Fundamentals of Electrodynamics

Fundamentals of

Electrodynamics

BORIS PODOLSKY

and

KAISER S. KUNZ

Research Center
New Mexico State University
Las Cruces, New Mexico

Dedicated to

B. Paula Podolsky

and

Ruth B. Kunz

Foreword

The main problem in preparing the final manuscript for publication was to bring the treatment up-to-date without undertaking such a thorough revision as to destroy the tone of the original manuscript. This was especially true of the revisions that have had to be made since Professor Podolsky's death.

The second author has been helped in this revision by the critical comments of Harry E. Denman, who read most of the original manuscript, suggesting improvements in logic or mathematical treatment. Section 28, moreover, is based on an article in *Journal of Mathematics and Physics* (1955), B. Podolsky and H. Denman.

The main purpose of the book is to treat classical electrodynamics in a logically consistent way so as to prepare the student for quantum electrodynamics, elementary particle physics, and quantum field theory. For this purpose one could make the treatment more modern by phrasing many of the arguments in group theoretic language, but this would have demanded too complete a revision of the original treatment.

As it stands the text should be suitable for the one-semester course in classical electrodynamics given in most universities as a part of a core curriculum in graduate physics.

While the treatment could no doubt be made more appealing to the expert by reducing the mathematical details given, I feel that the student will appreciate the time he is able to save by having these details spelled out clearly. This accounts also for our burdening many sentences with references to the specific equations that are needed to make a given step in the development. The extensive mathematical appendices should make the book more nearly self-contained and thus help to make it more readily comprehended by the average graduate student.

I would like to thank the following people at the Research Center, New Mexico State University, for their help in a difficult typing task: Beverly Landry, Barbara Edmonds, and Lou Ann Youngblood. It is a pleasure to acknowledge the help of my son Karl S. Kunz in proofreading the final copy.

Finally I would like to express my deep appreciation to my wife Ruth for her constant encouragement and for the long hours she spent helping to proofread the many versions of the manuscript.

KAISER S. KUNZ

July, 1969

Contents

Fundamentals of Electrodynamics

Chapter I

Introduction—General Dynamics

1. Space and Time of Classical Electrodynamics

A large part of our experience is embodied in our concepts of space and time. Part of this experience is subjective and is intimately connected with the observer's state of mind. Thus it is a matter of common experience that our judgment as to the lapse of time is very much influenced by what happens to us during this time.

In dealing with space, the necessity of using only objective measurements as our data is not questioned. Likewise the need for an objective measure of time has given rise to clocks, but we still have a direct subjective experience of time, which may or may not agree with the time as measured by our clocks. It has been found necessary in physics to deal only with clock time, and to exclude psychological time as being unsuited for the description and the understanding of objective reality. Thus *space* and *time* are to the physicist the results of measurements with ***rigid bodies*** and *clocks.**

Clocks are structures undergoing a change of condition in such a way that a certain configuration is indefinitely repeated. To some one occurrence of this repeated configuration we may assign the number 0; then starting from this configuration, we may assign the numbers 1, 2, 3, . . . to consecutive recurrences of it. These numbers can then be used to designate the time of each recurrence. To any phenomenon, occurring in the

* The reader familiar with quantum mechanics will recognize that the assumption that these measurements can be carried out with any desired degree of accuracy is not generally valid.

immediate neighborhood of one such recurrence and simultaneous with it, is assigned the same time as to the recurrence itself. The meaning of the terms *immediate neighborhood* and *simultaneous* in the immediate neighborhood are assumed to be known intuitively. It is also assumed that a sufficiency of recurrences exists to make possible a sufficiently accurate assignment of time to each phenomenon occurring in the neighborhood of a clock. To bring this about, it may be necessary to use secondary recurrent configurations obtained by subdividing the original unit of time and to assign to some phenomena fractional times. The time of phenomena not in the immediate neighborhood of each other must be determined by means of different clocks. Thus any desired number of sufficiently similar clocks are usually assumed to be available.

Most objects of physics are considered as divisible into parts. These parts are divisible into still smaller parts, and so on. The division is usually stopped when no purpose is apparently served by continuing it, which, of course, depends upon the problem at hand. The smallest part thus obtained is called a *particle*, a *material point*, or simply a *point*.* A phenomenon occurring in the immediate neighborhood of a point, at a definite time, is called an *event*.

The intuitive concept of an *absolutely rigid* body would require that if two such bodies were brought into contact, and if it were found that points A and B on one of them coincide respectively with points A' and B' on the other, then it would be possible to achieve such double coincidence of the same points whenever desired. A concept of the distance between a pair of points, as a quantity expressing their relationship, could then be introduced. In particular, the distance between A and B would be said to be equal to the distance between A' and B'. The concept of absolute rigidity can thus be said to consist in requiring that the distances between all points remain unchanged.

Completely rigid bodies, however, in the sense just defined, do not exist. Actually, only if the comparison of distances is made each time under the same standard conditions is a practical constancy of distances achieved. Accordingly, by a *rigid body* we shall mean a body with the property that the distances between its points do not depend on the body's history, but only on its physical condition. Equal distances are determined by a comparison made under standard conditions. Further, since it is essential

* As examples of particles we may take electrons, protons, etc. On the other hand, for some purposes, even a star may be considered as a particle. When the division is extended as far as our theory assumes possible, the ultimate parts are called *elementary particles*. It is doubtful whether elementary particles exist in any absolute sense.

that the standard conditions prevail over the entire rigid body, we shall consider that each rigid body is sufficiently small.* The introduction of standard conditions modifies, of course, the concept of distance.

One may make the concept of distance quantitative by choosing a suitable rigid body as a standard and on it making two marks. The distance between the marks is taken as a standard *unit of distance*, e.g., a *standard meter*, and the marks are called its *ends*. The distance between two points of a rigid body is said to be one unit if the two points can be made to coincide with the ends of the standard unit of distance when both bodies are under standard conditions. Similarly, by dividing the standard unit into equal parts, the distance between any two points of a rigid body may be measured.

By measuring the distances between the points of a rigid body and studying their various relationships, we may construct a *geometry* of such a body.

From this concrete geometry of individual rigid bodies we generalize to the concept of an abstract space in which the material points of all physical bodies are assigned points. This abstract space will be referred to as *position space* or simply as *space*.

If one chooses three points A, B, and C in a rigid body and measures the distances from these points to some other point P, then, in general, there exists no other point Q, in the immediate neighborhood of P, for which the corresponding distances to A, B, and C are the same. However, any two of the distances QA, QB, and QC may be equal to the corresponding distances from the point P. These relationships are analogous to those between points in a three-dimensional metric space of mathematics; therefore, this property of requiring three distances to locate a point in position space is expressed by saying that *space is three-dimensional*. It thus leads to the conclusion that each point of space may be characterized, in its relation to other points, by three suitably chosen numbers, called *coordinates*.

In order to have a convenient system of notation, it is very important that points that are in the immediate neighborhood of one another be given, in general, nearly equal coordinates, and that the distance between any two such points should be simply expressed in terms of the differences of their coordinates. The expression for the square of the distance

* Among the standard conditions are included the constancy and equality of speeds of the bodies compared, as well as the uniformity of the gravitational field over the two bodies. Thus, the extent of a body that may be considered as small will depend upon the degree of nonuniformity of certain fields and cannot well be discussed here.

between neighboring points in terms of their coordinates is called the *metric* of the coordinate space.*

This metric depends not only on the particular way of assigning three numbers to every point in space but also on the intrinsic geometry of space. Suppose we choose a point Q in the neighborhood of P and measure the distances PA, PB, PC, QA, QB, and QC; then the distance

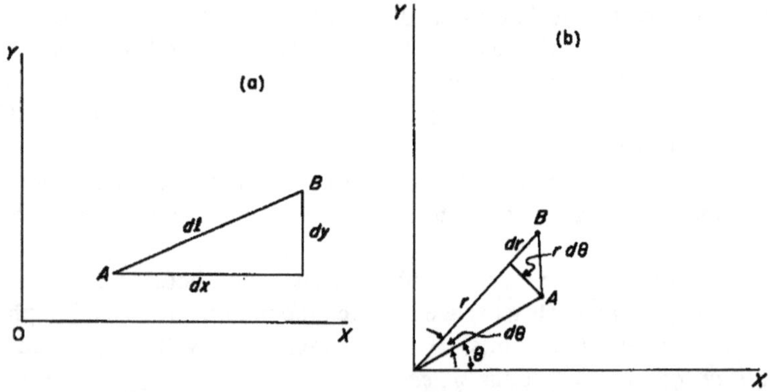

FIG. 1.1. Geometrical relationship of two neighboring points A and B in a plane for (a) Cartesian coordinates x, y and (b) polar coordinates r, θ.

PQ is not derivable from these distances without assuming a knowledge of the geometry of the space. Let us consider some subspaces of the three-dimensional physical space to illustrate the difference between a change in the metric due to a change in geometry and that due merely to a change in the way of associating three numbers with a point.

For example, if we are given a plane surface, we can assign Cartesian coordinates (x, y) to each point of the surface. These are pure numbers that describe the position of the point relative to the coordinate axes in terms of the unit of distance. Let the distance between two points A and B (see Figure 1.1(a)) very near one another be represented by the differential $d\ell$; then in Cartesian coordinates $(d\ell)^2$ is given by the *metric*

$$(d\ell)^2 = (dx)^2 + (dy)^2. \tag{1.1}$$

We could also have used polar coordinates (see Figure 1.1(b)) in which case

$$(d\ell)^2 = (dr)^2 + r^2 (d\theta)^2. \tag{1.2}$$

* The metric spaces of mathematics differ from physical spaces in that for the former the metric is just a number and has no necessary reference to a standard of physical length.

Obviously none of the geometrical properties of our space will be affected by this change of metric. In fact we can reduce the metric in Eq. (1.2) to that in Eq. (1.1) by the substitutions

$$r = (x^2 + y^2)^{1/2}$$
$$\theta = \tan^{-1}(y/x).$$

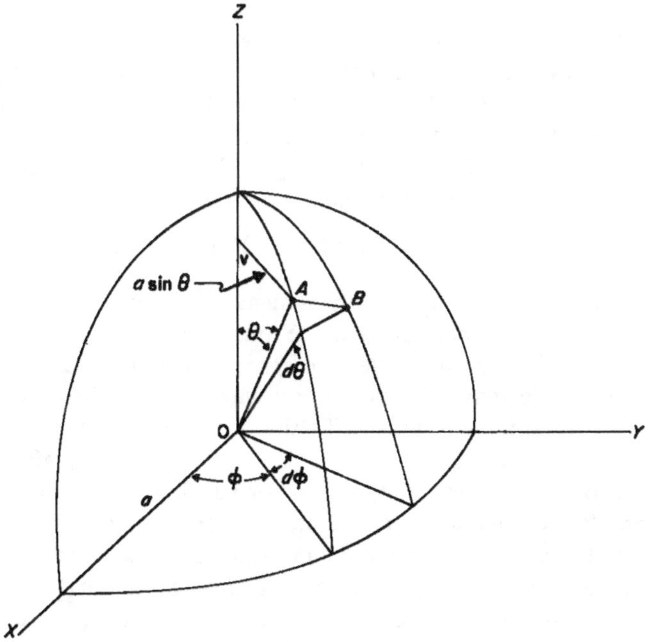

FIG. 1.2. Geometrical relationship of two neighboring points A and B on the surface of a sphere.

The multiple-valued nature of these definitions can be removed by requiring that $r \geq 0$ and $-\pi \leq \theta \leq \pi$.

To illustrate a change in the metric that involves a change in geometry, let us take the metric

$$(d\ell)^2 = a^2 (d\theta)^2 + a^2 \sin^2 \theta \, (d\phi)^2, \tag{1.3}$$

which represents geometrically the surface of a sphere (see Figure 1.2). We cannot by any transformation of coordinates θ and ϕ reduce this to the form of Eq. (1.1), a metric for a plane. This may be seen from the fact that the sum of the angles of a spherical triangle is not 180 degrees, and hence that Euclidean geometry cannot be applied to such a space.

From experience, the three-dimensional position space as abstracted from measurements on small rigid bodies is found to be Euclidean, i.e., it is possible to find a coordinate system for which the metric is of the form

$$(d\ell)^2 = (dx)^2 + (dy)^2 + (dz)^2. \tag{1.4}$$

In a space whose geometrical properties are determined by this metric it is well known that the formalism of vector analysis* is applicable. Moreover, since Eq. (1.4) is the metric of a Cartesian coordinate system with coordinates x, y, and z, we need not restrict the equation to infinitesimal distances; thus, if (x_1, y_1, z_1) and (x_2, y_2, z_2) are coordinates of any two points, the distance ℓ between them† is given by

$$\ell^2 = (x_2 - x_1)^2 + (y_2 - y_1)^2 + (z_2 - z_1)^2. \tag{1.5}$$

The theater in which physical phenomena take place involves not only the three-dimensional space but also the time as measured by clocks. This four-dimensional continuum is termed *space-time* and points within it are called *events*. We may label the events in space-time by a set of four numbers: three numbers to specify the location in space and one to specify the time. The numbering, in order to be convenient, must be made according to some system that, except for particular locations in space-time,‡ provides events in the immediate neighborhood of each other with nearly equal numbers. Such a system of numbering is called a *coordinate system*, or a *frame of reference*, and the ordered set of four numbers corresponding to an event are called its *coordinates*. To avoid confusion, the three numbers characterizing the point at which an event is located are called the *instantaneous coordinates* of the point. It is evident that the

* See Appendices A1 to A5. In what follows, the knowledge of vector analysis will be assumed.

† Strictly speaking, ℓ only represents the distance between the points based on a choice of the unit of distance, but it is often convenient to say simply that ℓ is the distance between the points. A thorough discussion of this point would bring out the two viewpoints one can take on the meaning of physical dimensions. See, in particular, D. C. Ipsen, *Units, Dimensions, and Dimensionless Numbers*, McGraw-Hill, 1960, Chap. 3.

‡ In two dimensions the polar coordinate system assigns points in the immediate neighborhood of one another nearly equal coordinates, with the exception of points surrounding the origin. In the immediate neighborhood of the origin the coordinates of points are (ε, r), where $\varepsilon \ll 1$ and $-\pi < \theta \leq \pi$. These points differ widely in their second coordinate.

system of numbering events is not unique because, instead of any given coordinate system, we could take another system in which the new coordinates would be any set of four independent continuous functions of the old. They should be continuous, except at certain singular points, in order that events in the immediate neighborhood of one another will again, as a rule, be assigned nearly equal coordinates.

We now make four assumptions about the properties of space and time. They are generalizations of the properties found for small rigid bodies and clocks and, within the limitations of electrodynamics, hold with great accuracy.* We shall discuss them in turn.

Assumption 1. Space is static.

This means that it is possible to choose a coordinate system in such a way that points in space having constant instantaneous coordinates also have constant distances between them. Because of this constancy, the distances referred to may be measured independently of time. A coordinate system satisfying the requirements of Assumption 1 is said to be *static*. After a static coordinate system is chosen, all points whose instantaneous coordinates do not vary in time, as well as bodies consisting entirely of such points, are said to be *at rest*. In what follows, only static coordinate systems will be considered.

Assumption 2. Clocks are synchronizable.

Suppose at some one time any two good clocks are made to read the same time when they are placed in the immediate neighborhoods of each other and have no relative motion. Then these clocks are said to be *synchronizable* if they can be adjusted on that occasion so as to read the same time whenever† they are again brought together under the same conditions. Assumption 2 requires, in principle, that any number of good clocks are available that are mutually synchronizable. In this discussion it is understood that the clocks are moved very *slowly*, i.e., each finite change of the coordinates of the clock is made during a long time. In what follows, we shall assume all the clocks to be synchronized, unless the contrary is obviously required by the context.

* We here neglect effects of gravitation on space and time, i.e., we set the gravitational constant γ equal to zero. This constant is defined by the formula $F = \gamma m_1 m_2/r^2$ for the magnitude of the force of attraction F between two particles with masses m_1 and m_2 that are r units apart.

† The words *when* and *whenever* are here unambiguous, since the two clocks show the same time.

With clocks synchronized, simultaneity of phenomena occurring not in the immediate neighborhood of each other acquires a meaning. Two phenomena are said to be *simultaneous* if they occur at the same time, the time for each being measured by the clock near it.

The measurement of the distance between points not within the same rigid body (but, of course, still at distances within the limitation of possible rigid bodies) can now be made precise. This distance is measured by making the two points coincide simultaneously with two points of a rigid body. *The rigid body and the clocks concerned are required to be at rest in some frame of reference.* The distance between the two points at the time of the coincidences is defined to be the same as the distance between the corresponding points of the rigid body.

In the definition of synchronization of clocks, as well as in the definition of the distance between points not in the same rigid body, the measuring instruments were required to be at *rest.* The term *rest* was, however, defined with respect to the coordinate system chosen. The choice of a static coordinate system is not unique; therefore synchronization of clocks and measurement of distances, i.e., the entire time-space description of phenomena, may depend upon the choice of the coordinate system.

Assumption 3. Space is Euclidean.

This statement acquires a meaning only for a static space, for which space and time can be considered independently. It means that Cartesian coordinate systems exist for the entire space, and that ordinary vector analysis is generally applicable. Unless the contrary is stated, we shall always assume a Cartesian coordinate system.

Thus, we assume the validity of Eqs. (1.4) and (1.5) and, as a consequence, the homogeneity and isotropy of space; space has the same properties at all points, and in all directions relative to these points. The limitation on the size of possible rigid bodies becomes unnecessary, and the distance between any two points becomes meaningful. A space-time system satisfying Assumptions 1 to 3 is said to be *flat.* Coordinates for which no material point exists are said to locate a point in empty space—that is to say, a possible place for a particle.

Assumption 4. Time is homogeneous.

By this we mean that the behavior of any isolated physical system can be described completely in terms of time intervals or time differences and does not change if there is a displacement in time of the initial state of this system. Thus, for instance, physical phenomena that are subject to experiment are assumed to appear exactly the same to twentieth-century physicists as to nineteenth-century physicists. The meaning of this assumption becomes more precise later when we have defined what is meant by an "isolated system" and by "initial state."

2. Equations of Motion

In considering geometrical properties of bodies, we regarded the bodies as consisting of points, but made no use of the material character of points. In fact, after a coordinate system has been constructed, no difference appears, geometrically, between empty and nonempty points. On the other hand, many aspects of the behavior of a physical object can be resolved into the behavior of its points. Thus, a nonempty point can be considered as an elementary physical object, the motion of which is one of the main concerns of physics. In what follows, a material point, considered in this way, shall be called a *particle*, while the term *point* shall be used only in the geometrical sense, for empty and nonempty points indiscriminately.

A fundamental property of interactions between physical systems, which may be called the *principle of separability*, may be stated as follows: Any physical system may be so far removed from all other matter that its motion, with any desired degree of accuracy, is independent of the existence of this matter. A system so removed is called an *isolated* system. The possibility of isolating a system, at least approximately, is a fundamental requirement for the existence of science; without this possibility the behavior of the entire universe would have to be known before anything could be predicted about any of its parts.

We shall often make use of the idea of a hypothetical displacement of a system of particles, in which the positions and motions of the particles relative to each other are unchanged. Such a displacement will be termed a *virtual displacement of the system as a whole* and the system obtained a *displaced system*. If B is a displaced system and A the original system, it is obviously possible to find a new coordinate system, obtained by translation and rotation of the old, in which system B has the same coordinates as A has in the original coordinate system.

We suppose that for an isolated system a Cartesian coordinate system can be chosen for which space is *mechanically homogeneous* and *isotropic;* that is, no mechanical experiment could distinguish one place, or direction, from another. We shall call such a coordinate system an *inertial* coordinate system. The assumption that such coordinate systems exist may be regarded as clarifying the meaning of the expression "independent of the existence of all other matter" used in defining an isolated system.

The inability to distinguish by an experiment one direction or place from another means, in particular, that all displaced systems behave in exactly the same way as the original system; that is, the laws governing the behavior of a displaced system, in its displaced coordinate system, must be the same as those of the original system in the original coordinate system.

Moreover, for such an isolated system, we have another general principle, the *principle of causality*, which may be stated as follows: The entire motion of an isolated system is determined by the *state* of the system at any one instant of time. This means that, if at any given time the state of the system is assigned, the entire motion of the system is determined; that is, it is calculable by means of the laws of motion. A little reflection shows that this principle is, to some extent, a definition of the term *state.* It can be reworded as follows: *State* is that which must be known at any given instant of time in order that the entire motion may be calculable. It must not be thought, however, that the principle of causality is therefore unimportant. Here the importance lies in the implication that *state* and *entire motion* are not logically equivalent.

The motion of a particle as a whole comprises its entire possible behavior. Any structural changes or rotation about its own axis can have a meaning only if the particle is considered as consisting of smaller parts; this, however, is contrary to its definition.*

When a particle moves, its coordinates change with the time, and can be considered as functions of the time. Designating these coordinates vectorially by the position vector $\mathbf{r}$, we have

$$\mathbf{r} = \mathbf{f}(t), \tag{2.1}$$

and this function completely determines all the properties of the motion.

* An electron with spin would thus not, ordinarily, be considered a particle in this formulation of classical electrodynamics because of the need to describe the relative motion of its parts. One could perhaps circumvent this objection by insisting that the angular momentum is an intrinsic property of the point particle and not associated with any describable motion.

Equation (2.1) is an assertion that, *a priori*, need not be true. In fact, it acquires a meaning only if this function can be experimentally determined, i.e., if a measurement of position can be performed at any time, with any desired degree of accuracy, without disturbing the motion. We assume this possibility as the *principle of determinateness.**

If we assume that it is possible to consider the particles that make up a group of bodies—this group being sufficiently removed from other bodies—as an isolated system,† then the motion of the system is completely determined by giving the positions of the particles as functions of the time. We shall use numerical or literal subscripts to designate the particles of a system. Thus, for example, (x_1, y_1, z_1) and (x_s, y_s, z_s) are the coordinates of the first and the sth particle, respectively, where s is some specific particle. We may slightly modify this idea and let (x_s, y_s, z_s) represent the coordinates of any one of the particles. Thus for a system of m particles s is any integer from 1 to m.

A transformation of coordinates

$$q_j = q_j(x_1, y_1, z_1;\ x_2, y_2, z_2;\ \ldots;\ x_m, y_m, z_m), \qquad j = 1, 2, \ldots, 3m, \tag{2.2a}$$

can be performed, where each of the $3m$ new coordinates q_j is a function of all the old coordinates. We shall let x_s, y_s, z_s represent the entire set of old coordinates and rewrite (2.2a) in the form

$$q_j = q_j(x_s, y_s, z_s), \qquad j = 1, 2, \ldots, 3m. \tag{2.2b}$$

If the q_j are independent parameters of the system of particles, the coordinates x_s, y_s, z_s, $s = 1, 2, \ldots, m$, of the individual particles can be expressed uniquely in terms of the coordinates q_j, $j = 1, 2, \ldots, 3m$. It may happen, however, that the particles of the system are so restricted in their motion that less than $3m$ numbers are sufficient to specify their positions. This happens, for instance, in the case of a rigid body. In such cases the $3m$ coordinates of the particles can be expressed as functions of a smaller number of independent parameters q_j; thus

$$x_s = x_s(q_j), \qquad y_s = y_s(q_j), \qquad z_s = z_s(q_j), \qquad s = 1, 2, \ldots, m, \tag{2.3}$$

where q_j stands for the entire set of parameters $q_1, q_2, \ldots, q_n$. Such a set of parameters that completely describe the configuration are called the

* We here neglect all quantum phenomena, i.e., put Planck's constant h equal to zero.

† This is not true in general, for, as we shall see in Chapter VI, the field of the particles should ordinarily be considered as a part of such a system.

generalized coordinates of the system, and their number n is called the number of *degrees of freedom* of the system.

Instead of Eq. (2.1) we now have

$$q_j = q_j(t), \qquad j = 1, 2, \ldots, n, \tag{2.4}$$

which may be called the equations of the *trajectory* of the system. If the functions in Eq. (2.4) are known, the entire behavior of the system is known; these functions therefore contain, on the basis of our present assumptions, all the information that we can have about the system. Thus, at any time $t = t_0$, the only physical data that can be given are contained in the set

$$q_j(t_0), \qquad \left(\frac{dq_j}{dt}\right)_{t=t_0}, \qquad \left(\frac{d^2q_j}{dt^2}\right)_{t=t_0}, \quad \ldots, \tag{2.5}$$

which must therefore be sufficient to specify a state.* But if all these quantities were necessary for such a specification, and could therefore be chosen arbitrarily, the term *state* would be equivalent to the term *motion*, and the principle of causality would become a tautology. Further, no law of motion would be possible; for such a law would be a condition on the functions in Eq. (2.4), and hence on the derivatives in (2.5), which then could not be arbitrarily chosen. Let us therefore suppose that the highest derivative required in specifying the state is the kth and that some p derivatives below it are not required. We conclude that the state of a system is determined, at any given time, by the q_j and the values of their $k - p$ derivatives for all the degrees of freedom of the system—where, of course, $p < k$. The state of the system is thus determined by $(k - p + 1)n$ of the quantities†

$$q_j, \quad \dot{q}_j \equiv \frac{dq_j}{dt}, \quad \ddot{q}_j \equiv \frac{d^2q_j}{dt^2}, \quad \ldots, \quad q_j^{(k)} \equiv \frac{d^kq_j}{dt^k}, \tag{2.6}$$

all taken for $t = t_0$.

Experience shows that, within the scope of the classical electrodynamics, a certain definite integral called the *action* has a minimum value for all sufficiently short trajectories. This property of trajectories is referred to as *Hamilton's principle*. It is in many ways the most important principle of classical theory.

* We shall not be concerned here with the mathematical question of whether or not these derivatives always exist.

† The symbol $\equiv$ here indicates a definition; it is also used for an identity. The context will indicate which usage is intended.

Suppose a and b designate two actual configurations of a system together with their respective times. The action W relative to any assumed motion connecting a and b is then given by

$$W = \int_a^b L\,dt, \tag{2.7}$$

where L is a function of the quantities, such as given in Eq. (2.6), defining the state of the system and in some cases also a function of the time. The function L is called the *Lagrangian*, and must be suitably chosen for each given system; it may be considered as the characteristic function of the system. Since W is defined for a large class of possible trajectories between a and b, it can be considered as the value of the functional* expressed by the integral in Eq. (2.7) relative to some set of acceptable functions $q_j(t), j = 1, 2, \ldots, n$, describing the intervening motion between a and b. According to Hamilton's principle the actual trajectory of the system between a and b must be such as to make W stationary, in the calculus of variation sense.† One may also say that for the actual trajectory the *variation* of W is zero; this is written

$$\delta W = 0. \tag{2.8}$$

For a sufficiently short time interval between a and b the trajectory taken by the system is actually that which minimizes W.

A variational equation, such as Eq. (2.8), as will be shown later, can be converted into a system of n differential equations. If L contains the kth derivative as the highest order derivative of the required functions q_j, the resulting Euler equations of the trajectory will in general be of $2k$th order. The trajectory will, therefore, require $2kn$ constants for its specification, $2k$ for each coordinate. On the other hand, as we have seen, the trajectory is determined by $(k - p + 1)n$ constants (see Eqs. (2.6)). It follows, therefore, that if for each given state the motion is to be uniquely determined,‡ one must have $2kn = (k - p + 1)n$. Now, since $p < k$ and both are positive integers, $k = 1$ and $p = 0$. Thus, a state of a system is determined by the coordinates and their time derivatives, called *velocities*.

* A functional is a mapping from some function space to the real (or complex) numbers. Thus the value of this functional is the real number W given by Eq. (2.7).

† See, for example, R. Courant and D. Hilbert, *Methods of Mathematical Physics*, Vol. I, Wiley-Interscience, New York, 1953, Chap. IV.

‡ One could take $2kn > (k - p + 1)n$ by assuming, in addition to Hamilton's principle, some appropriate auxiliary conditions. This would permit k to be greater than one, which would greatly complicate the theory. Fortunately, this expedient is not required in the usual formulation of classical electrodynamics.

Since only these occur in the Lagrangian, the equations of motion do not contain derivatives higher than the second. This important fact is responsible for the simplicity of classical electrodynamics.

If the system is an isolated system, then by virtue of Assumption 4 the value of the time coordinate t should not affect the behavior of the system. We here extend Assumption 4 by requiring that *the Lagrangian characterizing an isolated system can be so chosen as not to depend explicitly on the time t.*

Hamilton's principle for an isolated system of particles can thus be written

$$\delta W = \delta \int_a^b L(q_j, \dot{q}_j)\, dt = 0. \tag{2.9}$$

Since the variation of an integral with fixed limits of the independent variables is the integral of the variation of the integrand,

$$\delta W = \int_a^b \delta L\, dt = \int_a^b \left[\frac{\partial L}{\partial q_j} \delta q_j + \frac{\partial L}{\partial \dot{q}_j} \delta \dot{q}_j \right] dt, \tag{2.10}$$

where in agreement with the convention on repeated indices the summation is performed over all the values of j, from 1 to n. Further, since $\delta \dot{q}_j = d(\delta q_j)/dt$, we may integrate by parts the terms containing the $\delta \dot{q}_j$, obtaining finally

$$\delta W = \int_a^b \delta L\, dt = \left[\frac{\partial L}{\partial \dot{q}_j} \delta q_j \right]_a^b + \int_a^b \left[\frac{\partial L}{\partial q_j} - \frac{d}{dt} \frac{\partial L}{\partial \dot{q}_j} \right] \delta q_j\, dt. \tag{2.11}$$

Since the limits a and b are fixed, representing fixed values of the coordinates q_j, $\delta q_j = 0$ at these limits and thus the integrated part in Eq. (2.11) vanishes. On the other hand, the δq_j in the integrand are arbitrary and independent, which requires that the expressions in the brackets must vanish for each value of j; otherwise each δq_j could be chosen to have the same sign at each value of t as the sign of the brackets, and the variation of W would not vanish. We have, therefore, n Euler differential equations of the second order

$$\frac{d}{dt} \frac{\partial L}{\partial \dot{q}_j} = \frac{\partial L}{\partial q_j}, \qquad j = 1, 2, \ldots, n, \tag{2.12}$$

which in this application to mechanics are called the *equations of Lagrange.*

The quantity $p_j \equiv \partial L / \partial \dot{q}_j$ is called the *momentum* conjugate to the coordinate q_j and the quantity $F_j \equiv \partial L / \partial q_j$ the jth generalized *force*. The

Lagrange equations can therefore be written

$$\frac{dp_j}{dt} \equiv \dot{p}_j = F_j, \tag{2.13}$$

which is a generalized form of *Newton's second law of motion.*

Since $L = L(q_j, \dot{q}_j)$, we may write

$$dL = \frac{\partial L}{\partial q_j} dq_j + \frac{\partial L}{\partial \dot{q}_j} d\dot{q}_j = p_j \, d\dot{q}_j + F_j \, dq_j, \tag{2.14}$$

where, as in all that follows, a summation is implied with respect to each subscript occurring twice in the same term, unless the contrary is explicitly stated.* If, instead of the q_j and $\dot{q}_j$ as independent variables, which is implied by the form of Eq. (2.14), we wish to use the q_j and p_j, we may perform the transformation of Legendre by introducing a new function

$$H \equiv -L + \dot{q}_j p_j. \tag{2.15}$$

Then from Eqs. (2.13) and (2.14)

$$dH = -dL + p_j \, d\dot{q}_j + \dot{q}_j \, dp_j = \dot{q}_j \, dp_j - \dot{p}_j \, dq_j, \tag{2.16}$$

which indicates the differential dependence of H on the variables p_j and q_j.

In order to express H explicitly in terms of p_j and q_j, one solves the equations $p_j = \partial L/\partial \dot{q}_j$ for $\dot{q}_j$ in terms of p_j and q_j and substitutes these in Eq. (2.15). The function H, *when thus expressed as a function of q_j and p_j*, is called the *Hamiltonian function*, or simply the *Hamiltonian*, and plays a very important role in all advanced theory.

From Eq. (2.16) one can at once derive Hamilton's equations of motion

$$\frac{\partial H}{\partial p_j} = \dot{q}_j, \qquad \frac{\partial H}{\partial q_j} = -\dot{p}_j, \qquad j = 1, 2, \ldots, n. \tag{2.17}$$

They are also referred to as the *canonical form* of the equations of motion. There are twice as many equations in this set as we had in Eqs. (2.12), but this is accounted for by the introduction of new unknown functions p_j, a process common in passing from a set of second-order equations to a set of first order.†

* At times the summation sign will be employed for clarity or emphasis even though this rule would make it unnecessary.

† For an excellent treatment of the Lagrangian and Hamiltonian formulation of mechanics see C. Lanczos, *The Variation Principles of Mechanics*, 2nd ed., Univ. Toronto Press, Toronto, 1949.

The following equations are also sometimes useful and may be derived from the preceding equations:

$$\frac{\partial L}{\partial q_j} = -\frac{\partial H}{\partial q_j}, \qquad \frac{\partial L}{\partial \lambda} = -\frac{\partial H}{\partial \lambda}, \tag{2.18}$$

where λ is any parameter that may be contained in the Lagrangian function.

We now consider the definition of action, Eq. (2.7), in a different way. We regard the lower limit of integration as fixed, but the upper as variable. If, now, the equations of motion are taken as satisfied for all choices of the upper limit, the action is completely determined as soon as the coordinates and the time of the upper limit are given. The action can now be designated $W(q_k, t)$ where q_k and t stand for the generalized coordinates and the time at the upper limit. We can determine the functional dependence by repeating the variation of trajectory under these new conditions. Equation (2.11) still holds, but, since the set of trajectories considered in the variation are only those satisfying Lagrange's equations, the integral vanishes. On the other hand, since these trajectories necessarily differ at the upper limit b, the variations δq_j are not zero at b. The equation for the variation of $W(q_k, t)$ is thus given by

$$\delta W(q_k, t) = \frac{\partial L}{\partial \dot{q}_j}\, \delta q_j = p_j\, \delta q_j .$$

All the quantities are to be evaluated at b, which designates a configuration at time t given by the generalized coordinates q_j. This can be compared to the change

$$\delta W = \frac{\partial W}{\partial q_j}\, \delta q_j$$

computed directly for changes of δq_j in the variables of $W(q_k, t)$, and one concludes that

$$p_j = \frac{\partial W}{\partial q_j}. \tag{2.19}$$

The method of variation employed does not permit a variation in the time coordinate. One knows, however, since $W(q_k, t)$ is just the integral of L along the actual trajectory taken by the system, that

$$\frac{dW}{dt} \equiv \frac{\partial W}{\partial t} + \frac{\partial W}{\partial q_j}\frac{dq_j}{dt} = L, \tag{2.20}$$

where the total derivative includes the effect of changes in the q_j with time.

Combining Eq. (2.20) with Eqs. (2.15) and (2.19), we have

$$\frac{\partial W}{\partial t} = L - p_j \dot{q}_j = -H \tag{2.21}$$

If in this equation H is regarded as a function of the p_j and q_j, so that $H = H(p_j, q_j)$, then, making use of Eq. (2.19), we can write

$$H\left(\frac{\partial W}{\partial q_j}, q_j\right) + \frac{\partial W}{\partial t} = 0. \tag{2.22}$$

This important first-order partial differential equation for W is called the *Hamilton–Jacobi equation.*

Finally, from Eqs. (2.20) and (2.22), we obtain another useful relation

$$dW = \frac{\partial W}{\partial q_j} dq_j + \frac{\partial W}{\partial t} dt = p_j \, dq_j - H \, dt. \tag{2.23}$$

Although the Lagrangian of a system completely determines the motion, the motion of the system does not completely determine the Lagrangian. Thus, since the equations of motion are linear in L, we may multiply the Lagrangian by an arbitrary constant without affecting the motion. Moreover, we can add to L a term dR/dt, where R is any function of q_j and t. This is most readily seen from Eq. (2.7). In fact, the addition of dR/dt to L changes W by the amount

$$\Delta W = \int_a^b \frac{dR}{dt} dt = \int_a^b dR = R(b) - R(a) \tag{2.24}$$

which is independent of the trajectory, and therefore contributes nothing to δW.

We are interested at this point in determining the arbitrariness in the Lagrangian when we specify the equations of motion. That is, we want to know what form a Lagrangian L' must take if it is to yield the same* equations of motion as a given Lagrangian L. It is understood that the same generalized coordinates are to be used in these Lagrangians.

* If we require merely that L' gives us equivalent equations of motion the arbitrariness is very much greater. We could then have L' any Lagrangian arising in a canonical transform from the variables q_j, p_j in the original Hamiltonian H to the variables q_j and p_j' in a new Hamiltonian $H' = H + \partial F/\partial t$, where $F = F(p_j' - p_j, t)$ and $q_i = -(\partial F/\partial p_i) = (\partial F/\partial p_i')$. See H. Goldstein, *Classical Mechanics*, Addison-Wesley, Reading, Mass., 1950, Chap. 8.

We can go even further when L does not involve the time explicitly and require merely that

$$\dot{q}_j \frac{\partial L'}{\partial \dot{q}_j} - L' = \Phi\left(\dot{q}_j \frac{\partial L}{\partial \dot{q}_j} - L\right),$$

or in terms of the Hamiltonians $H' = \Phi(H)$. See F. J. Kennedy, Jr. and E. H. Kerner, *Amer. J. Phys.*, **33**, 463 (1965).

The requirement that the equations of motion are to be the same is taken to mean that

$$\frac{d}{dt}\frac{\partial L'}{\partial \dot{q}_i} - \frac{\partial L'}{\partial q_i} \equiv a\left(\frac{d}{dt}\frac{\partial L}{\partial \dot{q}_i} - \frac{\partial L}{\partial q_i}\right), \tag{2.25}$$

where a is some nonzero constant. Since the equations are linear, this becomes

$$\frac{d}{dt}\frac{\partial(L' - aL)}{\partial \dot{q}_i} - \frac{\partial(L' - aL)}{\partial q_i} \equiv 0. \tag{2.26}$$

Thus we need to solve the differential equations

$$\frac{d}{dt}\frac{\partial f}{\partial \dot{q}_i} - \frac{\partial f}{\partial q_i} \equiv 0, \tag{2.27}$$

where $f = L' - aL$. Expanding these equations, we get

$$\frac{\partial^2 f}{\partial \dot{q}_j\,\partial \dot{q}_i}\ddot{q}_j + \frac{\partial^2 f}{\partial q_j\,\partial \dot{q}_i}\dot{q}_j + \frac{\partial^2 f}{\partial t\,\partial \dot{q}_i} - \frac{\partial f}{\partial q_i} \equiv 0. \tag{2.28}$$

This, since $\ddot{q}_j$ enters each equation in only one place, shows that, for all j's and i's,

$$\frac{\partial^2 f}{\partial \dot{q}_j\,\partial \dot{q}_i} = 0. \tag{2.29}$$

On integrating this gives

$$\frac{\partial f}{\partial \dot{q}_i} = h_i(q_k, t)$$

and thus

$$f = \dot{q}_i h_i(q_k, t) + g(q_k, t). \tag{2.30}$$

Putting this f back into Eq. (2.28), we obtain the differential equation

$$\dot{q}_j\frac{\partial h_i(q_k, t)}{\partial q_j} + \frac{\partial h_i(q_k, t)}{\partial t} - \dot{q}_j\frac{\partial h_j(q_k, t)}{\partial q_i} - \frac{\partial g}{\partial q_i} \equiv 0$$

or

$$\dot{q}_j\left(\frac{\partial h_j}{\partial q_i} - \frac{\partial h_i}{\partial q_j}\right) + \left(\frac{\partial g}{\partial q_i} - \frac{\partial h_i}{\partial t}\right) \equiv 0; \tag{2.31}$$

and since this must be satisfied indentically in the $\dot{q}_j$ for all i

$$\frac{\partial h_j}{\partial q_i} - \frac{\partial h_i}{\partial q_j} = 0 \qquad \text{and} \qquad \frac{\partial g}{\partial q_i} - \frac{\partial h_i}{\partial t} = 0. \tag{2.32}$$

These, however, are just the conditions that must be satisfied in order for

$$\dot{q}_i h_i(q_k, t) + g(q_k, t)$$

to be a total derivative; hence

$$f = \frac{dR(q_k, t)}{dt} \tag{2.33}$$

and

$$L' = aL + \frac{d}{dt} R(q_k, t), \tag{2.34}$$

where R is an arbitrary function of the coordinates and the time.

It is convenient at this point to extend Assumption 3 on the homogeneity and isotropy of space. That assumption, as we have seen, leads to the requirement that the motion of an isolated system is equivalent to any of its displaced systems. We now assume *that at least one of the permissible Lagrangians for an isolated system must be of the form*

$$L = F(q_k, \dot{q}_k), \tag{2.35}$$

where F is invariant to any displacement or rotation of the system, as a whole. As will be seen later, this makes the concept of the total momentum precise. There still remains arbitrariness in the Lagrangian of the same form as that shown in Eq. (2.34) except that dR/dt must now be invariant to a displacement or rotation of the system. For a single particle this means that the arbitrariness dR/dt is merely an additive constant.

In the entire discussion of the Hamilton's principle and its consequences we have, until now, assumed that the system we have been considering was isolated. Let, now, such an isolated system be considered as consisting of two parts, say A and B. The equations of motion for the combined system are

$$\frac{d}{dt}\frac{\partial L}{\partial \dot{q}_i} - \frac{\partial L}{\partial q_i} = 0, \qquad i = 1, 2, \ldots, n,$$

showing that there is an equation corresponding to each coordinate q_i. Here L is a function of the q_i and the $\dot{q}_i$ only. If, since it is the Lagrangian of an isolated system, our coordinates are so chosen that some of them, say q_{A_i}, refer only to part A, while the rest of them, say q_{B_i}, refer to part B, then we can divide the equations of motion into those that correspond to the coordinates of part A, namely

$$\frac{d}{dt}\frac{\partial L}{\partial \dot{q}_{A_i}} - \frac{\partial L}{\partial q_{A_i}} = 0, \tag{2.36}$$

and those that correspond to coordinates of part B.

Although in form Eqs. (2.36) are those of an isolated system, the fact that L contains the q_B and $\dot{q}_B$ requires that the motion of part B be known in order that we may be able to solve these equations for the motion of part A. If the motion of part B is given—that is, its coordinates and velocities are known functions of time—we can substitute these for the q_B and $\dot{q}_B$ in the Lagrangian L, and obtain a function of the q_A, the $\dot{q}_A$, and t. With this Lagrangian the equations of motion (2.36) are solvable for the motion of part A.

We shall now show that another Lagrangian L' for the whole system can be chosen so that

$$L' = L'_A + L'_B + M'_{AB},$$

where L'_A and L'_B are Lagrangians for the parts A and B, respectively, when isolated, and M'_{AB} approaches zero when the two parts are separated.

We start by separating the parts A and B by a displacement of one of them to such a distance that the two parts can be considered as isolated systems and letting the new Lagrangian for the entire system be designated by L^*. We can, as seen above, use L^* to obtain the motion of part A, provided we consider the q_B and $\dot{q}_B$ to be given functions of time. However, it should not matter what these functions are, since the two parts are now isolated from each other. The same equations of motion for part A can therefore be obtained from some Lagrangian $L_A = L_A(q_A, \dot{q}_A)$, which does not involve the q_B and $\dot{q}_B$. This, by Eqs. (2.26), means

$$L^* - aL_A = h'_j(q_A, t)\,\dot{q}_{A_j} + g'(q_A, t), \tag{2.37}$$

where h'_j and g' must satisfy the conditions

$$\frac{\partial h'_i}{\partial q_{A_j}} = \frac{\partial h'_j}{\partial q_{A_i}} \tag{2.38}$$

and

$$\frac{\partial h'_i}{\partial t} = \frac{\partial g'}{\partial q_{A_i}}. \tag{2.39}$$

Since L^* and L_A, being Lagrangians of isolated systems, are independent of time, t should enter the right-hand side of Eq. (2.37) only through the functions $q_{B_i}(t)$ and their derivatives $\dot{q}_{B_i}(t)$. Changing to these variables by letting

$$h'_j(q_A, t) = h_j(q_A, q_B, \dot{q}_B)$$

and

$$g'(q_A, t) = g(q_A, q_B, \dot{q}_B),$$

Eq. (2.37) may be written in the form

$$L^* - aL_A = h_j(q_A, q_B, \dot{q}_B)\,\dot{q}_{A_j} + g(q_A, q_B, \dot{q}_B); \tag{2.40}$$

and the conditions (2.38) and (2.39) become

$$\frac{\partial h_i}{\partial q_{A_j}} = \frac{\partial h_j}{\partial q_{A_i}} \tag{2.41}$$

and

$$\frac{\partial h_i}{\partial q_{B_j}} \dot{q}_{B_j} + \frac{\partial h_i}{\partial \dot{q}_{B_j}} \ddot{q}_{B_j} = \frac{\partial g}{\partial q_{A_i}}, \tag{2.42}$$

respectively.

The general solution of the set of equations, Eqs. (2.41), is

$$h_i = \frac{\partial}{\partial q_{A_i}} R(q_A, q_B, \dot{q}_B), \tag{2.43}$$

where R is an arbitrary function of the variables indicated. Substituting this into the condition (2.42), we have

$$\frac{\partial}{\partial q_{A_i}}\left(\frac{\partial R}{\partial q_{B_j}} \dot{q}_{B_j} + \frac{\partial R}{\partial \dot{q}_{B_j}} \ddot{q}_{B_j}\right) = \frac{\partial g}{\partial q_{A_i}}, \tag{2.44}$$

or, integrating,

$$g = \frac{\partial R}{\partial q_{B_j}} \dot{q}_{B_j} + \frac{\partial R}{\partial \dot{q}_{B_j}} \ddot{q}_{B_j} + G(q_B, \dot{q}_B). \tag{2.45}$$

Equation (2.40) then becomes

$$L^* - aL_A = \frac{\partial R}{\partial q_{A_j}} \dot{q}_{A_j} + \frac{\partial R}{\partial q_{B_j}} \dot{q}_{B_j} + \frac{\partial R}{\partial \dot{q}_{B_j}} \ddot{q}_{B_j} + G(q_B, \dot{q}_B).$$

Since the Lagrangians L^* and L_A do not contain the $\ddot{q}_{B_j}$, being functions of the coordinates and velocities only, we have $\partial R/\partial \dot{q}_{B_j} = 0$. Thus $R = R(q_A, q_B)$ and our equation therefore becomes

$$\begin{aligned} L^* - aL_A &= \frac{\partial R}{\partial q_{A_j}} \dot{q}_{A_j} + \frac{\partial R}{\partial q_{B_j}} \dot{q}_{B_j} + G(q_B, \dot{q}_B) \\ &= \frac{dR}{dt} + G(q_B, \dot{q}_B) = \frac{dR'}{dt}, \end{aligned} \tag{2.46}$$

where

$$R' = R(q_A, q_B) + \int_0^t G(q_B, \dot{q}_B)\, dt + \text{constant}.$$

This is in agreement with Eq. (2.34), except that R in that equation is here replaced by R', a function whose dependence on time is expressed by means of the known functions $q_{B_i}(t)$ and $\dot{q}_{B_i}(t)$.

L^* must also give the same equations of motion for part B as some L_B, the q_A now being considered as given functions of t. Observe that aL_A does not contain the q_B or the $\dot{q}_B$, and that the total derivative dR/dt can be dropped without affecting equations of motion. It follows from Eq. (2.46) therefore that $G(q_B, \dot{q}_B)$, used as a Lagrangian of part B, will give the same equations of motion as L^*, and therefore the same as L_B. This requires that

$$G(q_B, \dot{q}_B) - bL_B(q_B, \dot{q}_B) = \frac{dS}{dt}, \tag{2.47}$$

where S is some function of q_B. Thus, finally, from Eqs. (2.46) and (2.47),

$$L^* = aL_A(q_A, \dot{q}_A) + bL_B(q_B, \dot{q}_B) + \frac{d(R+S)}{dt}. \tag{2.48}$$

Since L^* is just a limiting case of L, it is invariant to a displacement or a rotation of the combined system; therefore, the same must be true of $d(R + S)/dt$. Thus, this total derivative is within the arbitrariness allowed the Lagrangian of an isolated system, so that we can take instead of L the Lagrangian L' defined by

$$L' = L - \frac{d(R+S)}{dt} = aL_A + bL_B + M'(q_A, \dot{q}_A, q_B, \dot{q}_B), \tag{2.49}$$

which also defines M'. Comparison of this equation with Eq. (2.48) shows that as the two parts, A and B, are separated and L approaches L^*, L' approaches $aL_A + bL_B$, and M' approaches zero. Moreover, since aL_A and bL_B can be taken as the Lagrangians of parts A and B, respectively, when the systems are to be treated as isolated systems, we can write

$$L' = L'_A + L'_B + M'. \tag{2.50}$$

Thus one can always find a Lagrangian for a system of two parts which can be expressed as the sum of the Lagrangians of the parts plus terms having the property of approaching zero when the parts are separated. We shall consider, henceforth, that the adjustments necessary to achieve this are always made.

It is important to note that these adjustments require that we take for the Lagrangians of parts A and B of Eq. (2.49) particular functions L'_A and L'_B, with a and b chosen in a suitable way. This further restricts the concept of the Lagrangian.

We can obviously generalize our procedure and show that we can take as the total Lagrangian of a system of any number of parts the sum of the

Lagrangians of the parts plus terms that vanish when the parts are isolated from one another. It is this additiveness of the Lagrangians that makes the concept of the Lagrangian sufficiently definite to be useful.

Interactions between the various parts express themselves only through the additional terms, such as M' above, which may be called the *interaction function*, or simply the *interaction*, between the parts. Dropping the primes, the Lagrangian of a system of two parts becomes

$$L + L_A(q_A, \dot{q}_A) + L_B(q_B, \dot{q}_B) + M(q_A, \dot{q}_A, q_B, \dot{q}_B); \qquad (2.51)$$

and, assuming that we know q_B and $\dot{q}_B$ as functions of time, we can find the equations of motion of part A as a nonisolated system. They will be given by Eq. (2.36); from which we see that, since L_B will not enter the equations of motion, the same equations will result if we use, instead of L, the Lagrangian

$$L_A(q_A, \dot{q}_A, t) = L_A(q_A, \dot{q}_A) + M(q_A, \dot{q}_A, q_B(t), \dot{q}_B(t)). \qquad (2.52)$$

This is the general form of the Lagrangian of a nonisolated system and differs from the Lagrangian of an isolated system in that it generally contains the time t explicitly. It reduces to $L_A(q_A, \dot{q}_A)$ when part A is isolated. In some special cases, as in the case when part B is at rest, M may not involve time. Such cases are said to be *stationary*.

3. Laws of Conservation and the Principle of Relativity

We shall now consider certain consequences of the equations of motion that are known as the laws of conservation. They are of the form: *some expression involving coordinates and velocities = constant.* They are first integrals of the equations of motion, but they will be obtained here by using specific properties of the Lagrangian.

Consider first an isolated system. For such a system, as we have seen, L is a function of q_j and $\dot{q}_j$ only, these being sufficient to specify the state; therefore

$$\frac{dL}{dt} = \frac{\partial L}{\partial q_j}\dot{q}_j + \frac{\partial L}{\partial \dot{q}_j}\ddot{q}_j. \qquad (3.1)$$

Eliminating $\partial L/\partial q_s$ between this equation and the equations of motion (2.12), we obtain

$$\frac{dL}{dt} - \dot{q}_s \frac{d}{dt}\frac{\partial L}{\partial \dot{q}_s} - \frac{\partial L}{\partial \dot{q}_s}\frac{d\dot{q}_s}{dt} = 0. \tag{3.2}$$

The left side of this equation is evidently equal to

$$\frac{d}{dt}\left(L - \dot{q}_s \frac{\partial L}{\partial \dot{q}_s}\right),$$

so that, remembering the definitions of p_s and H, we may write

$$\frac{d}{dt}(L - \dot{q}_s p_s) = -\frac{dH}{dt} = 0. \tag{3.3}$$

On integration, we obtain the law of conservation

$$H = -L + \dot{q}_s p_s = \text{constant}. \tag{3.4}$$

For an nonisolated system we have, instead of Eq. (3.1), the more general equation

$$\frac{dL}{dt} = \frac{\partial L}{\partial t} + \frac{\partial L}{\partial q_s}\dot{q}_s + \frac{\partial L}{\partial \dot{q}_s}\cdot \tag{3.5}$$

Thus, in place of Eq. (3.2) we have

$$\frac{dL}{dt} - \dot{q}_s \frac{d}{dt}\frac{\partial L}{\partial \dot{q}_s} - \frac{\partial L}{\partial \dot{q}_s}\frac{d\dot{q}_s}{dt} = \frac{\partial L}{\partial t} \tag{3.6}$$

and finally that

$$\frac{dH}{dt} = -\frac{\partial L}{\partial t}\cdot \tag{3.7}$$

Thus if L does not explicitly contain the time, H is constant; hence for stationary cases we again have the law of conservation expressed by Eq. (3.4).

When the law of conservation $H = \textit{constant}$ holds, H is called the *energy* of the system, and the law is called the *law of conservation of energy*.*

Consider again an isolated system, and let Cartesian coordinates be used for all the particles. Let the radius vector $\mathbf{r}_s$ represent the three coordinates

* It is sometimes stated that H, even when it is constant, may nevertheless differ from the energy of the system. We consider that such a point of view introduces an unnecessary complication in the definition of energy and regard the definition given in the text as applying without exception.

of the sth particle; $s = 1, 2, \ldots, n$, where n is the number of particles. Space, in keeping with the assumptions made in Section 1, will be assumed to be homogeneous and isotropic, and the coordinate system will be assumed to be inertial. We have seen in the last section that the Lagrangian of an isolated system can be required to be invariant to a displacement of the system;* hence if we increase each $\mathbf{r}_s$ by the same infinitesimal vector $\boldsymbol{\epsilon}$, the change in the Lagrangian must be zero. Thus, using Eq. (A3.1),†

$$\delta L = \sum_s \frac{\partial L}{\partial \mathbf{r}_s} \cdot \delta \mathbf{r}_s = \sum_s \frac{\partial L}{\partial \mathbf{r}_s} \cdot \boldsymbol{\epsilon} = 0. \tag{3.8}$$

Using equations of motion (2.12) with the Cartesian coordinates of particles as the generalized coordinates, this is transformed as follows:

$$\boldsymbol{\epsilon} \cdot \frac{d}{dt} \sum_s \frac{\partial L}{\partial \dot{\mathbf{r}}_s} = 0. \tag{3.9}$$

Remembering that $\boldsymbol{\epsilon}$ is an arbitrary infinitesimal vector, and designating the momentum vector $\partial L / \partial \dot{\mathbf{r}}_s$ by $\mathbf{p}_s$, we obtain

$$\frac{d}{dt} \sum_s \mathbf{p}_s = 0 \qquad \text{or} \qquad \mathbf{p} = \sum_s \mathbf{p}_s = \textbf{constant}. \tag{3.10}$$

This is the *law of conservation of momentum*. Stated in words, it is as follows: For an isolated system the vector sum of the momenta of all the particles, which may be called the momentum of the system, is constant. Limitation to a Cartesian coordinate system, which was convenient for the purposes of derivation, may now be removed, by observing that Eq. (3.10) is a vector equation.

When the system is not isolated, δL will not be zero, but will have the form $\mathbf{F} \cdot \boldsymbol{\epsilon}$. Vector $\mathbf{F}$ is called the *resultant force* acting on the system. Instead of Eqs. (3.8) and (3.10) we will have

$$\delta L = \boldsymbol{\epsilon} \cdot \mathbf{F} \qquad \text{and} \qquad \frac{d\mathbf{P}}{dt} = \mathbf{F}. \tag{3.11}$$

* If we did not make this restriction on the Lagrangian for an isolated system, it would be possible to add an arbitrary total derivative that would change the expression for the momentum so that in general it would not be conserved. The equations of motion, of course, would not be affected.

† Equations marked with the letter *A* are to be found in the Appendix. Summation over the particles of a system will be explicitly shown, even though the convention of summing relative to repeated indices sometimes makes it unnecessary.

In the notation of this section Eq. (2.23) can be written

$$\delta W = \sum_s \mathbf{p}_s \cdot \delta \mathbf{r}_s - H\,\delta t. \tag{3.12}$$

Thus the change in W due to a uniform virtual displacement of the system by an infinitesimal amount $\boldsymbol{\epsilon}$ is given by

$$\delta W = \mathbf{P} \cdot \boldsymbol{\epsilon}. \tag{3.13}$$

This equation can be considered as defining the *total momentum of the system* in all cases.

Since, however, we often have the Lagrangian L given instead of the action function W, it is desirable to express the total momentum in terms of L. We have by Eq. (3.13), for a uniform virtual displacement $\delta \mathbf{r}_s = \boldsymbol{\epsilon}$,

$$\mathbf{P} \cdot \boldsymbol{\epsilon} = \delta W = \frac{\partial W}{\partial q_i}\,\delta q_i = \frac{\partial W}{\partial q_i} \sum_s \frac{\partial q_i}{\partial \mathbf{r}_s} \cdot \delta \mathbf{r}_s = \frac{\partial W}{\partial q_i} \sum_s \frac{\partial q_i}{\partial \mathbf{r}_s} \cdot \boldsymbol{\epsilon}.$$

Since $\partial W/\partial q_i$ is the component p_i of the momentum corresponding to the coordinate q_i and is thus equal to $\partial L/\partial \dot{q}_i$, our expression for the total momentum of the system becomes

$$\mathbf{P} = \frac{\partial L}{\partial \dot{q}_i} \sum_s \frac{\partial q_i}{\partial \mathbf{r}_s} = p_i \sum_s \frac{\partial q_i}{\partial \mathbf{r}_s}. \tag{3.14}$$

When the Cartesian coordinates of the particles are used as the generalized coordinates q_j of the system, Eq. (2.13) can be written

$$\frac{d\mathbf{p}_s}{dt} \equiv \dot{\mathbf{p}}_s = \mathbf{F}_s, \qquad s = 1, 2, \ldots, n,$$

where $\mathbf{F}_s = \partial L/\partial \mathbf{r}_s$ is termed the force acting on sth particle. Then the first of Eqs. (3.10) for an isolated system can be written

$$\sum_{s=1}^{n} \mathbf{F}_s = 0.$$

For two particles, $n = 2$, and the equation yields

$$\mathbf{F}_2 = -\mathbf{F}_1.$$

This is a special case of *Newton's third law of motion.*

When we have a single isolated particle, Eq. (3.8) requires, since $\boldsymbol{\epsilon}$ is an arbitrary vector, that $\partial L/\partial \mathbf{r} = 0$. This means that L is a function of the velocity $\dot{\mathbf{r}}$ only. Therefore, $\mathbf{p}$ which is equal to $\partial L/\partial \dot{\mathbf{r}}$ must also be a function of $\dot{\mathbf{r}}$ only. Equation (3.10) then states that a function of $\dot{\mathbf{r}}$ only is a constant, or

$$\dot{\mathbf{r}} = \textbf{constant} = \mathbf{v}. \tag{3.15}$$

This is *Newton's first law of motion* or the *law of inertia.*

When all the particles of a body move with the same constant velocity $\mathbf{v}$, their motion satisfies the equation of motion of isolated particles, Eq. (3.15), and the distances between them do not change. Such motion is therefore a possible motion of a rigid body. When a rigid body moves in this way, it is said to be in *translation by inertia*, with a velocity $\mathbf{v}$. Evidently a special case of such a motion is that for which $\mathbf{v} = 0$, the case of a body at rest.

At the beginning of the second section we introduced the concept of an inertial coordinate system, but at that time gave no way of constructing such a system. We now see that an inertial coordinate system is obtained by attaching a Cartesian coordinate system to a rigid body in translation by inertia;* this means choosing it so that the body is at rest in it. That such a coordinate system is an inertial one can be seen, for instance, from the fact that, the body being isolated, the only possible causes of the asymmetry of space, the velocity of the body and its time derivatives, are zero in this coordinate system. The use of various rigid bodies provides us with a method of constructing any desired number of inertial coordinate systems. Since a body at rest in one of these coordinate systems must have a constant velocity in another, and conversely, we say that the inertial coordinate systems move with constant velocities with respect to one another.

Experience shows that all inertial coordinate systems are equivalent. None of them possesses any unique quality, and physical laws can be stated in precisely the same form in all of them. This general statement is known as *Newton's principle of relativity.*

For any isolated system of moving particles it is possible to define the velocity of the system as a whole, in any given inertial coordinate system. To this end we make use of the law of conservation of momentum, Eq. (3.10). According to this law $\mathbf{P}$, the total momentum of the system, will

* To determine whether a rigid body is in translation by inertia we could observe whether the body changes its velocity or orientation relative to a set of isolated particles that have a common velocity. The latter would be evidenced by the particles' maintaining a constant distance between each pair of them.

As we observed before, an isolated system or an isolated particle is not achievable in an absolute sense, but this means only that an inertial coordinate system is a concept having only approximate realizations. The same, of course, holds for most concepts of physics.

remain constant during the motion; however, the vector **P** will be different in the various inertial systems, and it will be zero in one of them. In the inertial coordinate system in which **P** is zero, the system of particles as a whole is said to be at rest, and this coordinate system is known as the *proper coordinate system* for the system of particles. In any other inertial coordinate system the total momentum will no longer be zero, and the proper coordinate systems will appear in each such system as moving with some velocity **v**; this velocity is called the velocity of the system of particles as a whole in the given coordinate system.

Take once more an isolated system of particles, and consider the changes in the Lagrangian when the system is rotated through an infinitesimal angle. The changes in positions and velocities of the particles is given by

$$\delta\mathbf{r}_s = -\mathbf{r}_s \times \boldsymbol{\epsilon} \qquad \text{and} \qquad \delta\mathbf{v}_s = -\mathbf{v}_s \times \boldsymbol{\epsilon}, \tag{3.16}$$

where $\mathbf{v}_s = \dot{\mathbf{r}}_s$, and $\boldsymbol{\epsilon}$ is an infinitesimal vector along the axis of rotation. (See the first equations of Eqs. (A2.32) and (A2.33).) Since this is a rotation of the system as a whole, there should again be no change in the Lagrangian. The change in the Lagrangian is

$$\delta L = \sum_s \left(\frac{\partial L}{\partial \mathbf{r}_s} \cdot \delta\mathbf{r}_s + \frac{\partial L}{\partial \mathbf{v}_s} \cdot \delta\mathbf{v}_s\right);$$

therefore

$$\sum_s \left(\frac{\partial L}{\partial \mathbf{r}_s} \cdot \mathbf{r}_s \times \boldsymbol{\epsilon} + \frac{\partial L}{\partial \mathbf{v}_s} \cdot \mathbf{v}_s \times \boldsymbol{\epsilon}\right) = 0. \tag{3.17}$$

Substituting from the equation of motion (2.12), and remembering the definition of the momenta $\mathbf{p}_s$, one can write this equation as follows:

$$\boldsymbol{\epsilon} \cdot \sum_s \left(\mathbf{r}_s \times \frac{d\mathbf{p}_s}{dt} + \frac{d\mathbf{r}_s}{dt} \times \mathbf{p}_s\right) = 0. \tag{3.18}$$

Now $\boldsymbol{\epsilon}$ is an arbitrary infinitesimal vector, and the expression in parentheses is a total derivative of $\sum_s (\mathbf{r}_s \times \mathbf{p}_s)$; hence

$$\mathbf{M} = \sum_s (\mathbf{r}_s \times \mathbf{p}_s) = \textbf{constant}. \tag{3.19}$$

The vector **M** is called the total angular momentum of the system, and the fact that it remains unchanged in the motion of an isolated system is called the *law of conservation of angular momentum.*

Suppose a system is not isolated but its interaction with the surrounding matter is such that all directions about some axis are indistinguishable.

Then in keeping with the general principle we have been using the Lagrangian should show the same symmetry. In other words, a virtual rotation of the system as a whole about this axis will not change the Lagrangian. For definiteness, let this axis be the Z-axis of a Cartesian coordinate system. Then $\boldsymbol{\epsilon}$ in Eq. (3.18) has but one component different from zero, the component along the Z-axis. The corresponding component of $\mathbf{M}$ must therefore be constant. Thus, if as we have assumed, the Z-axis is an axis of symmetry of the system

$$M_z = \text{constant}. \tag{3.20}$$

If such a nonisolated system has an interaction Lagrangian with all other bodies that is independent of the orientation of the system, then the rotation vector $\boldsymbol{\epsilon}$ in Eq. (3.18) is again arbitrary and Eq. (3.19) holds. Thus angular momentum may be conserved even in a nonisolated system.

In general, however, when the system is not isolated, δL will not be zero, but will have the form $\boldsymbol{\epsilon} \cdot \mathbf{T}$. Instead of Eqs. (3.18) and (3.19) we then have

$$\delta L = \boldsymbol{\epsilon} \cdot \mathbf{T} \qquad \text{and} \qquad \frac{d\mathbf{M}}{dt} = \mathbf{T}. \tag{3.21}$$

The vector $\mathbf{T}$ is called the *resultant torque* acting on the system.

From Eqs. (3.12) and (3.16) we see that the change in W due to an infinitesimal rotation is given by

$$\delta W = -\sum_s \mathbf{p}_s \cdot \mathbf{r}_s \times \boldsymbol{\epsilon} = \sum_s \mathbf{r}_s \times \mathbf{p}_s \cdot \boldsymbol{\epsilon} = \mathbf{M} \cdot \boldsymbol{\epsilon}, \tag{3.22}$$

which can be regarded as the general definition of $\mathbf{M}$, the total angular momentum of any system.

As noted in the Appendix, Section A2, the so-called "position vector" $\mathbf{r}_s$ is not a true vector for transformations involving a change in the orgin. Thus, in general, the angular momentum and the torque depend on the choice of origin.

Problems

1. The metric in Eq. (1.3) can be put in form of that in Eq. (1.1) by setting

$$dx = a \sin\theta \, d\phi, \qquad dy = a \, d\theta.$$

Such a metric is said to be that of a *local Cartesian coordinate system* associated with particular values of θ and ϕ. What geometrical surface is described by such a metric and how is it related to the surface of the sphere?

2. Show that the necessary and sufficient condition that a space be static is that it is possible to express the distance ds between neighboring points (x_1, x_2, x_3) and $(x_1 + dx_1, x_2 + dx_2, x_3 + dx_3)$ in the form

$$(ds)^2 = \sum_{i=1}^{3} \sum_{j=1}^{3} g_{ij}\, dx_i\, dx_j,$$

where the g_{ij} are not functions of time.

3. Using vector analysis show that the differential distance $ds = |d\mathbf{r}|$ between the points

$$\mathbf{r} = \xi\mathbf{a} + \eta\mathbf{b} + \zeta\mathbf{c}$$

and

$$\mathbf{r} + d\mathbf{r} = (\xi + d\xi)\mathbf{a} + (\eta + d\eta)\mathbf{b} + (\zeta + d\zeta)\mathbf{c},$$

where **a**, **b**, and **c** are noncoplanar unit vectors, is given by

$$\begin{aligned}(ds)^2 = d\mathbf{r} \cdot d\mathbf{r} = {} & (d\xi)^2 + (d\eta)^2 + (d\zeta)^2 \\ & + 2\cos(\mathbf{a}, \mathbf{b})\, d\xi\, d\eta + 2\cos(\mathbf{b}, \mathbf{c})\, d\eta\, d\zeta \\ & + 2\cos(\mathbf{c}, \mathbf{a})\, d\zeta\, d\xi.\end{aligned}$$

Here $\cos(\mathbf{a}, \mathbf{b})$ stands for the cosine of the angle between the vectors **a** and **b**.

4. What is meant by the following statements?

(a) Space is static.
(b) Clocks are synchronizable.
(c) Space is three dimensional.
(d) Space is Euclidean.

5. Define the following: (a) objective time, (b) clock, (c) particle, (d) event, (e) rigid body, (f) absolutely rigid body, (g) coordinate system, (h) metric.

6. Give some reasons for disqualifying a laboratory room as an inertial coordinate system.

7. Can an electron ever be considered as an isolated system?

8. Explain how the Heisenberg uncertainty principle in quantum mechanics bears upon the validity of the assumption in Eq. (2.1).

9. What is meant by the "state" of a system?

10. How many generalized coordinates are needed to specify the configuration of a single free rigid body and that of a pendulum? What are the generalized momenta conjugate to these coordinates?

11. What is the physical significance of the interaction term between two systems as it occurs in the Hamiltonian of the combined system?

12. Show that

$$\delta \dot{q}_j = \frac{d}{dt}\,\delta q_j.$$

13. Derive the action function W for a particle of mass m in a uniform gravitation field of acceleration g. Assume the initial position of the particle is at the origin of a spherical coordinate system. Obtain the momentum, the Lagrangian, and the Hamiltonian using Eqs. (2.19) and (2.20).

14. The Lagrangian for a simple harmonic oscillator is

$$L = \tfrac{1}{2}m\dot{x}^2 - \tfrac{1}{2}kx^2.$$

Show that the equations of motion are unaffected if one adds a term

$$\frac{dR(x, t)}{dt}$$

to this Lagrangian.

15. What in general terms is a conservation law?

16. What is the *principle of relativity* and does it hold for Newton's three laws?

17. Knowing that

$$p^2\dot{q} + 2pq\dot{p} = 0,$$

one can conclude that some quantity is conserved. What quantity?

18. (a) Write down the expression for the angular momentum of a system of particles about a given axis, say the Z-axis. (b) The *moment of inertia* I of a rigid body about such an axis is defined by stating that the angular momentum of the body, when spinning with an angular velocity $\omega = \dot{\theta}$ about this axis, is $I\omega$. Derive a general expression for I.

19. If the Lagrangian is invariant to a displacement of a system, so are the equations of motion. Does the converse hold? *Hint:* Try $L = e^{aq}f(\dot{q})$.

20. If the Lagrangian of a system is invariant to a displacement, the total momentum is conserved. Is the converse true?

21. Derive an expression for the *torque* T, see Eq. (3.21), acting on a system of particles.

Chapter II

Newtonian Mechanics

4. Newtonian Space and Time

In a system consisting of two interacting particles, let one of the particles be suddenly affected by an external force, not acting directly on the other particle. Such a situation may arise, for example, when one of the particles collides with an object with which neither of the particles otherwise interacts. The motion of the particle so affected will suddenly change and, because of the interaction of the particles, sooner or later, the motion of the other particle will change also. The time elapsing between the change of the motion of the first particle and the consequent change of the motion of the second particle is said to be the *time of propagation of the interaction* between the particles. The interaction is considered as traveling the distance from the point occupied by the first particle when its motion is changed to the point occupied by the second particle when it is first affected by the interaction. The ratio of this interaction distance to the time of interaction is found to be a constant for empty space and is called *the speed of propagation of an interaction.*

The speed of propagation of an interaction should not be confused with the speed of propagation of a disturbance. The latter occurs when the system consists of a large number of particles near each other and is a secondary effect. A typical example of this is the propagation of sound. The process involved is the following: A number of particles are affected by some external forces. The effect is first propagated to all the other particles with the speed of propagation of the interaction between the particles. This is the primary effect and, in the case of sound, it is very small for all particles not very close to those originally disturbed. The originally disturbed particles now move and, when close enough to other particles, affect their motion. These move, and in turn affect still other particles. Thus a disturbance is propagated, and its speed is evidently smaller than the speed of propagation of interaction.

Evidently, interaction of matter may be used to transmit ideas by means of a prearranged code of signals. In principle, a signal can be transmitted as follows: The person sending the message changes the motion of a particle, while the person receiving it observes another particle near himself, but interacting with the first. Each change of the motion of the first particle may be called the *sending of a signal*, while the corresponding change of the motion of the second particle is the *arrival of the signal.* At each time t after the sending of a signal the signal can be received by an interacting particle located anywhere on a certain surface. In an inertial coordinate system, due to the isotropy of space, this surface will evidently be a sphere, the equation of which in Cartesian coordinates is

$$r = (x^2 + y^2 + z^2)^{1/2} = c(t - t_0). \tag{4.1}$$

Here c is the speed of propagation of the signal, which is the same as the speed of propagation of an interaction, and t_0 is the time at which it is sent. The point from which the signal is sent is taken as the origin.

Since c, the speed of propagation of an interaction, is very large* compared to ordinary speeds, the limiting case of $c \to \infty$ is a good approximation for certain phenomena treated under such subject titles as *electrostatics* and *magnetostatics.* We find it useful, partly because of the simplicity of the resulting structure, to develop this limiting case before passing on to the general case of a finite c treated in *electrodynamics.*

If, as we assume in electrostatics, the speed of propagation of the interaction c is infinite, then the equation of the signal, Eq. (4.1), becomes

$$t = t_0 = \text{constant} \qquad \text{(for all values of } \mathbf{r}). \tag{4.2}$$

In another inertial coordinate system in which the time is designated by t', the equation of the same signal will be

$$t' = t_0' = \text{constant}. \tag{4.3}$$

Since they designate the same event, t_0 corresponds to t_0', or by Eqs. (4.2)

* The speed of propagation of an interaction c is the speed of light in vacuum. According to recent measurements its value is $(2.997925 \pm 0.000003) \times 10^{10}$ cm/sec; *Phys. Today* **17**, 48 (1964).

and (4.3), to each time t of one inertial coordinate system corresponds one time t' of the other, independent of the values of x, y, and z, so that we must have

$$t' = f(t). \tag{4.4}$$

For infinitesimal increments this relation becomes

$$dt' = f'(t)\, dt. \tag{4.5}$$

On the other hand, for inertial coordinate systems, the relation between dt' and dt cannot depend on t, because all values of t are equivalent; thus, $f'(t) = a =$ constant. Integrating, we have $t' = at + b$. The constant b determines, merely, to what time t' shall correspond the initial time $t = 0$; we can therefore choose $b = 0$.

The constant a, while independent of $\mathbf{r}$ and t, may depend upon those parameters that characterize the coordinate system (x', y', z', t') and distinguish it from the coordinate system (x, y, z, t). We have seen that inertial coordinate systems are realized by attaching Cartesian coordinate systems to rigid bodies in translation by inertia, and that particles at rest in one inertial coordinate system are moving with a constant velocity in the other. Consider a particle P at rest in the primed coordinate system and having instantaneous coordinates expressed by the position vector $\mathbf{r}' = (x', y', z')$. In the unprimed system the particle P will have a position vector $\mathbf{r} = (x, y, z)$ whose rate of change with t is given by $\dot{\mathbf{r}} = (\dot{x}, \dot{y}, \dot{z}) =$ constant. The latter is clearly the velocity of the primed system in the unprimed system and will be designated by $\mathbf{v}$. The constant a may be a function of $\mathbf{v}$; but, since all directions in the coordinate system (x, y, z, t) are equivalent, a may depend only on the absolute value of the vector $\mathbf{v}$, but not on its direction. We thus have

$$t' = a(v)t. \tag{4.6}$$

Consider now a third inertial coordinate system $(\bar{x}, \bar{y}, \bar{z}, \bar{t})$ the velocity of which relative to the unprimed system is $\bar{\mathbf{v}}$ and relative to the primed system is $\bar{\mathbf{v}}'$. Then, we must have

$$\bar{t} = a(\bar{v})t = a(\bar{v}')t'. \tag{4.7}$$

From Eqs. (4.6) and (4.7) it follows that

$$a(\bar{v}) = a(\bar{v}')a(v). \tag{4.8}$$

It is clearly possible to change the magnitude of $\bar{\mathbf{v}}$ without changing the magnitudes of $\mathbf{v}$ and $\bar{\mathbf{v}}'$ by merely changing the relative direction of $\bar{\mathbf{v}}'$ with respect to $\mathbf{v}$. The right-hand side of Eq. (4.8) thus remains the same even though $\bar{\mathbf{v}}$ changes. This requires that

$$a(\bar{v}) = \text{constant}. \tag{4.9}$$

In particular, when $\bar{\mathbf{v}} = 0$, the third system coincides with the first, so that the value of the constant is unity.* Thus, by Eq. (4.6),

$$t' = t. \tag{4.10}$$

This means that time is the same for all inertial coordinate systems. We express this fact by saying that *time is absolute.*

In a somewhat similar way one can show that the distance between any two particles, *at any given time*, is the same for all inertial coordinate systems, i.e., *space is absolute.*

For coordinate systems that are not inertial, in general, time and space will not be absolute. It can be shown, however, that space and time remain absolute to the extent that the rigid bodies used as the standards of distance and the clocks are not affected by external forces. Such rigid bodies have been previously called absolutely rigid; such clocks may be called *perfect clocks.* With perfect clocks and absolutely rigid measuring rods space and time are absolute for all coordinate systems. Such a space–time system was assumed by Newton and is therefore called *Newtonian.* In Section 10, however, we shall see that the absoluteness of space and time can no longer be maintained when account is taken of the finiteness of the speed of propagation of interaction.

When space and time are absolute, we can add together vectors of displacement drawn in different inertial coordinate systems. Thus, if the position of a particle P at the time $t' = t$ in a primed system is designated by the position vector $\mathbf{r}'$ (see Fig. 4.1) and if the origin of this system moves with the velocity $\mathbf{v}$ with respect to an unprimed coordinate system, in which P's initial position is given by the vector $\mathbf{r}_0$; then the position vector of the particle at t with respect to the second coordinate system is evidently given by

$$\mathbf{r} = \mathbf{r}_0 + \mathbf{r}' + \mathbf{v}t. \tag{4.11}$$

It is customary to choose the coordinate systems in such a way that the origins coincide for $t = 0$; then $\mathbf{r}_0 = 0$ and Eq. (4.11) can be replaced by

$$\mathbf{r}' = \mathbf{r} - \mathbf{v}t. \tag{4.12}$$

* It is, of course, understood that the units in which space and time are to be measured are to be taken the same in each coordinate system.

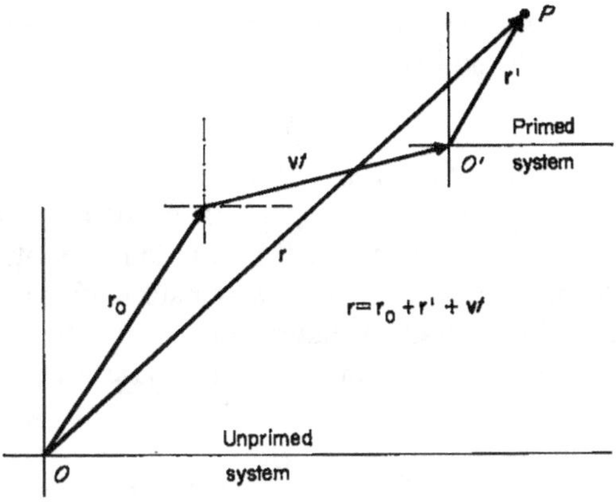

FIG. 4.1

This equation and Eq. (4.10) are referred to as the *Galilean transformation.* Differentiating Eq. (4.12) or Eq. (4.11) with respect to time gives the law of addition of velocities:

$$\mathbf{u}' = \mathbf{u} - \mathbf{v}. \tag{4.13}$$

Here $\mathbf{u} = \dot{\mathbf{r}}$ and $\mathbf{u}' = \dot{\mathbf{r}}'$ are the velocities of the particle P in the two coordinate systems. In the special case that the primed system moves in the positive X-axis of the unprimed system, $\mathbf{v}$ has only an X component and the Galilean transformation in terms of suitably chosen Cartesian coordinates becomes

$$x' = x - vt, \qquad y' = y, \qquad z' = z, \qquad t' = t. \tag{4.14}$$

5. Newtonian Particle

The Lagrangian characterizing a free particle, in an inertial coordinate system based on Newtonian space and time, cannot contain the coordinates of the particles, because, as we have seen, the Lagrangian of an isolated system is assumed not changed by a virtual displacement of the system. It must therefore be a function only of the velocity $\mathbf{v}$ of the particle. Further, since space is isotropic and the Lagrangian of an isolated system is assumed not to be changed by rotating the system, the direction of the velocity cannot play any role. The Lagrangian is therefore a function of $\mathbf{v}^2 = v^2$, and we may write

$$L = \varphi(v^2). \tag{5.1}$$

To determine further the form of this function we may use Newton's principle of relativity, that is, the equivalence of all inertial coordinate systems. In this case it means that the equation of motion arising from Hamilton's principle, Eq. (2.12), should remain unchanged in form when we perform a Galilean transformation of coordinates.* This requires, as we saw in Eq. (2.34), that in the new coordinate system the Lagrangian be given by

$$L' = aL + \frac{dR}{dt}(\mathbf{r}, t; \delta\mathbf{v}),$$

where a is a constant and $\delta\mathbf{v}$ a constant vector characterizing the transformation. From Eq. (5.1), therefore,

$$\varphi(|\mathbf{v} + \delta\mathbf{v}|^2) = a\varphi(v^2) + \frac{dR}{dt},$$

which to the first order in δv requires that

$$\varphi(v^2 + 2\mathbf{v} \cdot \delta\mathbf{v}) = a\,\delta(v^2) + \frac{dR}{dt};$$

hence, by the application of Eq. (A3.59) and Taylor's series,

$$\delta(v^2) + 2\mathbf{v} \cdot \delta\mathbf{v}\,\varphi'(v^2) = a\varphi(v^2) + \mathbf{v} \cdot \frac{\partial R}{\partial \mathbf{r}} + \frac{\partial R}{\partial t}. \tag{5.2}$$

Since the left-hand side does not involve $\mathbf{r}$ and t, we conclude that $a = 1$, $\partial R/\partial t = 0$, and

$$2\varphi'(v^2)\,\delta\mathbf{v} = \frac{\partial R}{\partial \mathbf{r}}. \tag{5.3}$$

From this equation, since R is not a function of $\mathbf{v}$, we conclude that $2\varphi'$ is a constant. This constant is designated by m and is called the *mass* of the particle. Integrating both sides of the equation $2\varphi' = m$, keeping in mind that the variable is v^2, we get for Eq. (5.1)

$$L = \tfrac{1}{2}mv^2. \tag{5.4}$$

A constant could, of course, be added, but, since this is true for all

* Requirement that L is unchanged in form under this transformation is too severe and leads to the erroneous result $L =$ constant.

Lagrangians occurring in electrostatics, it seems desirable to have this arbitrariness in the Lagrangian understood and to save oneself the inconvenience of adding an arbitrary constant to each Lagrangian. It is evident that the value of the constant m is immaterial, as long as we are dealing with a single particle. When, however, as we saw in Section 2, we are dealing with several interacting particles, the individual Lagrangians can be added only if the ratios of their masses are suitably chosen—these being the multipliers of the separate Lagrangians. Experimentally, the ratios of masses may be determined by the use of the law of conservation of momentum given in Eq. (3.10).

Action, in our case of one particle, is

$$W = \frac{1}{2} m \int_a^b v^2 \, dt. \tag{5.5}$$

From this it is evident that, by making the trajectory zigzag, the integral in Eq. (5.5) can be made as large as desired; therefore, W will have no minimum even for a short trajectory if m is negative. Thus, Hamilton's principle (see Section 2) requires that the masses of all particles be positive.*

The momentum, energy, and the angular momentum of a free particle are easily calculated and are

$$\mathbf{p} = m\mathbf{v}, \tag{5.6}$$

$$H = p^2/2m, \tag{5.7}$$

$$\mathbf{M} = m\mathbf{r} \times \mathbf{v}. \tag{5.8}$$

Problems

1. Show that a simple harmonic oscillator (or pendulum) will have the same period and hence register the same time in two inertial coordinate systems whose coordinates are related by Eq. (4.12). Assume a Lagrangian of the form

$$L = \tfrac{1}{2}m\dot{x}^2 - \tfrac{1}{2}kx^2.$$

* It may be argued that this is trivial, since by demanding that action should have a maximum instead of a minimum the masses would be made negative. The point we wish to make is that all masses are of the same sign, so that, choosing some mass as the unit, all masses are a positive number of times the unit mass.

2. Show that the transformations of Galilei corresponding to all finite values of the relative velocity **v** of the coordinate systems constitute a continuous mathematical group.

3. If a system of particles has zero angular momentum in one coordinate system, show that it must have zero angular momentum in all coordinate systems related to it by a Galilean transformation.

4. Why is space considered "absolute"? Are all spaces absolute?

5. What change would have to be made in Hamilton's principle if one decided to assign negative values to the masses of all particles?

Chapter III

The Electrostatic Field

6. Definition and Fundamental Properties

When a particle is a part of a system, its Lagrangian, in accordance with the general theory of nonisolated systems given in Section 2, will be $\frac{1}{2}mv^2 - U(\mathbf{r}, \mathbf{v}, t)$. Experience shows, however, that within the limitations of electrostatics this *interaction function U is independent of the velocity* **v**. The Lagrangian function for the particle will thus have the form

$$L = \tfrac{1}{2}mv^2 - U(\mathbf{r}, t). \tag{6.1}$$

The Hamiltonian corresponding to this Lagrangian can be immediately calculated and is

$$H = \frac{p^2}{2m} + U(\mathbf{r}, t). \tag{6.2}$$

When U is independent of t, H is the *energy* of the particle. The first term of the right-hand member is called its *kinetic* energy, and the second its *potential* energy.

Experience shows that the action of other particles upon a given particle is separable into *fields:* gravitational, electrostatic, etc. For each field there is a term in the interaction function U of the form $e\varphi(\mathbf{r}, t)$, where e is a constant depending only upon the properties of the particle, and φ is a function depending only on the rest of the system.

We shall consider here only the electrostatic field. In this case e is called the *electric charge* of the particle and φ the *potential* of the field acting on it. The possibility of writing

$$U = e\varphi(\mathbf{r}, t) \tag{6.3}$$

means that if, at any instant of time, a particle should be replaced by another, the ratios of the U's for the two particles will be a constant equal to the ratio of their electric charges, and therefore independent of the field. The values of the charges are thus determined as soon as the charge of one of the particles is assigned a value. This is customarily done by defining a *unit charge*, or selecting a *system of units*. It is to be remembered that the potential of a field, like the Lagrangian function, may be changed by the addition of an arbitrary constant. It is usually convenient to choose it in such a way that the value of the potential energy is zero at points infinitely removed from the system, i.e., when the physical interaction is presumably zero.

The equation of motion of the particle derived from the Lagrangian in Eq. (6.1) can be written

$$\frac{d}{dt}(m\mathbf{v}) = -\frac{\partial U}{\partial \mathbf{r}} = -e\,\nabla\varphi = e\boldsymbol{\mathscr{E}}, \tag{6.4}$$

where

$$\boldsymbol{\mathscr{E}} = -\nabla\varphi \tag{6.5}$$

is called the *electrostatic field strength.* When dealing with problems of electrostatics, we shall refer to it as the *field strength* or, simply, the *field.*

Since Eq. (6.5) is valid for any electrostatic field regardless of its source, it may be used to derive equations expressing universal properties of such a field. Among these equations the fundamental role is played by

$$\nabla \times \boldsymbol{\mathscr{E}} = 0. \tag{6.6}$$

This follows from Eq. (6.5) if one takes the *curl* of both sides and uses the fact that $\nabla \times \nabla\varphi = 0$. Note that, since the equation of motion (6.4) contains directly $\boldsymbol{\mathscr{E}}$, and not φ, $\boldsymbol{\mathscr{E}}$ is the primary measurable field quantity. From this point of view Eq. (6.6) is the necessary and sufficient condition for the existence of a potential field satisfying Eq. (6.5). For if Eq. (6.6) holds, then by Eq. (A4.18)

$$\oint \boldsymbol{\mathscr{E}} \cdot d\mathbf{r} = \int \nabla \times \boldsymbol{\mathscr{E}} \cdot d\mathbf{S} = 0, \tag{6.7}$$

where the first integral is taken along a closed curve, and $d\mathbf{S}$ is an element of some surface bounded by the curve. Since any two paths between the points A and B can be considered as parts of a closed curve, we have

$$\int_{APB} \boldsymbol{\mathscr{E}} \cdot d\mathbf{r} + \int_{BP'A} \boldsymbol{\mathscr{E}} \cdot d\mathbf{r} = 0,$$

where P and P' indicate points on the different paths. This can be written

$$\int_{APB} \mathscr{E} \cdot d\mathbf{r} = \int_{AP'B} \mathscr{E} \cdot d\mathbf{r}, \tag{6.8}$$

showing that the line integral of $\mathscr{E} \cdot d\mathbf{r}$ is independent of the path of integration.

This means that $\mathscr{E} \cdot d\mathbf{r}$ is a total differential of some function, say $-\varphi$; hence

$$\mathscr{E} \cdot d\mathbf{r} = -d\varphi = -d\mathbf{r} \cdot \nabla\varphi, \tag{6.9}$$

which must hold for arbitrary $d\mathbf{r}$. In order that Eq. (6.9) hold for arbitrary $d\mathbf{r}$ we must have

$$\mathscr{E} = -\nabla\varphi,$$

showing that, if $\nabla \times \mathscr{E} = 0$, then a scalar function φ exists satisfying Eq. (6.5). Moreover, using Eq. (6.9), we get

$$\int_{\mathbf{a}}^{\mathbf{b}} \mathscr{E} \cdot d\mathbf{r} = -\int_{\mathbf{a}}^{\mathbf{b}} d\varphi = \varphi(\mathbf{a}) - \varphi(\mathbf{b}). \tag{6.10}$$

If one keeps the lower limit fixed and varies the upper limit, one obtains from Eq. (6.10) an equation defining φ in terms of $\mathscr{E}$, namely,

$$\varphi(\mathbf{r}) = -\int_{\mathbf{a}}^{\mathbf{r}} \mathscr{E} \cdot d\mathbf{r} + \varphi(\mathbf{a}). \tag{6.11}$$

Example. If $\mathscr{E} = b\mathbf{r}/r^3$, what is the expression for the potential φ?

First we test whether $\nabla \times \mathscr{E} = 0$ in order to make sure a potential φ exists. We have

$$\nabla \times \mathscr{E} = \frac{b}{r^3}\nabla \times \mathbf{r} - b\mathbf{r} \times \nabla\left(\frac{1}{r^3}\right) = -b\mathbf{r} \times \frac{3\mathbf{r}}{r^5} = 0; \tag{6.12}$$

therefore φ exists and from Eq. (6.11) is given by

$$\varphi(\mathbf{r}) = -\int_{\mathbf{a}}^{\mathbf{r}} \frac{b\mathbf{r}'}{(r')^3} \cdot d\mathbf{r}' + \varphi(\mathbf{a}). \tag{6.13}$$

By making use of the relation

$$\mathbf{r}' \cdot d\mathbf{r}' = \tfrac{1}{2}\, d(\mathbf{r}' \cdot \mathbf{r}') = \tfrac{1}{2}\, d(r'^2) = r'\, dr' \tag{6.14}$$

and assuming that the potential is taken to be zero at $\mathbf{r} = \mathbf{a}$, we have from Eq. (6.13)

$$\varphi(\mathbf{r}) = -b\int_{\mathbf{a}}^{\mathbf{r}} \frac{dr'}{(r')^2} = b\left[\frac{1}{r'}\right]_{\mathbf{a}}^{\mathbf{r}} = b\left(\frac{1}{r} - \frac{1}{a}\right), \tag{6.15}$$

which is the desired potential.

7. Electrostatic Field Equations

In order to investigate how a system of particles determines the field we assume that we may regard the system as consisting of two parts, the particles and the field, and treat them in much the same way as we would two interacting mechanical systems.* From this point of view the Lagrangian for a system of interacting particles consists of three parts and can be written

$$L = L_p + L_f + L_{pf}, \tag{7.1}$$

where L_p is the Lagrangian for the particles alone, without the field and hence without the interaction, L_f is the Lagrangian of the field, and L_{pf} is the interaction between the field and the particles. Thus, the interaction between the particles expresses itself only through the field.

We know already that

$$L_p = \sum_s \tfrac{1}{2} m_s v_s^2, \qquad s = \text{the particle number}, \tag{7.2}$$

the summation being extended over all the particles, and that the Lagrangian for the sth particle of the system is (see Eqs. (6.1) and (6.3))

$$\tfrac{1}{2} m_s v_s^2 - e_s \varphi(\mathbf{r}_s, t).$$

Since $\frac{1}{2} m_s v_s^2$ is the Lagrangian of the particle treated as an isolated system, we conclude that $-e_s\varphi(\mathbf{r}_s, t)$ is the interaction function of the sth particle and the field. Thus the total interaction function is

$$L_{pf} = -\sum_s e_s \varphi(\mathbf{r}_s, t). \tag{7.3}$$

* As will be seen later, this procedure is *necessary* in electrodynamics, because in this more general case the field must be treated as a separate entity.

The form of L_f remains to be determined; however, before this can be done, we must choose what we are to regard as the coordinates of the field. Any set of quantities uniquely describing the field can be used, but, since the field is determined at any time t when $\varphi(\mathbf{r}, t)$ is given for all points of space, a natural suggestion is to use as the coordinates the value of $\varphi(\mathbf{r}, t)$ at each point of space.* In this case for each $\mathbf{r}$ there will be a coordinate $\varphi(\mathbf{r}, t)$ of the field, and a description of the change in the field will be fully given by the changes in these coordinates. The vector $\mathbf{r}$ is here not a function of time, as it is when it gives the position of a particle, but an independent parameter describing the particular coordinate in question. It is thus analogous to the subscript s used in our treatment of a system of particles.

As there will be a coordinate $\varphi(\mathbf{r}, t)$ for each $\mathbf{r}$, a continuous triple infinity of coordinates will exist. We shall assume, however, that, aside from certain generalizations, the φ's can be treated just as we did the coordinates in our previous work. Thus Hamilton's principle and most of our other assumptions can be used but need to be generalized to take care of a continuous infinity of coordinates. We shall, however, try to develop these generalizations naturally as we go along and not attempt an explicit statement of the mathematical assumptions thus entailed.

Since the behavior of a system of particles is considered to be given completely by a description of the motion of the individual particles, the field of the particles has physical significance only insofar as it affects this motion of the particles. Thus, although the $\varphi(\mathbf{r}, t)$ are taken as the coordinates of the field, the field is considered to be fully described by giving $\mathscr{E}(\mathbf{r}, t)$ throughout space. Since $\mathscr{E} = -\nabla\varphi$, the transformation $\varphi \rightarrow \varphi + k$, where k is a constant, leaves the field unchanged.

The Lagrangian of the field should depend only on those quantities that determine the state of the field, and hence φ should enter only through $\mathscr{E}$ and its derivatives. Referring to the definition of $\mathscr{E}$, Eq. (6.5), we see that this means that $\dot{\varphi}$ can enter only through $\dot{\mathscr{E}}$. If, on the other hand, L_f contained $\dot{\mathscr{E}}$, the equations of motion would contain $\ddot{\mathscr{E}}$. This would mean that the initial values of $\mathscr{E}$ and $\dot{\mathscr{E}}$ would have to be given before such equations could be solved.

When the states of the particles are given (say by giving $\mathbf{r}_s$ and $\dot{\mathbf{r}}_s$ for all

* One might consider that the $\mathscr{E}(\mathbf{r}, t)$ corresponding to different $\mathbf{r}$ could be taken as the coordinates of the field; but in that case we would have to impose the condition given by Eq. (6.6) on them. They would thus not satisfy the condition (see Section 2) that the generalized coordinates be independent of one another.

the particles), their interactions, and therefore $\mathscr{E}(\mathbf{r}_s)$, must be determined; because, otherwise, the forces acting upon the particles, and thus their subsequent behavior, will not be determined by their state.* We thus conclude that $\dot{\mathscr{E}}$, and therefore also $\dot{\varphi}$, does not occur in L_f.

If the Lagrangian L of a system does not contain $\dot{q}_j$, where q_j is a coordinate labeled by the discrete variable j, then the conjugate momentum

$$p_j \equiv \frac{\partial L}{\partial \dot{q}_j}$$

is clearly identically zero. Since $\varphi(\mathbf{r}, t) \equiv \varphi(x, y, z, t)$ is to be considered as a coordinate of the field labeled by the continuous variables x, y, and z, one is forced to adapt the definition of the conjugate momentum to a continuous set of coordinates. One replaces the partial derivative by a *variational* (or *functional*) derivative† and writes

$$p \equiv \frac{\delta L_f}{\delta \dot{\varphi}} = 0. \tag{7.4}$$

Thus the absence of $\dot{\varphi}$ in the Lagrangian L_f implies that the field in electrostatics carries no momentum. In accordance with Eq. (3.10), therefore, the total momentum $\mathbf{P}$ of the system, particles plus their field, is given by

$$\mathbf{P} = \mathbf{P}_{\text{particles}} = \sum_s \mathbf{p}_s = \sum_s m_s \mathbf{v}_s. \tag{7.5}$$

If a Galilean transformation of coordinates (see Eqs. (4.12) and (4.13)) is performed with the old positions $\mathbf{r}_s$ given in terms of the new by

$$\mathbf{r}_s = \mathbf{r}_s' + \mathbf{v}t, \tag{7.6}$$

$\mathbf{v}$ being the velocity of the origin of the primed system in the unprimed, then the corresponding velocities are related by

$$\mathbf{v}_s = \mathbf{v}_s' + \mathbf{v}. \tag{7.7}$$

The total momentum $\mathbf{P}$ of the system in the old coordinate system, by Eq. (7.5), is seen to be

$$\mathbf{P} = \sum_s m_s(\mathbf{v}_s' + \mathbf{v}) = \mathbf{P}' + \mathbf{v} \sum_s m_s, \tag{7.8}$$

* This discussion emphasizes the fact that in electrostatics we treat the field as a *mathematical fiction*, convenient for a description of the interactions of the particles.

† For a definition of the variational derivative see I. M. Gelfand and S. V. Fomin, *Calculus of Variations*, Prentice-Hall, Englewood Cliffs, N.J., 1963, p. 29. See also G. Wentzel, *Quantum Theory of Fields*, Wiley-Interscience, New York, 1949, pp. 1–4.

where $\mathbf{P}'$ is the total momentum of the system in the new coordinate system. If this coordinate system is chosen in such a way that $\mathbf{P}' = 0$, then $\mathbf{v}$ is called the velocity of the system of particles as a whole. With this choice

$$\mathbf{P} = \mathbf{v} \sum_s m_s = M\mathbf{v}, \tag{7.9}$$

where $M = \sum_s m_s$ and is called the *total mass* of the system. Thus, the field has no mass, and the relation between the total momentum, total mass, and the velocity of the system as a whole is the same as for a single particle. The result of the above considerations can be expressed by saying that in electrostatics *mass is additive.*

From Eqs. (7.5) and (7.9) it follows at once that

$$\mathbf{v} = \frac{\sum_s m_s \mathbf{v}_s}{\sum_s m_s} = \frac{d}{dt} \frac{\sum_s m_s \mathbf{r}_s}{\sum_s m_s}, \tag{7.10}$$

so that $\mathbf{v}$ can be considered as the velocity of a point having the position vector

$$\mathbf{r} = \frac{\sum_s m_s \mathbf{r}_s}{\sum_s m_s}. \tag{7.11}$$

This point is called the *center of mass*, or the *center of inertia*, of the system.

Returning to the consideration of the Lagrangian of Eq. (7.1), we see that the corresponding Hamiltonian can be easily calculated, and is

$$H = \sum_s \frac{p_s^2}{2m_s} + U, \tag{7.12}$$

where

$$U = -L_{pf} - L_f. \tag{7.13}$$

It is evident that U is the total contribution to the energy due to the interactions among the particles and may be called the *potential energy* of the system.

We must now make use of the *principle of superposition.* This expresses an important property of electromagnetic fields, namely: that *the total field produced by a system of particles is obtained by adding the fields of the individual particles.* It is expressed mathematically by the equation

$$\mathscr{E} = \sum_s \mathscr{E}^{(s)} \tag{7.14}$$

for the field strength $\mathscr{E}$, or the analogous equation

$$\varphi = \sum_s \varphi^{(s)} \tag{7.15}$$

for the potential. Here $\varphi^{(s)}$ and $\mathscr{E}^{(s)}$ are the potential and field strength, respectively, of the sth particle. This means, of course, that the addition of a particle to a system merely adds its field to the fields of the other particles, without changing these fields in any way.

Consider now the way in which the principle of superposition restricts the form of the Lagrangian of the system, and thus the form of the potential energy U. We have seen that for a single particle the Lagrangian may be taken as $L = \frac{1}{2}mv^2$, so that $U = 0$. For two particles U will result from their interaction, and may be designated by U_{12} or U_{21}. By the principle of superposition, as other particles are added, each will introduce the interaction of itself with all the other particles, without changing the interactions already existing. Thus, the total potential energy will be*

$$U = \frac{1}{2}\sum_{j}\sum_{\substack{k \\ (j \neq k)}} U_{jk}, \tag{7.16}$$

where the factor $\frac{1}{2}$ appears because in the double sum each term occurs twice. Thus, for example, besides U_{12} the sum will contain the term U_{21}, which is the same as U_{12}.

On the other hand, as we have seen, L_f must be a function of the field $\mathscr{E}$.† We assume that each portion of the space dV will contribute to L_f some amount $F(\mathscr{E})\,dV$ independently of the field in other portions of space. Then L_f will be

$$L_f = \int F(\mathscr{E})\,dV + \text{constant}. \tag{7.17}$$

The constant, which of course makes no difference in the field equations, should be chosen in such a way as to make Eq. (7.16) consistent with Eq. (7.13).

Remembering that L_f, being the Lagrangian for the field without the particles, cannot depend upon the direction of the field in an isotropic space, we see that F must be of the form

$$F(\mathscr{E}) = f(\mathscr{E}^2) = f(\mathscr{E}_x^2 + \mathscr{E}_y^2 + \mathscr{E}_z^2). \tag{7.18}$$

We can now determine the form of the function $f(\mathscr{E}^2)$. In fact, let the

* Since the field is used merely to express the interaction between the particles, we do not include the U_{jj} terms.

† The higher derivatives of φ, in terms such as $\nabla \cdot \mathscr{E}$, do not enter for reasons similar to those used to exclude from L the higher derivatives of the coordinates q_j of the particles.

value of $\mathscr{E}$, Eq. (7.14), be substituted into Eq. (7.17) using Eq. (7.18). Then, its right-hand side must separate into a sum of terms, none of which can contain more than two different $\mathscr{E}^{(s)}$. This is because Eq. (7.17) must be consistent with Eq. (7.16), and the latter shows that no term of U can contain the effect of more than two particles. Mathematically this can be stated as follows:

$$\frac{\partial^3 f}{\partial \mathscr{E}_i^{(m)}\, \partial \mathscr{E}_j^{(n)}\, \partial \mathscr{E}_k^{(p)}} = 0, \qquad \begin{cases} m, n, p = 1, 2, \ldots, n, \\ i, j, k = 1, 2, 3, \\ m \neq n \neq p \neq m, \end{cases} \tag{7.19}$$

where the superscript designates the particle producing the field and the subscript the component of that field—it being understood that $\mathscr{E}_1 = \mathscr{E}_x$, $\mathscr{E}_2 = \mathscr{E}_y$, and $\mathscr{E}_3 = \mathscr{E}_z$. Now

$$\sum_s \mathscr{E}_k^{(s)} = \mathscr{E}_k;$$

therefore

$$\frac{\partial f}{\partial \mathscr{E}_k^{(s)}} = \frac{\partial f}{\partial \mathscr{E}_j} \frac{\partial \mathscr{E}_j}{\partial \mathscr{E}_k^{(s)}} = \frac{\partial f}{\partial \mathscr{E}_k},$$

so that

$$\frac{\partial^3 f}{\partial \mathscr{E}_i\, \partial \mathscr{E}_j\, \partial \mathscr{E}_k} = 0. \tag{7.20}$$

In particular this means that

$$\frac{\partial^3 f}{\partial \mathscr{E}_x^3} = 0$$

or

$$f(\mathscr{E}^2) = \tfrac{1}{2} a \mathscr{E}_x^2 + b \mathscr{E}_x + c, \tag{7.21}$$

where a, b, and c are independent of $\mathscr{E}_x$. Since Eq. (7.20) also states (taking $i = 2$ and 3, and $j = k = 1$) that

$$\frac{\partial^3 f}{\partial \mathscr{E}_y\, \partial \mathscr{E}_x^2} = \frac{\partial^3 f}{\partial \mathscr{E}_z\, \partial \mathscr{E}_x^2} = 0,$$

a must be a constant. Finally, we see from Eq. (7.21) that, since $f = f(\mathscr{E}^2)$, f should contain $\mathscr{E}_x$ always as $\mathscr{E}_x^2$; hence $b = 0$. Moreover, since f is symmetric in $\mathscr{E}_x^2$, $\mathscr{E}_y^2$, and $\mathscr{E}_z^2$, being a function of their sum,

$$f(\mathscr{E}^2) = \tfrac{1}{2} a \mathscr{E}_x^2 + \tfrac{1}{2} a \mathscr{E}_y^2 + \tfrac{1}{2} a \mathscr{E}_z^2 = \tfrac{1}{2} a \mathscr{E}^2. \tag{7.22}$$

An arbitrary constant, which could be added, has been included in the expression for L_f, Eq. (7.17).

The value of the constant a depends upon the choice of the units. The sign of it could be, *a priori*, either positive or negative. For electrostatic field it turns out to be positive.* We choose $a = 1/4\pi$, which corresponds to the so-called *electrostatic system of units*. If we should choose $a = 1$, we would be employing rationalized units as one does in the MKS system of units.† Thus, with our choice of units,

$$L_f = \frac{1}{8\pi} \int \mathscr{E}^2 \, dV + \text{constant}. \tag{7.23}$$

We are now in a position to derive the field equations. However, purely for mathematical convenience, we wish first to introduce the concept of *charge density*, which we shall designate by the letter ρ. We consider the charge as being distributed in space, and ρ is defined as such a function of position that the charge contained within any portion of space V is equal to

$$Q_V = \int_V \rho(\mathbf{r}) \, dV. \tag{7.24}$$

The function ρ corresponding to a particle having a charge e_s and located at the point $\mathbf{r}_s$ is written

$$\rho_s(\mathbf{r}) = e_s \, \delta(\mathbf{r} - \mathbf{r}_s). \tag{7.25}$$

The function $\delta(\mathbf{r} - \mathbf{r}_s)$ (see Section A8) when integrated over any volume containing the point $\mathbf{r}_s$ must give unity; on the other hand, when the volume of integration does not include the point $\mathbf{r}_s$, the result is zero.‡ In other words, the charge density corresponding to a particle is everywhere zero, except at the point at which the particle is located; and at this point the density is infinite. The δ function has further the property

$$\int_V f(\mathbf{r}) \, \delta(\mathbf{r} - \mathbf{r}_0) \, dV = f(\mathbf{r}_0), \tag{7.26}$$

where the integral is taken over any volume containing the point $\mathbf{r}_0$.

* In electrodynamics it will be shown that the positive sign is demanded by Hamilton's principle (see Section 16). In a gravitational field the constant corresponding to a is negative, in which case like particles *attract* each other, instead of repelling.

† For a summary of the relationship between the Gaussian system of units (electrostatic units for the charges) and MKS system see Section A10 in the appendices.

‡ This is not a legitimate mathematical function but it is widely used in mathematical physics and has a counterpart in the theory of generalized functions.

Introducing the charge density of a system of point charges, we may transform Eq. (7.3) as follows:

$$L_{pf} = -\sum_s e_s\, \varphi(\mathbf{r}_s) = -\sum_s e_s \int \varphi(\mathbf{r})\, \delta(\mathbf{r} - \mathbf{r}_s)\, dV$$

$$= -\int \varphi(\mathbf{r}) \sum_s \rho_s(\mathbf{r})\, dV = -\int \rho\varphi\, dV, \quad (7.27)$$

where the integration is performed over the whole space, and

$$\rho = \sum_s \rho_s. \quad (7.28)$$

The potential energy U of Eq. (7.13) can now, with the help of Eqs. (7.23) and (7.27), be written in a form convenient for further use:

$$U = \int \left(\rho\varphi - \frac{\mathscr{E}^2}{8\pi}\right) dV - \text{constant}. \quad (7.29)$$

The Lagrangian of the field when interacting with the particles is $L_{pf} + L_f$; therefore the equation of motion of the field is obtained by setting

$$\delta \int (L_{pf} + L_f)\, dt = -\int \delta U\, dt = 0. \quad (7.30)$$

By Eq. (7.29)

$$\delta U = \delta \int \left(\rho\varphi - \frac{\mathscr{E}^2}{8\pi}\right) dV = \int \left(\rho\delta\varphi - \frac{\mathscr{E} \cdot \delta\mathscr{E}}{4\pi}\right) dV.$$

We shall consider that we are dealing with a finite system of charges, and, since the interaction between such a system and one at infinity is zero, we take φ at infinity to be zero. This means that in our variation we do not change the potential at infinity, in other words, at infinity $\delta\varphi = 0$. Further, since

$$\nabla \cdot (\mathscr{E}\, \delta\varphi) = \delta\varphi\, \nabla \cdot \mathscr{E} + \mathscr{E} \cdot \nabla\delta\varphi = \delta\varphi\, \nabla \cdot \mathscr{E} - \mathscr{E} \cdot \delta\mathscr{E}, \quad (7.31)$$

we see that

$$\delta U = \int \left(\rho - \frac{1}{4\pi} \nabla \cdot \mathscr{E}\right) \delta\varphi\, dV + \frac{1}{4\pi} \int \nabla \cdot (\mathscr{E}\, \delta\varphi)\, dV.$$

The last integral vanishes, as can be seen by transforming it by Gauss' theorem into a surface integral at infinity, where $\delta\varphi$ vanishes.

Equation (7.30) thus becomes

$$\int \delta U \, dt = \iint \left(\rho - \frac{1}{4\pi} \nabla \cdot \mathscr{E} \right) \delta\varphi \, dV \, dt = 0 \tag{7.32}$$

and, since this must hold for arbitrary $\delta\varphi$, the expression in the parentheses must be zero. This gives us the field equation

$$\nabla \cdot \mathscr{E} = 4\pi\rho. \tag{7.33}$$

This is the equation giving the connection between the charge distribution function ρ and the field $\mathscr{E}$ produced by it. Equations (6.6) and (7.33) are the fundamental equations of electrostatics. Taken with the boundary conditions specifying each particular problem, they suffice to determine $\mathscr{E}$ in each given case.

Of equal significance is the equation obtained from Eq. (7.33) by substitution of the value of $\mathscr{E}$ from Eq. (6.5),

$$\nabla^2\varphi = -4\pi\rho, \tag{7.34}$$

which is called *Poisson's equation*. An important special case of Eq. (7.34),

$$\nabla^2\varphi = 0, \tag{7.35}$$

holds in those regions of space in which $\rho = 0$ and is called *Laplace's equation.*

Equation (7.33) has a meaning at a given point $\mathbf{r}$ only if the first-order partial derivatives involved in $\nabla \cdot \mathscr{E}$ exist. For the very important case of point charges, the charge density being given by Eqs. (7.25) and (7.28), the field has singularities at the charges and these partial derivatives do not exist.

This difficulty may be overcome by defining $\nabla \cdot \mathscr{E}$ as follows:

$$\nabla \cdot \mathscr{E} = \lim_{V \to 0} \frac{1}{V} \int_S \mathscr{E} \cdot d\mathbf{S},$$

the volume V being taken around the point at which $\nabla \cdot \mathscr{E}$ is desired. For cases where the usual definition of $\nabla \cdot \mathscr{E}$ applies, the new definition gives the same result. This definition is equivalent to making Gauss' theorem (see Appendix, Eq. (A4.8)) universally valid.

Finally, we must observe an important fact, that L contains time only through v_s^2. This means that the Lagrangian is unchanged by a transformation $t \to -t$, and that, therefore, *in electrostatics there is no distinction between past and future.*

8. Coulomb's Law

The importance of Laplace's equation arises from the fact that, when we are dealing with a system of particles, ρ is zero everywhere except at the points occupied by the particles. Thus, except at these isolated points, the potential must satisfy Laplace's equation. In dealing with a particular problem we must, of all the possible solutions of Laplace's equation, choose the solution that has the right kind of singularities at the particles.

One way of exhibiting the nature of these singularities is by integrating $\mathscr{E}$ over a closed surface S enclosing a volume V containing a number of charged particles. We thus find that

$$\int_S \mathscr{E} \cdot d\mathbf{S} = \int_V \nabla \cdot \mathscr{E}\, dV = 4\pi \int_V \rho\, dV = 4\pi \sum_s \int_V \rho_s\, dV = 4\pi \sum_s e_s, \tag{8.1}$$

the sum being extended over all the particles included within the surface. This shows that the value of the integral is independent of the form of the surface and depends only upon the total charge contained within it.

If the surface in Eq. (8.1) is taken in such a way as to include only one particle of charge e, the integral is equal to $4\pi e$, however small the surface may be. The singularity of the field in the vicinity of each particle depends thus only upon the charge of the particle and may be investigated independently of the presence of other particles.

Consider therefore the field of a single point particle located at the origin. Its charge density is given by $\rho(\mathbf{r}) = e\, \delta(\mathbf{r})$, where $\delta(\mathbf{r})$ is the three-dimensional Dirac delta function introduced in the previous section. Since all directions from the origin are physically indistinguishable, we assume that the potential φ is a function of just the radial distance r. Thus $\varphi = \varphi(r)$ and by Eq. (A3.43)

$$\nabla\varphi = \frac{d\varphi}{dr} \nabla r = \frac{\mathbf{r}}{r} \varphi'(r) \tag{8.2}$$

and by Eq. (A3.49)

$$\nabla^2\varphi = \nabla \cdot \nabla\varphi = \mathbf{r} \cdot \nabla\left(\frac{\varphi'}{r}\right) + \frac{\varphi'}{r} \nabla \cdot \mathbf{r} = \frac{1}{r^2} \frac{d}{dr} (r^2\varphi'). \tag{8.3}$$

If we exclude the origin ($\mathbf{r} = 0$), then $\rho(\mathbf{r}) = 0$ and the potential satisfies Laplace's equation. Thus from Eqs. (7.35) and (8.3)

$$\frac{1}{r^2}\frac{d}{dr}\left(r^2\frac{d\varphi}{dr}\right) = 0,$$

the general solution of which is

$$\varphi = \frac{b}{r} + a.$$

The constant a is determined by the fact that we wish to apply as a boundary condition the requirement that the potential be zero at infinity; this makes $a = 0$. The field corresponding to this potential by Eq. (8.2) is

$$\mathcal{E} = -\nabla\left(\frac{b}{r}\right) = \frac{b\mathbf{r}}{r^3}. \tag{8.4}$$

The constant b can now be determined by the use of Eq. (8.1). Since

$$\int \frac{b\mathbf{r}}{r^3} \cdot d\mathbf{S} = 4\pi b, \tag{8.5}$$

we see that $b = e$, the charge of the particle. The potential is thus

$$\varphi = \frac{e}{r}, \tag{8.6}$$

which shows the nature of the singularity of the potential at a charge e. If other charges are present, then by the principle of superposition one needs to add the potential due to them. In the immediate neighborhood of the first particle this additional potential is finite and therefore the singularity at the origin is fully described by Eq. (8.6).

We have obtained the potential $\varphi(\mathbf{r})$ due to a point charge e located at the origin, one due to a charge density $\rho(\mathbf{r}) = e\,\delta(\mathbf{r})$. This corresponds to finding a solution of Poisson's equation (7.34) for this charge density and one concludes that

$$\nabla^2\left(\frac{e}{r}\right) = -4\pi e\,\delta(\mathbf{r}).$$

We have obtained thus a useful relation:

$$\nabla^2\left(\frac{1}{r}\right) = -4\pi\,\delta(\mathbf{r}). \tag{8.7}$$

When in the field of a charge e there is another charge e', the force acting on it, according to Eq. (6.4), will be

$$\mathbf{F} = e'\mathscr{E} = \frac{ee'\mathbf{r}}{r^3}. \tag{8.8}$$

This law of force is known as *Coulomb's law*, and the field $\mathscr{E} = e\mathbf{r}/r^3$ is called the Coulomb field of the charge e. The more familiar form of the law is obtained by taking the absolute magnitude of the force:

$$F = \frac{ee'}{r^2}. \tag{8.9}$$

This equation holds in the electrostatic system of units we selected in Section 7 by taking a in Eq. (7.22) to be $1/4\pi$. In this system the charges, such as e and e', are expressed in electrostatic units (esu), distances, such as r, in centimeters, and forces in dynes.

If the charge in Eq. (8.6) is the charge contained in an element of volume $dV' = dx'\,dy'\,dz'$ about the point $\mathbf{r}'$, then by this equation the potential at a point r due to this charge is

$$d\varphi(\mathbf{r}) = \frac{\rho(\mathbf{r}')\,dV'}{R}, \tag{8.10}$$

where R is the distance between the points $\mathbf{r}$ and $\mathbf{r}'$. From the linear character of Poisson's equation it is now evident that, in the general case, its solution can be compounded of the elementary solutions given by Eq. (8.10); therefore,

$$\varphi(\mathbf{r}) = \int \frac{\rho(\mathbf{r}')\,dV'}{R}. \tag{8.11}$$

While from its derivation the solution given by Eq. (8.11) might be expected to hold only for points outside of the charge distribution ρ, direct substitution shows that it is generally valid. Thus,

$$\begin{aligned}\nabla^2\varphi(\mathbf{r}) = \nabla^2\int \frac{\rho(\mathbf{r}')\,dV'}{R} &= \int \rho(\mathbf{r}')\,\nabla^2\left(\frac{1}{R}\right)\,dV' \\ &= -4\pi\int \rho(\mathbf{r}')\,\delta(\mathbf{R})\,dV' = -4\pi\rho(\mathbf{r}),\end{aligned}$$

where we have made use of Eqs. (8.7), (7.26), and where $\mathbf{R} = \mathbf{r} - \mathbf{r}'$.

For a system of particles, with a charge density given by Eqs. (7.28) and

(7.25), one generalizes Eq. (8.11) to

$$\varphi(\mathbf{r}) = \sum_s \frac{e_s}{R_s}, \tag{8.12}$$

where $\mathbf{R}_s = \mathbf{r} - \mathbf{r}_s$, and the summation is extended over all the particles present. This result could, of course, be obtained directly from Eq. (8.6) and the principle of superposition.

From Eq. (8.6) it follows directly that the interaction energy of any two particles having the charges e_j and e_k is, in accordance with Eq. (6.3),

$$U_{jk} = e_j \varphi_k(\mathbf{r}_j) = e_k \varphi_j(\mathbf{r}_k) = \frac{e_j e_k}{R_{jk}}, \tag{8.13}$$

where $\varphi_j(\mathbf{r}_k)$ is the potential produced by the charge e_j at the point $\mathbf{r}_k$, and R_{jk} is the distance between the charges. We see that this interaction energy is symmetrical in the two charges, which was, of course, to be expected of a correct theory. Collecting together our results from Eqs. (7.1), (7.2), (7.13), (7.16), and (8.13) we can write the Lagrangian for a system of particles electrically interacting with each other as follows:

$$L = \sum_s \frac{1}{2} m_s v_s^2 - \frac{1}{2} \sum_{j,k}{}' \frac{e_j e_k}{R_{jk}}, \tag{8.14}$$

where the prime after the symbol of summation indicates, as is customary, that terms with $j = k$ are omitted.

It is interesting to consider the value of the constant introduced in Eq. (7.17). As was indicated, it has to be chosen in such a way that the two expressions for U, Eqs. (7.29) and (7.16), are consistent, or so that

$$U = \int \rho\varphi \, dV - \frac{1}{8\pi} \int \mathscr{E}^2 \, dV - \text{constant} = \frac{1}{2} \sum_{j,k}{}' U_{jk}. \tag{8.15}$$

We observe first, however, that

$$\rho\varphi = \frac{\varphi \nabla \cdot \mathscr{E}}{4\pi} = \nabla \cdot \left(\frac{\varphi \mathscr{E}}{4\pi}\right) - \frac{\mathscr{E}}{4\pi} \cdot \nabla\varphi = \nabla \cdot \left(\frac{\varphi \mathscr{E}}{4\pi}\right) + \frac{\mathscr{E}^2}{4\pi}, \tag{8.16}$$

and hence

$$\int \rho\varphi \, dV = \frac{1}{4\pi} \int \mathscr{E}^2 \, dV + \frac{1}{4\pi} \int \nabla \cdot (\varphi \mathscr{E}) \, dV. \tag{8.17}$$

The second integral vanishes because the integrand is the divergence of $\varphi\mathscr{E}$ and by Eqs. (8.4) and (8.6) the magnitude of $\varphi\mathscr{E}$ behaves as $1/r^3$ at large distances. This is seen by replacing the volume integral over a sphere of radius R and observing that the integral drops off as $1/R$. Thus in the limit $R \to \infty$ the integral vanishes. We obtain therefore from Eqs. (8.15) and (8.17) the rather remarkable result that

$$\frac{1}{8\pi}\int \mathscr{E}^2\, dV = \frac{1}{2}\int \rho\varphi\, dV = U + \text{constant}. \tag{8.18}$$

We see from this equation that the potential energy aside from an added constant may be expressed as the volume integral corresponding either to the energy density $\mathscr{E}^2/8\pi$, distributed over the whole space and dependent only upon the field at each point, or to the energy density $\frac{1}{2}\rho\varphi$, which is associated with the charge, and is zero where ρ is zero. In particular, for particles, the first interpretation locates the energy entirely outside the particles, while the second interpretation locates it entirely within the particles. Neither can be considered as having greater reality than the other.

To determine the constant in Eq. (8.18) we evaluate $\int \rho\varphi\, dV$ for point particles. Substitution from Eqs. (7.28) and (7.25) into Eq. (8.18) gives

$$\begin{aligned} U + \text{constant} &= \frac{1}{2}\sum_j e_j \int \delta(\mathbf{r} - \mathbf{r}_j)\, \varphi(\mathbf{r})\, dV \\ &= \frac{1}{2}\sum_j e_j \varphi(\mathbf{r}_j). \end{aligned} \tag{8.19}$$

On substituting for $\varphi(r_j)$ from Eq. (8.12) we have

$$U + \text{constant} = \frac{1}{2}\sum_{j,k} \frac{e_j e_k}{R_{jk}} = \frac{1}{2}\sum_{j,k} U_{jk}, \tag{8.20}$$

so that, using Eq. (7.16), we may solve for the value of the constant:

$$\text{constant} = \frac{1}{2}\sum_{j,k} U_{jk} - \frac{1}{2}\sum_{j,k}{}' \, U_{jk} = \frac{1}{2}\sum_k U_{kk}. \tag{8.21}$$

The quantity $\frac{1}{2}U_{kk}$ can be considered as the potential energy of a single particle. We originally assumed the potential energy of a single free particle to be zero and took its Lagrangian to be $\frac{1}{2}mv^2$. This was permissible because the addition of a constant to the Lagrangian does not affect the equations of motion. Our treatment is here seen to eliminate all the self-energy terms $\frac{1}{2}U_{kk}$ in the potential energy U (see Eq. (8.15)).

If one takes an extended charge distribution for an individual particle, in place of a point charge, the self-energy term is just the Coulomb energy of the distribution. Taking the point charge of our treatment as the limit as the dimensions shrink to zero of an extended charge distribution, one finds that these self-energy terms are infinite. Their absence from the potential energy U is therefore very pleasing. Unfortunately, the matter is not so easily treated in the general case of electrodynamics (see Section 24).

Finally, it is to be noted that U, in either of the forms given in Eq. (8.18), is not suitable for substitution in the equation $-\int \delta U \, dt = 0$ of Section 7, since Eq. (8.18) already assumes the results of such a variation, namely, $\nabla \cdot \mathscr{E} = 4\pi\rho$.

Equation (8.14) gives the Lagrangian of a system of interacting particles with the field variables eliminated. It can be used to obtain the equations of motion of the particles. Thus, applying Eq. (2.12), we obtain for each particle ($s = 1, 2, \ldots, n$)

$$\frac{d}{dt} m_s \mathbf{v}_s = \sum_{k \neq s} \frac{e_s e_k (\mathbf{r}_s - \mathbf{r}_k)}{|\mathbf{r}_s - \mathbf{r}_k|^3}, \tag{8.22}$$

the right-hand side of the equation being merely the sum of the Coulomb forces on the sth particle due to all the other particles. This possibility of eliminating the field from our equations permits one to consider the field as merely a convenient mathematical fiction.

9. System of Charges at Rest

In this section we shall consider some approximate formulas for the field and the potential energy of a system of charges at rest. These arise from multipole-type expansions. Thus the potential given by Eq. (8.12), which can be written

$$\varphi(\mathbf{r}) = \sum_s \frac{e_s}{|\mathbf{r} - \mathbf{r}_s|}. \tag{9.1}$$

can be expanded in an infinite series that represents consecutively higher powers of $\mathbf{r}_s/r$. This is done by applying Taylor's expansion in three dimensions to the individual terms of Eq. (9.1).

Taylor's series in one variable can be written

$$f(x+a) = \left[1 + a\frac{d}{dx} + \frac{1}{2!}\left(a\frac{d}{dx}\right)^2 + \frac{1}{3!}\left(a\frac{d}{dx}\right)^3 + \cdots\right] f(x)$$
$$= \exp\left(a\frac{d}{dx}\right) f(x).$$

This generalizes in three dimensions to

$$f(x + a_x, y + a_y, z + a_z)$$
$$= \exp\left(a_x \frac{\partial}{\partial x}\right) f(x, y + a_y y, z + a_z z)$$
$$= \exp\left(a_x \frac{\partial}{\partial x}\right)\exp\left(a_y \frac{\partial}{\partial y}\right)\exp\left(a_z \frac{\partial}{\partial z}\right) f(x, y, z)$$

which, letting $\mathbf{r} = (x, y, z)$ and $\mathbf{a} = (a_x, a_y, a_z)$, can be written*

$$f(\mathbf{r} + \mathbf{a}) = \exp\left(a_x \frac{\partial}{\partial x} + a_y \frac{\partial}{\partial y} + a_z \frac{\partial}{\partial z}\right) f(x, y, z)$$
$$= \exp(\mathbf{a}\cdot\nabla) f(\mathbf{r}),$$
$$= \left[1 + \mathbf{a}\cdot\nabla + \frac{1}{2!}(\mathbf{a}\cdot\nabla)^2 + \frac{1}{3!}(\mathbf{a}\cdot\nabla)^3 + \cdots\right] f(\mathbf{r}). \quad (9.2)$$

If we take the origin of our coordinate system (base of the position vectors) at some suitable point C within the system of charges, then r can be treated as the distance from the system to the field point $\mathbf{r}$. At a distance r large compared to the r_s one can expand the individual terms of Eq. (9.1) using the Taylor's expansion of Eq. (9.2); thus

$$\varphi(\mathbf{r}) = \sum_s e_s\left[\frac{1}{r} - \mathbf{r}_s\cdot\nabla\frac{1}{r} + \frac{1}{2!}(\mathbf{r}_s\cdot\nabla)^2\frac{1}{r} - \cdots\right]. \quad (9.3)$$

In this expansion the main interest centers on the first nonvanishing term, which can often be used as a sufficiently accurate approximation for $\varphi(\mathbf{r})$ for large r. The first term in the expansion can be written

$$\varphi_0(\mathbf{r}) = \frac{1}{r}\sum_s e_s = \frac{Q}{r} \quad (9.4)$$

and can thus be used as an approximation for the field when the total charge $Q = \sum_s e_s$ is not zero. For this case the potential at large distances is the same as if all the charge were concentrated at the origin (point C).

* See J. M. H. Olmsted, *Real Variables: An Introduction to the Theory of Functions*, Appleton-Century-Crofts, New York, 1956, p. 391.

When the total charge is zero, we need to examine the next term

$$\varphi_1(\mathbf{r}) = -\sum_s e_s\mathbf{r}_s \cdot \nabla\left(\frac{1}{r}\right) = \frac{\not{p} \cdot \mathbf{r}}{r^3} = \frac{\not{p}}{r^2}\cos\theta, \tag{9.5}$$

where

$$\not{p} = \sum_s e_s\mathbf{r}_s \tag{9.6}$$

is called the *electric dipole moment* of the system. Thus, while $\varphi_0(\mathbf{r})$ is spherically symmetrical about the origin and drops off as $1/r$, $\varphi_1(\mathbf{r})$ depends as $\cos\theta$ on the spherical coordinate θ taken relative to a polar axis along the dipole $\not{p}$, and drops off as $1/r^2$.

The electric dipole moment $\not{p}$ being expressed by Eq. (9.6) in terms of position vectors $\mathbf{r}_s$ is not a true vector and thus will generally depend on the choice of a center C for the system of charges. If the origin is shifted to another point C' displaced by the vector $\mathbf{a}$ relative to C, then the new position vectors will be given by $\mathbf{r}_s' = \mathbf{r}_s - \mathbf{a}$ and the moment will be given by

$$\not{p}' = \sum_s e_s\mathbf{r}_s' = \sum_s (e_s\mathbf{r}_s - e_s\mathbf{a}) = \not{p} - Q\mathbf{a}. \tag{9.7}$$

Only, therefore, if $Q = 0$, is the dipole moment independent of the choice of a center C for the system.

This independence of the choice of an origin, for $Q = 0$, can be made explicit by breaking up the summation in Eq. (9.6) and writing

$$\not{p} = \sum_s e_s^+\mathbf{r}_s^+ - \sum_t e_t^-\mathbf{r}_t^- \tag{9.8}$$

and

$$Q = \sum_s e_s^+ - \sum_t e_t^- = 0, \tag{9.9}$$

where e_s^+, e_t^- are the magnitudes of the positive and negative charges, respectively, and $\mathbf{r}_s^+$, $\mathbf{r}_t^-$ their position vectors. The first summation with respect to s is over the positive charges and second summation with respect to t is over the negative charges. Let

$$Q^+ = \sum_s e_s^+ = \sum_t e_t^- \tag{9.10}$$

and

$$\mathbf{r}^+ = \frac{1}{Q^+}\sum_s e_s^+\mathbf{r}_s^+, \qquad \mathbf{r}^- = \frac{1}{Q^+}\sum_t e_t^-\mathbf{r}_t^- ; \tag{9.11}$$

then from Eq. (9.8)

$$\mathscr{p} = Q^+(\mathbf{r}^+ - \mathbf{r}^-). \tag{9.12}$$

Since the e_s^+ and e_t^- are all positive, $\mathbf{r}^+$ and $\mathbf{r}^-$ represent sort of mean position vectors for the positive and negative charges and can thus be considered the *electric centers* of the positive and negative charges, respectively. Equation (9.12) shows, therefore, that the dipole moment has the direction of the directed line segment connecting the center $\mathbf{r}^-$ of the negatives charges with the center $\mathbf{r}^+$ of the positive charges and a magnitude equal to the product of the total positive charge Q^+ and the separation $|\mathbf{r}^+ - \mathbf{r}^-|$ between the centers.

Two equal charges of opposite sign, e and $-e$, with position vectors $\mathbf{r}_1$ and $\mathbf{r}_2$ are easily seen to have a dipole moment given by

$$\mathscr{p} = e(\mathbf{r}_1 - \mathbf{r}_2) \tag{9.13}$$

and is often said to represent a *finite electric dipole*. It can be readily shown that such a pair of charges has also higher electric moments and hence its potential field is not just $\varphi_1(\mathbf{r})$ as defined by Eq. (9.5). This potential field $\varphi_1(\mathbf{r})$, in fact, satisfies Laplace's equation everywhere except at $\mathbf{r} = 0$ and thus requires a *point dipole*, one obtained by letting $\mathbf{r}_2 \to \mathbf{r}_1$ while keeping the product $|\mathscr{p}| = e|\mathbf{r}_1 - \mathbf{r}_2|$ constant. This requires that e become infinite.

We saw earlier that the charge density ρ of a unit charge at $\mathbf{r}_0$ is given by the Dirac delta function $\delta(\mathbf{r} - \mathbf{r}_0)$. A finite dipole formed by charges $e_1 = 1/a$ and $e_2 = -1/a$ located at $\mathbf{a}/2$ and $-\mathbf{a}/2$ has a unit dipole moment given by

$$\mathscr{p} = \frac{\mathbf{a}}{a} \tag{9.14}$$

and a charge density

$$\rho(\mathbf{r}) = \frac{1}{a}\,\delta\left(\mathbf{r} - \frac{1}{2}\mathbf{a}\right) - \delta\left(\mathbf{r} + \frac{1}{2}\mathbf{a}\right). \tag{9.15}$$

Letting $a \to 0$ in Eq. (9.15), we obtain the charge distribution

$$\rho(\mathbf{r}) = -\nabla_{\mathbf{a}}\,\delta(\mathbf{r}) \tag{9.16}$$

of a *unit point dipole* at the origin. Again such a "function" is not a true mathematical function and can only be handled rigorously by the theory of distributions.*

* For a discussion of distributions from a physicist's viewpoint see I. Stakgold, *Boundary Value Problems of Mathematical Physics*, Vol. I, Macmillan, New York, 1967, pp. 28–54.

The electric fields $\mathscr{E}_0$ and $\mathscr{E}_1$ corresponding to the monopole and dipole potentials $\varphi_0(\mathbf{r})$ and $\varphi_1(\mathbf{r})$ are given by

$$\mathscr{E}_0 = -\nabla\varphi_0 = -Q\,\nabla\left(\frac{1}{r}\right) = \frac{Q\mathbf{r}}{r^3}, \tag{9.17}$$

the same as for a single charge Q at the origin, and

$$\mathscr{E}_1 = -\nabla\varphi_1 = -\nabla\left(\frac{\not{p}\cdot\mathbf{r}}{r^3}\right) = -(\not{p}\cdot\mathbf{r})\nabla\left(\frac{1}{r^3}\right) - \frac{\nabla(\not{p}\cdot\mathbf{r})}{r^3}$$

$$= \frac{3(\not{p}\cdot\mathbf{r})\mathbf{r} - r^2\not{p}}{r^5}. \tag{9.18}$$

Here we have made use of Eqs. (A3.28), (A3.43), and (A3.52) to set

$$\nabla\left(\frac{1}{r^3}\right) = \frac{-3\mathbf{r}}{r^5} \tag{9.19}$$

and

$$\nabla(\not{p}\cdot\mathbf{r}) = \not{p}. \tag{9.20}$$

From Eq. (9.18) the field off the end of the dipole ($\mathbf{r}$ parallel to $\not{p}$) is given by $\mathscr{E}_1 = 2\not{p}/r^3$ and in the equatorial plane of the dipole ($\mathbf{r}$ perpendicular to $\not{p}$) it is given by $\mathscr{E}_1 = -\not{p}/r^3$.

Setting $\mathbf{u} = \mathbf{r}/r$, we may write

$$\mathscr{E}_1 = \frac{\not{p}}{r^3}\,[3(\not{p}\cdot\mathbf{u})\mathbf{u} - \not{p}], \tag{9.21}$$

which shows that the direction of $\mathscr{E}_1$ does not depend on the radial distance r. As this distance increases in any given direction from the origin (u fixed), the direction of $\mathscr{E}_1$ is fixed and its magnitude $\mathscr{E}_1$ drops off exactly as $1/r^3$. In particular, using spherical coordinates (r, θ, ϕ) with the polar axis along the dipole, we have

$$\mathscr{E}_1 = \frac{\not{p}}{r^3}(3\cos\theta - 1). \tag{9.22}$$

The next approximation, the *quadrupole approximation*, becomes important when the total charge Q and the electric dipole moment $\not{p}$ are both zero. From Eq. (9.3) it is seen to be

$$\varphi_2(\mathbf{r}) = \frac{1}{2!}\sum_s e_s(\mathbf{r}_s\cdot\nabla)^2\frac{1}{r} = \frac{1}{2!}\,\frac{\mathbf{r}\cdot\mathbf{Q}\cdot\mathbf{r}}{2r^3}, \tag{9.23}$$

where $\mathbf{Q}$, the *electric quadrupole moment dyadic*, is given by

$$\mathbf{Q} = \sum_s e_s(3\mathbf{r}_s\mathbf{r}_s - r_s^2\mathbf{1}). \tag{9.24}$$

Here $\mathbf{1}$ is the unit dyadic (see Appendix A7). In tensor or component form Eq. (9.23) becomes

$$\varphi_2(\mathbf{r}) = \frac{1}{2!}\frac{Q_{jk}x_jx_k}{r^5}, \tag{9.25}$$

where by Eq. (9.24)

$$Q_{jk} = \sum_s e_s(3x_j^s x_k^s - r_s^2\,\delta_{jk}). \tag{9.26}$$

In general, the $(n + 1)$th term in the multipole expansion of Eq. (9.3) represents the contribution of the *2^n-pole electric moment tensor* $Q'_{k_1k_2\cdots k_n}$ and is given by

$$\varphi_n(\mathbf{r}) = \frac{1}{n!}\frac{1}{r^{2n+1}} Q_{k_1k_2\cdots k_n}x_{k_1}x_{k_2}\cdots x_{k_n}. \tag{9.27}$$

In this expression the summation convention applies to each of the n subscripts $k_1 \cdots k_n$, which thus separately take on the values 1, 2, 3.

Consider next the increase in the potential energy U of a system of stationary charges when it is placed in an external field described by the potential $\varphi(\mathbf{r})$. It is assumed for this discussion that the distribution of charges that produces the potential $\varphi(\mathbf{r})$ is unaffected by the presence of the system. Let the center of the system, however defined, be at $\mathbf{r}$ and the individual charges be at $\mathbf{r} + \mathbf{r}_s$; by Eqs. (6.3) and (9.2) this potential energy is

$$\begin{aligned} U &= \sum_s e_s\varphi(\mathbf{r} + \mathbf{r}_s) \\ &= \sum_s e_s\left[\varphi(\mathbf{r}) + \mathbf{r}_s\cdot\nabla\varphi(\mathbf{r}) + \frac{1}{2!}(\mathbf{r}_s\cdot\nabla)^2\varphi(\mathbf{r}) + \cdots\right]. \end{aligned} \tag{9.28}$$

The first term in the expansion,

$$U_0 \equiv \sum_s e_s\varphi(\mathbf{r}) = Q\varphi(\mathbf{r}), \tag{9.29}$$

represents the monopole contribution due to the net charge Q of the system. In this approximation the force acting on the system is

$$\mathbf{F}_0 \equiv -\nabla U_0 = -Q\,\nabla\varphi(\mathbf{r}) = Q\mathscr{E}(\mathbf{r}). \tag{9.30}$$

The next term

$$U_1 \equiv \sum_s e_s\mathbf{r}_s\cdot\nabla\varphi(\mathbf{r}) = \boldsymbol{p}\cdot\nabla\varphi(\mathbf{r}) = -\boldsymbol{p}\cdot\mathscr{E}(\mathbf{r}) \tag{9.31}$$

represents the dipole moment contribution, and the force on the system due to this dipole is

$$\mathbf{F}_1 \equiv -\nabla U_1 = \nabla(\boldsymbol{p}\cdot\boldsymbol{\mathscr{E}}) = (\boldsymbol{p}\cdot\nabla)\boldsymbol{\mathscr{E}}(\mathbf{r}). \tag{9.32}$$

Use has been made of Eqs. (A3.28), (9.31), and (6.6) and the fact that $\boldsymbol{p}$ is constant, the charges being at rest.

In Section 3 we defined the resultant torque $\mathbf{T}$ acting on a system by the equation

$$\delta L = \boldsymbol{\epsilon}\cdot\mathbf{T} \tag{9.33}$$

(see Eq. (3.21)), where $\boldsymbol{\epsilon}$ represents an infinitesimal rotation of the system. Since the charges are at rest and the rotation does not affect the field or the total charge, to a first approximation

$$\delta L = -\delta U = -\delta(U_0 + U_1) = -\varphi\,\delta Q + \boldsymbol{\mathscr{E}}\cdot\delta\boldsymbol{p} = \delta\boldsymbol{p}\cdot\boldsymbol{\mathscr{E}} \tag{9.34}$$

The change in a vector such as $\boldsymbol{p}$ due to an infinitesimal rotation $\boldsymbol{\epsilon}$ (see Eq. (A2.32)) is $\boldsymbol{\epsilon}\times\boldsymbol{p}$; therefore, from Eqs. (9.33) and (9.34)

$$\boldsymbol{\epsilon}\cdot\mathbf{T} = \delta\boldsymbol{p}\cdot\boldsymbol{\mathscr{E}} = \boldsymbol{\epsilon}\times\boldsymbol{p}\cdot\boldsymbol{\mathscr{E}} = \boldsymbol{\epsilon}\cdot\boldsymbol{p}\times\boldsymbol{\mathscr{E}}.$$

Since this is to hold for an arbitrary $\boldsymbol{\epsilon}$, we have

$$\mathbf{T} = \boldsymbol{p}\times\boldsymbol{\mathscr{E}}. \tag{9.35}$$

The next higher term in Eq. (9.28)

$$U_2 \equiv \frac{1}{2!}\sum_s e_s(\mathbf{r}_s\cdot\nabla)^2\varphi(\mathbf{r}) = \frac{1}{2}\sum_s e_s\left(x_i^s\frac{\partial}{\partial x_i}\right)\left(x_j^s\frac{\partial}{\partial x_j}\right)\varphi(\mathbf{r})$$

represents the quadrupole moment contribution to the potential energy of the system. Thus from Eq. (9.26)

$$U_2 = \frac{1}{6}\left[Q_{ij}\frac{\partial^2\varphi(\mathbf{r})}{\partial x_i\,\partial x_j} + \sum_s r_s^2\,\delta_{ij}\frac{\partial^2\varphi(\mathbf{r})}{\partial x_i\,\partial x_j}\right].$$

Since the potential $\varphi(\mathbf{r})$ is assumed produced by charges remote from the system,

$$\delta_{ij}\frac{\partial^2\varphi(\mathbf{r})}{\partial x_i\,\partial x_j} = \frac{\partial^2\varphi(\mathbf{r})}{\partial x_i^2} = \nabla^2\varphi = 0,$$

and hence

$$U_2 = \frac{1}{3!}Q_{ij}\frac{\partial^2\varphi(\mathbf{r})}{\partial x_i\,\partial x_j} \equiv \frac{1}{3!}\mathbf{Q}:\nabla\nabla\varphi(\mathbf{r}). \tag{9.36}$$

Consider next the interaction potential energy $U(\mathbf{r}-\mathbf{r}')$ between a system S composed of charges e_s with position vectors $\mathbf{r}+\mathbf{r}_s$ and a second system S' composed of charges e_t' having position vectors $\mathbf{r}'+\mathbf{r}_t$, where $\mathbf{r}$ and $\mathbf{r}'$ are the position vectors of the centers C and C' of the two systems. We may take $U(\mathbf{r}-\mathbf{r}')$ as defined by

$$U(\mathbf{r}-\mathbf{r}') \equiv \sum_{s,t} \frac{e_s e_t'}{|(\mathbf{r}+\mathbf{r}_s)-(\mathbf{r}'+\mathbf{r}_t)|}, \tag{9.37}$$

the sum of the Coulomb energies of all those pairs composed of a charge e_s from S and a charge e_t' from S'.

Applying Taylor's expansion in three dimensions, as given by Eq. (9.2), to each term of the right-hand side of Eq. (9.37), we have

$$U(\mathbf{r}-\mathbf{r}') = \sum_s e_s \exp(\mathbf{r}_s \cdot \nabla)\, \varphi(\mathbf{r}), \tag{9.38}$$

where

$$\varphi(\mathbf{r}) = \sum_t \frac{e_t}{|\mathbf{r}-(\mathbf{r}'+\mathbf{r}_t)|} \tag{9.39}$$

is the potential at $\mathbf{r}$ due to the system S' at $\mathbf{r}'$. This potential can be expanded as in Eq. (9.3) to give

$$\begin{aligned}\varphi(\mathbf{r}) &= \sum_t e_t \exp(\mathbf{r}_t \cdot \nabla') \frac{1}{|\mathbf{r}-\mathbf{r}'|} \\ &= \varphi_0(\mathbf{r}) + \varphi_1(\mathbf{r}) + \varphi_2(\mathbf{r}) + \cdots,\end{aligned} \tag{9.40}$$

where, setting $\mathbf{R}=\mathbf{r}-\mathbf{r}'$ and using the operator relation $\nabla' = -\nabla$ of Eq. (A3.68),

$$\begin{aligned}\varphi_n(\mathbf{r}) &= \frac{1}{n!}\sum_t e_t(\mathbf{r}_t \cdot \nabla')^n \frac{1}{|\mathbf{r}-\mathbf{r}'|} \\ &= \frac{1}{n!}\sum_t e_t(-\mathbf{r}_t \cdot \nabla)^n \frac{1}{R}.\end{aligned} \tag{9.41}$$

Next we can expand the exponential operator in Eq. (9.38) as in Eq. (9.28) to obtain the double series expansion

$$\begin{aligned}U(\mathbf{R}) = U_{00} &+ U_{01} + U_{02} + \cdots \\ &+ U_{10} + U_{11} + U_{12} + \cdots \\ &+ U_{20} + U_{21} + U_{22} + \cdots \\ &+ \cdots,\end{aligned} \tag{9.42}$$

where

$$U_{mn} \equiv \frac{1}{m!} \sum_s e_s(\mathbf{r}_s \cdot \boldsymbol{\nabla})^m \varphi_n(\mathbf{r})$$

$$= \frac{1}{m!}\frac{1}{n!} \sum_{s,t} e_s e_t (\mathbf{r}_s \cdot \boldsymbol{\nabla})^m (\mathbf{r}_t \cdot \boldsymbol{\nabla}')^n \frac{1}{R}. \tag{9.43}$$

If we interchange the above roles of the two systems, then $\mathbf{r}$ and $\mathbf{r}'$, $\boldsymbol{\nabla}$ and $\boldsymbol{\nabla}'$, and the subscripts s and t are interchanged. We thus see that U_{mn} goes over into U_{nm}. This, as required, does not affect the interaction energy $U(\mathbf{R})$ between the two systems, as can be seen from Eq. (9.42).

If neither of the total charges $Q = \sum_s e_s$ and $Q' = \sum_t e_t$ vanishes, a useful approximation for the interaction energy is the *monopole–monopole interaction* term

$$U_{00} = \sum_s \sum_t e_s e_t \frac{1}{R} = \frac{QQ'}{R}. \tag{9.44}$$

If $Q = 0$, then from Eq. (9.43) all the interactions terms U_{0n} are zero and the lowest-order term (first order in the $\mathbf{r}_s$ and $\mathbf{r}_t$) that can contribute is the *dipole–monopole* interaction term

$$U_{10} = \sum_{s,t} e_s e_t (\mathbf{r}_s \cdot \boldsymbol{\nabla}) \frac{1}{R} = Q'(\boldsymbol{p} \cdot \boldsymbol{\nabla}) \frac{1}{R} = -Q' \frac{\boldsymbol{p} \cdot \mathbf{R}}{R^3}. \tag{9.45}$$

If $Q = Q' = 0$, then the terms U_{m0} are also zero and the lowest-order term (second order in the $\mathbf{r}_s$ and $\mathbf{r}_t$) that can contribute is the *dipole–dipole interaction* term

$$U_{11} = \sum_{s,t} e_s e_t (\mathbf{r}_s \cdot \boldsymbol{\nabla})(-\mathbf{r}_t \cdot \boldsymbol{\nabla}) \frac{1}{R}$$

$$= -(\boldsymbol{p} \cdot \boldsymbol{\nabla})(\boldsymbol{p}' \cdot \boldsymbol{\nabla}) \frac{1}{R}$$

$$= \boldsymbol{p} \cdot \boldsymbol{\nabla}\left(\frac{\boldsymbol{p}' \cdot \mathbf{R}}{R^3}\right) = \frac{R^2 \boldsymbol{p} \cdot \boldsymbol{p}' - 3(\boldsymbol{p} \cdot R)(\boldsymbol{p}' \cdot \mathbf{R})}{R^5}. \tag{9.46}$$

In obtaining Eq. (9.46) use has been made of Eqs. (9.43), (A3.65), and (A3.57).

Problems

1. In the expression $U = e\phi(\mathbf{r}, t)$ of Eq. (6.3) is e a derivable quantity or can it be determined only by experiment?

2. Can a $\phi(x, y, z)$ exist such that

$$\text{(a) } \mathscr{E} = x\mathbf{i} + xy\mathbf{j} + xyz\mathbf{k}, \qquad \text{(b) } \mathscr{E} = yz\mathbf{i} + xz\mathbf{j} + xy\mathbf{k},$$

$$\text{(c) } \mathscr{E} = \tfrac{1}{2}yz^2\mathbf{i} + \tfrac{1}{2}xz^2\mathbf{j} + xyz\mathbf{k}?$$

3. Show that the Hamiltonian given in Eq. (6.2) follows from the Lagrangian given in Eq. (6.1). What is the Hamiltonian corresponding to $L = a\dot{q}^2 + b\dot{q} + cq$?

4. What choice of a in Eq. (7.22) is required when one used the MKS system of units?

5. Determine the differential expression for the divergence of a vector $\mathbf{A}$ expressed in spherical coordinates by (a) transformation of coordinates from a Cartesian coordinate system and (b) using the definition in terms of an integral.

6. Obtain an expression for Laplace's equation in cylindrical coordinates.

7. Explain why an electric field $\mathscr{E}$ described by a potential function $\phi(\mathbf{r}, t)$ that satisfies Laplace's equation throughout a region V must necessarily be *solenoidal* (divergence zero) and *irrotational* (curl zero) everywhere throughout V.

8. What is the principle of superposition? Express it mathematically in two different ways and show their equivalence.

9. Express the Dirac delta function $\delta(\mathbf{r} - \mathbf{r}_0)$ in spherical coordinates. What difficulties occur when $\mathbf{r}_0$ is at the origin of this coordinate system? How can they be overcome?

10. Show that the integral definition for the divergence of a vector **A** agrees with the usual definition in terms of derivatives when **A** possesses the required partial derivatives at the point.

11. Show that the electrostatic energy U of a uniformly charged sphere of radius a as computed from the two formulas

$$U = \frac{1}{8\pi}\int \mathscr{E}^2\,dV \quad \text{and} \quad U = \frac{1}{2}\int \rho\phi\,dV,$$

where the integral is over all space, is the same. Set this energy equal to the rest energy mc^2 of the electron and obtain the classical electron radius $a = e^2/2mc^2$.

12. Generalize problem 11 by letting the charge density $\rho(r)$ be an arbitrary function of the radial distance r. What expression for the classical electron radius does one obtain when all the charge is assumed to be concentrated at the surface, i.e., $\rho(r) = (q/4\pi a^2)\delta(r - a)$?

13. Show that the field of a charge distribution given by $\rho = e^{-r}$ is $\mathscr{E} = 4\pi[2 - (r^2 + 2r + 2)e^{-r}]\mathbf{r}/r^3$.

14. By taking the appropriate limiting case of a volume distribution, show that the potential $\phi(\mathbf{r})$ due to a surface S having a surface density of charge $\sigma(\mathbf{r}')$ is

$$\phi(\mathbf{r}) = \int \frac{\sigma(\mathbf{r}')}{|\mathbf{r} - \mathbf{r}'|}\,dS,$$

where dS is the element of surface area.

15. Using the same approach as in problem 14, show that the potential $\phi(\mathbf{r})$ due to an arbitrary curve C of linear charge density $\tau(\mathbf{r}')$ is

$$\phi(\mathbf{r}) = \int \frac{\tau(\mathbf{r}')}{|\mathbf{r} - \mathbf{r}'|}\,ds,$$

where s is the distance along C.

16. Using the results of problem 15, find the potential in the equatorial plane of a line segment of length L with a uniform linear charge density τ. What happens as $L \to \infty$?

17. Find the potential of a uniformly charged line using Gauss's theorem.

18. Show that Taylor's series expansion can be written formally

$$f(x + \varepsilon_1) = e^{\varepsilon_1 \partial/\partial x} f(x),$$

which in three dimensions generalizes to

$$f(\mathbf{r} + \boldsymbol{\epsilon}) = \exp(\boldsymbol{\epsilon} \cdot \boldsymbol{\nabla}) f(\mathbf{r})$$

The expansion of the exponential term in the latter equation leads then to Eq. (9.2).

19. Show directly that the dipole field given by Eq. (9.18) is a conservative field and that its divergence vanishes everywhere except possibly at $r = 0$. Discuss the divergence at $r = 0$.

20. The position of a charge e along the X-axis is given by

$$x = A \cos \omega t.$$

Show that its dipole moment is the same as that of two charges $\frac{1}{2}e$ and $-\frac{1}{2}e$ at x and the image point $x' = -x$, respectively. Explain physically, by subtracting their charge distributions, why these distributions have the same dipole moment. How do their quadrupole moments differ?

21. Determine the dipole interaction energy of two point dipoles in the following orientations:

(a) Dipoles are parallel to each other and to the line joining them.
(b) Dipoles are antiparallel and perpendicular to the line joining them.
(c) Dipoles are perpendicular to each other and to the line joining them.

22. A finite electric dipole has an electric dipole moment given by Eq. (9.13). Find the first nonvanishing higher multipole moment.

Chapter IV

Relativistic Mechanics

10. Relativistic Space and Time

In the last two chapters we have considered the consequences of the assumption that the speed of propagation of interaction of all matter, which we shall designate by c, is infinite. Actually, however, this speed is not infinite, although it is very large (approximately 3×10^{10} cm/sec) in comparison with most speeds of everyday life. The preceding theory is therefore an approximate theory, holding with a great degree of accuracy in those cases when the speeds we are dealing with are small in comparison with c; it could be obtained from the more exact theory by going over to the limit that the exact theory approaches as c is made to approach infinity.

In discussing electrostatics we made free use of the equivalence of all inertial coordinate systems, that is, of *the principle of relativity*. This made no difference in our usual concepts of space and time; in fact, the principle of relativity, together with the assumption of the infinite speed of propagation of interaction, led us to the absoluteness of space and time. However, when the principle of relativity is combined with the experimental fact that the speed of propagation of interaction is finite, we are led to revise completely our concepts of space and time, and hence of energy, mass, etc. The resulting theory is called the *special theory of relativity*.

While Newton's laws were known to be invariant for a Galilean transformation and thus were considered valid in any inertial coordinate system, most scientists prior to the twentieth century believed in a *luminiferous ether* in which light waves propagated and through which bodies moved without resistance. Thus a body was absolutely at rest if it were not moving with respect to the ether. All attempts, however, to measure this motion of bodies relative to the ether failed. The most crucial of the experiments performed was that by Michelson and Morley* in 1887 in which they showed by means of an interferometer that the time for light to travel to a mirror and back was independent of the orientation relative to motion of the earth about the sun. The orbital speed of the earth is about 3×10^6 cm/sec, or about one ten thousandth of the velocity of light, and would have been detectable as a shift in the interference fringes.

The history of the thought and experiments prior to Einstein's elevation of the principle of relativity to a universal principle is both interesting and instructive. Since we shall, with Einstein, take the principle of relativity as a fundamental assumption, it is very important that the reader here supplement the text by reading some of the very excellent treatments† of the experimental basis of the special theory of relativity.

To see in 1905 that the principle of relativity, which maintains the complete equivalence of all inertial frames of reference, could be taken as a universally valid assumption took the greatness of Einstein's physical insight. Moreover, it took his courage of thought to undertake the resulting revision of the established and unquestioned concepts of space and time.‡ Following him it is now easy to see what changes need to be made; changes that bring unity and beauty into electrodynamics, which is otherwise little more than a collection of empirical formulas.

When one combines the well-established finiteness of the speed of propagation of interaction c, which one also refers to as the *speed of light*, with the principle of relativity, one concludes that this speed is the same in all inertial coordinate systems. Otherwise, there would be some inertial coordinate system, or perhaps a set of such systems, in which this velocity would be a minimum, and this would make the inertial systems no longer equivalent. Also, of course, this speed cannot depend on the direction of propagation because of the assumed isotropy of space.

* A. A. Michelson and E. W. Morley, *Am. J. Sci.*, **34**, 333 (1887). This was a more refined experiment than that report by Michelson in 1881: see A. A. Michelson, *Am. J. Sci.*, **22**, 20 (1881).

† Some of these are: W. Pauli, *Theory of Relativity*, Pergamon, New York, 1958; C. Møller, *The Theory of Relativity*, Clarendon Press, Oxford, 1952; G. Stephenson and C. W. Kilmister, *Special Relativity for Physicists*, Longmans Green, London and New York, 1958; A. Einstein, *The Meaning of Relativity*, Princeton Univ. Press, Princeton, N.J., 1953; and H. Arzeliès, *Relativistic Kinematics*, Pergamon, New York, 1966.

‡ A. Einstein, *Ann. Phys.*, **17**, 891 (1905).

The discussion of Section 4, leading to the conclusion that time is absolute, must now be modified. As in Section 4, we consider the equation of a signal propagated with the speed of light, taking for simplicity the time of sending the signal as zero. Instead of Eq. (4.1) we now have

$$x^2 + y^2 + z^2 = c^2t^2. \tag{10.1}$$

Let another coordinate system, moving with the velocity $\mathbf{v}$ relatively to the first, be chosen in such a way that the origins of the two coordinate systems coincide at the time $t = 0$, and let the moment of coincidence be chosen in the second coordinate system as the initial moment $t' = 0$. The equation of the signal in the second coordinate system will be

$$x'^2 + y'^2 + z'^2 = c^2t'^2. \tag{10.2}$$

Equations (10.1) and (10.2) show that, whenever the expression

$$-s^2 = x^2 + y^2 + z^2 - c^2t^2 \tag{10.3}$$

is zero, so is the expression

$$-s'^2 = x'^2 + y'^2 + z'^2 - c^2t'^2. \tag{10.4}$$

Let us consider the way in which the primed coordinates are related to the unprimed. In general, each of the coordinates (x', y', z', t') is a function of all the coordinates (x, y, z, t) of the first coordinate system; therefore,

$$\begin{aligned}
dx' &= \frac{\partial x'}{\partial x}dx + \frac{\partial x'}{\partial y}dy + \frac{\partial x'}{\partial z}dz + \frac{\partial x'}{\partial t}dt,\\
dy' &= \frac{\partial y'}{\partial x}dx + \frac{\partial y'}{\partial y}dy + \frac{\partial y'}{\partial z}dz + \frac{\partial y'}{\partial t}dt,\\
dz' &= \frac{\partial z'}{\partial x}dx + \frac{\partial z'}{\partial y}dy + \frac{\partial z'}{\partial z}dz + \frac{\partial z'}{\partial t}dt,\\
dt' &= \frac{\partial t'}{\partial x}dx + \frac{\partial t'}{\partial y}dy + \frac{\partial t'}{\partial z}dz + \frac{\partial t'}{\partial t}dt.
\end{aligned} \tag{10.5}$$

These differential relations cannot, however, depend explicitly upon the coordinates, because all points of space and time are equivalent, i.e., a given displacement (dx, dy, dz, dt) in the first coordinate system must correspond to a constant displacement (dx', dy', dz', dt') in the second coordinate system, irrespective of the coordinates of the initial event (x, y, z, t). This means that all the differential coefficients $\partial x'/\partial x$,

$\partial x'/\partial y, \ldots, \partial t'/\partial t$ are independent of the coordinates; therefore

$$\begin{aligned} dx' &= a_{11}\,dx + a_{12}\,dy + a_{13}\,dz + a_{14}\,dt, \\ &\;\;\vdots \\ dt' &= a_{41}\,dx + a_{42}\,dy + a_{43}\,dz + a_{44}\,dt, \end{aligned} \tag{10.6}$$

where the coefficients $a_{11}, a_{12}, \ldots, a_{44}$ may depend only on $\mathbf{v}$. Integrating Eqs. (10.6) and taking into account the convention that initially the origins of the two coordinate systems coincide, we have

$$\begin{aligned} x' &= a_{11}x + a_{12}y + a_{13}z + a_{14}t, \\ &\;\;\vdots \\ t' &= a_{41}x + a_{42}y + a_{43}z + a_{44}t. \end{aligned} \tag{10.7}$$

We make use of these results to determine the relation between s and s'. If we use Eq. (10.7) to substitute for x', y', z', and t' in Eq. (10.4), we obtain for s'^2 a second degree polynomial in x, y, z, and t. Now since this polynomial must be zero whenever $s^2 = 0$, and vice versa, these polynomials must differ only by a constant multiplier; that is,

$$s'^2 = A(v)s^2. \tag{10.8}$$

This equation indicates that the multiplier may still be a function of the magnitude of the relative velocity of the coordinate systems. The direction of the velocity will not enter due to the isotropy of space. We have here a situation paralleling that in Section 4, Eq. (10.8) being analogous to Eq. (4.6). In fact the argument there needs only to be amended by a substitution of s^2 for t, s'^2 for t', and $A(v)$ for $a(v)$. The final equation (4.10) is thus replaced by the equation

$$s^2 = s'^2. \tag{10.9}$$

This shows that in all inertial coordinate system having a common origin (0, 0, 0, 0), corresponding to any given event A, the quantity

$$s^2 = c^2t^2 - x^2 - y^2 - z^2, \tag{10.10}$$

relating to an event B at (x, y, z, t), is independent of the coordinate system used to assign coordinates to this event. We may express this by writing

$$s^2 = \text{invariant}. \tag{10.11}$$

The restriction of the inertial coordinate systems to those having a common origin can be removed by taking the coordinates of the event A, used before as the common origin, to be (x_1, y_1, z_1, t_1) and the coordinate of the event B to be (x_2, y_2, z_2, t_2). We then have that

$$s^2 = c^2(t_2 - t_1)^2 - (x_2 - x_1)^2 - (y_2 - y_1)^2 - (z_2 - z_1)^2 \quad (10.12)$$

is an invariant. The quantity s is called the *interval* between the two events A and B. The differential form of Eq. (10.12) is

$$(ds)^2 = c^2(dt)^2 - (dx)^2 - (dy)^2 - (dz)^2. \quad (10.13)$$

Quantities that depend upon the choice of a coordinate system are said to be *relative* while those that are independent of the coordinate system are said to be *absolute*. Thus, within the range of validity of the assumptions made in Section 4 the time interval and the space interval between two events are both absolute. On our present assumptions they are no longer absolute; nevertheless, the sum of their squares, or the interval s as expressed by Eq. (10.12), is still absolute.

Equation (10.12) can be put in the form

$$s^2 = (t_2 - t_1)^2\left[c^2 - \frac{(\mathbf{r}_2 - \mathbf{r}_1)^2}{(t_2 - t_1)^2}\right]. \quad (10.14)$$

When the two events are the position of the same particle at two different times, the quantity $(\mathbf{r}_2 - \mathbf{r}_1)^2/(t_2 - t_1)^2$ is the square of the average velocity of the particle between the points $\mathbf{r}_1$ and $\mathbf{r}_2$. Since, as we shall see in Section 12, the velocity of a particle is always less than c, this quantity must be less than c^2. Equation (10.14) then shows that the interval between any two events corresponding to the same particle is real.

For two events at the same place (same position coordinates) the interval is merely

$$s = c|t_2 - t_1|, \quad (10.15)$$

which is proportional to the time elapsed between the two events. For this reason a real interval is called *timelike*. On the other hand, for two events occurring at the same time the interval is the distance between the two events, multiplied by $i = \sqrt{-1}$. Accordingly, an imaginary interval is called *spacelike*. If the first event is the sending of a light signal from some point and the second event is its reception at some other point, then, as in Eqs. (10.1) and (10.2), the interval between the events is zero. A zero interval between distinct events is thus said to be *lightlike*.

If we set $t_2 = t_1$ in Eq. (10.12), we have either a zero or a pure imaginary interval s. Thus for a timelike interval (s real and nonzero) we must have $t_2 \neq t_1$. Let us suppose that in some particular coordinate system $t_2 > t_1$. Since s^2 is positive for a timelike interval, Eq. (10.14) shows that from the point $\mathbf{r}_1$ at the time t_1 a signal can be sent which, moving with a speed smaller than c, could arrive at the point $\mathbf{r}_2$ at the time t_2. Since in no coordinate system could a signal arrive before it is sent, the event $(\mathbf{r}_1, t_1)$ must be earlier than the event $(\mathbf{r}_2, t_2)$ in all coordinate systems. Thus, for a timelike interval one and the same end is earlier than the other in all coordinate systems. For a spacelike interval this is no longer true and, as we shall see, by a suitable choice of the coordinate system either end of the interval can be made to be earlier than the other.

Consider two events occurring at same particle. If the events are sufficiently close together in time, they can be referred to an inertial coordinate system in which the particle is momentarily at rest. The time difference between these events is called the *proper time* difference. It is the time difference that would be measured by a good clock attached to the particle. It is also the time that determines the speeds of all processes (physical, chemical, or biological) in a system moving with the particle. Since in this coordinate system in which the particle is at rest

$$dx' = dy' = dz' = 0, \qquad dt' = dt_p, \tag{10.16}$$

where dt_p is the proper time difference, Eq. (10.13) states that

$$ds = c\,dt' = c\,dt_p. \tag{10.17}$$

Accordingly, in any other coordinate system with coordinates (x, y, z, t), in which the particle moves with the velocity $\mathbf{v}$,

$$dt_p = \frac{1}{c}\,ds = \frac{1}{c}\sqrt{c^2(dt)^2 - (dx)^2 - (dy)^2 - (dz)^2} = \sqrt{1 - \frac{v^2}{c^2}}\,dt; \tag{10.18}$$

therefore

$$dt_p \leq dt. \tag{10.19}$$

This shows that, in a coordinate system in which a particle moves with a velocity $\mathbf{v} \neq 0$, a clock moving with the particle (or merely a clock moving with a velocity $\mathbf{v} \neq 0$) will appear to be losing time; that is, it will appear to be going slower than a clock at rest.

At first sight this may seem to contradict the principle of relativity by offering a means of knowing which clock is really moving. A little thought shows that this is not the case. The relation between the coordinate system moving with the particle and the coordinate system in which the particle has the velocity $\mathbf{v}$ is symmetrical. Thus, to an observer moving with the particle, the clock at rest in the other coordinate system appears slow. The asymmetry in the positions of dt and dt_p in Eq. (10.18) arises from the fact that in the proper coordinate system we take two reading of time *at the same place* and can thus refer to a *single clock*, in the unprimed coordinate system the same two events require time readings *at different places* and must therefore be referred to *different clocks.*

Integrating Eq. (10.18), we obtain

$$t_p = \int_{t_1}^{t_2} \sqrt{1 - \frac{v^2}{c^2}}\, dt \leq t, \qquad (t_2 > t_1). \tag{10.20}$$

This equation leads to some interesting conclusions. Suppose that a traveler, originally at rest in a coordinate system (any given inertial coordinate system), moved away from us and, after a lengthy journey at speeds near the speed of light, returned and again came to rest in our coordinate system. The clock that he carried with him can be compared with the clocks that remained with us and will show a time different from the time shown by them. In fact, his clock will be behind by the amount $t - t_p$. The traveler will thus return younger than a person who was born at the same time as he, but who remained at rest in our coordinate system.* Again, there is no contradiction with the principle of relativity. The traveler's motion was not an inertial motion; for, otherwise, he could not leave us, if he moved with the same velocity as we, or else he could not return to us and come to rest in our coordinate system. Our analysis merely showed that the time difference between any two events is greater for an observer moving along an inertial path than along any other trajectory through the two events. In our case the two events are the two comparisions of the clocks before and after the trip.

Equation (10.20) shows also that for any fixed t the proper time t_p approaches zero, as v approaches c. To make the physical meaning of this clearer, let us suppose that in some coordinate system we have two points $\mathbf{r}_1$ and $\mathbf{r}_2$ in space very far from each other. We can always choose two events $(\mathbf{r}_1, 0)$ and $(\mathbf{r}_2, t)$ at these locations with a timelike interval between them by making t sufficiently large. Since the interval is timelike, it is possible for a particle to travel between the two events by moving with the average speed $|\mathbf{r}_2 - \mathbf{r}_1|/t = v$ that is less than c. The length of the trip from $\mathbf{r}_1$ to $\mathbf{r}_2$ measured in the proper time of a uniformly moving particle

* This is the famous *twin paradox*. There is a very extensive literature on it. See, for example, references given in G. Holton, *Am. J. Phys.* **30,** 462 (1962).

is by Eq. (10.20)

$$t_p = \sqrt{1 - \frac{v^2}{c^2}}\, t. \tag{10.21}$$

It is now evident that by reducing t, thus making v approach c, we can make t_p as small as we please. In other words, the distance between any two points of space can be traveled in an arbitrarily short proper time, by traveling with a speed sufficiently near c.

It is often convenient to introduce a new variable, proportional to time,

$$\tau = ict, \tag{10.22}$$

where $i = \sqrt{-1}$. With the help of this new variable Eq. (10.13) can be written in the more symmetrical form

$$-(ds)^2 = (dx)^2 + (dy)^2 + (dz)^2 + (d\tau)^2. \tag{10.23}$$

This equation shows that $-(ds)^2$ is a generalization to four dimensions of the square of a vector of displacement in three dimensions, the four-dimensional position vector will be designated x_α and has components

$$x_1 = x, \qquad x_2 = y, \qquad x_3 = z, \qquad x_4 = \tau.$$

Accordingly, the entire history of a particle can be represented in four dimensions by a line consisting of the events of its history; to each instant of time, and thus to each τ, there corresponds a position vector $\mathbf{r}$, which together with τ specifies a four-vector x_α. The terminus of this four-vector traces out a line in this four-dimensional space,* which is called the *world-line* of the particle.

11. The Lorentz Transformation

One of the fundamental questions of the theory of relativity is the following: *Given that in one inertial coordinate system an event has the*

* This is referred to as the *Minkowski space*. See H. Minkowski, *Raum und Zeit*, lecture delivered at the Congress of Scientists, Cologne, 21 Sept. 1908. Translation given in A. Einstein, H. A. Lorentz, H. Minkowski, and H. Weyl, *The Principle of Relativity*, Dover, New York (translation first published in 1923).

One can employ only real vectors by sticking with the metric in Eq. (10.13) and using contravariant and covariant vectors as in A. O. Barut, *Electrodynamics and Classical Theory of Fields and Particles*, Macmillan, New York, 1964, p. 6.

coordinates (x, y, z, t), *what will be its coordinates in another coordinate system, moving with the velocity* $\mathbf{v}$ *relatively to the first*?

Equations connecting quantities in the new coordinate system with the corresponding quantities in the old are called the *equations of transformation.* In particular, in Section 4, starting with the assumption of infinite speed of propagation of interaction, we found the Galilean equations of transformation. What conditions must the equations of transformation satisfy now? We have given an answer to this question in the previous section. They must be linear homogeneous equations of the type given in Eq. (10.7) and they must leave invariant s^2, see Eq. (10.10). These conditions are sufficient to determine the equations of transformation in each particular case.

Introducing the four-dimensional vector x_α, we can write Eq. (10.9) in the form

$$x_\alpha^2 = x^2 + y^2 + z^2 + \tau^2 = x'^2 + y'^2 + z'^2 + \tau'^2. \tag{11.1}$$

Mathematically considered, a transformation of coordinates of the type given by (10.7), leaving the square of a four-dimensional vector x_α invariant, is a rotation in four-dimensional space. This mathematical analogy (and it is no more than that) enables us to obtain the desired equations of transformation very simply.

Consider a transformation that leaves x_α^2 invariant and affects only the x and y coordinates, that is, such that $\tau' = \tau$ and $z' = z$. Equation (11.1) then demands that

$$x'^2 + y'^2 = x^2 + y^2; \tag{11.2}$$

and by analogy with what we would have in three-dimensional space this is termed a *rotation in the XY-plane* of our four-dimensional space. Moreover, since any transformation satisfying Eq. (11.2) can be achieved by a rotation of the spatial coordinates about the *Z*-axis, the transformation to the primed coordinates amounts to merely an ordinary rotation of the *X*-axis and *Y*-axis in space.

Let us take, however, a rotation in the *XT*-plane. The equations of transformation for such a rotation, from analytic geometry, are

$$\begin{aligned} x' &= x \cos\alpha + \tau \sin\alpha, \\ \tau' &= -x \sin\alpha + \tau \cos\alpha, \\ y' &= y, \qquad z' = z, \end{aligned} \tag{11.3}$$

where α is the angle of rotation. This angle, however, will turn out to be imaginary.

The corresponding equations involving the relative velocities of the coordinate systems are obtained by finding the physical analogs of $\sin \alpha$ and $\cos \alpha$. This can be done by considering some special case. It is convenient to take for this purpose the transformation for the origin of the second coordinate system, that is, the point $x' = y' = z' = 0$. This point moves with the velocity $\mathbf{v}$ with respect to the first coordinate system, so that its coordinates in the first coordinate system are

$$x = v_x t, \qquad y = v_y t, \qquad z = v_z t. \tag{11.4}$$

On the other hand, Eq. (11.3) gives, upon substituting $x' = y' = z' = 0$,

$$x = -\tau \tan \alpha = -(ic \tan \alpha)t, \qquad y = z = 0. \tag{11.5}$$

A comparison of Eqs. (11.4) and (11.5) shows that $\mathbf{v}$ must be directed along the X-axis; $v_z = v_y = 0$, $v_x = v = -ic \tan \alpha$. Therefore, we have

$$\tan \alpha = \frac{iv}{c}, \qquad \cos \alpha = \frac{1}{\sqrt{1 - v^2/c^2}}, \qquad \sin \alpha = \frac{iv}{c\sqrt{1 - v^2/c^2}}. \tag{11.6}$$

The choice of a positive sign for the radical in the expression for $\cos \alpha$ may be justified as follows: When $\mathbf{v}$ goes to zero, the primed system becomes identical with the unprimed system and hence $x' = x$. From the first equation of Eqs. (11.3) this is seen to require that when $v = 0$ we must have $\cos \alpha = 1$.

The equations of transformation (11.3) can thus be written in the form

$$\begin{aligned} x' &= \frac{x + iv\tau/c}{\sqrt{1 - v^2/c^2}}, & \tau' &= \frac{\tau - ivx/c}{\sqrt{1 - v^2/c^2}}, \\ y' &= y, & z' &= z, \end{aligned} \tag{11.7}$$

or, in terms of t and t', in the form

$$\begin{aligned} x' &= \frac{x - vt}{\sqrt{1 - v^2/c^2}}, & t' &= \frac{t - vx/c^2}{\sqrt{1 - v^2/c^2}}, \\ y' &= y, & z' &= z. \end{aligned} \tag{11.8}$$

These equations were obtained by performing a rotation in the XT-plane. They correspond to the case when the vector $\mathbf{v}$ is directed along the X-axis; however, since we can nearly always choose the X-axis in the direction of $\mathbf{v}$, the restriction is of little importance.

The equations of transformation (11.8) are known as the *Lorentz transformation equations.* They were obtained by Lorentz as early as

1904,* from a consideration of the simplicity that was introduced into the electrodynamics of moving systems with their help. To have done this before the introduction of the theory of relativity was a remarkable achievement. It must be noted, however, that the time t' was to Lorentz, in 1904, merely a convenient device, and not in any sense a true time.

Equations (11.8) may be solved for the unprimed variables in terms of the primed; the result is

$$x = \frac{x' + vt'}{\sqrt{1 - v^2/c^2}}, \qquad y = y', \qquad z = z', \qquad t = \frac{t' + vx'/c^2}{\sqrt{1 - v^2/c^2}}. \tag{11.9}$$

One sees from these equations that when $x = 0$, $x' = -vt'$, or that the first coordinate system is moving with the velocity $-\mathbf{v}$ with respect to the second. This is just what was to be expected on the Newtonian theory, or from a consideration of the symmetry between the coordinate systems. The only difference between the coordinate systems is that the prime system moves with a velocity $\mathbf{v}$ with respect to the unprimed system while the unprimed system moves with a velocity $-\mathbf{v}$ with respect to the primed system. Thus, Eqs. (11.9) follow from Eqs. (11.8) by merely replacing $\mathbf{v}$ by $-\mathbf{v}$. We also note that, for $c \to \infty$, Eqs. (11.8) approach the corresponding equations of the Galilean transformation (see Eqs. (4.14))

$$x' = x - vt, \qquad y' = y, \qquad z' = z, \qquad t' = t.$$

EXAMPLE 1. Determine the length of a uniformly moving rod relative to its length in a coordinate system in which it is at rest.

Let the coordinate system in which the rod is at rest be designated as the primed coordinate system, then with the conventions established above the rod will have a velocity $\mathbf{v}$ along the positive X-axis in the unprimed system. Let the spatial coordinates of the end points of the rod in the primed system be at $(0, 0, 0)$ and (a', b', c'). The spatial coordinates of these end points in the unprimed system at $t = 0$ (a given instance in the unprimed system) are obtained by setting $t = 0$ in Eq. (11.8). Thus the first end is at $(0, 0, 0)$ and the second end at (x, y, z) where

$$a' = \frac{x}{\sqrt{1 - v^2/c^2}}, \qquad b' = y, \qquad c' = z.$$

The position of the second end relative to the first is thus given by the

* H. A. Lorentz, *Proc. Amsterdam Acad. Sci.*, **6**, 809 (1904).

displacement vector

$$(x, y, z) = \left(a'\sqrt{1 - \frac{v^2}{c^2}}, b', c'\right). \tag{11.10}$$

We see, therefore, that the dimensions (b' and c') of the rod perpendicular to the direction of motion are unchanged but that the longitudinal dimension changes from a' to

$$x = a'\sqrt{1 - \frac{v^2}{c^2}} \tag{11.11}$$

as one goes from the primed coordinate system to the unprimed system in which the rod moves with a velocity $\mathbf{v}$.

Such a contraction is known as the *Lorentz–Fitzgerald contraction*, and was proposed by them to explain the experimental results obtained by Michelson. We see, however, that this is a perfectly symmetrical effect; a rod at rest in *either* of the two coordinates systems appears contracted to the observer in the other coordinate system.

To understand better the nature of apparent contraction of the rod in the direction of its motion one sees from Eqs. (11.8) that for $t = 0$

$$t' = \frac{-vx/c^2}{\sqrt{1 - v^2/c^2}} = \frac{-vx'}{c^2} \tag{11.12}$$

Thus to an observer moving with the rod the positions $(0, 0, 0)$ and (x', y', z') of the end points of the rods are taken at slightly different times— the first at $t' = 0$ and the second at $t' = va'/c^2$. He is therefore not surprised that an unprimed system observer gets the "wrong" length. Note, however, that the primed observer's measurements of lengths can be similarly criticized by an observer in the unprimed coordinate system.

EXAMPLE 2. A clock at rest at the origin of the primed coordinate system comes into coincidence, at the time t', with another clock, which is at rest in the unprimed coordinate system. What will be the reading of the second clock?

In the primed coordinate system the event corresponding to the coincidence has the coordinates $(0, 0, 0, t')$. The corresponding coordinates in the unprimed system, according to the Eqs. (11.9), are

$$t = \frac{t'}{\sqrt{1 - v^2/c^2}}, \qquad x = \frac{vt'}{\sqrt{1 - v^2/c^2}} = vt. \tag{11.13}$$

The expression for x is, of course, what was to be expected. The first equation shows that the reading of the clock in the unprimed system will be greater, in the ratio of 1 to $\sqrt{1 - v^2/c^2}$, than the reading of the clock in the primed system; in other words, a moving clock will appear to go slow. Conversely, a clock at rest in the unprimed system will appear to go slow when it is compared with the clocks that it passes in the primed system.

EXAMPLE 3. A particle moves in the XY-plane with a velocity $\mathbf{u} = (u_x, u_y, 0)$. What will be its velocity $\mathbf{u}'$ in a second coordinate system, moving in the X-direction with the velocity $\mathbf{v}$ with respect to the first?

We may take the first coordinate system to be the unprimed system and the second the primed system; thus Eqs. (11.8) and (11.9) are applicable. In terms of differentials, Eqs. (11.8) become

$$dx' = \frac{dx - v\,dt}{\sqrt{1 - v^2/c^2}}, \qquad dt' = \frac{dt - v\,dx/c^2}{\sqrt{1 - v^2/c^2}}, \qquad (11.14)$$
$$dy' = dy, \qquad dz' = dz.$$

Dividing by dt' and remembering that $u_x = dx/dt$, $u_x' = dx'/dt'$, etc., we have

$$u_x' = \frac{dx - v\,dt}{dt - (v\,dx/c^2)} = \frac{u_x - v}{1 - (vu_x/c^2)},$$
$$u_y' = \frac{u_y\sqrt{1 - v^2/c^2}}{1 - (vu_x/c^2)}, \qquad u_z' = \frac{u_z\sqrt{1 - v^2/c^2}}{1 - (vu_x/c^2)}. \qquad (11.15)$$

These equations are known as the equations for *addition of velocities*. In the special case when $\mathbf{u}$ is parallel to $\mathbf{v}$, they acquire the simple form

$$u' = \frac{u - v}{1 - (vu/c^2)} \qquad \text{or} \qquad u = \frac{u' + v}{1 + (vu'/c^2)}. \qquad (11.16)$$

When c is made to approach infinity, these equations, of course, pass into the corresponding equations for the Galilean transformation (see Eq. (4.13)).

An interesting special case arises when $u = c$. Equation (11.16) then shows that also $u' = c$. Thus, a particle moving with the maximum speed moves with the same speed in all coordinate systems.

If the motion with speed c is not parallel to the X-axis, but makes an angle β with it, then $u_x = c\cos\beta$, $u_y = c\sin\beta$. Equations (11.15) give in this case

$$u'_x = \frac{c^2\cos\beta - vc}{c - v\cos\beta}, \qquad u'_y = \frac{(c^2\sin\beta)\sqrt{1 - v^2/c^2}}{c - v\cos\beta}, \tag{11.17}$$

from which it follows that

$$(u'^2_x + u'^2_y)^{1/2} = c, \qquad \text{and} \qquad \tan\beta' = \frac{u'_y}{u'_x} = \frac{c\sin\beta\sqrt{1 - v^2/c^2}}{c\cos\beta - v}. \tag{11.18}$$

Thus, the speed remains c, but the direction of motion is different. This result applies equally well to a ray of light. The change in its direction as expressed by the second equation of Eqs. (11.18) is known as *optical aberration.*

For small values of v/c this equation becomes

$$\tan\beta' \simeq \frac{\sin\beta}{\cos\beta - (v/c)} \simeq \left(1 + \frac{v}{c}\sec\beta\right)\tan\beta.$$

Since this shows that $\Delta \equiv \beta' - \beta$ is small, we have also

$$\begin{aligned}\tan\beta' &= \frac{\tan\beta + \tan\Delta}{1 - \tan\beta\tan\Delta} \\ &\simeq (\Delta + \tan\beta)(1 + \Delta\tan\beta) \\ &\simeq \tan\beta + \Delta(1 + \tan^2\beta).\end{aligned}$$

Comparing these last two equations, we have

$$\Delta \simeq \frac{v}{c}\sin\beta, \tag{11.19}$$

which is the nonrelativistic formula for aberration.

In the theory of relativity it is often convenient to use a four-dimensional vector analysis. We have already encountered the four-vector

$$x_\alpha = (x, y, z, \tau) = (\mathbf{r}, \tau), \tag{11.20}$$

where x_α is meant to stand for the entire set of components

$$x_1 = x, \qquad x_2 = y, \qquad x_3 = z, \qquad x_4 = \tau \equiv ict. \tag{11.21}$$

In general, a four-vector will be a set of four components transforming like x_α under a Lorentz transformation. It will be designated by a letter bearing a Greek-letter subscript (A_β, u_α, w_γ, etc.).

If

$$A_\alpha = (A_1, A_2, A_3, A_4)$$
$$B_\alpha = (B_1, B_2, B_3, B_4)$$

are two four-vectors, then their *inner product* is defined as

$$A_\alpha B_\alpha \equiv \sum_{\alpha=1}^{4} A_\alpha B_\alpha.$$

Thus the summation convention is extended to mean that a repeated Greek-letter subscript is to be summed from 1 to 4. We shall designate summation from 1 to 3 by using Latin letters, thus

$$A_j B_j \equiv \sum_{j=1}^{3} A_j B_j.$$

An important four-vector is the differential

$$dx_\alpha = (dx, dy, dz, d\tau) = (d\mathbf{r}, d\tau). \tag{11.22}$$

By taking the inner product of two four-vectors we obtain a *scalar*, a quantity that is invariant under a Lorentz transformation of the four-vectors, for example,

$$(dx_\alpha)^2 \equiv dx_\alpha\, dx_\alpha = (d\mathbf{r})^2 - c^2(dt)^2 = -(ds)^2. \tag{11.23}$$

Dividing each component of a vector by a scalar we obtain another vector. Thus, starting with dx_α and using Eq. (10.18), we can define the velocity four-vector

$$u_\alpha \equiv \frac{dx_\alpha}{ds} = \left(\frac{\mathbf{v}}{c\sqrt{1 - v^2/c^2}}, \frac{i}{\sqrt{1 - v^2/c^2}}\right). \tag{11.24}$$

It has a constant magnitude, since

$$u_\alpha^2 = \frac{(dx_\alpha)^2}{(ds)^2} = -1. \tag{11.25}$$

Similarly, we may define a four-dimensional acceleration

$$g_\alpha = \frac{du_\alpha}{ds} = \frac{d^2 x_\alpha}{ds^2}. \tag{11.26}$$

By differentiating Eq. (11.25) we have

$$u_\alpha g_\alpha = 0. \tag{11.27}$$

Two four-vectors, such as these, whose inner product is zero, are said to be *orthogonal* to each other.

The principle of relativity requires that every physical law be basically independent of the choice of a particular inertial coordinate system. Tensor analysis (see Appendix A5) is a mathematical formalism permitting the expression of physical laws in a form that readily reflects its independence of coordinate system.

Let a general Lorentz transformation

$$x'_\alpha = a_{\alpha\beta} x_\beta \tag{11.28}$$

represents a change from the coordinates x_α of one inertial frame to the coordinates x'_α of any other inertial frame. Then a four-dimensional tensor $A_{\alpha_1\alpha_2\ldots\alpha_n}$ is an 4^n-component object that transforms under this Lorentz transformation in the following way:

$$A'_{\alpha_1\alpha_2\cdots\alpha_n} = a_{\alpha_1\beta_1} a_{\alpha_2\beta_2} \cdots a_{\alpha_n\beta_n} A_{\beta_1\beta_2\cdots\beta_n} \tag{11.29}$$

where $\alpha_1, \alpha_2, \ldots, \alpha_n$ can each have any one of the four values 1, 2, 3, 4 and each of the β's are summed over these same values. This tensor is said to be of rank n.

In this terminology a scalar A is a tensor of rank 0 because its transformation equation can be written

$$A' = A. \tag{11.30}$$

A vector A_α is a tensor of rank 1 and hence transforms thus

$$A'_\alpha = a_{\alpha\beta} A_\beta . \tag{11.31}$$

A tensor $A_{\alpha\beta}$ of rank 2 under a Lorentz transformation becomes

$$A'_{\alpha\beta} = a_{\alpha\gamma} a_{\beta\delta} A_{\gamma\delta} . \tag{11.32}$$

One can make the form of a physical law independent of the inertial coordinate system employed by expressing it in the *covariant form*

$$\text{tensor } A = \text{tensor } B, \tag{11.33}$$

where the tensors are of the same rank. On transforming to a new (primed) coordinate system the law becomes simply

$$\text{tensor } A' = \text{tensor } B',$$

where the primes designate the transformed tensors. Also, since the tensors are of the same rank, they can be subtracted to give a tensor C.

The physical law can then also be written in the covariant form

$$\text{tensor } C = 0, \tag{11.34}$$

where 0 is the zero tensor of that rank. One sees from Eq. (11.29) that the zero tensor, which is required to have all its components equal to zero in one coordinate system, will also have all of them zero in any other inertial coordinate system.

We shall require as a principle that *each physical law be capable of being written in covariant form*, as in Eq. (11.33) or (11.34). This we shall call the *principle of invariance* and leave undetermined the extent to which it is a consequence of the principle of relativity.

EXAMPLE 4. Given that in one inertial coordinate system an event has coordinates (x, y, z, t); what will be its coordinates in another inertial coordinate system, moving with the velocity $\mathbf{v}$ relatively to the first?

In this example we shall not restrict the direction of $\mathbf{v}$, as was done in the special *Lorentz transformation* previously considered. For this reason the equations that we shall derive connecting (x, y, z, t) with (x', y', z', t') will be called the *general Lorentz transformation.* We shall assume however, for the sake of simplicity, that the origins of the two coordinate systems coincide at $t = t' = 0$.

The transformation will thus again be a rotation in the four-dimensional Minkowski space. As in driving the special Lorentz transformation we are only interested in rotations in planes passing through the T-axis. We have seen that a rotation in the XT-plane corresponds to having the new coordinate system moving with the velocity $\mathbf{v}$ along the X-axis of the old coordinate system. Since the velocity $\mathbf{v} = (v_x, v_y, v_z)$ specifies a direction in the spatial dimensions, we could either rotate our spatial axes so that the new X-axis is in the direction of $\mathbf{v}$ and then rotate in the new XT-plane or, what is clearly the same, we could leave the X-axis unchanged and rotate in this same plane. By choosing the latter we need merely to rotate in the plane determined by the T-axis and the four-dimensional vector $(v_x, v_y, v_z, 0)$ determined by $\mathbf{v}$.

Four-dimensionally these directions are specified by the unit vectors

$$z_\alpha = (0, 0, 0, 1) \qquad \text{along the } T\text{-axis} \tag{11.35}$$

and

$$w_\alpha = \left(\frac{v_x}{v}, \frac{v_y}{v}, \frac{v_z}{v}, 0\right) \qquad \text{along } \mathbf{v}. \tag{11.36}$$

These vectors are orthogonal to each other because by inspection

$$z_\alpha w_\alpha = 0. \tag{11.37}$$

We thus have the situation discussed in Appendix A5 of a general rotation in the plane determined by two orthogonal unit vectors (see Eq. (A5.34) and following equations).

Our transformation will therefore be

$$x'_\alpha = a_{\alpha\beta} x_\beta, \tag{11.38}$$

where $a_{\alpha\beta}$ is given by Eq. (A5.45) with $u_\alpha = z_\alpha$ and $v_\alpha = w_\alpha$, namely,

$$a_{\alpha\beta} = \delta_{\alpha\beta} - (z_\alpha z_\beta + w_\alpha w_\beta)(1 - \cos\theta) + (z_\alpha w_\beta - z_\beta w_\alpha)\sin\theta. \tag{11.39}$$

With the values of z_α and w_α given by Eqs. (11.35) and (11.36) we have

$$\begin{aligned} a_{jk} &= \delta_{jk} - \frac{1}{v^2} v_j v_k (1 - \cos\theta), \\ a_{j4} &= \frac{-v_j}{v} \sin\theta = -a_{4j}, \\ a_{44} &= \cos\theta, \end{aligned} \tag{11.40}$$

where i, j, and k each take on one of the values 1, 2, and 3.

If a particle is at rest in the primed coordinate system, it is moving with the velocity $\mathbf{v}$ in the unprimed coordinate system. Let the particles path in the unprimed coordinate system be given by $x_k(t)$, $k = 1, 2, 3$. Then the events on the world line of this particle satisfy the differential conditions

$$dx'_j = 0, \qquad \frac{dx_k}{dt} = v_k. \tag{11.41}$$

On the other hand, from the differential form of Eq. (11.38)

$$dx'_j = a_{j\beta}\,dx_\beta = a_{jk}\,dx_k + a_{j4} ic\,dt = \left(a_{jk}\frac{dx_k}{dt} + ica_{j4}\right) dt.$$

Thus for events on the world line, by virtue of Eqs. (11.41),

$$a_{jk} v_k + ica_{j4} = 0. \tag{11.42}$$

Substitution from Eqs. (11.40) now gives

$$\left[\delta_{jk} - \frac{1}{v^2} v_j v_k (1 - \cos\theta)\right] v_k - \frac{icv_j}{v} \sin\theta = 0,$$

which reduces to

$$\tan\theta = \frac{-iv}{c}. \tag{11.43}$$

We require that when $\mathbf{v} = 0$, the two coordinate systems exactly coincide and thus that $a_{\alpha\beta} = \delta_{\alpha\beta}$. From the last equation of Eqs. (11.40) this is seen to require that when $v = 0$ we have $\cos\theta = 1$. With this requirement and that of Eq. (11.43) we have

$$\begin{aligned} \cos\theta &= \frac{1}{\sqrt{1 - v^2/c^2}} \equiv \gamma, \\ \sin\theta &= \frac{-iv/c}{\sqrt{1 - v^2/c^2}} = \frac{-i\gamma v}{c}. \end{aligned} \tag{11.44}$$

Equations (11.38) and (11.40) then yield the transformation equations

$$\begin{aligned} x' &= x + \frac{1}{v^2}(\gamma - 1) v_k x_k v_x - \gamma v_x t, \\ y' &= y + \frac{1}{v^2}(\gamma - 1) v_k x_k v_y - \gamma v_y t, \\ z' &= z + \frac{1}{v^2}(\gamma - 1) v_k x_k v_z - \gamma v_z t, \\ t' &= \gamma\left(t - \frac{v_k x_k}{c^2}\right), \end{aligned} \tag{11.45}$$

remembering of course that $x_4 = \tau = ict$. In vector form these equations are

$$\begin{aligned} \mathbf{r}' &= \mathbf{r} + \frac{1}{v^2}(\gamma - 1)\mathbf{v} \cdot \mathbf{r}\, \mathbf{v} - \gamma \mathbf{v} t \\ t' &= \gamma\left(t - \frac{\mathbf{v} \cdot \mathbf{r}}{c^2}\right), \end{aligned} \tag{11.46}$$

where as above

$$\gamma = \frac{1}{\sqrt{1 - v^2/c^2}}.$$

12. Relativistic Particle

We must now investigate the properties of a particle in relativistic mechanics. As before, we start with the case of a free particle. In our investigation we shall be guided by the principle of invariance and the fact that, in the limit as c is made to approach infinity, these properties must become those of a Newtonian particle, discussed in Section 5.

Hamilton's principle (see Eq. (2.9)),

$$\delta W = \delta \int_a^b L\, dt = 0, \tag{12.1}$$

becomes invariant if the differential

$$dW = L\, dt \tag{12.2}$$

is a four-dimensional invariant, that is a scalar, and if the limits of integration are specified in an invariant way, as for example by specifying the two events between which the integration takes place. Due to the fact that L is a function of the position and velocity of the particle, dW must be a function of x_α and of the various dx_β; but for a free particle, since its behavior cannot depend on its location in space and time, it must be a function of these dx_β alone. The only independent invariant function of this kind is $(dx_\beta)^2 = -(ds)^2$. This is evident from the fact that any change in the dx_β not affecting $(ds)^2$ can be produced by a rotation of coordinates; hence, since space is assumed to be isotropic, the value of an invariant function dW must be independent of such changes and thus depend only on the value of $(ds)^2$, i.e., it must be a function of $(ds)^2$.

On the other hand, Eq. (12.2) shows that dW must be a homogeneous function of the first degree in the differentials. Thus, we must have

$$dW = -B\, ds, \qquad \text{or} \qquad W = -B \int ds, \tag{12.3}$$

where B is some constant. Comparing this with Eqs. (10.17) and (10.20), we see that action is proportional to the proper time of the particle, measured from the event corresponding to the lower limit of integration in Eq. (12.3). We saw in Section 10 that for inertial motion the proper time has its maximum value. This is in agreement with Eq. (12.1), and shows that W will have a minimum if B is positive. This is, of course, the reason for introducing the minus sign in the expression for W.

Comparing Eqs. (12.2) and (12.3), we see that

$$L = -B\frac{ds}{dt} = -B\sqrt{c^2 - v^2}, \tag{12.4}$$

which, at first sight, may seem to be entirely unlike the expression $\frac{1}{2}mv^2$ obtained in Section 5.* However, when v is very small in comparison with c, we can write Eq. (12.4) in the form

$$L \simeq -Bc + \frac{1}{2}\frac{Bv^2}{c}, \tag{12.5}$$

with neglect of the higher terms. The first term, being a constant, could be dropped; the second term becomes equal to $\frac{1}{2}mv^2$ if $B = mc$. We therefore choose this value of B, obtaining

$$W = -mc\int_a^b ds \tag{12.6}$$

and

$$L = -mc^2\sqrt{1 - v^2/c^2}. \tag{12.7}$$

It is to be noted that the requirement of invariance of W makes L quite definite; no arbitrary additive constant is now allowed.

Let us next calculate δW, which from Eq. (12.6) is

$$\delta W = -mc\,\delta\int_a^b ds = -mc\int_a^b \delta\, ds. \tag{12.8}$$

* The assumption of invariance of dW means that

$$L(\mathbf{v})\,dt = L(\mathbf{v}')\,dt',$$

which for Newtonian space and time ($dt = dt'$) would reduce to the invariance of L, namely,

$$L(\mathbf{v}) = L(\mathbf{v}').$$

Applying to this the argument of Section 5, since now $\delta L = 0$, one obtains $\varphi' = 0$ and $L(\mathbf{v}) =$ *constant*, which is not satisfactory. However, as was proved in Section 2, with $t = t'$ we need no longer insist on $L(\mathbf{v}) = L(\mathbf{v}')$, but can assume a less restrictive condition,

$$L(\mathbf{v}') = L(\mathbf{v}) + \frac{dR}{dt}.$$

Thus in Section 5 a suitable Lagrangian is obtained but the L is, of course, not invariant.

Now, since $-(ds)^2 = (dx_\alpha)^2$, we have

$$-2\, ds\, \delta\, ds = 2\, dx_\alpha\, d\delta x_\alpha,$$

and thus

$$\delta\, ds = -u_\alpha\, d\delta x_\alpha. \tag{12.9}$$

Substitution of this result into Eq. (12.8) and integration by parts gives, finally,

$$\delta W = mc\int_a^b u_\alpha\, d\delta x_\alpha = [mcu_\alpha\, \delta x_\alpha]_a^b - mc\int_a^b \delta x_\alpha\, du_\alpha. \tag{12.10}$$

The equations of motion are obtained, as in Section 2, by setting δW equal to zero under the condition that $\delta x_\alpha = 0$ at the end points a and b of the path. Thus from Eq. (12.10) we must have

$$-mc\int_a^b \delta x_\alpha\, du_\alpha = 0$$

for arbitrary δx_α. This requires that

$$du_\alpha = 0, \tag{12.11a}$$

or in integral form

$$u_\alpha = \text{constant four-vector.} \tag{12.11b}$$

This is the four-dimensional, or covariant, form of the nonrelativistic equation

$$\mathbf{v} = \textbf{constant},$$

which we obtained in Section 3. We may observe, in passing, that we now have four equations, instead of three. The fourth equation, however, is not independent of the other three, since the four components of u_α must satisfy Eq. (11.25).

As in Section 2, we may define the momentum to be $\mathbf{p} = \partial L/\partial \mathbf{v}$, which by Eq. (12.7) is

$$\mathbf{p} = \frac{m\mathbf{v}}{\sqrt{1 - v^2/c^2}}. \tag{12.12}$$

By Eq. (2.15) the Hamiltonian is

$$H = \mathbf{p}\cdot\mathbf{v} - L = \frac{mc^2}{\sqrt{1 - v^2/c^2}} = \sqrt{c^2p^2 + m^2c^4} \tag{12.13}$$

and, since by Eq. (12.11) it is conserved, it is referred to as the energy of the particles and written

$$E = \frac{mc^2}{\sqrt{1 - v^2/c^2}}. \tag{12.14}$$

One obtains a covariant treatment by taking paths satisfying the equations of motion, Eqs. (12.11), and letting W be a function of the upper limit b. From Eq. (12.10), since $du_\alpha = 0$,

$$\delta W = mcu_\alpha \, \delta x_\alpha,$$

where the right-hand side is evaluated at this upper limit. Making use of the fact that W is then a function only of the x_α defining the upper limit, we have also

$$\delta W = \frac{\partial W}{\partial x_\alpha} \delta x_\alpha.$$

Comparing these equations and defining the momentum components p_α in keeping with Eqs. (2.19), we have

$$p_\alpha = \frac{\partial W}{\partial x_\alpha} = mcu_\alpha, \tag{12.15}$$

where $\alpha = 1, 2, 3, 4$. The three equations corresponding to $\alpha = 1, 2, 3$ are the relativistic form of Eq. (2.19). The fourth equation ($\alpha = 4$) defines the fourth component p_4 of the momentum four-vector. Thus by Eqs. (2.21), (12.13), and (12.14) we have

$$p_4 = \frac{\partial W}{\partial x_4} = \frac{1}{ic}\frac{\partial W}{\partial t} = \frac{i}{c} H = \frac{i}{c} E. \tag{12.16}$$

We may readily check that Eqs. (12.15) and (12.16) are consistent with Eqs. (12.12) and (12.14).

From Eqs. (12.12) and (12.14) it follows that

$$\mathbf{p} = \frac{E}{c^2}\mathbf{v}. \tag{12.17}$$

On account of this equation, which is analogous to the equation of the Newtonian theory, $\mathbf{p} = m\mathbf{v}$, the quantity

$$\frac{E}{c^2} = \frac{m}{\sqrt{1 - v^2/c^2}}$$

is sometimes called the mass of the particle. We prefer, however, to preserve the name *mass* for the quantity m, which characterizes the inertial properties of the particle, in all kinds of motion, leaving without a special name the quantity E/c^2 which appertains to one speed only.

It is evident that the momentum $\mathbf{p}$, Eq. (12.12), approaches $m\mathbf{v}$ when v is small in comparison with c. On the other hand, in this case

$$E = mc^2 + \tfrac{1}{2}mv^2, \tag{12.18}$$

with neglect of the higher terms. This differs by a term mc^2 from the Newtonian expression for the kinetic energy of the particle. We shall call the quantity mc^2 *the rest energy* of the particle.

One sees from Eq. (12.14) that as $v \to c$ the energy E approaches infinity. Thus, it would require an infinite energy to bring a particle to speed c, while the energy of a particle with greater speed is imaginary. We therefore conclude that all particles of finite mass move with speeds smaller than c.

The speed of interaction is equal to c and we have now shown that a particle must move with a speed less than c. Since the only other type of signal, a disturbance, must, as we saw in Section 4, travel with a speed less than the speed of interaction, we conclude that a signal cannot be sent with a speed greater than c.

The square of the four-vector p_α is the scalar

$$p_\alpha^2 = m^2c^2u_\alpha^2 = -m^2c^2, \tag{12.19}$$

in accordance with Eqs. (12.15) and (11.25). In the three-dimensional form this relation is

$$E^2 = m^2c^4 + c^2p^2, \tag{12.20}$$

which is a basic equation relating the energy to the magnitude of the momentum.

Finally, we note that since $(p_1, p_2, p_3, iE/c)$ are the components of a four-vector p_α, they transform as the components of x_α, i.e., in accordance with Eqs. (11.7). Thus, the Lorentz transformations that apply to the energy and the components of the momentum are

$$\begin{aligned} p_x' &= \frac{p_x - Ev/c^2}{\sqrt{1 - v^2/c^2}}, \qquad p_y' = p_y, \qquad p_z' = p_z, \\ E' &= \frac{E - vp_x}{\sqrt{1 - v^2/c^2}}. \end{aligned} \tag{12.21}$$

13. Bodies in Relativistic Theory

In the theory of relativity absolutely rigid bodies are impossible, because such bodies could be given a motion that would contradict the predictions of the theory. This is seen at once by noting that if one edge of a body is set in motion the far edge would have to respond at once to maintain the dimensions of the body. One could thus propagate a signal across the body with infinite velocity. This clearly contradicts the fundamental assumption in relativity of a finite velocity of propagation c, valid for all inertial coordinate systems.

This in no way affects the possibility of what we called rigid bodies, bodies that return to the same shape whenever they are brought to standard conditions (see Section 1).

We have defined a particle as a part of a physical object, a part so small that no further purpose is served by continuing the process of division. This situation may arise, either because we limit ourselves to problems that do not require a more detailed analysis of the structure of the object, or because a more detailed analysis is not possible within the limitations of classical electrodynamics. In the latter case we shall say that the particle is a *classical elementary particle*.* Such an elementary particle cannot be treated as having a finite radius, because a finite radius would imply that the particle is either absolutely rigid, or possesses certain elastic properties. In the former case it contradicts the theory of relativity, as has just been shown; in the latter case it would possess an infinite number of degrees of freedom, behaving as a complicated structure and not as an elementary particle. We see, therefore, that *within the limitations of classical theory an elementary particle must be considered as a point*. A typical example of a classical elementary particle is an *electron*. It is a particle having a mass of $(9.1091 \pm 0.0004) \times 10^{-28}$ grams and a negative charge of $(4.80298 \pm 0.00020) \times 10^{-10}$ electrostatic units.† Thus, as an elementary particle an electron cannot properly be treated as having a finite radius. We shall return to this question in Section 24.

Let us now consider a body as a system of particles. The energy and

* By this terminology we hope to ignore the problem of what an elementary particle is in a more general theory, say in quantum field theory.

† *Phys. Today*, **17**, 48 (1964).

momentum of a system were defined in Section 3. If $\boldsymbol{\epsilon}$ in Eq. (3.13) has the components $(\delta x_1, \delta x_2, \delta x_3)$, Eq. (3.13) implies the following:

$$P_1 = \frac{\partial W}{\partial x_1}, \qquad P_2 = \frac{\partial W}{\partial x_2}, \qquad P_3 = \frac{\partial W}{\partial x_3}. \tag{13.1}$$

If we write, as the fourth component,

$$P_4 = \frac{\partial W}{\partial x_4} = \frac{\partial W}{\partial \tau} = \frac{iE}{c}, \tag{13.2}$$

the energy-momentum four-vector P_α for the system will behave exactly as the four-vector p_α for a single particle. If we take the X-axis and X'-axis parallel and the latter in the direction of the velocity $\mathbf{v}$ of the primed coordinate system in the unprimed system, then as seen from Eqs. (12.21) (with, of course, v replaced by $-v$) the transformation equations are

$$\begin{aligned} P_x &= \frac{P'_x + E'v/c^2}{\sqrt{1 - v^2/c^2}}, \qquad P_y = P'_y, \qquad P_z = P'_z, \\ E &= \frac{E' + vP'_x}{\sqrt{1 - v^2/c^2}}. \end{aligned} \tag{13.3}$$

Let us choose the proper coordinate system—that is, the system in which the total momentum of the body is zero—as the primed coordinate system. The velocity $\mathbf{v}$ of the system as a whole in the unprimed coordinate system is defined, as before, as the velocity in it of this primed coordinate system. The energy of the body in the proper coordinate system is called the proper energy of the body and is denoted by E_0. Substituting $P'_x = P'_y = P'_z = 0$ and $E' = E_0$ in Eq. (13.3) and writing the first three equations as a single vector equation, we have

$$\mathbf{P} = \frac{E_0\mathbf{v}/c^2}{\sqrt{1 - v^2/c^2}}, \qquad E = \frac{E_0}{\sqrt{1 - v^2/c^2}}, \tag{13.4}$$

where $\mathbf{v}$ is in the X-direction. Since the first equation is a vector equation and holds in one three-dimensional coordinate system, it holds in all three-dimensional coordinate systems; thus Eqs. (13.4) are not dependent upon the X-axis being taken parallel to $\mathbf{v}$.

The dependence of these quantities upon $\mathbf{v}$ is exactly the same as for a particle of mass $M = E_0/c^2$; M is therefore called the *mass of the body*. Being defined in terms of the proper coordinate system it is invariant. This invariance can be seen also from the relation

$$P_\alpha^2 = P^2 - \frac{E^2}{c^2} = -c^2M^2 = \text{invariant}, \tag{13.5}$$

which is the analog of Eq. (12.19). This equation can be used, moreover, as an invariant definition of the mass of the body. Incidentally, as this equation shows, the mass of an isolated body is conserved because, as we saw in Section 3, the energy and the total momentum of an isolated system are conserved.

It is, perhaps, necessary to warn against confusing invariance with conservation. A quantity is invariant if it is unchanged when a transformation is made from one coordinate system to another; a quantity is conserved if it remains constant in a given coordinate system. Thus, the energy of an isolated system is conserved, but it is not invariant, being proportional to the fourth component of a four-vector; on the other hand, for a nonisolated system the mass, although still an invariant, is not, as we shall see, conserved.

Let us now consider the energy relations involved when a body, or a system of particles, spontaneously separates into two systems, say parts 1 and 2. As seen from the proper coordinate system for the body, we have

$$E_0 = Mc^2 = E_1 + E_2, \tag{13.6}$$

since the energy is conserved. Here E_0 is the total energy before the parts separate and E_1 and E_2 the energies of the parts after they separate. When the two parts separate so far that their interaction can be neglected, Eq. (13.4) becomes applicable to each part, and thus

$$\begin{aligned} E_1 &= \frac{M_1c^2}{\sqrt{1 - v_1^2/c^2}} \geqslant M_1c^2, \\ E_2 &= \frac{M_2c^2}{\sqrt{1 - v_2^2/c^2}} \geqslant M_2c^2, \end{aligned} \tag{13.7}$$

where v_1, v_2 are the speeds, and M_1, M_2 the masses of the two parts after separation. Substitution from these inequalities into Eq. (13.6) now gives

$$M \geqslant M_1 + M_2. \tag{13.8}$$

This means that the mass of the system as a whole is necessarily greater than the sum of the masses of the parts into which it can divide spontaneously.

When the opposite inequality is true, the system cannot spontaneously separate into these two parts and is therefore stable with respect to such a separation. This generalizes as follows: If a system is regarded as consisting of parts 1, 2, . . . , n, the masses of which after separation would be M_1, M_2, . . . , M_n, and if the quantity

$$\sum_s M_s - M \tag{13.9}$$

is positive, then the system is stable with respect to the possibility of such a division. The quantity (13.9) is called the *mass defect* of the system. In particular, when the parts considered are the individual particles of the system, the quantity

$$\sum_s m_s - M \tag{13.10}$$

is called to *total mass defect* of the system. It may be positive, negative, or zero, depending on the stability properties of the system. As a general rule, we would expect that

$$M \neq \sum_s m_s. \tag{13.11}$$

Problems

1. What are the fundamental assumptions in the special theory of relativity? How do these assumptions differ from those made in classical theory (electrostatics)?

2. Which of the following quantities are *relative* and which are *absolute*? (Explain):

(a) Time between two events
(b) Distance between two events
(c) Velocity of light
(d) Velocity of a particle
(e) Frequency of a light wave
(f) $|\mathbf{r} - \mathbf{r}'|^2 - c^2(t - t')^2$

3. A *light year* is defined as the distance traveled by a light signal in one year. How many kilometers (or miles) does it represent? How far, approximately, is the closest star, the most remote galaxy?

4. Can one prove that it is impossible to choose an inertial coordinate system in which a signal arrives before it is sent? Explain.

5. Explain what is meant by the assertion that the interval between two events is generally made up partly of a time interval and partly of a space interval. When is the total interval *timelike* and when is it *spacelike*?

6. A space traveler A moves at a uniform velocity $\mathbf{v}$ from a fixed point Q in some inertial coordinate system S to a distant point P at a speed near c and then returns to Q at a velocity $-\mathbf{v}$. A second traveler B starts from Q at the same time with a velocity $-\mathbf{v}$ and travels to a point P' diametrically opposite P and then returns to Q at a velocity $\mathbf{v}$. Compare their total elapsed times for the trip and the elapsed time for an observer remaining at Q and one in a coordinate system S' moving with a velocity $\mathbf{v}$ with respect to Q.

7. If particles A and B move along the X-axis with velocities of $c/2$ and $-c/2$ respectively, what is their relative velocity?

8. A gunner A at $x = -d_1$ is moving along the X-axis at a velocity v_1 and a target B at $y = d_2$ is moving along the Y-axis at a velocity v_2. If the velocity of the bullet is v, what direction must A shoot to hit B?

9. Obtain Eqs. (11.9) directly from Eqs. (11.8).

10. What would be the effect, if any, of having the first clock in Example 2, p. 81, at some point other than the origin of the primed coordinate system?

11. An astronaut leaves the Earth, travels a distance d to Pluto, and returns. If he travels at uniform velocities $\mathbf{v}$ and $-\mathbf{v}$ going and returning, respectively, except for a very short interval when he is making the necessary accelerations, what time difference when he returns will there be between his clock and the clocks on Earth? Can one show that the effect of the accelerations must become relatively less and less important as the length of the journey increases?

12. Obtain an expression for the correction to the Galilean transformation of velocities that will apply when $v \ll c$. Would it be appreciable for astronauts traveling at around 20,000 miles per hour?

13. Show that the general Lorentz transformation of Eqs. (11.46) can be expressed by

$$\begin{bmatrix} r' \\ \tau' \end{bmatrix} = \begin{bmatrix} \Phi & b \\ b^\dagger & \gamma \end{bmatrix} \begin{bmatrix} r \\ \tau \end{bmatrix},$$

where Φ is a 3×3 matrix, b and r are 3×1 matrices, and $b^\dagger$ is the Hermitian conjugate $[b^\dagger = (b^T)^*]$. Find Φ and b.

14. The length of a rod moving with a velocity v along the positive X-axis of a given coordinate system can also be determined by observing the time t it takes for it to pass a stationary point, say the origin of the given coordinate system. Does this method give the same Fitzgerald contraction as that obtained in Example 1, p. 81?

15. A space ship A, having a velocity v, as shown in the figure, will just go through a gap in a screen B when $v \ll c$. Will the Fitzgerald contraction

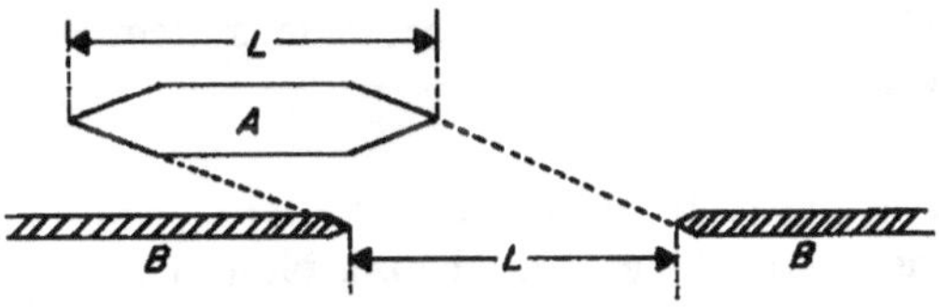

of the ship's length L when v approaches c make it go through with room to spare, or must one take the viewpoint of an observer on A that the size L of the hole in B is contracted and that therefore A will not go through? *Hint:* Will there be also an apparent turning of space ship or screen?

16. Generalize the concept of a solid angle to four-dimensional space.

17. A unit four-dimensional cube can be specified as the set of all points (x_1, x_2, x_3, x_4) such that $0 \leq x_k \leq 1$, where $k = 1, 2, 3, 4$. It is bounded by the hyperplanes of the form $(0, x_2, x_3, x_4)$, $(1, x_2, x_3, x_4)$, etc. How many such hypersurfaces are there? What is their shape? How many two-dimensional surfaces, such as $(0, x_2, 1, x_4)$, are needed to bound these hypersurfaces? How many edges and how many vertices of the unit four-dimensional cube are there?

18. At what speed is the energy of a free particle equal to twice its rest energy?

19. If a particle were started from rest in a constant gravitational field equal to that on the surface of the Earth, how long would it take to acquire a speed of $c/2$?

20. Express the velocity $\mathbf{v}$ of a free particle in terms of its momentum $\mathbf{p}$ and its rest mass m.

21. Express the speed v of a particle in terms of its energy E and rest energy mc^2. Find the nonrelativistic limit valid for $T \ll mc^2$, where $T = E - mc^2$, and the limit for $T \gg mc^2$.

22. Express p, the magnitude of the momentum of a free particle, in terms of the energy $T = E - mc^2$ it has in excess of its rest energy. Also express T in terms of p.

23. By analogy with the results of Section 11 obtain the general Lorentz transformation of the momentum $\mathbf{p}$ and the energy E for a relative velocity $\mathbf{v}$ of a primed coordinate system with respect to an unprimed system.

24. What is the relationship between energy and momentum for a particle of mass zero? Does it hold for a photon?

25. What is the total mass defect of a helium atom? Does this explain its stability? How is the mass defect related to the energy released in fusion reactions?

26. A free particle of mass m_1 moving with a velocity $\mathbf{v}_1$ strikes a second particle of mass m_2 and velocity $\mathbf{v}_2$. If there is a very inelastic collision and the particles stay together, what is the mass m and velocity $\mathbf{v}$ of the combined particles after impact? Compare your answer in the nonrelativistic limit of $v \ll c$ with the predictions of Newtonian mechanics.

27. Is an unexcited hydrogen atom heavier or lighter than the combined mass of an isolated proton and an isolated electron? How large is the mass defect in this case?

Chapter V

The Electromagnetic Field

14. Electromagnetic Forces

Let us consider again interactions between particles, noting however one essential difference between what we had in Sections 6 and 7 and the situation we have now. This difference is that, while previously the interaction was determined by the state of the system of particles at the time considered, now, since the interaction is propagated with finite speed, it is determined at each moment in part by previous states of the system. Consider for simplicity a system consisting of just two particles. In electrostatics their interaction could be expressed not only by means of the field, but also directly in terms of their positions at any given moment, for example, by Eq. (8.13) or (8.22). Thus, the behavior of the two particles at any given moment was sufficient to determine their interaction. Now, on the contrary, with the interaction assumed to be expressed only in terms of a field, a given motion of one of the particles will cause a corresponding field to act on the other particle only after a lapse of time, the time of propagation of the interaction. In the interim the field may be said to be moving from the one particle to the other. It is this phenomenon of a moving field, quite alien to the considerations of the first four chapters, which justifies the name *electrodynamic field*, as distinguished from the electrostatic field previously discussed.

At first sight the situation confronting us seems to contradict hopelessly the principle of causality, which demands that the state of the system *at any given moment* determines the subsequent behavior of the system. We can get around the difficulty, however, by introducing coordinates describing the field, as was done in Section 7. One thus treats the field as a separate entity. Indeed, if the state of the field at any given moment is included in specifying the state of the system, then the subsequent behavior of the system is completely determined. Thus, while in electrostatics the introduction of the field was merely a convenience, but not a necessary mathematical device, its introduction in electrodynamics is essential.

We must now consider more critically the process of exploring a field with the help of a particle, the process upon which the procedure of Section 6 depends. When we write $U = e\varphi$, Eq. (6.3), we imply that it is possible, in principle, to determine φ at each instant of time and each point of space, up to an arbitrary constant. The fact that not U itself, but the force $\mathbf{F} = e\mathscr{E} = -e\,\nabla\varphi$, is directly measurable presents no difficulty; because, knowing $\mathscr{E}$, we can, with the help of Eq. (6.11), construct φ. The difficulty arises when we wish to know $\mathscr{E}$ at different places at nearly the same time. In Section 6 this difficulty was not emphasized, because it could there be resolved in a trivial manner, by merely moving the test particle from one point to another with a sufficiently high speed. In the present case we cannot do this since the speed of the test particle must be smaller than c.

Therefore, we suppose that, in addition to the interacting particles constituting our system, we have any desired number of test particles distributed throughout space. It is assumed that they may be thickly enough distributed to permit a determination of the variation of the field with time at any desired point. In order that the test particles themselves should produce only negligibly small fields, the test particles are assumed to have infinitesimal charges. This procedure would be justifiable if it were possible to have arbitrarily small charges. Such charges, however, do not exist. The smallest known charge is the charge of the electron; therefore, *electrodynamics has limits of applicability determined by the indivisibility of this electronic charge.* What, more precisely, these limitations are will be discussed in Section 24.

We proceed now to generalize the expression $dW = -mc\,ds$, that arose in Eqs. (12.3) and (12.6), to make it applicable to the case when the particle is acted upon by an electrodynamic field. This means, of course, that we must add to this expression terms corresponding to the interaction between the field and the particle. These will be expressed through an electrodynamic potential. We require, by analogy with electrostatics, that to a constant potential should correspond an absence of field;* this

* We hereby disregard any nonelectromagnetic forces that might be introduced by the addition of terms like $\psi(x_\alpha)\,ds$ to dW, ψ being a scalar.

requirement may be regarded as a part of the definition of the potential. In our generalization of dW this requirement will express itself as follows: the interaction term to be added to $-mc\,ds$ must, when the potential is constant, reduce to a total differential. Only under this condition will the motion corresponding to different constant potentials be the same.

When the potential is constant, the total differential resulting from the additional terms cannot contain the coordinates explicitly. Otherwise, the action function for an isolated particle in the absence of a field would involve the coordinates, contrary to our basic assumptions in Sections 1 and 2.

The only total differentials remaining to us in this case are dx_α. These can be multiplied by constants and added, the sum remaining a total differential. Since the terms to be added to $-mc\,ds$ must constitute an invariant, they must be of the form $a\varphi_\alpha\,dx_\alpha$, where a is an arbitrary constant, and φ_α is an arbitrary four-dimensional vector. This vector is a natural generalization of the potential φ of Section 6. Since the interaction is proportional to e, it is in keeping with our system of units to set $a = e/c$, which still leaves $a\varphi_\alpha$ completely arbitrary. The generalized expression for dW becomes, therefore,

$$dW = -mc\,ds + \frac{e}{c}\,\varphi_\alpha\,dx_\alpha. \tag{14.1}$$

To facilitate a comparison with electrostatics it is convenient to introduce the following notation:

$$\varphi_4 = i\varphi, \qquad (\varphi_1, \varphi_2, \varphi_3) = \mathbf{A}. \tag{14.2}$$

φ is called the *scalar potential*, and $\mathbf{A}$ is called the *vector potential*. It is evident that φ is not an invariant in the four-dimensional sense. Like the energy, which is also proportional to the fourth component of a four-vector, it remains unchanged only for transformations of coordinates not involving time. With this notation Eq. (14.1) becomes

$$dW = -mc\,ds - e\varphi\,dt + \frac{e}{c}\,\mathbf{A}\cdot d\mathbf{r}, \tag{14.3}$$

and the corresponding Lagrangian is

$$L = \frac{dW}{dt} = -mc^2\sqrt{1 - v^2/c^2} - e\varphi + \frac{e}{c}\,\mathbf{A}\cdot\mathbf{v}. \tag{14.4}$$

Expanding the square root in powers of v/c and keeping only the first two terms, we have

$$L \simeq -mc^2 + \frac{1}{2} mv^2 - e\varphi + \frac{e}{c} \mathbf{A} \cdot \mathbf{v}. \tag{14.5}$$

Ignoring for the moment the last term, we have just the Lagrangian of electrostatic theory (see Eqs. (6.1) and (6.2)) minus a constant mc^2 representing the rest mass of the particle. We conclude that the scalar potential φ is to be identified with the electrostatic potential of the nonrelativistic theory, as we have anticipated by using the same symbol.

The theory of relativity, however, requires the additional term $e\mathbf{A} \cdot \mathbf{v}/c$ in Eq. (14.5), which enters even at nonrelativistic velocities ($v \ll c$). Thus, even though terms in the expansion of $\sqrt{1 - v^2/c^2}$ higher than those retained in Eq. (14.5) may be negligible, the term $e\mathbf{A} \cdot \mathbf{v}/c$, being of first power in v/c, may be important. It relates to the presence of a magnetic field.

The momentum corresponding to the Lagrangian of Eq. (14.4) is

$$\mathbf{p} = \frac{\partial L}{\partial \mathbf{v}} = \frac{m\mathbf{v}}{\sqrt{1 - v^2/c^2}} + \frac{e}{c} \mathbf{A}, \tag{14.6}$$

which differs from the momentum of a free particle by the term $e\mathbf{A}/c$. We recall that in electrostatics the expression for the momentum of a particle in terms of its velocity did not depend upon the field. Lagrange's equations for a particle can be written in the vector form

$$\frac{d\mathbf{p}}{dt} = \nabla L, \tag{14.7}$$

which by Eqs. (14.4) and (14.6) becomes

$$\frac{d}{dt}\left(\frac{m\mathbf{v}}{\sqrt{1 - v^2/c^2}} + \frac{e}{c} \mathbf{A}\right) = \nabla\left(-e\varphi + \frac{e}{c} \mathbf{v} \cdot \mathbf{A}\right). \tag{14.8}$$

To arrive at a more convenient form of this equation we shall now perform a number of mathematical transformations. We have, by Eq. (A3.60),

$$\frac{d\mathbf{A}}{dt} = \frac{\partial \mathbf{A}}{\partial t} + (\mathbf{v} \cdot \nabla)\mathbf{A}. \tag{14.9}$$

Further, since the operator ∇ in Eq. (14.8) involves only partial differentiations with respect to the coordinates, the velocity is treated as a constant.

By Eq. (A3.28), therefore, we have

$$\nabla(\mathbf{v}\cdot\mathbf{A}) = (\mathbf{v}\cdot\nabla)\mathbf{A} + \mathbf{v}\times(\nabla\times\mathbf{A}). \tag{14.10}$$

It will also be convenient to introduce the notation

$$\mathbf{P} = \frac{m\mathbf{v}}{\sqrt{1-v^2/c^2}}, \quad \text{and} \quad T = \frac{mc^2}{\sqrt{1-v^2/c^2}}, \tag{14.11}$$

and call these quantities the *kinetic momentum* and the *kinetic energy* of the particle, respectively. Since these are just the momentum and energy of a free particle (no field), they satisfy the relations

$$T^2 = c^2P^2 + m^2c^4 \tag{14.12}$$

and

$$\mathbf{v} = \frac{c^2\mathbf{P}}{T} \tag{14.13}$$

derived before, in Eqs. (12.19) and (12.17), for such a particle.

On substitution from Eqs. (14.9) and (14.10) into Eq. (14.8) the terms involving $(\mathbf{v}\cdot\nabla)\mathbf{A}$ cancel, and we are left with

$$\frac{d\mathbf{P}}{dt} = e\left(-\nabla\varphi - \frac{1}{c}\frac{\partial\mathbf{A}}{\partial t}\right) + \frac{e}{c}\mathbf{v}\times(\nabla\times\mathbf{A}) = e\mathscr{E} + \frac{e}{c}\mathbf{v}\times\mathscr{H}, \tag{14.14}$$

where

$$\mathscr{E} = -\nabla\varphi - \frac{1}{c}\frac{\partial\mathbf{A}}{\partial t} \tag{14.15}$$

and

$$\mathscr{H} = \nabla\times\mathbf{A}. \tag{14.16}$$

The vectors $\mathscr{E}$ and $\mathscr{H}$ are called the *electric field* and the *magnetic field*, respectively. When the vector potential **A** is zero, or even when it is merely independent of time, the expression for the electric field reduces to the old expression of electrostatics. The right-hand member of Eq. (14.14) is called the *Lorentz force;* it is separated into the *electric force* $e\mathscr{E}$ and the *magnetic force* $e\mathbf{v}\times\mathscr{H}/c$. It is evident that these are not forces in the usual sense, because the left-hand member of Eq. (14.14) is not $d\mathbf{p}/dt$. For small v, however, Eq. (14.14) can be written, approximately, as

$$\frac{d}{dt}(m\mathbf{v}) = e\mathscr{E} + \frac{e}{c}\mathbf{v}\times\mathscr{H}, \tag{14.17}$$

which is the original form in which the equations of motion were obtained by Lorentz. The historical reason for calling the right-hand side of

Eq. (14.17) a *force* is based on the similarity of this equation to Newton's equations of motion (see Section 3).

The Hamiltonian corresponding to the Lagrangian of Eq. (14.4), calculated in the usual way, is

$$H = -L + \mathbf{p} \cdot \mathbf{v} = T + e\varphi. \tag{14.18}$$

This expression is independent of **A**. Since, however, both φ and **A** are contained in L, there is, in general, no conservation of energy unless φ and **A** are both independent of t (see Eq. (3.7)).

It is of interest to obtain the equations of motion also in an invariant form. To accomplish this we shall find an invariant form of δW and set it equal to zero. From Eq. (14.1) we have

$$W = \int_a^b \left(-mc\,ds + \frac{e}{c}\,\varphi_\alpha\,dx_\alpha\right); \tag{14.19}$$

therefore, with the help of Eq. (12.9), we obtain for the variation

$$\delta W = \int_a^b \left(mcu_\alpha\,d\,\delta x_\alpha + \frac{e}{c}\,\varphi_\alpha\,d\,\delta x_\alpha + \frac{e}{c}\frac{\partial\varphi_\alpha}{\partial x_\beta}\,\delta x_\beta\,dx_\alpha\right).$$

Here we have made use of the fact that, since the φ_α are assumed to be given functions of the coordinates x_β,

$$\delta\varphi_\alpha = \frac{\partial\varphi_\alpha}{\partial x_\beta}\,\delta x_\beta. \tag{14.20}$$

Integration by parts now gives

$$\delta W = \left[\left(mcu_\alpha + \frac{e}{c}\,\varphi_\alpha\right)\delta x_\alpha\right]_a^b - \int_a^b \delta x_\alpha\,d\left(mcu_\alpha + \frac{e}{c}\,\varphi_\alpha\right) + \int_a^b \frac{e}{c}\frac{\partial\varphi_\beta}{\partial x_\alpha}\,\delta x_\alpha\,dx_\beta, \tag{14.21}$$

where, in the last term, summation indices α and β have been interchanged. If the limits are fixed, δx_α is zero at the limits, and the integrated part vanishes. In this case, equating δW to zero we obtain, since the δx_α are arbitrary,

$$d\left(mcu_\alpha + \frac{e}{c}\,\varphi_\alpha\right) = \frac{e}{c}\frac{\partial\varphi_\beta}{\partial x_\alpha}\,dx_\beta,$$

or

$$mc\,du_\alpha = \frac{e}{c}\left(\frac{\partial\varphi_\beta}{\partial x_\alpha} - \frac{\partial\varphi_\alpha}{\partial x_\beta}\right) dx_\beta = \frac{e}{c} F_{\alpha\beta}\, dx_\beta, \tag{14.22}$$

where

$$F_{\alpha\beta} = \frac{\partial\varphi_\beta}{\partial x_\alpha} - \frac{\partial\varphi_\alpha}{\partial x_\beta}. \tag{14.23}$$

In Eq. (14.22) the quantity $F_{\alpha\beta}$ is multiplied by an arbitrary four-vector dx_β and gives a four-vector; therefore $F_{\alpha\beta}$ is a tensor of second rank. From its definition in Eq. (14.23) it is evident that $F_{\alpha\beta}$ is an antisymmetric tensor, that is, that

$$F_{\alpha\beta} = -F_{\alpha\beta}. \tag{14.24}$$

Comparison of Eq. (14.23) with Eqs. (14.15) and (14.16), by means of the relations in Eqs. (14.2), shows that

$$\begin{aligned} &F_{12} = -F_{21} = \mathscr{H}_z, \qquad &&F_{23} = -F_{32} = \mathscr{H}_x, \qquad &&F_{31} = -F_{13} = \mathscr{H}_y, \\ &F_{41} = -F_{14} = i\mathscr{E}_x, &&F_{42} = -F_{24} = i\mathscr{E}_y, &&F_{43} = -F_{34} = i\mathscr{E}_z, \end{aligned} \tag{14.25}$$

the other components being zero. The electric and the magnetic fields behave as three-dimensional vectors as long as the coordinate transformations do not involve time. They are not vectors, nor components of vectors, in a four-dimensional sense—unlike, for example, p, the components of which are three of the four components of a four-dimensional vector p_α. In any case the electric and the magnetic fields determine the six independent components of a four-dimensional antisymmetric tensor of second rank, $F_{\alpha\beta}$. For this reason the tensor $F_{\alpha\beta}$ is called the *electromagnetic-field tensor.*

With the help of Eq. (11.24) the equation obtained from Eq. (14.22), for the special case of $\alpha = 1$, becomes

$$dP_x = e\mathscr{E}_x\, dt + \frac{e}{c}(\mathscr{H}_z\, dy - \mathscr{H}_y\, dz),$$

with similar expressions for the cases when $\alpha = 2$ and 3. Together these give Eq. (14.14). On the other hand, for $\alpha = 4$ we get

$$dT = e(\mathscr{E}_x\, dx + \mathscr{E}_y\, dy + \mathscr{E}_z\, dz),$$

which can be written

$$\frac{dT}{dt} = e\,\mathbf{v}\cdot\mathscr{E}. \tag{14.26}$$

This equation is, of course, a consequence of the others; it can be easily obtained from Eq. (14.12), with the help of Eqs. (14.13) and (14.14), in the following manner:

$$\frac{dT}{dt} = c^2 \frac{\mathbf{P} \cdot \dot{\mathbf{P}}}{T} = \mathbf{v} \cdot \frac{d\mathbf{P}}{dt} = e\,\mathbf{v} \cdot \mathscr{E} + \frac{e}{c}\,\mathbf{v} \cdot \mathbf{v} \times \mathscr{H} = e\,\mathbf{v} \cdot \mathscr{E}.$$

If the lower limit in Eq. (14.21) is fixed, and the upper limit is variable, and if, for each upper limit, the trajectory is chosen so that it satisfies the equations of motion, Eqs. (14.22); then

$$dW = \left(mcu_\alpha + \frac{e}{c}\,\varphi_\alpha\right) dx_\alpha,$$

or

$$p_\alpha = \frac{\partial W}{\partial x_\alpha} = mcu_\alpha + \frac{e}{c}\,\varphi_\alpha. \qquad (14.27)$$

If we refer to Eq. (11.24), we see that the first three components of Eq. (14.27) are equivalent to Eq. (14.6). The fourth component gives Eq. (14.18), provided that $p_4 = iH/c$, which is in agreement with Eq. (12.13).

Another interesting relation is obtained by combining Eq. (14.27) with Eq. (11.25):

$$\left(p_\alpha - \frac{e}{c}\,\varphi_\alpha\right)^2 = m^2c^2u_\alpha^2 = -m^2c^2, \qquad (14.28)$$

or in another form,

$$(H - e\varphi)^2 = c^2\left(p - \frac{e}{c}\,A\right)^2 + m^2c^4. \qquad (14.29)$$

An approximate form of this relation, applicable when v is small in comparison with c, is often useful. It can be obtained by remembering that $(\mathbf{P} - e\mathbf{A}/c) = \mathbf{P}$, and its square approaches m^2v^2 for small v. Thus, the first term of the right-hand member of Eq. (14.29) is, in this case, small in comparison with the other term. Thus, writing Eq. (14.29) in the form

$$H = e\varphi + \sqrt{m^2c^4 + c^2(\mathbf{P} - e\mathbf{A}/c)^2},$$

we have approximately

$$H = mc^2 + e\varphi + \frac{1}{2m}\left(\mathbf{p} - \frac{e}{c}\,\mathbf{A}\right)^2. \qquad (14.30)$$

The sign of the square root is determined by the fact that in the absence of the field $H = E = mc^2 + p^2/2m$, approximately.

EXAMPLE 1. Consider the motion of a particle in a constant homogeneous electric field. We may take $\mathbf{A} = 0$; then $\mathscr{E} = -\nabla\varphi$, and $\mathbf{P} = \mathbf{p}$. Let us choose the X-axis in the direction of $\mathscr{E}$, and the Y-axis in such a way that the initial velocity lies in the XY plane; the motion will then be found to be in this plane. The equations of motion, Eqs. (14.14), in this case become

$$\dot{p}_x = \frac{d}{dt}\frac{mv_x}{\sqrt{1 - v^2/c^2}} = e\mathscr{E}, \qquad \dot{p}_y = 0, \qquad \dot{p}_z = 0.$$

Integrating, we have

$$p_x = e\mathscr{E}t + \text{constant}, \qquad p_y = \text{constant}, \qquad p_z = 0. \tag{14.31}$$

To simplify our solution we shall measure time from the instant at which $p_x = 0$ and the space coordinates from the position of the particle at $t = 0$. The boundary conditions are then

$$x = y = z = 0, \quad p_x = 0 \qquad \text{at} \quad t = 0.$$

Then, $p_x = e\mathscr{E}t$ and, according to Eq. (14.12),

$$T^2 = m^2c^4 + c^2p^2 = m^2c^4 + c^2p_y^2 + c^2e^2\mathscr{E}^2t^2.$$

The initial value of T, which we shall designate by T_0, is

$$T_0 = \sqrt{m^2c^4 + c^2p_y^2}.$$

Since by Eqs. (14.13) and (14.31)

$$\dot{x} = v_x = \frac{c^2p_x}{T} = \frac{c^2e\mathscr{E}t}{\sqrt{T_0^2 + c^2e^2\mathscr{E}^2t^2}},$$

we have, on integrating and applying the boundary condition $x = 0$ at $t = 0$,

$$x = \frac{1}{e\mathscr{E}}(\sqrt{T_0^2 + c^2e^2\mathscr{E}^2t^2} - T_0). \tag{14.32}$$

Likewise, we have

$$\dot{y} = \frac{c^2p_y}{T} = \frac{c^2p_y}{\sqrt{T_0^2 + c^2e^2\mathscr{E}^2t^2}}$$

and

$$y = \frac{cp_y}{e\mathscr{E}}\sinh^{-1}\frac{ce\mathscr{E}t}{T_0}. \tag{14.33}$$

The equation of the curve described by the particle is obtained by the elimination of t from Eqs. (14.32) and (14.33); thus

$$x = \frac{T_0}{e\mathscr{E}}\left(\cosh\frac{e\mathscr{E}y}{cp_y} - 1\right) = \frac{2T_0}{e\mathscr{E}}\sinh^2\left(\frac{e\mathscr{E}y}{2cp_y}\right), \tag{14.34}$$

which shows that the curve is a *catenary*.

EXAMPLE 2. Consider a particle moving in a constant homogeneous magnetic field. Let the Z-axis be chosen in the direction of the field. The equations of motion from Eqs. (14.14) and (14.26) are

$$\frac{dP_x}{dt} = \frac{e\mathscr{H}}{c}\frac{dy}{dt}, \qquad \frac{dP_y}{dt} = -\frac{e\mathscr{H}}{c}\frac{dx}{dt}, \qquad \frac{dP_z}{dt} = 0, \qquad \frac{dT}{dt} = 0.$$

Integration of these equations gives

$$P_x = \frac{e\mathscr{H}}{c}(y - a), \qquad P_y = -\frac{e\mathscr{H}}{c}(x - b),$$

$$P_z = \text{constant}, \qquad T = \text{constant},$$

where $e\mathscr{H}a/c$ and $e\mathscr{H}b/c$ are constants of integration. By a proper choice of the spatial origin these equations become just

$$P_x = \frac{e\mathscr{H}y}{c}, \qquad P_y = -\frac{e\mathscr{H}x}{c}, \qquad P_z = \text{constant}, \qquad T = \text{constant}. \tag{14.35}$$

Further, since $T = \text{constant}$, Eq. (14.12) shows that $P^2 = \text{constant}$; consequently,

$$P_z^2 + \frac{e^2\mathscr{H}^2}{c^2}(x^2 + y^2) = P^2 = \text{constant}.$$

This shows that $x^2 + y^2 = \text{constant}$ and that therefore the motion takes place on a circular cylinder, with its axis along the Z-axis. The radius of the circular cross section is

$$R = (x^2 + y^2)^{1/2} = \frac{c}{e\mathscr{H}}(P^2 - P_z^2)^{1/2} = \text{constant}.$$

From Eq. (14.13)

$$v_z = \frac{c^2P_z}{T} = \text{constant}, \qquad \text{and} \qquad v = \frac{c^2P}{T} = \text{constant};$$

consequently,

$$v_x^2 + v_y^2 = v^2 - v_z^2 = \text{constant}.$$

This shows that the particle moves with a constant speed v_z along the cylinder and with a constant angular velocity ω around the cylinder, where

$$|\omega| = \frac{1}{R}\sqrt{v_x^2 + v_y^2}. \tag{14.36}$$

The particle will thus move along a helix. Since from Eqs. (14.35)

$$\omega = \left(\frac{v_y}{x}\right)_{y=0} = \left(\frac{c^2 P_y}{xT}\right)_{y=0} = -\frac{ec\mathscr{H}}{T}, \tag{14.37}$$

the angular velocity ω is opposite in sign to e when viewed from the positive Z-axis (direction of the magnetic field); therefore, for a positive e and v_z the particle will follow the threads of a left-handed screw. The pitch of the helix in this case will be

$$\lambda = \frac{2\pi}{|\omega|}v_z = \frac{2\pi c P_z}{e\mathscr{H}}. \tag{14.38}$$

Suppose $v_z = 0$, then by Eq. (14.37) the particle moves in a circle of radius

$$R = \frac{v}{|\omega|} = \frac{c^2 P}{T\,|\omega|} = \frac{cP}{|e|\,\mathscr{H}}. \tag{14.39}$$

This formula provides us with a means of measuring the momentum of a particle of known charge e by measuring the radius R in a known magnetic field. If $v_z \neq 0$, the formula applies to the magnitude of the transverse momentum.

For nonrelativistic velocities, $v \ll c$,

$$\omega \simeq -\frac{e\mathscr{H}}{mc} \tag{14.40}$$

and $|\omega|/2\pi$ is referred to as the *cyclotron frequency* of the particle in the magnetic field.

Example 3. The equations of motion of a particle in a combined electric and magnetic field can be written

$$du_\alpha = \frac{e}{mc^2}F_{\alpha\beta}\,dx_\beta,$$

or

$$\frac{d^2 x_\alpha}{ds^2} = \frac{e}{mc^2} F_{\alpha\beta} \frac{dx_\beta}{ds}. \tag{14.41}$$

This is a system of ordinary linear homogeneous differential equations and can be solved in the usual way. The result will be of the form $x_\alpha = f_\alpha(s)$, but it is of no particular interest to us, because of its complexity. We shall consider only the nonrelativistic case* $v \ll c$.

We take $\mathscr{H}$ along the Z-axis and $\mathscr{E}$ in the plane XZ, and use Eqs. (14.17) to obtain as equations of motion

$$m\frac{dv_x}{dt} = e\mathscr{E}_x + \frac{e}{c}\mathscr{H}v_y, \qquad m\frac{dv_y}{dt} = -\frac{e}{c}\mathscr{H}v_x, \qquad m\frac{dv_z}{dt} = e\mathscr{E}_z. \tag{14.42}$$

The last equation is directly integrable and gives

$$v_z = \frac{e}{m}\mathscr{E}_z t + \text{constant}. \tag{14.43}$$

For the other equations we make a substitution

$$v_x' = v_x, \qquad \text{and} \qquad v_y' = v_y + \frac{c\mathscr{E}_x}{\mathscr{H}}. \tag{14.44}$$

The new variables v_x' and v_y' will now satisfy the same equations as v_x and v_y would satisfy without the electric field. The solution is therefore obtained immediately from the results of the previous example by taking the limiting case of small v. The resultant motion will be a superposition of the following motions: An accelerated motion in the direction of the Z-axis, with the speed v_z given by Eq. (14.43); a uniform rotation in the XY-plane with an angular velocity $\omega = -\mathscr{H}/mc$ about a circle of radius

$$\begin{aligned} R &= \frac{mc}{|e|\,\mathscr{H}}(v_x'^2 + v_y'^2)^{1/2} \\ &= \frac{mc}{|e|\,\mathscr{H}}\left[v_x^2 + \left(v_y + \frac{e\mathscr{E}_x}{\mathscr{H}}\right)^2\right]^{1/2}; \end{aligned} \tag{14.45}$$

and, finally, a uniform motion of the center of the circle in the negative Y-direction with the speed $c\mathscr{E}_x/\mathscr{H}$. Another way of describing the motion is by saying that the motion is along a cycloid in a plane parallel to the

* This does not mean that the motion in this case will be the same as in electrostatics. We now have the presence of the magnetic field, the origin of which does not concern us in this problem.

XY-plane and that this plane is moving with a uniform acceleration in the Z direction.

A few words of criticism of our procedure are necessary. While it is true, as was assumed in the beginning of Sections 13 and 14, that in the absence of a field a particle is an isolated system, this statement is strictly applicable only to particles of zero charge; for a charged particle there is always its own field, from which it cannot be isolated. An assumption as to the behavior of a charged particle isolated from its own field is not a physical but a metaphysical question. It is to be noted that, if we accept the metaphysical assumption that in the absence of the field a particle is an isolated system, $\mathscr{E}$ and $\mathscr{H}$ in Eq. (14.14) *must include the field of the particle itself*, that is, they are *total fields*.

Another possible physical assumption would be: *A charged particle isolated with its field behaves as an uncharged particle.* This amounts to saying that the behavior of a particle is not affected by its own field. The line of reasoning employed in obtaining Eq. (14.14) then applies, and this same equation results, but now $\mathscr{E}$ and $\mathscr{H}$ are merely the *external fields.* A particle that would satisfy such an assumption is precisely what we had previously called a test particle. It can be used, with the help of Eq. (14.14), to explore the external fields. If the fields assumed in the preceding examples are external fields into which a particle is assumed to be placed, the results obtained are applicable strictly only to such test particles.

Actual charged particles interact with one another and with themselves and, as the concept of the field is introduced for the purpose of accounting for these interactions, one cannot assume that the fields of all particles are negligible without destroying the physical meaning of the field altogether. Thus, in classical electrodynamics one falls back upon the metaphysical assumption referred to above.

To what extent and under what conditions one can expect actual charged particles to behave in accordance with the equations thus obtained will be discussed in Section 24.

15. The Lorentz Transformation of the Field

We shall now find how the field quantities $\mathscr{E}$ and $\mathscr{H}$ transform when we perform the Lorentz transformation. To this end it is easier to consider first the transformation of the second-order tensor $F_{\alpha\beta}$.

The general rule of transformation of such tensors, when coordinates are transformed according to the formula*

$$x'_\alpha = a_{\alpha\beta} x_\beta , \tag{15.1}$$

is (see Eq. (A5.9))

$$F'_{\alpha\beta} = a_{\alpha\lambda} a_{\beta\mu} F_{\lambda\mu} , \tag{15.2}$$

where a double summation, with respect to λ and μ, is implied. However, with the help of an artifice, we can avoid the use of these complicated sums. We note that, if any particular coordinate is unchanged, for example, if $x'_2 = x_2$, Eq. (15.1) implies that $a_{2\beta} = \delta_{2\beta}$, in other words,

$$a_{22} = 1, \qquad a_{2\beta} = 0 \quad \text{if} \quad \beta \neq 2.$$

If this result is substituted into Eq. (15.2), one has

$$F'_{2\beta} = \delta_{2\lambda} a_{\beta\mu} F_{\lambda\mu} = a_{\beta\mu} F_{2\mu} . \tag{15.3}$$

Thus, components of a tensor, one of the indices of which refers to the coordinate remaining unchanged, transform as the components of a vector with respect to the other index.

Now, in the Lorentz transformation of Eqs. (11.7) coordinates x_2 and x_3 (i.e., y and z) remain unchanged. This means, first, that F_{23} (and, of course, F_{32}) remains unchanged, since the indices refer only to the unchanged axes. Further, $F_{2\beta}$ and $F_{3\beta}$ will be transformed as vectors. To transform F_{14} we have the following considerations: Since the transformation matrix $a_{\alpha\beta}$ contains no nonzero elements between the two-dimensional subspace formed by the coordinates y and z and that formed by the coordinates x and τ, we may treat the transformations within these two subspaces separately. The invariant expression $F_{\alpha\beta} F_{\alpha\beta}$ is thus also invariant if α and β are restricted to the values 1 and 4 corresponding to the second subspace. Thus, since $F_{\beta\alpha} = -F_{\alpha\beta}$, we have

$$\sum_{\alpha,\beta=1 \text{ and } 4} F_{\alpha\beta} F_{\alpha\beta} = F_{14}^2 + F_{41}^2 = 2F_{14}^2 = \text{invariant},$$

and therefore that F_{14} is invariant in this transformation. Collecting

* Throughout this book we consider only transformations between orthogonal Cartesian coordinates, so that no distinction need be made between covariant and contravariant tensors.

results, we have, using Eqs. (11.7),

$$F'_{23} = F_{23}, \qquad F'_{14} = F_{14},$$
$$F'_{12} = \frac{F_{12} + ivF_{42}/c}{\sqrt{1 - v^2/c^2}}, \qquad F'_{13} = \frac{F_{13} + ivF_{43}/c}{\sqrt{1 - v^2/c^2}}, \tag{15.4}$$
$$F'_{42} = \frac{F_{42} - ivF_{12}/c}{\sqrt{1 - v^2/c^2}}, \qquad F'_{43} = \frac{F_{43} - ivF_{13}/c}{\sqrt{1 - v^2/c^2}}.$$

It is easy to rewrite these equations, with the help of Eqs. (14.25), in terms of the components of $\mathscr{E}$ and $\mathscr{H}$. We then obtain as transformation equations

$$\mathscr{E}'_x = \mathscr{E}_x, \qquad \mathscr{H}'_x = \mathscr{H}_x,$$
$$\mathscr{E}'_y = \frac{\mathscr{E}_y - v\mathscr{H}_z/c}{\sqrt{1 - v^2/c^2}}, \qquad \mathscr{H}'_y = \frac{\mathscr{H}_y + v\mathscr{E}_z/c}{\sqrt{1 - v^2/c^2}}, \tag{15.5}$$
$$\mathscr{E}'_z = \frac{\mathscr{E}_z + v\mathscr{H}_y/c}{\sqrt{1 - v^2/c^2}}, \qquad \mathscr{H}'_z = \frac{\mathscr{H}_z - v\mathscr{E}_y/c}{\sqrt{1 - v^2/c^2}}.$$

Let us consider what changes in our equations result from a reversal in the direction of time. This corresponds to the transformation of coordinates

$$x' = x, \qquad y' = y, \qquad z' = z, \qquad t' = -t. \tag{15.6}$$

This, of course, is not a rotation of the axes; and equations invariant with respect to rotations may not be invariant to this transformation. The velocity will, of course, change its sign. The form of the Lagrangian, Eq. (14.4), will be the same in the new coordinate system as in the old if we put*

$$\varphi' = \varphi \qquad \text{and} \qquad \mathbf{A}' = -\mathbf{A}, \tag{15.7}$$

because $\mathbf{v}' = -\mathbf{v}$. From this it follows at once that

$$\mathscr{E}' = \mathscr{E} \qquad \text{and} \qquad \mathscr{H}' = -\mathscr{H}. \tag{15.8}$$

* It is not obvious, of course, that one should require that the Lagrangian be invariant for this transformation. One can see, however, that the acceleration should not change sign and hence that the force $\partial L/\partial \mathbf{r}$ should be invariant. Since the position vector $\mathbf{r}$ is unchanged by the transformation, this requires that L remain the same. The action function $W = \int L\,dt$, however, must switch sign.

In keeping with this analysis is the fact that the transformation $\mathbf{v} \rightarrow -\mathbf{v}$, $\mathscr{E} \rightarrow \mathscr{E}$, and $\mathscr{H} \rightarrow -\mathscr{H}$ leaves the Lorentz force unchanged.

We see that, unlike the Lorentz transformation, the transformation of Eqs. (15.6) leaves our equations invariant only in the absence of a magnetic field; if a magnetic field is present, we must change its sign.

Finally, it is interesting to consider what combinations of $\mathscr{E}$ and $\mathscr{H}$ remain unchanged when we perform a Lorentz transformation. An obvious one is the scalar

$$\tfrac{1}{2}F_{\alpha\beta}F_{\alpha\beta} = \mathscr{H}^2 - \mathscr{E}^2. \tag{15.9}$$

Another invariant is

$$\tfrac{1}{8}i\ \varepsilon_{\alpha\beta\gamma\delta}F_{\alpha\beta}F_{\gamma\delta} = \mathscr{E}\cdot\mathscr{H} = \sqrt{-\,|F_{\alpha\beta}|}\,, \tag{15.10}$$

where $\varepsilon_{\alpha\beta\gamma\delta}$ has the following values: zero when any two or more of the subscripts are equal to each other; $+1$ when the subscripts are an even permutation of the numbers 1, 2, 3, and 4; and -1 when they are an odd permutation of these numbers. The expression $|F_{\alpha\beta}|$ is the determinant

$$|F_{\alpha\beta}| = \begin{vmatrix} 0 & \mathscr{H}_z & -\mathscr{H}_y & -i\mathscr{E}_x \\ -\mathscr{H}_z & 0 & \mathscr{H}_x & -i\mathscr{E}_y \\ \mathscr{H}_y & -\mathscr{H}_x & 0 & -i\mathscr{E}_z \\ i\mathscr{E}_x & i\mathscr{E}_y & i\mathscr{E}_z & 0 \end{vmatrix}. \tag{15.11}$$

It can be shown that there are no other independent invariants that can be constructed from the components of $F_{\alpha\beta}$.

EXAMPLE. We can, in most cases, as will be shown in the example of Section 17, choose a coordinate system in which the electric and the magnetic fields at any given point are parallel. Let the magnitudes of the electric and the magnetic fields in this coordinate system be $\mathscr{E}_0$ and $\mathscr{H}_0$, respectively; and let the direction of the two fields be chosen as the X-axis. The tensor $F_{\alpha\beta}$ will then have the form

$$F_{\alpha\beta} = \begin{vmatrix} 0 & 0 & 0 & -i\mathscr{E}_0 \\ 0 & 0 & \mathscr{H}_0 & 0 \\ 0 & -\mathscr{H}_0 & 0 & 0 \\ i\mathscr{E}_0 & 0 & 0 & 0 \end{vmatrix}. \tag{15.12}$$

The values of $\mathscr{E}_0$ and $\mathscr{H}_0$ can be found with the help of the invariants in Eqs. (15.9) and (15.10); thus

$$\mathscr{E}_0^2 - \mathscr{H}_0^2 = \mathscr{E}^2 - \mathscr{H}^2 \quad \text{and} \quad \mathscr{E}_0\cdot\mathscr{H}_0 = \mathscr{E}_0\mathscr{H}_0 = \mathscr{E}\cdot\mathscr{H}. \tag{15.13}$$

The right-hand members of these equations are supposed to be known in some coordinate system; being invariants they are equal to the corresponding expressions in any other coordinate system, and in particular in the coordinate system in which the electric and the magnetic fields are parallel.

Equations (15.13) are sufficient to determine $\mathscr{E}_0$ and $\mathscr{H}_0$, except in the case when, in the original coordinate system, $\mathscr{E} = \mathscr{H}$ and $\mathscr{E} \cdot \mathscr{H} = 0$. In this case both invariants are equal to zero, and the electric and magnetic fields are equal and perpendicular in all coordinate systems.

16. Electromagnetic Field Equations

So far, in considering an electromagnetic field, our chief concern has been with the effect of the field on a particle. We shall consider in this chapter the equations that must be satisfied by the field quantities, that is, the question of the way in which the field itself is determined.

First, of course, we have the equations of definition, Eqs. (14.15) and (14.16),

$$\mathscr{E} = -\nabla\varphi - \frac{1}{c}\frac{\partial \mathbf{A}}{\partial t}, \qquad \mathscr{H} = \nabla \times \mathbf{A}, \tag{16.1}$$

or, in the invariant form, Eq. (14.23),

$$F_{\alpha\beta} = \frac{\partial \varphi_\beta}{\partial x_\alpha} - \frac{\partial \varphi_\alpha}{\partial x_\beta}. \tag{16.2}$$

If the field is given, these equations may be regarded as defining the potentials φ and $\mathbf{A}$; however, the potentials are not completely determined in this way. This is because the expression for dW, Eq. (14.1), can be changed by an addition of a total differential of an arbitrary function of coordinates and time, without affecting the equations of motion. Thus, we could add

$$\frac{e}{c}\,d\chi = \frac{e}{c}\frac{\partial \chi}{\partial x_\alpha}\,dx_\alpha,$$

where χ is an arbitrary function of the x_α's. This means that the potential φ_α may be changed into

$$\varphi'_\alpha = \varphi_\alpha + \frac{\partial \chi}{\partial x_\alpha}, \tag{16.3}$$

as is evident from Eq. (14.1), without affecting the equations of motion or the tensor $F_{\alpha\beta}$. This can be seen also by a direct substitution of φ'_α instead of φ_α into Eq. (16.2), whereupon the terms containing derivatives of χ cancel.* In the three-dimensional form Eq. (16.3) becomes

$$\varphi' = \varphi - \frac{1}{c}\frac{\partial \chi}{\partial t}, \qquad \mathbf{A}' = \mathbf{A} + \nabla\chi. \tag{16.4}$$

One of the methods of fixing the potentials is to require them to vanish at infinity, but this requirement is not sufficient. This is evident from the fact that we can change the potentials without changing their values at infinity by choosing χ in such a way that its derivatives vanish at infinity.

We have seen that only the electric and the magnetic fields enter into the equations of motion. This means, of course, that potentials are auxiliary quantities, introduced for mathematical convenience. It is therefore desirable to eliminate the potentials from Eqs. (16.1), obtaining relations involving only the fields $\mathscr{E}$ and $\mathscr{H}$. This is easily done. We take the *curl* of the first of the two equations, substitute for $\nabla \times \mathbf{A}$ from the second, and make use of the identity $\nabla \times \nabla\varphi = 0$. Thus, we obtain

$$\nabla \times \mathscr{E} = -\frac{1}{c}\frac{\partial \mathscr{H}}{\partial t}. \tag{16.5}$$

Also, from the second of the two equations,

$$\nabla \cdot \mathscr{H} = 0. \tag{16.6}$$

This pair of equations is usually called the *first pair* of the *Maxwell–Lorentz equations*. It is the first of the two pairs of equations that were set up by Maxwell and were later modified by Lorentz to apply to elementary processes.

We can easily obtain the integral form of these equations. Multiplying the first equation by $\cdot\, d\mathbf{S}$, and integrating over any surface S bounded by some curve C, we have

$$\frac{1}{c}\frac{\partial}{\partial t}\int_S \mathscr{H} \cdot d\mathbf{S} = \int_S \nabla \times \mathscr{E} \cdot d\mathbf{S} = \int_C \mathscr{E} \cdot d\mathbf{r}. \tag{16.7}$$

The integral $\int_S \mathscr{H} \cdot d\mathbf{S}$ represents the *magnetic flux* through the surface S. The last expression is called the *electromotive force around the closed contour C*. Contrary to what we had in electrostatics, it is in general not

* The invariance of equations of electrodynamics with respect to the transformation (16.3) is called *gauge invariance* or *Eichinvarianz*.

zero. This is, of course, a consequence of the fact that, while in electrostatics we had $\nabla \times \mathscr{E} = 0$, this is not generally true in electrodynamics.

Integrating $\mathscr{H} \cdot d\mathbf{S}$ over a closed surface, and using Eq. (16.6), we obtain the integral form of Eq. (16.6),

$$\int \mathscr{H} \cdot d\mathbf{S} = \int \nabla \cdot \mathscr{H} \, dV = 0, \tag{16.8}$$

which, on comparison with Eq. (8.1), shows that there are no magnetic charges analogous to the electric charges.

Elimination of the potentials may be also easily carried out using Eq. (16.2). In fact, we have

$$\frac{\partial F_{\alpha\beta}}{\partial x_\gamma} + \frac{\partial F_{\beta\gamma}}{\partial x_\alpha} + \frac{\partial F_{\gamma\alpha}}{\partial x_\beta} = 0, \tag{16.9}$$

as may be easily verified by substitution from Eq. (16.2). It can also be shown that when Eq. (16.9) holds and one requires that $F_{\alpha\beta} = -F_{\beta\alpha}$, then the quantities $F_{\alpha\beta}$ must be of the form given by Eq. (16.2). On the other hand, by giving α, β, and γ various values, we see that this equation is equivalent to Eqs. (16.5) and (16.6). Thus, for example, if $\alpha = 1$, $\beta = 2$, $\gamma = 4$, Eq. (16.9) becomes, with the help of Eq. (14.25),

$$\frac{\partial \mathscr{H}_z}{\partial \tau} + \frac{\partial(-i\mathscr{E}_y)}{\partial x} + \frac{\partial(i\mathscr{E}_x)}{\partial y} = 0,$$

which, upon simplification, reduces to the z component of Eq. (16.5). Similarly, for $\alpha = 1$, $\beta = 2$, $\gamma = 3$, we obtain Eq. (16.6). Thus, Eq. (16.9) is the invariant form of the first pair of the Maxwell–Lorentz equations.

We have seen that this pair is a consequence of the definition of the field, and is independent of the way in which the field is produced. The second pair, which we shall now proceed to obtain, gives us the relations between the field and the system producing it.

We first construct the action function for the field. In electrostatics we had the form

$$W_f = \int L_f \, dt, \qquad L_f = \int F(\mathscr{E}) \, dV, \qquad F(\mathscr{E}) = \frac{\mathscr{E}^2}{8\pi}, \tag{16.10}$$

according to Eqs. (7.17) and (7.22). We can make use of this information, because in the limit of $\mathscr{H} = 0$ and a constant electric field $\mathscr{E}$ the electrodynamic equations must reduce to these. From the first two of these equations we have

$$W_f = \iint F\, dt\, dV = \frac{1}{ic} \int F\, d\Omega, \qquad (16.11)$$

where $d\Omega = dx_1\, dx_2\, dx_3\, dx_4$ is the four-dimensional volume element. Further, since in the present theory we wish W_f to be an invariant, F must now also be an invariant, which $\mathscr{E}^2/8\pi$ is not. Moreover, the principle of superposition again requires that F should be a quadratic polynomial in the field quantities.

The only two independent quadratic invariants available can be shown to be the expressions $\frac{1}{2}F_{\alpha\beta}F_{\alpha\beta} = \mathscr{H}^2 - \mathscr{E}^2$ and $\sqrt{-|F_{\alpha\beta}|} = \mathscr{E} \cdot \mathscr{H}$. The first, in the absence of a magnetic field, reduces to $-\mathscr{E}^2$, which is proportional to $F(\mathscr{E})$ of Eq. (16.10); the second reduces to zero. Thus, the expression

$$F = -\frac{1}{16\pi} F_{\alpha\beta}F_{\alpha\beta} + b\mathscr{E} \cdot \mathscr{H}, \qquad (16.12)$$

where b is any constant, will be invariant, dimensionally homogeneous, and will reduce to the right limit in the absence of a magnetic field. It turns out, however, that the choice of the constant b in no way alters the field equations. When Eq. (16.1) is taken into account we can put b equal to zero, without any loss of generality. Our expression for W_f is therefore

$$W_f = -\frac{1}{16\pi ic} \int F_{\alpha\beta}F_{\alpha\beta}\, d\Omega = \frac{1}{8\pi} \iint (\mathscr{E}^2 - \mathscr{H}^2)\, dV\, dt. \qquad (16.13)$$

The limits of integration in Eq. (16.11) must be fixed in an invariant way. Formerly we integrated over the whole space and from the time t_1 to time t_2. Four-dimensionally speaking, we integrated over a four-dimensional region bounded on two sides by the three-dimensional hyperplanes $t_1 =$ constant and $t_2 =$ constant, but otherwise unbounded. Since the time involved refers to one particular choice of a coordinate system, this way of stating the limits is not invariant. In another coordinate system the hyperplane $t_1 =$ constant will be inclined to the time axis and will be merely one of the spacelike hyperplanes. The term *spacelike* applied to a hypersurface means that the interval between any two points of it is spacelike.

We shall generalize the type of region in four dimensions that we integrate over to find W_f by requiring only that it be bounded by two nonintersecting hypersurfaces on which the potential are to be given and unvaried. In the spatial dimensions, between these hypersurfaces, the region will again extend to infinity, where the potentials will also be assumed to be unvaried (usually zero).

By taking the variation of W_f, Eq. (16.13), and setting it equal to zero, we would obtain the equations of the field in the absence of particles. If, in a particular coordinate system, the potentials are given at the initial and final moments of time, and **A** is made to vary rapidly with time, then $\mathscr{E}^2$, which depends upon the rate of change of **A**, will be large and will increase with the rapidity of the change in **A**. On the other hand, $\mathscr{H}^2$ is independent of the time rate of change of **A**, depending only upon its spatial derivatives, which may be kept small. Thus the expression for W_f in Eq. (16.13) can be made as large as one chooses by taking **A** to be a sufficiently rapidly varying function of time. Thus W_f has no maximum. Since Hamilton's principle requires that the action be a minimum, at least over very short time intervals, the sign could not be reversed from that given in Eq. (16.13). This is the same reasoning we used in specifying that the a in Eq. (7.22) had to be a positive constant.

To obtain general field equations, we must add together W_f of Eq. (16.13) and W_{pf} of Section 14, generalizing the latter to many particles, and put the variation of the sum equal to zero. From Eq. (14.1) W_{pf} for a single particle interacting with the field is given in differential form by

$$dW_{pf} = \frac{e}{c}\,\varphi_\alpha(\mathbf{r}, t)\,dx_\alpha.$$

For many particles we therefore have

$$\begin{aligned} W_{pf} &= \int \sum_s \frac{e_s}{c}\,\varphi_\alpha(\mathbf{r}_s, t)\,dx_\alpha \\ &= \int \left[-\sum_s e_s\varphi(\mathbf{r}_s, t) + \sum_s \frac{e_s}{c}\,\mathbf{A}(\mathbf{r}_s, t)\cdot\mathbf{v}_s \right] dt. \end{aligned} \qquad (16.14)$$

We further transform this formula by means of the relation

$$\sum_s e_s\phi(\mathbf{r}_s, t) = \int \rho(\mathbf{r}, t)\varphi(\mathbf{r}, t)\,dV = \int \rho\varphi\,dV, \qquad (16.15)$$

which we had in Eq. (7.27), and a similar one

$$\sum_s e_s\mathbf{A}(\mathbf{r}_s, t)\cdot\mathbf{v}_s = \int \rho\mathbf{v}\cdot\mathbf{A}\,dV, \qquad (16.16)$$

which is easily obtained in the same way. Thus we have

$$W_{pf} = \int\!\!\int [-\rho\varphi + \rho\mathbf{v}\cdot\mathbf{A}/c]\,dV\,dt = \frac{1}{ic}\int s_\alpha\varphi_\alpha\,d\Omega, \qquad (16.17)$$

where

$$s_\alpha = \left(\frac{\rho v_x}{c}, \frac{\rho v_y}{c}, \frac{\rho v_z}{c}, i\rho\right). \tag{16.18}$$

Since $s_\alpha \varphi_\alpha$ is a scalar, and φ_α is a four-vector, s_α is a four-vector called the *current-density four-vector.** The field equations are to be obtained from the equation

$$\delta(W_f + W_{pf}) = \delta \frac{1}{ic} \int \left[s_\alpha \varphi_\alpha - \frac{1}{16\pi} F_{\gamma\beta} F_{\gamma\beta} \right] d\Omega = 0. \tag{16.19}$$

We shall now perform the indicated variation. Since we are looking for the equations of motion for the field, with the motion of particles supposed known, we vary the potentials φ_α, leaving s_α fixed. It will be convenient to designate the coordinates of points on the two hypersurfaces bounding the region Ω by ξ_α and x_α, respectively. Thus, we shall write

$$\delta W = \delta(W_f + W_{pf}) = \frac{1}{ic} \delta \int_{\xi_\alpha}^{x_\alpha} \left[s_\beta \varphi_\beta - \frac{1}{16\pi} F_{\gamma\beta} F_{\gamma\beta} \right] d\Omega. \tag{16.20}$$

Instead of keeping x_α fixed, we shall generalize our treatment, for future use, by allowing the upper limit to undergo a displacement δx_α. This displacement, of course, must be set equal to zero to obtain the equations of motion. We thus have

$$ic\,\delta W = \int_{\xi_\alpha}^{x_\alpha + \delta x_\alpha} \left[s_\beta(\varphi_\beta + \delta\varphi_\beta) - \frac{1}{16\pi} (F_{\gamma\beta} + \delta F_{\gamma\beta})^2 \right] d\Omega$$
$$- \int_{\xi_\alpha}^{x_\alpha} \left[s_\beta \varphi_\beta - \frac{1}{16\pi} F_{\gamma\beta}^2 \right] d\Omega.$$

Considering all variations as being small, and neglecting terms of second degree in the variations, this becomes

$$\begin{aligned} ic\,\delta W &= \int_{\xi_\alpha}^{x_\alpha} \left(s_\beta\,\delta\varphi_\beta - \frac{1}{8\pi} F_{\gamma\beta}\,\delta F_{\gamma\beta} \right) d\Omega + \int_{x_\alpha}^{x_\alpha+\delta x_\alpha} \left(s_\beta \varphi_\beta - \frac{1}{16\pi} F_{\gamma\beta}^2 \right) d\Omega \\ &= \int_{\xi_\alpha}^{x_\alpha} \left(s_\beta\,\delta\varphi_\beta - \frac{1}{8\pi} F_{\gamma\beta}\,\delta F_{\gamma\beta} \right) d\Omega \\ &\quad + \int_{x_\alpha}^{x_\alpha+\delta x_\alpha} \left(s_\beta \varphi_\beta - \frac{1}{16\pi} F_{\gamma\beta}^2 \right) \delta x_\alpha\, dS_\alpha . \end{aligned}$$

* It is shown in Appendix A9 that s_α is also a four-vector when $\rho = \Sigma_s e_s \delta(\mathbf{r} - \mathbf{r}_s)$, corresponding to a system of point charges.

Here the four-dimensional integral over the thin layer between x_α and $x_\alpha + \delta x_\alpha$ has been replaced by the surface integral over the hypersurface x_α. This is done by taking for the volume element $d\Omega$ the scalar product $\delta x_\alpha \, dS_\alpha$ and by replacing one integration by multiplication by δx_α.

Now,

$$F_{\gamma\beta}\,\delta F_{\gamma\beta} = F_{\gamma\beta}\,\delta\left(\frac{\partial \varphi_\beta}{\partial x_\gamma} - \frac{\partial \varphi_\gamma}{\partial x_\beta}\right) = F_{\gamma\beta}\,\frac{\partial \delta\varphi_\beta}{\partial x_\gamma} - F_{\gamma\beta}\,\frac{\partial \delta\varphi_\gamma}{\partial x_\beta},$$

which, after the interchange of the summation indices in the second term, becomes

$$F_{\gamma\beta}\,\delta F_{\gamma\beta} = (F_{\gamma\beta} - F_{\beta\gamma})\,\frac{\partial \delta\varphi_\beta}{\partial x_\gamma} = -2F_{\beta\gamma}\,\frac{\partial \delta\varphi_\gamma}{\partial x_\beta};$$

therefore,

$$\begin{aligned} ic\,\delta W &= \int_{\xi_\alpha}^{x_\alpha}\left(s_\beta\,\delta\varphi_\beta + \frac{1}{4\pi}\,F_{\beta\gamma}\,\frac{\partial \delta\varphi_\beta}{\partial x_\gamma}\right) d\Omega + \int_{x}\left(s_\beta \varphi_\beta - \frac{1}{16\pi}\,F_{\gamma\beta}^2\right)\delta x_\alpha\, dS_\alpha \\ &= \int_\Omega\left(s_\beta - \frac{1}{4\pi}\,\frac{\partial F_{\beta\gamma}}{\partial x_\gamma}\right)\delta\varphi_\beta\, d\Omega + \frac{1}{4\pi}\int_\Omega \frac{\partial}{\partial x_\gamma}\,(F_{\beta\gamma}\,\delta\varphi_\beta)\, d\Omega \\ &\quad + \int_{x_\alpha}\left(s_\beta\varphi_\beta - \frac{1}{16\pi}\,F_{\gamma\beta}^2\right)\delta x_\alpha\, dS_\alpha. \end{aligned} \tag{16.21}$$

Now the integrand in the second integral is a four-dimensional divergence, so that the integral reduces to an integral over the boundary; and, since we assume $\delta\varphi_\beta$ to be zero over the boundary, the integral vanishes. In the present case the last integral is also zero, because the boundary is kept fixed, and so $\delta x_\alpha = 0$. Since $\delta\varphi_\beta$ in the first integral is arbitrary, δW will be zero if and only if

$$\frac{\partial F_{\beta\gamma}}{\partial x_\gamma} = 4\pi s_\beta. \tag{16.22}$$

These are the required field equations in an invariant form.

Written out in the three-dimensional form, with the help of Eqs. (14.25) and (16.18), these field equations are

$$\nabla \times \mathscr{H} - \frac{1}{c}\frac{\partial \mathscr{E}}{\partial t} = \frac{4\pi\rho\mathbf{v}}{c}, \tag{16.23}$$

for $\beta = 1, 2,$ and 3; and

$$\nabla \cdot \mathscr{E} = 4\pi\rho, \tag{16.24}$$

for $\beta = 4$. We have thus obtained the *second pair of Maxwell–Lorentz equations*.

It is surprising that as far back as the middle of the last century Maxwell was able to derive essentially these equations by a process of generalization from the experimental data then available.

For convenience of reference let us collect the field equations. From Eqs. (16.5), (16.6), (16.23), and (16.24)

$$\nabla \times \mathscr{E} + \frac{1}{c}\frac{\partial \mathscr{H}}{\partial t} = 0, \qquad \nabla \cdot \mathscr{H} = 0, \tag{16.25}$$

$$\nabla \times \mathscr{H} - \frac{1}{c}\frac{\partial \mathscr{E}}{\partial t} = \frac{4\pi\rho\mathbf{v}}{c}, \qquad \nabla \cdot \mathscr{E} = 4\pi\rho. \tag{16.26}$$

In addition we have, for each particle, from Eqs. (14.11) and (14.14) the equation of motion

$$\frac{d}{dt}\frac{m\mathbf{v}}{\sqrt{1 - v^2/c^2}} = e\mathscr{E} + \frac{e}{c}\mathbf{v} \times \mathscr{H}. \tag{16.27}$$

As a consequence of Eqs. (16.26) we have also the *continuity equation*

$$\nabla \cdot \rho\mathbf{v} + \frac{\partial \rho}{\partial t} = 0 \tag{16.28}$$

for the charge. It is the mathematical expression of the requirement of *charge conservation*, i.e., that the net charge in the universe stays the same (see the discussion in Example 2 of Appendix A4).

If now we know the coordinates and velocities of all the particles at any moment of time (and thus ρ and $\mathbf{v}$ for all points), as well as $\mathscr{E}$ and $\mathscr{H}$ throughout space at this time, the first of Eqs. (16.25) and of Eqs. (16.26) determine the rates of change of the fields $\mathscr{E}$ and $\mathscr{H}$ while Eq. (16.27) determines the rate of change of $\mathbf{v}$. Thus, the state of the system, as we initially required, is determined by the state of the particles and the field.

In specifying the initial state of a system we need therefore to specify not only the positions and velocities of all the particles but an initial field that satisfies the equations

$$\nabla \cdot \mathscr{H} = 0, \qquad \nabla \cdot \mathscr{E} = 4\pi\rho.$$

It is important to note, however, that these equations, with given ρ, are not sufficient to determine the initial values of the field. Thus, any $\mathscr{E}$ and $\mathscr{H}$ that satisfy these equations may be replaced by $\mathscr{E}' = \mathscr{E} + \nabla \times \mathbf{\Phi}$ and $\mathscr{H}' = \mathscr{H} + \nabla \times \boldsymbol{\psi}$, where $\mathbf{\Phi}$ and $\boldsymbol{\psi}$ are arbitrary vector functions of coordinates, and these equations will still be satisfied, because the divergence of a curl is zero. Thus, *the field is not merely a property of the particles, but is an object in itself*, which cannot be eliminated.

That the equations

$$\nabla \cdot \mathscr{H} = 0 \qquad \text{and} \qquad \nabla \cdot \mathscr{E} = 4\pi\rho$$

are truly initial conditions that must be imposed upon $\mathscr{E}$ and $\mathscr{H}$ may be seen from the fact that, if they are satisfied at any one moment of time, they will be satisfied at all times, by virtue of the remaining equations. Thus, putting

$$P \equiv \nabla \cdot \mathscr{H} \qquad \text{and} \qquad Q \equiv \nabla \cdot \mathscr{E} - 4\pi\rho,$$

we find from Eq. (16.25)

$$\frac{\partial P}{\partial t} = \nabla \cdot \frac{\partial \mathscr{H}}{\partial t} = \nabla \cdot (-c\,\nabla \times \mathscr{E}) = -c\,\nabla \cdot \nabla \times \mathscr{E} = 0,$$

so that if $P = 0$ for $t = 0$, it is zero for all t. Similarly, from Eqs. (16.26) and (16.28)

$$\frac{\partial Q}{\partial t} = \nabla \cdot \frac{\partial \mathscr{E}}{\partial t} - 4\pi \frac{\partial \rho}{\partial t} = \nabla \cdot (c\,\nabla \times \mathscr{H} - 4\pi\rho v) - 4\pi \frac{\partial \rho}{\partial t}$$
$$= -4\pi\left(\nabla \cdot \rho \mathbf{v} + \frac{\partial \rho}{\partial t}\right) = 0.$$

Thus $Q = 0$ for all t if it is zero for $t = 0$.

We know that, as $c \to \infty$, the equations of electrodynamics must approach the equations of electrostatics. In this sense electrostatics is a limiting case of electrodynamics. There are, however, cases when the equations of electrostatics hold exactly. Thus, when all particles are at rest, and all fields are independent of time, the equations of electrodynamics become

$$\nabla \times \mathscr{E} = 0, \qquad \nabla \cdot \mathscr{H} = 0, \qquad \nabla \times \mathscr{H} = 0, \qquad \nabla \cdot \mathscr{E} = 4\pi\rho. \qquad (16.29)$$

The second and third of these are satisfied by $\mathscr{H} = 0$. The other two equations are just the equations of electrostatics. Thus, we see that any field obtained in electrostatics is also a possible field of electrodynamics.* In this sense *electrostatics is not only a limiting case of electrodynamics, but also a special case of it.*

* The same could, of course, be said of the fields obtained in magnetostatics—those satisfying the equations $\mathbf{E} = 0$, $\nabla \times \mathbf{H} = 0$, and $\nabla \cdot \mathbf{H} = 0$.

Equation (16.24) is the same as in electrostatics; therefore, as in electrostatics, we have for any closed surface

$$\int \mathscr{E} \cdot d\mathbf{S} = 4\pi Q, \tag{16.30}$$

where Q is the total charge within the surface.

Taking the four-dimensional divergence of both sides of Eq. (16.22), we obtain

$$\frac{\partial^2 F_{\alpha\beta}}{\partial x_\alpha \partial x_\beta} = 4\pi \frac{\partial s_\alpha}{\partial x_\alpha}. \tag{16.31}$$

The left-hand member of Eq. (16.31) can be shown to vanish as follows: by interchanging first the summation indices and then the order of differentiation we have

$$\frac{\partial^2 F_{\alpha\beta}}{\partial x_\alpha \partial x_\beta} = \frac{\partial^2 F_{\beta\alpha}}{\partial x_\beta \partial x_\alpha} = \frac{\partial^2 F_{\beta\alpha}}{\partial x_\alpha \partial x_\beta};$$

therefore, since $F_{\beta\alpha} = -F_{\alpha\beta}$, it is equal to its own negative and must consequently be zero. We thus have

$$\frac{\partial s_\alpha}{\partial x_\alpha} = 0, \tag{16.32}$$

which is the covariant form of the continuity equation (16.28).

It may be useful to write down the form that the second pair of Maxwell–Lorentz equations takes when we have but a single particle situated at the origin:

$$\nabla \times \mathscr{H} - \frac{1}{c}\frac{\partial \mathscr{E}}{\partial t} = \frac{4\pi}{c} e\mathbf{v}\, \delta(\mathbf{r}), \tag{16.33}$$

$$\nabla \cdot \mathscr{E} = 4\pi e\, \delta(\mathbf{r}). \tag{16.34}$$

Suppose, now, that we found $\mathscr{E}$ and $\mathscr{H}$ satisfying Eqs. (16.25) and (16.26), with given ρ and $\mathbf{v}$. Other solutions of these equations, which we may represent by $\mathscr{E} + \mathbf{e}$ and $\mathscr{H} + \mathbf{h}$, must satisfy the equations

$$\frac{1}{c}\frac{\partial(\mathscr{H} + \mathbf{h})}{\partial t} = -\nabla \times (\mathscr{E} + \mathbf{e}), \qquad \nabla \cdot (\mathscr{H} + \mathbf{h}) = 0; \tag{16.35}$$

$$\frac{1}{c}\frac{\partial(\mathscr{E} + \mathbf{e})}{\partial t} = \nabla \times (\mathscr{H} + \mathbf{h}) - \frac{4\pi\rho\mathbf{v}}{c}, \qquad \nabla \cdot (\mathscr{E} + \mathbf{e}) = 4\pi\rho. \tag{16.36}$$

Since, however, $\mathscr{E}$ and $\mathscr{H}$ satisfy Eqs. (16.25) and (16.26), these equations become

$$\frac{1}{c}\frac{\partial \mathbf{h}}{\partial t} = -\nabla \times \mathbf{e}, \qquad \nabla \cdot \mathbf{h} = 0; \tag{16.37}$$

$$\frac{1}{c}\frac{\partial \mathbf{e}}{\partial t} = \nabla \times \mathbf{h}, \qquad \nabla \cdot \mathbf{e} = 0. \tag{16.38}$$

These equations are exactly of the form that Eqs. (16.25) and (16.26) take for the case when $\rho = 0$. We thus see that any solution of the equations for empty space may be added to the solutions originally found. This is, of course, what could be immediately predicted on the basis of the theory of differential equations. Since Eqs. (16.25) and (16.26) are linear non-homogeneous differential equations, their general solution consists of the sum of any special solution and the general solution of the same equations with the inhomogeneous terms $4\pi\rho\mathbf{v}/c$ and $4\pi\rho$ removed.

Substituting from Eqs. (14.15) and (14.16) into Eq. (16.26), we obtain

$$\nabla \times (\nabla \times \mathbf{A}) + \frac{1}{c}\frac{\partial}{\partial t}\left(\nabla\varphi + \frac{1}{c}\frac{\partial \mathbf{A}}{\partial t}\right) = \frac{4\pi\rho\mathbf{v}}{c},$$

and

$$\nabla \cdot \left(\nabla\varphi + \frac{1}{c}\frac{\partial \mathbf{A}}{\partial t}\right) = -4\pi\rho;$$

or, since

$$\nabla \times (\nabla \times \mathbf{A}) = \nabla\nabla \cdot \mathbf{A} - \nabla^2\mathbf{A},$$

$$\nabla^2\mathbf{A} - \frac{1}{c^2}\frac{\partial^2 \mathbf{A}}{\partial t^2} = \frac{-4\pi\rho\mathbf{v}}{c} + \nabla\left(\nabla \cdot \mathbf{A} + \frac{1}{c}\frac{\partial \varphi}{\partial t}\right)$$

and

$$\nabla^2\varphi - \frac{1}{c^2}\frac{\partial^2 \varphi}{\partial t^2} = -4\pi\rho - \frac{1}{c}\frac{\partial}{\partial t}\left(\nabla \cdot \mathbf{A} + \frac{1}{c}\frac{\partial \varphi}{\partial t}\right).$$

We have seen that potentials are not completely determined by the fields. The equations above show that it is convenient to restrict the potential further by requiring that

$$\nabla \cdot \mathbf{A} + \frac{1}{c}\frac{\partial \varphi}{\partial t} = 0. \tag{16.39}$$

This is known as the *Lorentz relation* and is allowable by virtue of the arbitrariness in the potentials expressed by the gauge transformation in

Eq. (16.4). With this choice of gauge the above equations reduce to

$$\nabla^2 \mathbf{A} - \frac{1}{c^2}\frac{\partial^2 \mathbf{A}}{\partial t^2} = \frac{-4\pi\rho\mathbf{v}}{c},$$
$$\nabla^2 \varphi - \frac{1}{c^2}\frac{\partial^2 \varphi}{\partial t^2} = -4\pi\rho. \tag{16.40}$$

A relativistic form for these equations are obtained by substitution from Eq. (14.23) into Eq. (16.22). Thus

$$\frac{\partial F_{\alpha\beta}}{\partial x_\beta} = \frac{\partial}{\partial x_\beta}\left(\frac{\partial \varphi_\beta}{\partial x_\alpha} - \frac{\partial \varphi_\alpha}{\partial x_\beta}\right) = \frac{\partial^2 \varphi_\beta}{\partial x_\alpha\, \partial x_\beta} - \frac{\partial^2 \varphi_\alpha}{\partial x_\beta\, \partial x_\beta} = 4\pi s_\alpha,$$

which on introducing the Lorentz relation

$$\frac{\partial \varphi_\beta}{\partial x_\beta} = 0 \tag{16.41}$$

becomes

$$\frac{\partial^2 \varphi_\alpha}{\partial x_\beta^2} = -4\pi s_\alpha. \tag{16.42}$$

Equations (16.41) and (16.42) are equivalent to Eqs. (16.39) and (16.40), respectively.

We need to show that the Lorentz condition can be imposed on the potentials describing an arbitrary field. Thus assume that we have initially a potential four-vector φ_β which does not satisfy Eq. (16.41) and thus

$$\frac{\partial \varphi_\beta}{\partial x_\beta} = f(x_\gamma),$$

where $f(x_\gamma)$ is an arbitrary function. By making a gauge transformation

$$\varphi_\beta' = \varphi_\beta + \frac{\partial \chi}{\partial x_\beta},$$

we have

$$\frac{\partial \varphi_\beta'}{\partial x_\beta} = f(x_\beta) + \frac{\partial^2 \chi}{\partial x_\beta^2} = 0,$$

provided χ can be chosen so as to satisfy the inhomogeneous wave equation

$$\frac{\partial^2 \chi}{\partial x_\beta^2} = \nabla^2 \chi - \frac{1}{c^2}\frac{\partial^2 \chi}{\partial t^2} = -f(\mathbf{r}, t).$$

This equation is of the same form as Eq. (19.1), which will be shown to have a solution given by Eq. (19.14). Thus the Lorentz relation applies to the new potential φ'_β.

Even with the restriction imposed by Eq. (16.41) the potentials are not completely determined by the field. Requiring that φ_β and φ'_β both satisfy Eq. (16.41) leads to

$$\frac{\partial^2 \chi}{\partial x_\beta^2} \equiv \nabla^2 \chi - \frac{1}{\chi^2}\frac{\partial^2 \chi}{\partial t^2} = 0. \tag{16.43}$$

Thus, not only is Eq. (16.41) a permissible restriction, but the transformation (16.3) is still permissible, provided χ satisfies the wave equation (16.43).

When no charges are present $s_\alpha = 0$. It is then often convenient to put the scalar potential φ equal to zero. This is within the permissible arbitrariness of φ_α. For, suppose we have a solution of Eqs. (16.41) and (16.42) for which $\varphi_4 = i\varphi \neq 0$. We can then go over to φ'_α for which

$$\varphi' = \varphi - \frac{1}{c}\frac{\partial \chi'}{\partial t} = 0,$$

provided a χ' can be found satisfying this differential equation. Such a solution always exists and is given by

$$\chi' = c\int_0^t \varphi \, dt + \chi_0, \tag{16.44}$$

where χ_0 is independent of time.

Suppose the original potentials satisfy the Lorentz relation. Under what conditions will the primed potentials obtained above satisfy this relation? By our previous discussion, one answer is that χ' must satisfy Eq. (16.43). Since φ satisfies the second equation of Eqs. (16.40) with $\rho = 0$, it can easily be shown that χ' will satisfy Eq. (16.43) provided

$$\nabla^2 \chi_0 = \frac{1}{c}\left(\frac{\partial \varphi}{\partial t}\right)_{t=0}.$$

This equation is of the form of Poisson's equation (see Eq. (7.34)) corresponding to $\rho = -(1/4\pi c)(\partial\varphi/\partial t)_{t=0}$ and is therefore always solvable by Eq. (8.11). Thus, the assumption $\varphi = 0$ is always permissible in the absence of charges even for potentials satisfying the Lorentz relation. Our Eqs. (16.39) and (16.40) then become (with $\varphi = 0$)

$$\nabla^2 \mathbf{A} - \frac{1}{c^2}\frac{\partial^2 \mathbf{A}}{\partial t^2} = 0 \quad \text{and} \quad \nabla \cdot \mathbf{A} = 0. \tag{16.45}$$

The remaining arbitrariness in **A** can be found by using Eqs. (16.4) and requiring that $\varphi' = \varphi = 0$. These equations show that χ must be independent of time, and that

$$\mathbf{A}' = \mathbf{A} + \nabla\chi. \tag{16.46}$$

We thus see that $\mathbf{A}'$ will satisfy Eq. (16.45) provided

$$\nabla^2\chi = 0. \tag{16.47}$$

Thus, any solution of Laplace's equation independent of time can be used for χ in Eqs. (16.46) without violating any of the requirements.

EXAMPLE 1. Find the field produced by a charge e moving uniformly with a velocity $\mathbf{v}$.

In the coordinate system moving with the particle, the particle will be permanently at rest. Let us designate by primes all quantities measured in this coordinate system, and suppose the particle to be at the origin. In this coordinate system the field produced by the particle will be independent of time, and the field equations (16.25) and (16.26) reduce to

$$\begin{aligned} \nabla \cdot \mathscr{E}' &= 4\pi e\,\delta(\mathbf{r}'), & \nabla \times \mathscr{E}' &= 0, \\ \nabla \cdot \mathscr{H}' &= 0, & \nabla \times \mathscr{H}' &= 0. \end{aligned} \tag{16.48}$$

The last two equations are satisfied by

$$\mathscr{H}' = 0, \tag{16.49}$$

while the first two are the same as the corresponding equations of electrostatics and give

$$\mathscr{E}' = \frac{e\mathbf{r}'}{r'^3}.$$

Now, in order to obtain the field in the coordinate system in which the particle moves with the velocity $\mathbf{v}$, we need only to perform a Lorentz transformation to this coordinate system. Choosing, for convenience, the direction of $\mathbf{v}$ as the direction of X-axis, we have, in accordance with Eqs. (15.5)* and (11.8),

$$\mathscr{E}_x = \mathscr{E}'_x = \frac{ex'}{r'^3} = \frac{e(1 - v^2/c^2)(x - vt)}{[(x - vt)^2 + (1 - v^2/c^2)(y^2 + z^2)]^{3/2}},$$

* Since the transformation is from the primed to the unprimed system the inverse of Eq. (15.5) should be used. It is obtained from Eq. (15.5) by interchanging the primed and unprimed quantities and substituting $-\mathbf{v}$ for $\mathbf{v}$.

since

$$r' = (x'^2 + y'^2 + z'^2)^{1/2} = \left[\frac{(x - vt)^2}{1 - v^2/c^2} + y^2 + z^2\right]^{1/2}.$$

Likewise

$$\mathscr{E}_y = \frac{\mathscr{E}'_y}{\sqrt{1 - v^2/c^2}} = \frac{e(1 - v^2/c^2)y}{[(x - vt)^2 + (1 - v^2/c^2)(y^2 + z^2)]^{3/2}},$$

with a similar expression for $\mathscr{E}_z$. If we choose the coordinate system in such a way that $t = 0$ at that moment at which we wish to know the field, then the above equations simplify to the single vector equation

$$\mathscr{E} = \frac{e(1 - v^2/c^2)\mathbf{r}}{[x^2 + (1 - v^2/c^2)(y^2 + z^2)]^{3/2}}. \tag{16.50}$$

Equation (16.50) shows that for a moving charge, as for a charge at rest, the electric field everywhere is directed radially away from the charge. The magnitude of $\mathscr{E}$, however, is not independent of the direction. This can best be seen by writing it in the form

$$\mathscr{E} = \frac{e(1 - v^2/c^2)}{[1 - (v^2/c^2)\sin^2\theta]^{3/2}r^2}, \tag{16.51}$$

where θ is the angle between the directions of $\mathbf{r}$ and $\mathbf{v}$. $\mathscr{E}$ has a maximum value for $\theta = \pi/2$ given by

$$\mathscr{E}_{\max} = \frac{e}{r^2\sqrt{1 - v^2/c^2}}; \tag{16.52}$$

the minimum value occurs for $\theta = 0$, and $\theta = \pi$—that is, in the direction of $\pm\mathbf{v}$—and is

$$\mathscr{E}_{\min} = \frac{e(1 - v^2/c^2)}{r^2}. \tag{16.53}$$

Again, using Eq. (11.8) and the inverse of Eqs. (15.5), we obtain

$$\mathscr{H}_x = 0, \qquad \mathscr{H}_y = \frac{-v\mathscr{E}'_z/c}{\sqrt{1 - v^2/c^2}}, \qquad \mathscr{H}_z = \frac{v\mathscr{E}'_y/c}{\sqrt{1 - v^2/c^2}}. \tag{16.54}$$

The result is best expressed in spherical coordinates. Taking the polar axis along the path traced by the particle, we have

$$\mathscr{H}_r = \mathscr{H}_\theta = 0, \qquad \mathscr{H}_\varphi = \frac{e(1 - v^2/c^2)v\sin\theta/c}{(1 - v^2\sin^2\theta/c^2)^{3/2}r^2}. \tag{16.55}$$

This shows that the magnetic field lines are a family of circles with their planes perpendicular to **v** and having their centers on the polar axis.

EXAMPLE 2. Determine the behavior of the electric and the magnetic fields at the boundary between two regions, assuming a steady state.*

It is the purpose of this example to account quantitatively for the seeming discontinuities of the electric and magnetic fields that are often observed at boundary surfaces. To do this we make use of the common assumption by physicists that physical quantities change continuously, and therefore that the boundary surfaces, which appear to be surfaces of discontinuity, are in actuality very thin transition regions in which $\mathscr{E}$ and $\mathscr{H}$ undergo perhaps a rapid, but nevertheless a continuous change. If $\mathscr{E}$ and $\mathscr{H}$ undergo an extremely rapid change in this thickness, they may have quite different values on the two sides of the transition region, and the fields would appear to have changed discontinuously. It is these total changes of the fields and not a detailed picture of the changes taking place within the transition layer with which we are here concerned.

Consider a very thin layer of variable charge density $\rho(\mathbf{r})$ in a space free of other charges. Since the potential in this case is given by (see Eq. (8.11))

$$\varphi(\mathbf{r}) = \int_{V'} \frac{\rho(\mathbf{r}')}{|\mathbf{r} - \mathbf{r}'|} \, dV',$$

the electric field is

$$\mathscr{E}(\mathbf{r}) = -\nabla\varphi(\mathbf{r}) = \int_{V'} \frac{(\mathbf{r} - \mathbf{r}')\,\rho(\mathbf{r}')}{|\mathbf{r} - \mathbf{r}'|^3} \, dV'. \tag{16.56}$$

* The usual treatment of this problem is to apply Gauss' theorem to a cylinder lying in the transition layer with its elements perpendicular to this layer, and having one face in each of the two regions separated by the layer; and next to apply Stoke's theorem to a circuit passing through the transition region, and then parallel in each of the main regions to the transition layer.

In the usual consideration of the limiting case when the thickness of the transition layer goes to zero, it is customary to say that the contribution to the integrals of the portion of the surface integral taken inside the layer, and of the portion of the circuit inside the layer, approaches zero. This, however, is not true unless the fields in the layer remain finite. Since the charge density and the current density often become infinite in this layer, it is not at all obvious that the fields do not also become infinite.

The treatment given here, although somewhat lengthy, has the advantage of stating the conditions a physical transition layer and the external fields must satisfy in order that the equations for the discontinuities of the electric and magnetic fields at boundaries between regions may be applicable.

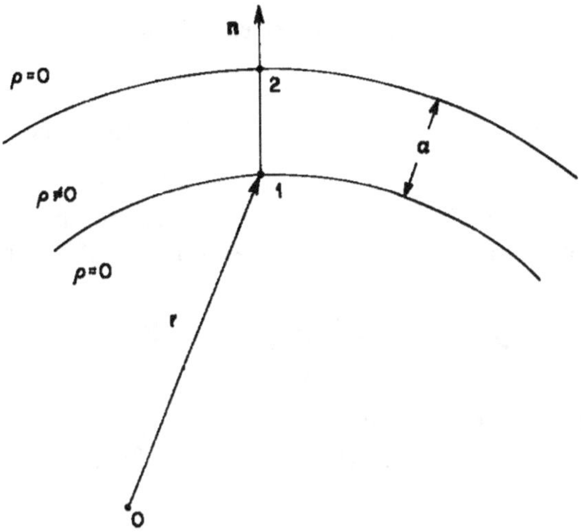

FIG. 16.1. Transition region between two media.

The difference between the fields at two points 1 and 2 on the opposite sides of the layer (see Fig. 16.1) is then

$$\mathscr{E}_2 - \mathscr{E}_1 = \mathscr{E}(\mathbf{r} + \alpha\mathbf{n}) - \mathscr{E}(\mathbf{r})$$
$$= \int_{V'} \left(\frac{\mathbf{r} - \mathbf{r}' + \alpha\mathbf{n}}{|\mathbf{r} - \mathbf{r}' + \alpha\mathbf{n}|^3} - \frac{\mathbf{r} - \mathbf{r}'}{|\mathbf{r} - \mathbf{r}'|^3} \right) \rho(\mathbf{r}')\, dV', \qquad (16.57)$$

where $\mathbf{n}$ is the unit vector normal to the surface of the layer and α is the thickness of the layer. We wish to calculate the approximate value of this difference under certain restrictive conditions, namely:

1. The layer is so thin and uniform that one normal $\mathbf{n}$ may be used with sufficient accuracy for both boundary surfaces near $\mathbf{r}$.

2. $\rho(\mathbf{r}')$ varies so slowly in any direction parallel to the layer that for distances from $\mathbf{r}$ less than some a, where a is large in comparison with the thickness of the layer α, $\mathbf{n}$ may be regarded as being parallel to $\nabla\rho$. That is, for

$$\alpha \ll |\mathbf{r} - \mathbf{r}'| < a,$$

we may assume

$$\mathbf{n} \times \nabla\rho(\mathbf{r}') = 0. \qquad (16.58)$$

3. The curvature of the layer near $\mathbf{r}$ is so small that any radius of curvature near $\mathbf{r}$ is large in comparison with a. Thus, if b is the smallest

radius of curvature near $\mathbf{r}$, we must be able to find a length a such that

$$b \gg a \gg \alpha,$$

and for this a, the condition expressed in Eq. (16.58) must hold.

Assuming that the above conditions are fulfilled, we may proceed as follows. First, we introduce for convenience the notation $\mathbf{R} = \mathbf{r} - \mathbf{r}'$, and estimate the contribution to the integral in Eq. (16.57), which can now be written as

$$\mathscr{E}_2 - \mathscr{E}_1 = \int_{V'} \left\{ \frac{\mathbf{R} + \alpha\mathbf{n}}{|\mathbf{R} + \alpha\mathbf{n}|^3} - \frac{\mathbf{R}}{R^3} \right\} \rho(\mathbf{r}')\, dV', \tag{16.59}$$

due to that part of V' for which $|\mathbf{r} - \mathbf{r}'| = R > a$.

Now $dV' = dh\, dS$, where dS is an element of surface and h is measured perpendicular to the surface. Thus

$$I \equiv \left| \int_{R>a} \left\{ \frac{\mathbf{R} + \alpha\mathbf{n}}{|\mathbf{R} + \alpha\mathbf{n}|^3} - \frac{\mathbf{R}}{R^3} \right\} dS \int_0^h \rho(\mathbf{r}')\, dh \right|$$

is the magnitude of the contribution of that portion of V' for which $R > a$. The vector in the braces is a function of R, and since R for the region $R > a$ does not differ appreciably for $\mathbf{r}'$'s lying on opposite sides of the transition layer, this vector can be considered as approximately constant for the integration with respect to h. Thus

$$\begin{aligned} I &\simeq \left| \int_{R>a} \left\{ \frac{\mathbf{R} + \alpha\mathbf{n}}{|\mathbf{R} + \alpha\mathbf{n}|^3} - \frac{\mathbf{R}}{R^3} \right\} \sigma\, dS \right| \\ &\lesssim |\sigma|_{\max} \int_{R>a} \left| \frac{\mathbf{R} + \alpha\mathbf{n}}{|\mathbf{R} + \alpha\mathbf{n}|^3} - \frac{\mathbf{R}}{R^3} \right| dS, \end{aligned}$$

where

$$\sigma = \int_0^h \rho(\mathbf{r}')\, dh$$

is equal to the charge per unit area of the layer. Moreover to the first power in α/R,

$$\begin{aligned} \left| \frac{\mathbf{R} + \alpha\mathbf{n}}{|\mathbf{R} + \alpha\mathbf{n}|^3} - \frac{\mathbf{R}}{R^3} \right| &\simeq \left[\left(\frac{\alpha\mathbf{n}}{R^3} - \frac{3\alpha\mathbf{n} \cdot \mathbf{R}\mathbf{R}}{R^5} \right)^2 \right]^{1/2} \\ &\simeq \left[\frac{\alpha^2}{R^6} + \frac{3\alpha^2(\mathbf{n} \cdot \mathbf{R})^2}{R^8} \right]^{1/2} \\ &\leq \left[\frac{4\alpha^2}{R^6} \right]^{1/2} = \frac{2\alpha}{R^3}\,; \end{aligned} \tag{16.60}$$

thus

$$I \lesssim 2\alpha|\sigma|_{\max} \int_{R>a} \frac{dS}{R^3}. \tag{16.61}$$

The integral

$$\int_{R>a} \frac{dS}{R^3}$$

of course depends on the shape of the transition layer.* Suppose the layer is sufficiently regular to permit one to speak of the distance u of a point from $\mathbf{r}$, as measured along the surface, and to draw on the surface closed curves of constant u. Consider now a strip of the surface lying between two adjacent curves of constant u, and designate by $R_{\min}$ the minimum value of R on this strip. Let

$$R_{\min} = G(u)\, u,$$

where $G(u)$ is the minimum ratio of the actual distance of a point on the surface from r to the distance between the same points measured along the surface. For a plane

$$G(u) = 1;$$

for all ordinary surfaces

$$G(u) \simeq 1. \tag{16.62}$$

Also, if $\ell(u)$ is the length of the strip,

$$\ell(u) = 2\pi R_{\min}\, F(u),$$

where $F(u) \simeq 1$. Thus, if du is the width of the strip,

$$\int_{R>a} \frac{dS}{R^3} = \int_a^\infty \frac{\ell(u)\, du}{R^3} \leq 2\pi \int_a^\infty \frac{F(u)\, du}{u^2 G^2(u)};$$

and hence

$$\int_{R>a} \frac{dS}{R^3} \leq \frac{2\pi F_{\max}}{G^2_{\min}} \int_a^\infty \frac{du}{u^2} = \frac{2\pi F_{\max}}{G^2_{\min}} \frac{1}{a}. \tag{16.63}$$

Putting this in the inequality (16.61), we have the result that the absolute value of the contribution to the integral in Eq. (16.59), due to the part of the layer for which $R > a$, is less than

$$4\pi|\sigma|_{\max} \frac{F_{\max}}{G^2_{\min}} \left(\frac{\alpha}{a}\right),$$

* The mathematical surface over which this integral is to be taken is, of course, indefinite to the extent of the thickness of the transition layer.

which approaches zero as $\alpha/a \to 0$. Thus, neglecting quantities of the order of α/a,

$$\mathscr{E}_2 - \mathscr{E}_1 = \int_{V_0} \left\{ \frac{\mathbf{R} + \alpha\mathbf{n}}{|\mathbf{R} + \alpha\mathbf{n}|^3} - \frac{\mathbf{R}}{R^3} \right\} \rho(\mathbf{r}')\, dV', \tag{16.64}$$

where V_0 is the portion of the layer for which $R < a$.

Let us now introduce a cylindrical coordinate system (u, θ, z), where u and z are defined by the equation

$$\mathbf{r}' - \mathbf{r} = -\mathbf{R} = z\mathbf{n} + u\mathbf{s},$$

$\mathbf{s}$ being perpendicular to $\mathbf{n}$. Thus the origin is at $\mathbf{r}$ and the Z-axis is along $\mathbf{n}$. Therefore

$$\mathbf{R} = -z\mathbf{n} - u\mathbf{s}, \qquad R = [z^2 + u^2]^{1/2}; \tag{16.65}$$

and

$$\begin{aligned} \mathbf{R} + \alpha\mathbf{n} &= (\alpha - z)\mathbf{n} - u\mathbf{s}, \\ |\mathbf{R} + \alpha\mathbf{n}| &= [(\alpha - z)^2 + u^2]^{1/2}. \end{aligned} \tag{16.66}$$

For V_0 the condition (16.58), $\mathbf{n} \times \nabla\rho = 0$, holds. But $\mathbf{n} = \nabla z$, so that $\nabla z \times \nabla\rho = 0$, or ρ is a function of z alone (see Theorem 1, Appendix A3). We therefore write $\rho(\mathbf{r}') = \rho(z)$. By virtue of condition 3 we can regard V_0 as being a flat disc of thickness α and radius a. Thus, using Eqs. (16.65) and (16.66) and the fact that $dV' = u\, du\, d\theta\, dz$,

$$\mathscr{E}_2 - \mathscr{E}_1 \simeq \int_0^{2\pi} d\theta \int_0^{\alpha} \rho(z)\, dz \int_0^a \left\{ \frac{(\alpha - z)\mathbf{n} - u\mathbf{s}}{[u^2 + (\alpha - z)^2]^{3/2}} + \frac{z\mathbf{n} + u\mathbf{s}}{[u^2 + z^2]^{3/2}} \right\} u\, du. \tag{16.67}$$

For the normal component of this difference we now have

$$\begin{aligned} (\mathscr{E}_2 - \mathscr{E}_1)_n &= \mathbf{n} \cdot (\mathscr{E}_2 - \mathscr{E}_1) \\ &\simeq \int_0^{2\pi} d\theta \int_0^{\alpha} \rho(z)\, dz \int_0^a \left\{ \frac{\alpha - z}{[u^2 + (\alpha - z)^2]^{3/2}} + \frac{z}{[u^2 + z^2]^{3/2}} \right\} u\, du. \end{aligned} \tag{16.68}$$

The integral

$$\begin{aligned} &\int_0^a \left\{ \frac{\alpha - z}{[u^2 + (\alpha - z)^2]^{3/2}} + \frac{z}{[u^2 + z^2]^{3/2}} \right\} u\, du \\ &\qquad = 2 - \frac{\alpha - z}{[a^2 + (\alpha - z)^2]^{1/2}} - \frac{z}{[a^2 + z^2]^{1/2}}. \end{aligned}$$

Since, however, $z \leq \alpha \ll a$,

$$[a^2 + (\alpha - z)^2]^{-1/2} = a^{-1}\left[1 + \frac{(\alpha - z)^2}{a^2}\right]^{-1/2} \simeq a^{-1}\left[1 - \frac{(\alpha - z)^2}{2a^2}\right]$$

and

$$[a^2 + z^2]^{-1/2} = a^{-1}\left[1 + \frac{z^2}{a^2}\right]^{-1/2} = a^{-1}\left[1 - \frac{z^2}{2a^2}\right].$$

Thus

$$\int_0^a \left\{\frac{\alpha - z}{[u^2 + (\alpha - z)^2]^{3/2}} + \frac{z}{[u^2 + z^2]^{3/2}}\right\} u\, du$$

$$\simeq 2 - \frac{\alpha - z}{a}\left[1 - \frac{(\alpha - z)^2}{2a^2}\right] - \frac{z}{a}\left[1 - \frac{z^2}{2a^2}\right]$$

$$\simeq 2 - \frac{\alpha - z}{a} - \frac{z}{a} = 2 - \frac{\alpha}{a}. \tag{16.69}$$

Therefore, with neglect of the term in α/a, Eq. (16.68) becomes

$$(\mathscr{E}_2 - \mathscr{E}_1)_n = 2\int_0^{2\pi} d\theta \int_0^\alpha \rho(z)\, dz = 4\pi \int_0^\alpha \rho(z)\, dz = 4\pi\sigma, \tag{16.70}$$

where

$$\sigma = \int_0^\alpha \rho(z)\, dz = \text{surface density of charge} \tag{16.71}$$

is equal to the charge per unit area of the layer.

Let $\mathbf{t}$ be a constant unit vector tangential to the layer and making an angle θ with the vector $\mathbf{s}$. Then $\mathbf{t} \cdot \mathbf{s} = \cos\theta$ and $\mathbf{t} \cdot \mathbf{n} = 0$; therefore, by Eq. (16.67),

$$(\mathscr{E}_2 - \mathscr{E}_1) \cdot \mathbf{t}$$

$$= \int_0^{2\pi} d\theta \int_0^\alpha \rho(z)\, dz \int_0^a \left\{\frac{1}{[u^2 + z^2]^{3/2}} - \frac{1}{[u^2 + (\alpha - z)^2]^{3/2}}\right\} \mathbf{t} \cdot \mathbf{s}\, u^2\, du$$

or

$$(\mathscr{E}_2 - \mathscr{E}_1)_t$$

$$= \int_0^\alpha \rho(z)\, dz \int_0^a \left\{\frac{1}{[u^2 + z^2]^{3/2}} - \frac{1}{[u^2 + (\alpha - z)^2]^{3/2}}\right\} u^2\, du \int_0^{2\pi} \cos\theta\, d\theta.$$

Thus, since

$$\int_0^{2\pi} \cos\theta\, d\theta = 0,$$

$$(\mathscr{E}_2 - \mathscr{E}_1)_t = 0 \tag{16.72}$$

with neglect of terms of the order of α/a. Since **t** is an arbitrary unit vector perpendicular to **n**, this equation states that there is no appreciable change in the tangential component of $\mathscr{E}$ across the transition layer.

We wish next to consider the behavior of the magnetic field $\mathscr{H}$ at boundaries. In order to arrive at an expression for $\mathscr{H}$ analogous to Eq. (16.56) for $\mathscr{E}$, we use the differential equation for **A**, Eq. (16.40), to find the expression for **A**. Since we are assuming a steady state,

$$\nabla^2\mathbf{A} = -4\pi\mathbf{j}, \qquad \text{where} \quad \mathbf{j} = \frac{\rho\mathbf{v}}{c},$$

and since each component of this differential equation is of the form of Poisson's equation, a solution is*

$$\mathbf{A} = \int \frac{\mathbf{j}(\mathbf{r}')}{|\mathbf{r} - \mathbf{r}'|}\, dV'. \tag{16.73}$$

Hence

$$\mathscr{H} = \nabla \times \mathbf{A} = -\int \left[\mathbf{j} \times \nabla\left(\frac{1}{R}\right)\right] dV'$$

or

$$\mathscr{H} = \int \frac{\mathbf{j} \times \mathbf{R}}{R^3}\, dV'. \tag{16.74}$$

The conditions that the transition layer must satisfy are the same as those given in the treatment of the electric field, except that the condition $\mathbf{n} \times \nabla\rho = 0$ is replaced by a condition on **j**, to be given later.

The difference between the magnetic field on one side of the transition layer and the field on the other side, due to the transition layer, is thus given by

$$\mathscr{H}_2 - \mathscr{H}_1 = \int \mathbf{j} \times \left\{\frac{\mathbf{R} + \alpha\mathbf{n}}{|\mathbf{R} + \alpha\mathbf{n}|^3} - \frac{\mathbf{R}}{R^3}\right\} dV'.$$

* The **A** so given is only a particular solution of the differential equation; to it could be added an arbitrary external field corresponding to a solution of the homogeneous equation. We assume, however, that this external field does not change appreciably in a distance α and hence can be neglected. For a discussion of the reason for considering the integral in Eq. (16.73) as the field produced by the layer, see Section 19.

Moreover, the absolute value of the contribution to this difference of the portion of the layer lying outside the sphere $R = a$ is less than

$$I_2 \equiv \left| \int_{R>a} \mathbf{j} \times \left\{ \frac{\mathbf{R} + \alpha\mathbf{n}}{|\mathbf{R} + \alpha\mathbf{n}|^3} - \frac{\mathbf{R}}{R^3} \right\} dV' \right|$$

$$\simeq \left| \int_{R>a} \int_0^h \mathbf{j}\, dh \times \left\{ \frac{\mathbf{R} + \alpha\mathbf{n}}{|\mathbf{R} + \alpha\mathbf{n}|^3} - \frac{\mathbf{R}}{R^3} \right\} dS \right|,$$

and, letting

$$\mathbf{J} = \int_0^h \mathbf{j}\, dh,$$

$$I_2 < |\mathbf{J}|_{\max} \int_{R>a} \left| \frac{\mathbf{R} + \alpha\mathbf{n}}{|\mathbf{R} + \alpha\mathbf{n}|^3} - \frac{\mathbf{R}}{R^3} \right| dS$$

or

$$I_2 < 2\alpha |\mathbf{J}|_{\max} \int_{R>a} \frac{dS}{R^3}.$$

This being the same as (16.61), except that $|\mathbf{J}|_{\max}$ now takes the place of $|\sigma|_{\max}$, it vanishes as α/a approaches zero. Thus, again neglecting terms of the order of α/a,

$$\mathscr{H}_2 - \mathscr{H}_1 = \int_{V_0} \mathbf{j} \times \left\{ \frac{\mathbf{R} + \alpha\mathbf{n}}{|\mathbf{R} + \alpha\mathbf{n}|^3} - \frac{\mathbf{R}}{R^3} \right\} dV', \tag{16.75}$$

where as before V_0 is the portion of the layer for which $R \leq a$.

Introducing the cylindrical coordinate system used for the electric field, we can now specify the condition to be imposed on $\mathbf{j}$; it is

$$\mathbf{j}(\mathbf{r}') = \mathbf{j}(z) \qquad \text{for} \quad R < a \ll b, \tag{16.76}$$

where b is the minimum radius of curvature near $\mathbf{r}$. Making use of this condition and the equation of continuity,

$$\nabla \cdot \rho\mathbf{v} + \frac{\partial \rho}{\partial t} = 0,$$

we have, for a steady state,

$$\nabla \cdot \mathbf{j} = \frac{\partial j_x}{\partial x} + \frac{\partial j_y}{\partial y} + \frac{\partial j_z}{\partial z} = \nabla \cdot \frac{\rho\mathbf{v}}{c} = 0$$

or, by Eq. (16.76)

$$\frac{\partial j_z}{\partial z} = 0, \qquad j_z = \text{constant}.$$

Since j_z is zero at the surfaces of the transition layer this constant is zero; thus

$$\mathbf{j} \cdot \mathbf{n} = 0. \tag{16.77}$$

The difference $\mathscr{H}_2 - \mathscr{H}_1$ (see Eq. (16.75)) when expressed in terms of cylindrical coordinates becomes, with use of Eqs. (16.65) and (16.66),

$$\mathscr{H}_2 - \mathscr{H}_1 \cong \int_0^{2\pi} d\theta \int_0^{\alpha} dz \int_0^{a} u\, du\, \mathbf{j}(z) \times \left\{ \frac{(\alpha - z)\mathbf{n} - u\mathbf{s}}{[u^2 + (\alpha - z)^2]^{3/2}} + \frac{z\mathbf{n} + u\mathbf{s}}{[u^2 + z^2]^{3/2}} \right\}. \tag{16.78}$$

The normal component of this difference, since $\mathbf{v} \times \mathbf{n} \cdot \mathbf{n} = 0$, is

$$(\mathscr{H}_2 - \mathscr{H}_1) \cdot \mathbf{n}$$
$$\cong \int_0^{\alpha} dz \int_0^{a} \left\{ \frac{1}{[u^2 + z^2]^{3/2}} - \frac{1}{[u^2 + (\alpha - z)^2]^{3/2}} \right\} u^2\, du \int_0^{2\pi} \mathbf{j} \cdot \mathbf{s} \times \mathbf{n}\, d\theta.$$

Since $\mathbf{j}$ in the last integral is independent of θ, it will make a constant angle with $\mathbf{t}$ in the integration over θ. This, together with the fact that $\mathbf{s} \times \mathbf{n}$ is at right angles to $\mathbf{s}$—the vectors $\mathbf{j}$, $\mathbf{t}$, and $\mathbf{s} \times \mathbf{n}$, being perpendicular to $\mathbf{n}$—makes it possible to take the angle between $\mathbf{j}$ and $\mathbf{s} \times \mathbf{n}$ to be $\theta + \varphi$, where φ is constant. Using these properties, we have

$$\int_0^{2\pi} \mathbf{j} \cdot \mathbf{s} \times \mathbf{n}\, d\theta = |\mathbf{j}| \int_0^{2\pi} \cos(\theta + \varphi)\, d\theta = 0,$$

and thus

$$\mathbf{n} \cdot (\mathscr{H}_2 - \mathscr{H}_1) = (\mathscr{H}_2 - \mathscr{H}_1)_n = 0. \tag{16.79}$$

Because $\mathbf{s} \times \mathbf{t}$ is parallel to $\mathbf{n}$, Eq. (16.77) can be replaced by the equation

$$\rho\mathbf{v}(z) \cdot \mathbf{s} \times \mathbf{t} = \rho\mathbf{v}(z) \times \mathbf{s} \cdot \mathbf{t} = 0.$$

Making use of this fact, we have

$$(\mathscr{H}_2 - \mathscr{H}_1) \cdot \mathbf{t}$$
$$= \int_0^{2\pi} d\theta \int_0^{\alpha} (\mathbf{j} \times \mathbf{n} \cdot \mathbf{t})\, dz \int_0^{a} \left\{ \frac{\alpha - z}{[u^2 + (\alpha - z)^2]^{3/2}} + \frac{z}{[u^2 + z^2]^{3/2}} \right\} u\, du.$$

The last integral has been evaluated (see Eq. (16.69)); therefore neglecting quantities of the order of magnitude of α/a,

$$(\mathscr{H}_2 - \mathscr{H}_1) \cdot \mathbf{t} = \frac{2}{c} \int_0^{2\pi} d\theta \int_0^{\alpha} \rho\mathbf{v}(z)\, dz \times \mathbf{n} \cdot \mathbf{t}$$
$$= (4\pi\mathbf{J} \times \mathbf{n}) \cdot \mathbf{t},$$

where

$$J = \frac{1}{c}\int_0^{\alpha} \rho\mathbf{v}(z)\, dz. \tag{16.80}$$

Moreover, since by Eq. (16.79) $\mathscr{H}_2 - \mathscr{H}_1$ is perpendicular to $\mathbf{n}$, as are also $\mathbf{J} \times \mathbf{n}$ and $\mathbf{t}$, they are all in the same plane; and since $\mathbf{t}$ is an arbitrary vector in this plane, the total change in $\mathscr{H}$ at the boundary is

$$\mathscr{H}_2 - \mathscr{H}_1 = 4\pi\mathbf{J} \times \mathbf{n}, \tag{16.81}$$

with neglect of terms of the order of α/a.

EXAMPLE 3. The magnetic field produced by a uniform current in an infinite cylindrical conductor (vector method).

The equations to be satisfied by $\mathscr{H}$ are

$$\nabla \cdot \mathscr{H} = 0, \tag{16.82}$$

$$\nabla \times \mathscr{H} = J\mathbf{k}, \tag{16.83}$$

where

$$J = \begin{Bmatrix} 4\pi\rho v/c, & \text{inside} \\ 0, & \text{outside} \end{Bmatrix} \text{the conductor,}$$

and $\mathbf{k}$ is a unit vector parallel to the axis of the cylinder and in the direction of the current.

Let Q be the point of observation described by the position vector $\mathbf{r}$ drawn from a point on the axis of the conductor (see Fig. 16.2) and let

$$\mathbf{p} = \mathbf{k} \times \mathbf{r}, \tag{16.84}$$

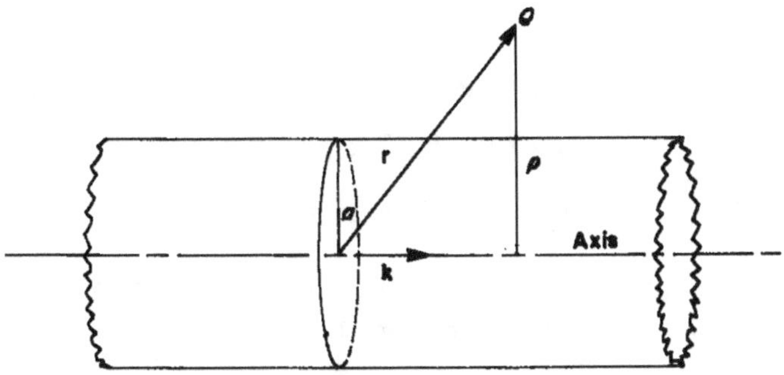

FIG. 16.2. Cylindrical conductor carrying a uniform current.

which means that $p = |\mathbf{k} \times \mathbf{r}|$ is the distance of Q from the axis. Then by Eq. (A3.27)

$$\nabla \times \mathbf{p} = \nabla \times (\mathbf{k} \times \mathbf{r}) = \mathbf{k}\, \nabla \cdot \mathbf{r} - \mathbf{k} \cdot \nabla \mathbf{r},$$

so that by Eqs. (A3.48) and (A3.52)

$$\nabla \times \mathbf{p} = 2\mathbf{k}. \tag{16.85}$$

Also by Eqs. (A3.28) and (A3.29)

$$\begin{aligned} \nabla \mathbf{p} \cdot \mathbf{p} &= 2\, \mathbf{p} \cdot \nabla \mathbf{p} + 2\, \mathbf{p} \times (\nabla \times \mathbf{p}), \\ \nabla p^2 &= 2\, \mathbf{p} \cdot \nabla(\mathbf{k} \times \mathbf{r}) + 4\, \mathbf{p} \times \mathbf{k}, \\ 2p\, \nabla p &= 2\, \mathbf{k} \times (\mathbf{p} \cdot \nabla \mathbf{r}) + 4\, \mathbf{p} \times \mathbf{k} = 2\, \mathbf{k} \times \mathbf{p} + 4\, \mathbf{p} \times \mathbf{k} = 2\, \mathbf{p} \times \mathbf{k}; \end{aligned}$$

hence

$$\nabla p = \frac{\mathbf{p} \times \mathbf{k}}{p}, \tag{16.86}$$

and by Eq. (A3.26)

$$\nabla \cdot \mathbf{p} = \nabla \cdot (\mathbf{k} \times \mathbf{r}) = -\mathbf{k} \cdot \nabla \times \mathbf{r} = 0. \tag{16.87}$$

Since

$$\mathbf{k} \cdot \mathbf{p} \times (\mathbf{p} \times \mathbf{k}) = -|\mathbf{p} \times \mathbf{k}|^2 = -p^2 \neq 0,$$

except on the axis, every vector, and in particular $\mathscr{H}$, can be expressed in the form $A\mathbf{p} + B\mathbf{p} \times \mathbf{k} + C\mathbf{k}$, again with the exception of the axis. Further, since the magnetic field is cylindrically symmetrical, the coefficients A, B, and C must be functions of p alone. Thus

$$\mathscr{H} = A(p)\mathbf{p} + B(p)\mathbf{p} \times \mathbf{k} + C(p)\mathbf{k}, \tag{16.88}$$

and Eq. (16.82) becomes, using Eq. (A3.23),

$$\begin{aligned} \nabla \cdot \mathscr{H} = A\, \nabla \cdot \mathbf{p} + \mathbf{p} \cdot \nabla A + B\, \nabla \cdot (\mathbf{p} \times \mathbf{k}) \\ + \mathbf{p} \times \mathbf{k} \cdot \nabla B + \mathbf{k} \cdot \nabla C = 0. \end{aligned} \tag{16.89}$$

In order to simplify this we note that by Eqs. (A3.26), (16.85), and (16.86)

$$\nabla \cdot (\mathbf{p} \times \mathbf{k}) = \mathbf{k} \cdot \nabla \times \mathbf{p} = 2k^2 = 2, \tag{16.90}$$

$$\nabla A = A'\, \nabla p = A' \frac{\mathbf{p} \times \mathbf{k}}{p}, \tag{16.91}$$

with similar equations for ∇B and ∇C. Substituting from Eqs. (16.87), (16.90), and (16.91) in Eq. (16.89), we have

$$\nabla \cdot \mathscr{H} = 2B + B' \frac{|\mathbf{p} \times \mathbf{k}|^2}{p} = 2B + B'p = 0.$$

Thus

$$2Bp + B'p^2 = \frac{d}{dp}(Bp^2) = 0,$$

or

$$B = \frac{d}{p^2}, \tag{16.92}$$

where d is a constant.

Equation (16.83) becomes, by virtue of Eqs. (16.88) and (A3.24),

$$\nabla \times \mathscr{H} = A\,\nabla \times \mathbf{p} - \mathbf{p} \times \nabla A + B\,\nabla \times (\mathbf{p} \times \mathbf{k}) - (\mathbf{p} \times \mathbf{k}) \times \nabla B - \mathbf{k} \times \nabla C = J\mathbf{k}. \tag{16.93}$$

Moreover, by Eqs. (A3.27), (16.86), (16.87), (16.91), and (A2.25)

$$\nabla \times (\mathbf{p} \times \mathbf{k}) = \mathbf{k} \cdot \nabla \mathbf{p} - \mathbf{k}\nabla \cdot \mathbf{p} = 0,$$

$$\mathbf{p} \times \nabla A = A' \frac{\mathbf{p} \times (\mathbf{p} \times \mathbf{k})}{p} = A'\left(-\frac{p^2\mathbf{k}}{p}\right) = -A'p\mathbf{k},$$

$$(\mathbf{p} \times \mathbf{k}) \times \nabla B = B' \frac{(\mathbf{p} \times \mathbf{k}) \times (\mathbf{p} \times \mathbf{k})}{p} = 0,$$

and

$$\mathbf{k} \times \nabla C = C' \frac{\mathbf{k} \times (\mathbf{p} \times \mathbf{k})}{p} = C' \frac{\mathbf{p}}{p}.$$

These results, together with Eq. (16.85), when used in Eq. (16.93), give

$$\nabla \times \mathscr{H} = (2A + A'p)\mathbf{k} - C' \frac{\mathbf{p}}{p} = J\mathbf{k}. \tag{16.94}$$

Since $\mathbf{k}$ and $\mathbf{p}$ are perpendicular to each other, Eq. (16.94) requires

$$2A + A'p = J,$$

and

$$C' = 0.$$

The first equation may be integrated as follows:

$$2Ap + A'p^2 = \frac{d}{dp}(Ap^2) = Jp,$$

$$Ap^2 = \tfrac{1}{2}Jp^2 + b,$$

where b = constant.

Thus

$$A = \frac{1}{2} J + \frac{b}{p^2}, \tag{16.95}$$

and from the second equation,

$$C = \text{constant.} \tag{16.96}$$

Finally, therefore, Eq. (16.88) becomes

$$\mathscr{H} = \left(\frac{1}{2} J + \frac{b}{p^2}\right)\mathbf{p} + \frac{d}{p^2}\mathbf{p} \times \mathbf{k} + C\mathbf{k}. \tag{16.97}$$

As J is $4\pi\rho v/c$ inside the cylinder and zero outside, we have for the field inside

$$\mathscr{H}_i = \left(2\pi\rho\frac{v}{c} + \frac{b_i}{p^2}\right)\mathbf{p} + \frac{d_i}{p^2}\mathbf{p} \times \mathbf{k} + C_i\mathbf{k}, \tag{16.98}$$

and for the field outside

$$\mathscr{H}_e = \frac{b_e}{p^2}\mathbf{p} + \frac{d_e}{p^2}\mathbf{p} \times \mathbf{k} + C_e\mathbf{k}. \tag{16.99}$$

Since the magnitude of $\mathbf{p}$ is the distance of the point $\mathbf{r}$ from the axis of the cylinder and since the field must remain finite as $\mathbf{p} \to 0$, it is necessary that b_i and d_i in Eq. (16.98) be zero. Thus

$$\mathscr{H}_i = 2\pi\rho\frac{v}{c}\mathbf{p} + C_i\mathbf{k}, \tag{16.100}$$

$$\mathscr{H}_e = \frac{b_e}{p^2}\mathbf{p} + \frac{d_e}{p^2}\mathbf{p} \times \mathbf{k} + C_e\mathbf{k}. \tag{16.101}$$

As a final condition, these fields must have the same value at the surface of the cylinder, $p = a$ (see Example 2). This means, since $\mathbf{p}$, $\mathbf{k}$, and $\mathbf{p} \times \mathbf{k}$ are independent vectors, that the coefficients of $\mathbf{p}$, $\mathbf{k}$, and $\mathbf{p} \times \mathbf{k}$ in Eqs. (16.100) and (16.101) are respectively equal when $p = a$. Thus

$$2\pi\rho\frac{v}{c} = \frac{b_e}{a^2},$$

$$\frac{d_e}{a^2} = 0,$$

and

$$C_i = C_e = C.$$

Putting b_e, d_e, C_e, and C_i from these equations into Eqs. (16.100) and (16.101), we have

$$\mathscr{H}_i = 2\pi\rho \frac{v}{c} \mathbf{p} + C\mathbf{k} \tag{16.102}$$

$$\mathscr{H}_e = 2\pi\rho \frac{v}{c} \frac{a^2}{p^2} \mathbf{p} + C\mathbf{k} = \frac{2I}{p^2} \mathbf{p} + C\mathbf{k}, \tag{16.103}$$

where $I = \rho(v/c)\pi a^2$ = total current. In order to interpret Eqs. (16.102) and (16.103) properly, it is necessary to recall that **p** is directed around the wire and is equal in magnitude to the distance of the point from the axis. Thus the field, except for an arbitrary constant field in the k-direction, is circular both outside and inside the cylinder. Inside it is directly proportional to the distance from the axis and outside inversely proportional to the distance.

EXAMPLE 4. Find the field produced by a uniform current in an infinite cylindrical conductor, making use of Gauss' and Stokes' theorems.

As in Example 3,

$$\nabla \cdot \mathscr{H} = 0, \tag{16.104}$$

$$\nabla \times \mathscr{H} = J\mathbf{k}, \tag{16.105}$$

where J, **k**, and **p** are as there defined. Also, as before, from considerations of symmetry,

$$\mathscr{H} = A(p)\,\mathbf{p} + B(p)\,\mathbf{p} \times \mathbf{k} + C(p)\,\mathbf{k}. \tag{16.106}$$

Take a cylinder of arbitrary radius p, coaxial with the given cylinder and of length h (see Fig. 16.3), and integrate $\mathscr{H} \cdot d\mathbf{S}$ over the surface of this cylinder. Then

$$\int \mathscr{H} \cdot d\mathbf{S} = \int \nabla \cdot \mathscr{H}\, dV = 0$$

by the use of Gauss' theorem; or

$$\int_{S_3} \mathscr{H} \cdot \mathbf{k}\, dS + \int_{S_1} \mathscr{H} \cdot (-\mathbf{k})\, dS + \int_{S_2} \mathscr{H} \cdot \frac{\mathbf{p} \times \mathbf{k}}{p}\, dS = 0. \tag{16.107}$$

Since $\mathscr{H}$ is the same at the two ends S_1 and S_3 of the cylinder, the first two integrals cancel each other and Eq. (16.107) reduces to

$$\int_{S_2} \mathscr{H} \cdot \frac{\mathbf{p} \times \mathbf{k}}{p}\, dS = 0. \tag{16.108}$$

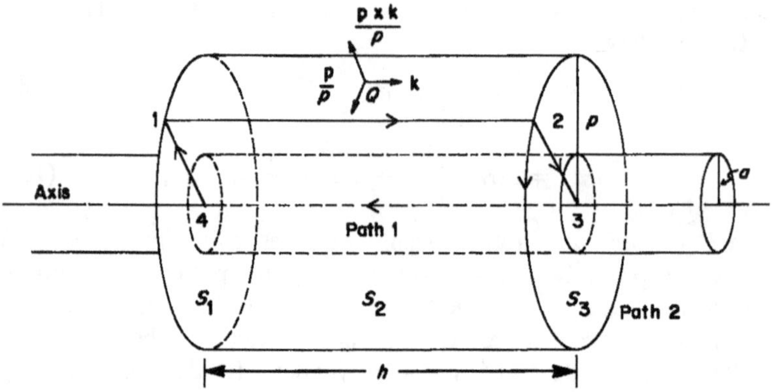

FIG. 16.3. Gaussian surface and Stokes' contours for obtaining the field produced by a uniform current in a cylindrical conductor.

Now

$$\begin{aligned}\mathscr{H} \cdot \mathbf{p} \times \mathbf{k} &= A(p)\, \mathbf{p} \cdot \mathbf{p} \times \mathbf{k} + B(p)\, (\mathbf{p} \times \mathbf{k})^2 + C(p)\, \mathbf{k} \cdot \mathbf{p} \times \mathbf{k} \\ &= B(p)\, [p^2 - (\mathbf{p} \cdot \mathbf{k})^2] = p^2\, B(p)\end{aligned}$$

and hence Eq. (16.108) becomes

$$\int_{S_2} pB(p)\, dS = pB(p) \int_{S_2} dS = 2\pi p^2 h B(p) = 0.$$

This requires that $B(p) = 0$, and thus Eq. (16.106) becomes

$$\mathscr{H} = A(p)\mathbf{p} + C(p)\mathbf{k}. \tag{16.109}$$

We now apply Stokes' theorem, with $\mathscr{H} \cdot d\mathbf{r}$ being integrated along path 1 in Fig. 16.3, and $d\mathbf{\Sigma}$ being an element of surface of the rectangle bounded by this path; thus,

$$\int_{\text{path 1}} \mathscr{H} \cdot d\mathbf{r} = \int \nabla \times \mathscr{H} \cdot d\mathbf{\Sigma} = \int J\mathbf{k} \cdot d\mathbf{\Sigma} = 0.$$

Here we have used Eq. (16.105) and the fact that $\mathbf{k}$ is perpendicular to $d\mathbf{\Sigma}$. Breaking up path 1 into its parts, we have

$$\int_1^2 \mathscr{H} \cdot d\mathbf{r} + \int_2^3 \mathscr{H} \cdot d\mathbf{r} + \int_3^4 \mathscr{H} \cdot d\mathbf{r} + \int_4^1 \mathscr{H} \cdot d\mathbf{r} = 0.$$

Since $\mathscr{H}$ is the same along 2–3 as along 1–4, the second and fourth integrals cancel. Moreover, $\mathscr{H}$ does not change as we move along either the side 1–2 or the side 3–4; therefore

$$\mathscr{H} \cdot \int_1^2 d\mathbf{r} + \mathscr{H}_0 \cdot \int_3^4 d\mathbf{r} = \mathscr{H} \cdot h\mathbf{k} + \mathscr{H}_0 \cdot (-h\mathbf{k}) = 0,$$

where $\mathscr{H}_0$ is the field along the axis of the given cylinder. Thus

$$\mathscr{H} \cdot \mathbf{k} = \mathscr{H}_0 \cdot \mathbf{k},$$

and, by Eq. (16.109), for arbitrary p

$$C(p) = C(0)$$

or

$$C = \text{constant}. \tag{16.110}$$

If the field is to vanish at infinity, we must require that

$$C = 0.$$

Next we apply Stokes' theorem to $\mathscr{H} \cdot d\mathbf{r}$ integrated around the closed path 2, which, as shown in Fig. 16.3, is a circular path of radius p about the axis. We then have

$$\int_{\text{path 2}} \mathscr{H} \cdot d\mathbf{r} = \int_{S_3} (\nabla \times \mathscr{H}) \cdot d\mathbf{S} = \int_{S_3} J\mathbf{k} \cdot \mathbf{k}\, dS = \int_{S_3} J\, dS. \tag{16.111}$$

Suppose, first, that $p < a$; then J is everywhere equal to $4\pi\rho v/c$, and hence

$$\int_{\text{path 2}} \mathscr{H} \cdot d\mathbf{r} = 4\pi\rho \frac{v}{c} \int_{S_3} dS = 4\pi^2 p^2 \rho \frac{v}{c}. \tag{16.112}$$

On the other hand, for $p > a$, since $J = 0$ outside the cylinder, the right-hand integral reduces to an integral over a cross section of the given cylinder. This means, of course, that this integral is the same for all paths in which $p > a$; thus

$$\int_{\text{path 2}} \mathscr{H} \cdot d\mathbf{r} = \begin{cases} 4\pi^2 p^2 \rho \dfrac{v}{c}, & \text{for } p < a \\ 4\pi^2 a^2 \rho \dfrac{v}{c}, & \text{for } p > a. \end{cases} \tag{16.113}$$

Now $d\mathbf{r} = (\mathbf{p}/p)\,dr$ on path 2, and therefore by Eq. (16.109)

$$\int_{\text{path 2}} \mathscr{H} \cdot d\mathbf{r} = \int_{\text{path 2}} A(p)p\,dr = A(p)p\,(2\pi p).$$

Putting this into Eq. (16.113)

$$\begin{aligned} A(p) &= 2\pi\rho\frac{v}{c}, \qquad \text{for} \quad p < a \\ A(p) &= 2\pi\frac{a^2}{p^2}\rho\frac{v}{c}, \qquad \text{for} \quad p > a. \end{aligned} \tag{16.114}$$

If we put these values of $A(p)$ together with $C = 0$, as given by Eq. (16.110), back in Eq. (16.109), we have as our final result

$$\mathscr{H}_i = 2\pi\rho\frac{v}{c}\mathbf{p}$$

and

$$\mathscr{H}_e = 2\pi\frac{a^2}{p^2}\rho\frac{v}{c}\mathbf{p}.$$

These equations agree with Eqs. (16.102) and (16.103) of Example 3 if we impose on the latter equations the condition that the field must approach zero as p approaches infinity.

EXAMPLE 5. A method of producing uniform magnetic fields.*

Suppose we have two infinite parallel cylinders A and B within each of which we have uniform current densities of magnitude $j = \rho v/c$; however, the direction of the current density vector in A is opposite to its direction in B. We may take $\mathbf{k}$ to be a unit vector in the direction of the current in A (out of the paper in Figs. 16.4a and 16.4b) and $-\mathbf{k}$ a unit vector in the direction of the current in B.

Furthermore, suppose that the two cylinders are separated by a distance less than the sum of the radii. Under these circumstances the two principal types of relationship between the cylinders A and B are illustrated in Fig. 16.4 by means of cross sections. In the region common to A and B we would have two currents running in opposite directions.

* I. I. Rabi, *Rev. Sci. Instr.*, **5**, 78 (1934).

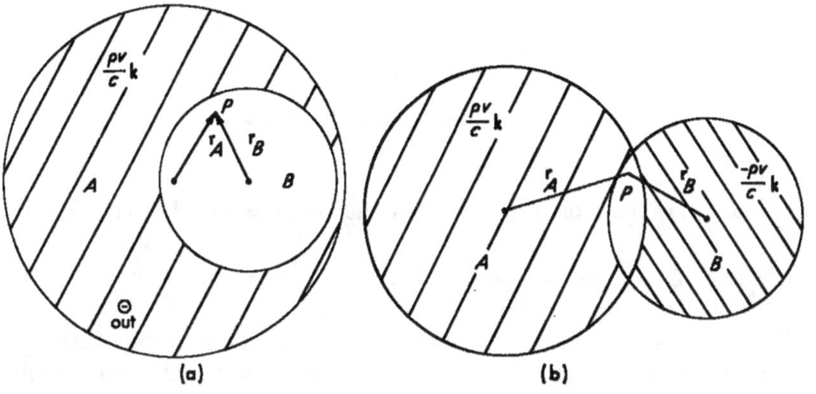

FIG. 16.4. Cross sections of cylindrical conductors that will produce a uniform magnetic field.

The field in the region common to A and B would be a superposition of the fields of both. Since this region is inside both A and B, the contribution of each is given by Eq. (16.102). Thus the total field at a point P in this region is given by

$$\begin{aligned}\mathscr{H} &= 2\pi\rho\frac{v}{c}\mathbf{k}\times\mathbf{r}_A + C\mathbf{k} + 2\pi\rho\frac{v}{c}(-\mathbf{k})\times\mathbf{r}_B + C(-\mathbf{k}) \\ &= 2\pi\frac{\rho v}{c}\mathbf{k}\times(\mathbf{r}_A - \mathbf{r}_B) = \frac{1}{2}J\,\mathbf{k}\times(\mathbf{r}_A - \mathbf{r}_B),\end{aligned} \tag{16.115}$$

where $\mathbf{r}_A$ and $\mathbf{r}_B$ are vectors from some point on the axis of A and some point on the axis of B, respectively, to the point P. Now $\mathbf{r}_A - \mathbf{r}_B$ is a vector drawn between two fixed points on the axes of the two cylinders, and hence is constant. Moreover, $\mathbf{k}\times(\mathbf{r}_A - \mathbf{r}_B)$ is obviously perpendicular to the plane of the axes of the two cylinders A and B, and equal in magnitude to the distance between the axes of A and B. Thus $\mathscr{H}$ in the region common to the cylinders is homogeneous, and directly proportional to the distance between the axes of A and B.

Since the region common to A and B has two current density vectors equal in magnitude but opposite in direction, the currents cancel each other and thus the conducting material in this region can be removed. A uniform magnetic field then exists in this region common to the two cylinders.

17. Energy and Momentum of the Field

The Lagrangian of empty space, in accordance with Eq. (16.13), is

$$L = L_f = -\frac{1}{16\pi}\int F_{\alpha\beta}F_{\alpha\beta}\,dV = \frac{1}{8\pi}\int (\mathscr{E}^2 - \mathscr{H}^2)\,dV. \tag{17.1}$$

We now wish to find the Hamiltonian corresponding to this Lagrangian. To this end we must first recall that the generalized coordinates describing the field were the potentials φ and $\mathbf{A}$. The potentials were connected with the field by the relations (16.1), i.e.,

$$\mathscr{E} = -\nabla\varphi - \frac{1}{c}\frac{\partial \mathbf{A}}{\partial t}, \qquad \mathscr{H} = \nabla \times \mathbf{A}. \tag{17.2}$$

The only generalized velocity entering the Lagrangian is $\dot{\mathbf{A}} = \partial\mathbf{A}/\partial t$. Since each component of $\mathbf{A}$ at each point of space is to be considered as an independent coordinate, there will be a momentum vector corresponding to $\mathbf{A}$ for each point, which may be designated by $\mathbf{p}_A$.

It is convenient* to regard the integral in Eq. (17.1) as the limit of the sum

$$L' = \frac{1}{8\pi}\sum_s (\mathscr{E}_s^2 - \mathscr{H}_s^2)\,\Delta V_s, \tag{17.3}$$

obtained by considering the space as being divided into small regions ΔV_s, where s is the numbering index, the field in each region being considered as essentially homogeneous. The momentum corresponding to each such region is then

$$\mathbf{p}_{A_s} = \frac{\partial L'}{\partial \dot{\mathbf{A}}_s} = -\frac{1}{4\pi c}\mathscr{E}_s\,\Delta V_s. \tag{17.4}$$

It is evident, since $\dot{\varphi}$ is not contained in Eq. (17.3), that the corresponding momentum component p_{φ_s} is zero. Therefore, in accordance with Eq. (2.15) the Hamiltonian is given by

$$H' = -L' + \sum_s \dot{\mathbf{A}}_s\cdot\mathbf{p}_{A_s} = \frac{1}{8\pi}\sum_s\left(-\mathscr{E}_s^2 + \mathscr{H}_s^2 - \frac{2}{c}\dot{\mathbf{A}}_s\cdot\mathscr{E}_s\right)\Delta V_s; \tag{17.5}$$

* The straightforward transformation involves the use of the calculus of functionals, which we wish to avoid for pedagogical reasons.

or, going to the limit, and then substituting for $\dot{A}$ from Eq. (17.2),*

$$H = E = \frac{1}{8\pi}\int\left(-\mathscr{E}^2 + \mathscr{H}^2 - \frac{2}{c}\dot{\mathbf{A}}\cdot\mathscr{E}\right)dV$$

$$= \frac{1}{8\pi}\int(\mathscr{E}^2 + \mathscr{H}^2 + 2\mathscr{E}\cdot\nabla\varphi)\,dV. \tag{17.6}$$

The last term of this expression can be transformed as follows:

$$\frac{1}{4\pi}\int\mathscr{E}\cdot\nabla\varphi\,dV = \frac{1}{4\pi}\int\nabla\cdot(\varphi\mathscr{E})\,dV - \frac{1}{4\pi}\int\varphi\nabla\cdot\mathscr{E}\,dV. \tag{17.7}$$

The first term of the right-hand member transforms into a surface integral taken over the boundary of the region. Since the region considered is the entire space, the boundary may be assumed to be an infinite sphere. Assuming as we shall that $\varphi\mathscr{E}$ vanishes at infinity faster than $1/r^2$, this surface integral over the infinite sphere vanishes. The last term in Eq. (17.7) also vanishes because, in accordance with Eq. (16.26), $\nabla\cdot\mathscr{E}$ vanishes in empty space. Thus, Eq. (17.6) becomes

$$E = \frac{1}{8\pi}\int(\mathscr{E}^2 + \mathscr{H}^2)\,dV, \tag{17.8}$$

which justifies considering the *energy density* to be given by

$$w = \frac{1}{8\pi}(\mathscr{E}^2 + \mathscr{H}^2). \tag{17.9}$$

It is to be noted, however, that any expression differing from w by the divergence of a vector vanishing at infinity faster than $1/r^2$ is equally well justified. Thus, the less convenient expression $(\mathscr{E}^2 + \mathscr{H}^2 + 2\,\mathscr{E}\cdot\nabla\varphi)/8\pi$ could also be used.

The rate of change of the energy density at any point can be obtained by partial differentiation of Eq. (17.9) with respect to time. Thus,

$$\frac{\partial w}{\partial t} = \frac{1}{4\pi}\left(\mathscr{E}\cdot\frac{\partial\mathscr{E}}{\partial t} + \mathscr{H}\cdot\frac{\partial\mathscr{H}}{\partial t}\right), \tag{17.10}$$

which, with the help of the field equations (16.25) and (16.26), becomes for empty space

$$\frac{\partial w}{\partial t} = \frac{c}{4\pi}(\mathscr{E}\cdot\nabla\times\mathscr{H} - \mathscr{H}\cdot\nabla\times\mathscr{E}) = -\frac{c}{4\pi}\nabla\cdot(\mathscr{E}\times\mathscr{H}). \tag{17.11}$$

* The letter E is used to represent the energy and is customarily used in place of H when the dynamical variable aspect of the Hamiltonian is to be emphasized.

Introducing, for convenience,

$$\boldsymbol{\gamma} = \frac{c}{4\pi} \boldsymbol{\mathscr{E}} \times \boldsymbol{\mathscr{H}}, \tag{17.12}$$

we have

$$-\frac{d}{dt}\int w\,dV = -\int \frac{\partial w}{\partial t}\,dV = \int \nabla \cdot \boldsymbol{\gamma}\,dV = \int \boldsymbol{\gamma} \cdot d\mathbf{S}, \tag{17.13}$$

The integrations here may be performed over the volume of any portion of space and its boundary surface, respectively, and therefore this equation shows that the rate of decrease of the energy contained within any closed surface is equal to an integral of $\boldsymbol{\gamma}$ over the surface. Thus, $\boldsymbol{\gamma}$ can be considered as having the direction of the energy flow, and its magnitude as giving the amount of the energy flowing per unit of transverse area per unit of time. This vector is usually called *Poynting's vector*. It is evident that Eq. (17.13) would also hold if we replaced $\boldsymbol{\gamma}$ by $\boldsymbol{\gamma}' = \boldsymbol{\gamma} + \nabla \times \boldsymbol{\Psi}$, where $\boldsymbol{\Psi}$ is an arbitrary vector function of $\mathbf{r}$ and t.

Let us now consider the momentum of the electromagnetic field. If we use Eq. (2.23) and regard time as unvaried, we can write

$$\delta W = p_j\,\delta q_j. \tag{17.14}$$

Going back to the device of dividing space into small regions ΔV_s, used at the beginning of this section, we can rewrite this equation in the form

$$\delta W' = \sum_s \mathbf{p}_{A_s} \cdot \delta \mathbf{A}_s. \tag{17.15}$$

Here use has been made of the fact $p_{\varphi_s} = 0$. Substitution from Eq. (17.4) gives

$$\delta W' = -\frac{1}{4\pi c}\sum_s \boldsymbol{\mathscr{E}}_s \cdot \delta \mathbf{A}_s \Delta V_s, \tag{17.16}$$

or, in the limit,

$$\delta W' = -\frac{1}{4\pi c}\int \boldsymbol{\mathscr{E}} \cdot \delta \mathbf{A}\,dV. \tag{17.17}$$

Suppose now that $\delta\mathbf{A}$ is a result of a displacement $\boldsymbol{\epsilon}$ of the entire system. Then, in accordance with Eq. (3.13),

$$\delta W = \mathbf{P} \cdot \boldsymbol{\epsilon}, \tag{17.18}$$

where $\mathbf{P}$ is the total momentum of the system. If we express $\delta\mathbf{A}$ in terms of $\boldsymbol{\epsilon}$, and compare Eqs. (17.17) and (17.18), we can obtain an expression for P.

When the entire system is displaced by the amount $d\mathbf{r} = \boldsymbol{\epsilon}$ the field at the point $\mathbf{r}$ after the displacement will be the same as the field at the point $\mathbf{r} - \boldsymbol{\epsilon}$ before the displacement. Thus,

$$\delta A = -\frac{\partial A}{\partial x_i}\,dx_i = -\boldsymbol{\epsilon}\cdot\nabla\mathbf{A}. \tag{17.19}$$

Substitution of this value into Eq. (17.17) gives

$$\begin{aligned}\delta W &= \frac{1}{4\pi c}\int \mathscr{E}\cdot(\boldsymbol{\epsilon}\cdot\nabla\mathbf{A})\,dV \\ &= \frac{\boldsymbol{\epsilon}}{4\pi c}\cdot\int[\mathscr{E}\times(\nabla\times\mathbf{A}) + \mathscr{E}\cdot\nabla\mathbf{A}]\,dV,\end{aligned} \tag{17.20}$$

where, in performing the last transformation, we made use of the formula (A3.30). Since $\boldsymbol{\epsilon}$ is an arbitrary small vector, a comparison of Eqs. (17.18) and (17.20) yields

$$\mathbf{P} = \frac{1}{4\pi c}\int \mathscr{E}\times\mathscr{H}\,dV + \frac{1}{4\pi c}\int \mathscr{E}\cdot\nabla\mathbf{A}\,dV. \tag{17.21}$$

The last integral may be transformed, using Eq. (A4.10), as follows:

$$\int \mathscr{E}\cdot\nabla\mathbf{A}\,dV = \int \mathbf{A}\,\mathscr{E}\cdot d\mathbf{S} - \int \mathbf{A}\,\nabla\cdot\mathscr{E}\,dV. \tag{17.22}$$

The first term on the right vanishes because the boundary is taken at infinity, where both $\mathbf{A}$ and $\mathscr{E}$ vanish; the second term vanishes because $\nabla\cdot\mathscr{E} = 0$ in empty space. Thus,

$$\mathbf{P} = \frac{1}{4\pi c}\int \mathscr{E}\times\mathscr{H}\,dV, \tag{17.23}$$

and we are justified in regarding

$$\mathbf{G} = \frac{\boldsymbol{\Upsilon}}{c^2} = \frac{1}{4\pi c}\,\mathscr{E}\times\mathscr{H} \tag{17.24}$$

as the *momentum density* of the electromagnetic field. It should be noted, however, that any vector differing from $\mathbf{G}$ by a vector the volume integral of which over the entire space vanishes is equally justified by this discussion; thus, for example, although with less convenience, we could have used $(\mathscr{E}\times\mathscr{H} + \mathscr{E}\cdot\nabla\mathbf{A})/4\pi c$ as the momentum density.

More powerful than the method we used above is the four-dimensional covariant method. We recall Eqs. (13.1) and (13.2), which can be summarized as follows:

$$\delta W = P_\alpha \, \delta x_\alpha , \tag{17.25}$$

where

$$(P_1, P_2, P_3) = \mathbf{P} \qquad \text{and} \qquad P_4 = \frac{iE}{c} . \tag{17.26}$$

It is further to be recalled that the variation of W implied in Eq. (17.25) was supposed to be produced by varying the upper limit of the integral $\int L \, dt$. In introducing a covariant form of Hamilton's principle we replaced this integral by an integral over a four-dimensional region Ω bounded by two nonintersecting three-dimensional hypersurfaces. The variation implied in Eq. (17.25) corresponds to a displacement of one of these hypersurfaces, that is, changing the coordinates of all points of the hypersurface by an amount δx_α, without in any other way altering the system. In particular, the distribution of the potentials φ_β on the hypersurface must be the same after the displacement as before.

We calculated such a variation of W in the previous section, obtaining Eq. (16.21). In the present case, however, we are dealing with empty space, so that $s_\beta = 0$. Since the field equations (16.22) are assumed to hold, Eq. (16.21) becomes

$$\delta W = \frac{1}{4\pi i c}\left[\int_\Omega \frac{\partial}{\partial x_\gamma}(F_{\beta\gamma} \, \delta\varphi_\beta) \, d\Omega - \frac{1}{4}\int_{x_\alpha} F_{\gamma\beta}^2 \, \delta x_\alpha \, dS_\alpha\right]. \tag{17.27}$$

The first integral, by a generalization of Gauss' theorem, transforms immediately into a surface integral over the boundary of the region Ω. However, since $\delta\varphi_\beta$ is zero over the undisplaced surface, the integration need be taken only over the hypersurface that undergoes the displacement, but in its undisplaced position. Thus,

$$\delta W = \frac{1}{4\pi i c}\left[\int_{x_\alpha} F_{\beta\gamma} \, \delta\varphi_\beta \, dS_\gamma - \frac{1}{4}\int_{x_\alpha} F_{\gamma\beta}^2 \, \delta x_\alpha \, dS_\alpha\right], \tag{17.28}$$

where the integrations are performed over the hypersurface, the coordinates of which are x_α. Our problem, then, is solved as soon as we determine the value of $\delta\varphi_\beta$ on this hypersurface.

We shall designate, for the moment, the unvaried potentials by a superscript 0, so that the potential distribution before the displacement, as function of coordinates, may be designated by $\varphi_\beta^0(x_\alpha)$. The required variation is then

$$\delta\varphi_\beta = \varphi_\beta(x_\alpha) - \varphi_\beta^0(x_\alpha), \tag{17.29}$$

where the varied potential is without the superscript. Since the potential φ_β treated as a field in our four-dimensional space is displaced along with the hypersurface we see that

$$\varphi_\beta(x_\alpha + \delta x_\alpha) = \varphi_\beta^0(x_\alpha), \tag{17.30}$$

and hence from Eq. (17.29)

$$\delta\varphi_\beta = \varphi_\beta(x_\alpha) - \varphi_\beta(x_\alpha + \delta x_\alpha); \tag{17.31}$$

therefore, to first order in the variations

$$\delta\varphi_\beta = -\frac{\partial\varphi_\beta}{\partial x_\alpha}\,\delta x_\alpha . \tag{17.32}$$

Thus, Eqs. (17.25) and (17.28) become

$$\delta W = P_\alpha\,\delta x_\alpha = \frac{-\delta x_\alpha}{4\pi i c}\left[\int_{x_\alpha} F_{\beta\gamma}\frac{\partial\varphi_\beta}{\partial x_\alpha}\,dS_\gamma + \frac{1}{4}\int_{x_\alpha} F_{\gamma\beta}^2\,dS_\alpha\right];$$

and, since δx_α is arbitrary, this requires that

$$P_\alpha = \frac{1}{4\pi i c}\left[\int_{x_\alpha} F_{\gamma\beta}\frac{\partial\varphi_\beta}{\partial x_\alpha}\,dS_\gamma - \frac{1}{4}\int_{x_\alpha} F_{\gamma\beta}^2\,dS_\alpha\right].$$

We shall now transform the first of the two integrals so as to express P_α in terms of only the field tensor $F_{\alpha\beta}$. This is done as follows:

$$\begin{aligned}\int_{x_\alpha} F_{\gamma\beta}\frac{\partial\varphi_\beta}{\partial x_\alpha}\,dS_\gamma &= \int_{x_\alpha} F_{\gamma\beta}\left(F_{\alpha\beta} + \frac{\partial\varphi_\alpha}{\partial x_\beta}\right)dS_\gamma \\ &= \int_{x_\alpha} F_{\alpha\beta}F_{\gamma\beta}\,dS_\gamma + \int_{x_\alpha}\frac{\partial}{\partial x_\beta}(F_{\gamma\beta}\varphi_\alpha)\,dS_\gamma - \int_{x_\alpha}\varphi_\alpha\frac{\partial F_{\gamma\beta}}{\partial x_\beta}\,dS_\gamma .\end{aligned} \tag{17.33}$$

The last term of the right-hand member vanishes by virtue of the field equations for empty space, Eq. (16.22) with $s_\beta = 0$. It will be shown below that the second integral also vanishes.

Let us consider together with this integral an integral designated by the subscript 0 involving the same integrand, but integrated over an instantaneous space, i.e., a hyperplane perpendicular to the time axis. Neglecting an integral over a cylindrical hypersurface at infinity, which is permissible when the field vanishes sufficiently rapidly at infinity, we have

$$-\int_0 \frac{\partial}{\partial x_\gamma}(F_{\gamma\beta}\varphi_\alpha)\,dS_\beta + \int_{x_\alpha}\frac{\partial}{\partial x_\gamma}(F_{\gamma\beta}\varphi_\alpha)\,dS_\beta = \int\frac{\partial^2(F_{\gamma\beta}\varphi_\alpha)}{\partial x_\beta\,\partial x_\gamma}\,d\Omega = 0,$$

the integration in the last integral being over the four-dimensional volume V between the two hypersurfaces.* The latter is zero as can be shown by using the same type of argument as was used in deriving Eq. (16.31). Thus

$$\int_{x_\alpha} \frac{\partial}{\partial x_\gamma}(F_{\gamma\beta}\varphi_\alpha)\,dS_\beta = \int_0 \frac{\partial}{\partial x_\gamma}(F_{\gamma 4}\varphi_\alpha)\,dS_4$$

$$= \int_0 \frac{\partial}{\partial x_k}(F_{k4}\varphi_\alpha)\,dV = \int_S F_{k4}\varphi_\alpha\,dS_k,$$

dS_k being an element of the surface S at infinity bounding the volume V. If $F_{k4}\varphi_\alpha$ goes to zero as the distance approaches infinity faster than the reciprocal of the square of the distance, then the last integral is zero. In this case

$$\int_{x_\alpha} \frac{\partial}{\partial x_\gamma}(F_{\gamma\beta}\varphi_\alpha)\,dS_\gamma = 0.$$

We have thus shown that

$$\int_{x_\alpha} F_{\gamma\beta}\frac{\partial\varphi_\beta}{\partial x_\alpha}\,dS_\gamma = \int_{x_\alpha} F_{\alpha\beta}F_{\gamma\beta}\,dS_\gamma.$$

Moreover, since the hypersurface designated by x_α is any hypersurface with spacelike intervals, we shall drop the subscript x_α on the integrals. Thus, by Eq. (17.33)

$$P_\alpha = \frac{1}{4\pi ic}\int\left(F_{\alpha\beta}F_{\gamma\beta}\,dS_\gamma - \frac{1}{4}F_{\gamma\beta}F_{\gamma\beta}\,dS_\alpha\right).$$

With a change of summation indices this becomes

$$P_\alpha = \frac{1}{4\pi ic}\int\left(F_{\alpha\gamma}F_{\beta\gamma}\,dS_\beta - \frac{1}{4}F_{\epsilon\gamma}F_{\epsilon\gamma}\,dS_\alpha\right) = \int T_{\alpha\beta}\,dS_\beta, \qquad (17.34)$$

where

$$T_{\alpha\beta} = \frac{1}{4\pi ic}\left(F_{\alpha\gamma}F_{\beta\gamma} - \frac{1}{4}\delta_{\alpha\beta}F_{\epsilon\gamma}F_{\epsilon\gamma}\right). \qquad (17.35)$$

The tensor $T_{\alpha\beta}$ is called the *energy–momentum tensor*. It is a symmetrical tensor, i.e.,

$$T_{\alpha\beta} = T_{\beta\alpha}. \qquad (17.36)$$

* In case the two hypersurfaces intersect, the four-dimensional volume between them is divided into two parts lying on each side of the surface of intersection. The equation above then holds for each of these portions, and the final result is the same.

Also, since

$$\delta_{\alpha\alpha} \equiv \sum_{\alpha=1}^{4} \delta_{\alpha\alpha} = 4,$$

we have

$$T_{\alpha\alpha} = \frac{1}{4\pi ic}\left[F_{\alpha\gamma}F_{\alpha\gamma} - \frac{1}{4}\,\delta_{\alpha\alpha}F_{\epsilon\gamma}F_{\epsilon\gamma}\right] = 0. \tag{17.37}$$

Thus the *trace* of the energy–momentum tensor vanishes.

The most important property of $T_{\alpha\beta}$ for empty space is the following:

$$\frac{\partial T_{\alpha\beta}}{\partial x_\beta} = 0, \tag{17.38}$$

which can be verified by direct substitution. Thus

$$\begin{aligned} 4\pi ic\,\frac{\partial T_{\alpha\beta}}{\partial x_\beta} &= \frac{\partial}{\partial x_\beta}\left[F_{\alpha\gamma}F_{\beta\gamma} - \frac{1}{4}\,\delta_{\alpha\beta}F_{\epsilon\gamma}F_{\epsilon\gamma}\right] \\ &= \frac{\partial F_{\alpha\gamma}}{\partial x_\beta}F_{\beta\gamma} + F_{\alpha\gamma}\frac{\partial F_{\beta\gamma}}{\partial x_\beta} - \frac{1}{4}\frac{\partial}{\partial x_\alpha}(F_{\epsilon\gamma}F_{\epsilon\gamma}). \end{aligned} \tag{17.39}$$

The second term vanishes in empty space by the second pair of Maxwell's equations, Eq. (16.22); therefore,

$$4\pi ic\,\frac{\partial T_{\alpha\beta}}{\partial x_\beta} = \frac{\partial F_{\alpha\gamma}}{\partial x_\beta}F_{\beta\gamma} - \frac{1}{2}\frac{\partial F_{\epsilon\gamma}}{\partial x_\alpha}F_{\epsilon\gamma} = F_{\beta\gamma}\left(\frac{\partial F_{\alpha\gamma}}{\partial x_\beta} - \frac{1}{2}\frac{\partial F_{\beta\gamma}}{\partial x_\alpha}\right).$$

Making use now of the first pair of Maxwell's equations, in the form of Eq. (16.9), we obtain

$$\begin{aligned} 4\pi ic\,\frac{\partial T_{\alpha\beta}}{\partial x_\beta} &= F_{\beta\gamma}\left[\frac{\partial F_{\alpha\gamma}}{\partial x_\beta} + \frac{1}{2}\left(\frac{\partial F_{\alpha\beta}}{\partial x_\gamma} + \frac{\partial F_{\gamma\alpha}}{\partial x_\beta}\right)\right] \\ &= \frac{1}{2}F_{\beta\gamma}\left(\frac{\partial F_{\alpha\gamma}}{\partial x_\beta} + \frac{\partial F_{\alpha\beta}}{\partial x_\gamma}\right) = \frac{1}{2}F_{\beta\gamma}\frac{\partial F_{\alpha\gamma}}{\partial x_\beta} - \frac{1}{2}F_{\gamma\beta}\frac{\partial F_{\alpha\beta}}{\partial x_\gamma}. \end{aligned} \tag{17.40}$$

The right-hand side of Eq. (17.40) vanishes since the two terms differ only in sign and the labeling of their summation indices. Thus Eq. (17.38) holds.

When, in any particular coordinate system, the hypersurface of Eq. (17.34) is a hyperplane perpendicular to the fourth coordinate axis, the hypersurface is an ordinary three-dimensional instantaneous space. The hypersurface element dS_β has, then, only $dS_4 = dV$ as a nonvanishing component, and Eq. (17.34) becomes

$$P_\alpha = \int T_{\alpha 4}\, dV. \tag{17.41}$$

Thus,

$$T_{\alpha 4} = \textit{density of momentum four-vector } P_\alpha. \tag{17.42}$$

Accordingly, since $P_4 = iE/c = (i/c)\int w\, dV$, by Eqs. (15.9) and (17.35)

$$w = \frac{cT_{44}}{i} = -\frac{1}{4\pi}\left[F_{4\gamma}F_{4\gamma} - \frac{1}{4}\delta_{44}F_{\epsilon\gamma}F_{\epsilon\gamma}\right] = \frac{1}{8\pi}(\mathscr{E}^2 + \mathscr{H}^2), \tag{17.43}$$

in agreement with Eq. (17.9). Letting $\alpha = k$, where i, j, and k are the indices 1, 2, 3 in cyclic order, in Eq. (17.41) and using Eq. (17.35), we have for the kth component of the momentum density G

$$G_k = T_{k4} = \frac{1}{4\pi ic}F_{k\gamma}F_{4\gamma} = \frac{1}{4\pi ic}(F_{ki}F_{4i} + F_{kj}F_{4j})$$

$$= \frac{1}{4\pi c}(\mathscr{H}_j\mathscr{E}_i - \mathscr{H}_i\mathscr{E}_j) = \frac{1}{4\pi c}(\mathscr{E} \times \mathscr{H})_k, \tag{17.44}$$

which is in agreement with Eq. (17.24).

Let us investigate closer the meaning of Eq. (17.38). For $\alpha = 4$, the equation becomes

$$\frac{\partial T_{44}}{\partial \tau} + \frac{\partial T_{4k}}{\partial x_k} = 0, \tag{17.45}$$

where, as is usual, we use the Latin summation index for the summation over the values $k = 1$, 2, and 3. On substituting from Eqs. (17.43) and (17.44), and remembering that $T_{\alpha\beta} = T_{\beta\alpha}$, we have

$$\frac{\partial w}{\partial t} + \nabla \cdot \gamma = 0, \tag{17.46}$$

which is just Eq. (17.11).

As we have seen, Eq. (17.46) expresses the conservation of energy within an isolated system, since it leads to Eq. (17.13), which states that the rate of diminution of the energy contained within a closed surface is equal to the energy flow through the surface. A similar equation should exist for the law of conservation of momentum, and, in fact, Eq. (17.38) for $\alpha = k = 1, 2, 3$ gives us the desired result. Thus, in this case

$$\frac{\partial T_{k4}}{\partial \tau} + \frac{\partial T_{k\ell}}{\partial x_\ell} = 0, \tag{17.47}$$

or

$$\frac{\partial G_k}{\partial t_\ell} + \frac{\partial}{\partial x_\ell}(icT_{k\ell}) = 0. \tag{17.48}$$

If, for convenience, we put

$$\Pi_{k\ell} = icT_{k\ell}, \tag{17.49}$$

then

$$\frac{\partial G_k}{\partial t} + \frac{\partial \Pi_{k\ell}}{\partial x_\ell} = 0. \tag{17.50}$$

From this it follows that for any portion of space

$$-\frac{d}{dt}\int G_k\, dV = \int \frac{\partial \Pi_{k\ell}}{\partial x_\ell}\, dV = \int \Pi_{k\ell}\, dS_\ell, \tag{17.51}$$

where dS_ℓ is now the ℓth component of the surface element dS of the boundary of that portion of space being considered. Equation (17.51) shows that $\Pi_{k\ell}$ is the density of flow of momentum; more precisely, it is the rate at which the momentum having the direction of X_k-axis flows in the direction of X_ℓ-axis per unit of transverse area per unit of time. The tensor $-\Pi_{k\ell}$ is sometimes called the *Maxwell stress tensor.*

Energy and momentum of field with particles present. Let us now review our derivation of Eq. (17.34) in order to see what modifications, if any, are necessary when our system contains particles. Since we are, at present, interested only in the momentum and the energy of the field, we must consider the motion of the particles as given, so that coordinates of the particles are not to be varied. We need, then, to use only $W_{pf} + W_f$, as was done in Eq. (16.19). Returning to the variation of this quantity computed in Section 16, and still regarding the field equations (16.22) as satisfied, Eq. (16.21) becomes

$$\delta W = \frac{1}{4\pi ic}\int_\Omega \frac{\partial}{\partial x_\gamma}(F_{\beta\gamma}\varphi_\beta)\, d\Omega + \frac{1}{ic}\int\left(s_\beta\varphi_\beta - \frac{1}{16\pi}F^2_{\gamma\beta}\right)\delta x_\alpha\, dS_\alpha. \tag{17.52}$$

This differs from the right-hand member of Eq. (17.27) only by the term

$$\frac{1}{ic}\int s_\beta\varphi_\beta\, \delta x_\alpha\, dS_\alpha,$$

which would contribute to P_α an additional term

$$\frac{1}{ic}\int s_\beta \varphi_\beta \, dS_\alpha . \tag{17.53}$$

Thus, P_α is now

$$P_\alpha = \frac{1}{4\pi ic}\left(\int F_{\gamma\beta}\frac{\partial \varphi_\beta}{\partial x_\alpha} dS_\gamma - \frac{1}{4}\int F_{\gamma\beta}^2 \, dS_\alpha + 4\pi\int s_\beta \varphi_\beta \, dS_\alpha\right). \tag{17.54}$$

The transformation of the first integral of this expression, given in Eq. (17.33), still holds. The middle term of the right-hand member of Eq. (17.33) vanishes as before, but the last term becomes

$$-\int \varphi_\alpha \frac{\partial F_{\gamma\beta}}{\partial x_\beta} dS_\gamma = -4\pi\int \varphi_\alpha s_\gamma \, dS_\gamma .$$

Thus, we will still have

$$P_\alpha = \int T_{\alpha\beta} \, dS_\beta , \tag{17.55}$$

provided we now set

$$T_{\alpha\beta} = \frac{1}{4\pi ic}\left(F_{\alpha\gamma}F_{\beta\gamma} - \frac{1}{4}\delta_{\alpha\beta}F_{\epsilon\gamma}F_{\epsilon\gamma}\right) + \frac{1}{ic}(\varphi_\gamma s_\gamma \delta_{\alpha\beta} - \varphi_\alpha s_\beta). \tag{17.56}$$

We see, therefore, that $T_{\alpha\beta}$ differs from the previous expression, Eq. (17.35), only by terms that vanish outside of the charges. Thus, for points outside of matter the old expression for $T_{\alpha\beta}$ is the same as that for a field without particles.

In order to gain further familiarity with Eq. (17.50), let us calculate the force acting on a body in otherwise empty space. Let us take the integration in Eq. (17.41) to be extended over the space outside of the body. The rate of change of the momentum in this space is the force acting upon the field outside the body. In accordance with Newton's third law, this force must be the negative of the force acting upon the body. Thus, by Eqs. (17.41) and (17.44), the force acting upon the body is

$$F_k = -\frac{d}{dt}\int G_k \, dV. \tag{17.57}$$

This rate of change of the momentum in the space considered may be regarded as consisting of two parts, the change due to the change of the momentum density and the change due to the motion of the boundary.

The first part is obviously $\int (\partial G_k/\partial t)\, dV$; the second is $\int G_k \mathbf{v} \cdot d\mathbf{S}$, since $\mathbf{v} \cdot d\mathbf{S}$ is the volume per unit of time swept out by the element of the boundary surface $d\mathbf{S}$. Here $\mathbf{v}$ is the velocity of this element of the boundary. Thus

$$F_k = -\int \frac{\partial G_k}{\partial t}\, dV - \int G_k \mathbf{v} \cdot d\mathbf{S}.$$

The surface element $d\mathbf{S}$ is here regarded as an element of the surface enclosing the space outside of the body, and is thus positive when directed into the body.* It is convenient to change the convention and to regard $d\mathbf{S}$ as an element of the surface of the body; to do this, we need only to change its sign. Thus, using Eq. (17.50) and Eq. (17.51) with the sign of dS_ℓ changed, we finally obtain

$$F_k = \int \frac{\partial \Pi_{k\ell}}{\partial x_\ell}\, dV + \int G_k \mathbf{v} \cdot d\mathbf{S} = \int (-\Pi_{k\ell} + G_k v_\ell)\, dS_\ell. \tag{17.58}$$

For a body at rest

$$-\Pi_{k\ell}\, dS_\ell = \text{force on an element of surface of the body in the direction of the } X_k\text{-axis.} \tag{17.59}$$

This relation accounts for $-\Pi_{k\ell}$ being referred to as a *stress tensor.*

Let us express $\Pi_{k\ell}$ for a charge-free region ($s_\alpha = 0$) in terms of $\mathscr{E}$ and $\mathscr{H}$. We have, in accordance with Eq. (17.49) and Eq. (17.56),

$$4\pi \Pi_{k\ell} = F_{k\gamma} F_{\ell\gamma} - \tfrac{1}{4}\delta_{k\ell} F_{\epsilon\gamma} F_{\epsilon\gamma}; \tag{17.60}$$

therefore,

$$4\pi \Pi_{11} = F_{1\gamma}^2 - \tfrac{1}{4} F_{\epsilon\gamma}^2 = \tfrac{1}{2}(\mathscr{E}^2 + \mathscr{H}^2) - \mathscr{E}_x^2 - \mathscr{H}_x^2.$$

Similarly, one has

$$4\pi \Pi_{12} = F_{1\gamma} F_{2\gamma} = F_{13} F_{23} + F_{14} F_{24} = -(\mathscr{E}_x \mathscr{E}_y + \mathscr{H}_x \mathscr{H}_y).$$

It is easy to see that, in general,

$$\Pi_{k\ell} = \frac{1}{8\pi} [(\mathscr{E}^2 + \mathscr{H}^2)\delta_{k\ell} - 2(\mathscr{E}_k \mathscr{E}_\ell + \mathscr{H}_k \mathscr{H}_\ell)]. \tag{17.61}$$

* We assume that the field drops off rapidly enough at large distances that there is no contribution from the surface at infinity.

EXAMPLE. Tensor $T_{\alpha\beta}$ may usually be brought to diagonal form.* In that coordinate system in which $T_{\alpha\beta}$ is diagonal,

$$G_k = T_{k4} = 0, \qquad \text{or} \qquad \mathbf{G} = \frac{1}{4\pi c} \mathscr{E} \times \mathscr{H} = 0, \tag{17.62}$$

which shows that $\mathscr{E}$ is parallel to $\mathscr{H}$. This is the case which we considered previously in the example of Section 15. Let, as before, the common direction of the electric and magnetic fields be chosen as the X-axis. Then, using the notation of that example, one has

$$T_{11} = \frac{\Pi_{11}}{ic} = \frac{1}{8\pi ic} [(\mathscr{E}_0^2 + \mathscr{H}_0^2) - 2(\mathscr{E}_{0x}^2 + \mathscr{H}_{0x}^2)]$$
$$= \frac{i}{8\pi c} (\mathscr{E}_0^2 + \mathscr{H}_0^2) = \frac{iw_0}{c}, \tag{17.63}$$

where w_0 is the energy density in this coordinate system.

Similarly we obtain

$$T_{22} = T_{33} = -\frac{iw_0}{c} \qquad \text{and} \qquad T_{44} = T_{11} = \frac{iw_0}{c}. \tag{17.64}$$

Since the principal values of energy–momentum tensor are equal in pairs, we may rotate our coordinate system in the YZ-plane and the XT-plane without spoiling the diagonality of the tensor.† A rotation in the YZ-plane has no effect because the electric and magnetic vectors are both along the X-axis. A rotation in the YZ-plane, which corresponds to shifting to a coordinate system moving uniformly in the direction of the X-axis, gives rise to the Lorentz transformation of Eqs. (15.5). One can quickly check that, since only $\mathscr{E}_x$ and $\mathscr{H}_x$ are different from zero, the primed fields obtained from this Lorentz transformation are the same as the unprimed fields.

Let us consider next the angular momentum of the electromagnetic field. In accordance with Eq. (3.19),

$$\mathbf{M} = \sum_s (\mathbf{r}_s \times \mathbf{p}_s). \tag{17.65}$$

Considering that each element of space dV contains the momentum $G\,dV$, we can generalize Eq. (17.65) into

$$\mathbf{M} = \int \mathbf{r} \times \mathbf{G}\,d\mathbf{V}, \tag{17.66}$$

* The exceptional case when this transformation is not possible is discussed in the example of Section 15. Further light is also thrown on the subject by the discussion in Section 23.

† See discussion of the reduction of a symmetrical tensor to diagonal form in Appendix 5.

for any portion of space. Written in terms of components this becomes

$$M_{k\ell} = \int (x_k G_\ell - x_\ell G_k)\, dV = \int (x_k T_{\ell 4} - x_\ell T_{k4})\, dS_4, \qquad (17.67)$$

where

$$M_{12} = M_z, \qquad M_{23} = M_x, \qquad M_{31} = M_y. \qquad (17.68)$$

Equation (17.67) is readily generalized into a tensor equation

$$M_{\alpha\beta} = \int_P (x_\alpha T_{\beta\gamma} - x_\beta T_{\alpha\gamma})\, dS_\gamma, \qquad (17.69)$$

where integration is over some spacelike hyperplane P. The three components of **M** are then expected to be given by the independent space components of this antisymmetrical tensor of second rank.

In a coordinate system in which the hyperplane P is perpendicular to the time axis and thus represents a three-dimensional volume, the space components of $M_{\alpha\beta}$, as given by Eq. (17.69), reduce to those of Eq. (17.67). On the other hand $M_{\alpha\beta}$ is an antisymmetric tensor, and its law of transformation is the same as that of $F_{\alpha\beta}$, as given by Eq. (15.4). Since the components of **M** are conserved in all coordinate systems, the equations of transformation lead to the conclusion that the remaining components of $M_{\alpha\beta}$ must also be conserved.

This conclusion, derived by a series of generalizations, must be verified directly before it can be accepted. This can be easily done. Thus, consider the difference in the values of $M_{\alpha\beta}$ for two parallel hypersurfaces, say $M^2_{\alpha\beta} - M^1_{\alpha\beta}$, where the superscripts refer to the two hypersurfaces. The two hypersurfaces, together with a hypersurface at infinity, may be considered as the boundaries of the portion of the four-dimensional space between the two hypersurfaces. Since all the quantities involved are considered as vanishing sufficiently fast at infinity, we can write

$$M^2_{\alpha\beta} - M^1_{\alpha\beta} = \int (x_\alpha T_{\beta\gamma} - x_\beta T_{\alpha\gamma})\, dS_\gamma, \qquad (17.70)$$

the integral being taken over the entire boundary of the region Ω included between the two hypersurfaces. By a generalization of Gauss' theorem this becomes

$$M^2_{\alpha\beta} - M^1_{\alpha\beta} = \int \frac{\partial}{\partial x_\gamma} (x_\alpha T_{\beta\gamma} - x_\beta T_{\alpha\gamma})\, d\Omega$$

$$= \int \left(\frac{\partial x_\alpha}{\partial x_\gamma} T_{\beta\gamma} - \frac{\partial x_\beta}{\partial x_\gamma} T_{\alpha\gamma} \right) d\Omega + \int \left(x_\alpha \frac{\partial T_{\beta\gamma}}{\partial x_\gamma} - x_\beta \frac{\partial T_{\alpha\gamma}}{\partial x_\gamma} \right) d\Omega.$$

By Eq. (17.40) and the symmetry of $T_{\alpha\beta}$

$$M^2_{\alpha\beta} - M^1_{\alpha\beta} = \int (\delta_{\alpha\gamma} T_{\beta\gamma} - \delta_{\beta\gamma} T_{\alpha\gamma})\, d\Omega$$

$$= \int (T_{\beta\alpha} - T_{\alpha\beta})\, d\Omega = 0; \tag{17.71}$$

thus,

$$M^2_{\alpha\beta} = M^1_{\alpha\beta} = \text{constant.} \tag{17.72}$$

This, for α and β equal to 1, 2, and 3 and the hypersurfaces perpendicular to the T-axis, leads to the conservation of the angular momentum.

The other three equations contained in Eqs. (17.72) in the coordinate system in which the hyperplane P is perpendicular to the time axis lead to the conclusion that

$$M_{k4} = \int (x_k T_{44} - x_4 T_{k4})\, dS_4$$

$$= \frac{i}{c}\left(\int w x_k\, dV - c^2 t \int G_k\, dV\right) = \text{constant.}$$

This can also be written

$$\int w\mathbf{r}\, dV - c^2 t\mathbf{P} = \textbf{constant.} \tag{17.73}$$

If we introduce

$$\mathbf{R} = \frac{\int w\mathbf{r}\, dV}{\int w\, dV} = \frac{1}{E}\int w\mathbf{r}\, dV, \tag{17.74}$$

then Eq. (17.73), upon being divided by E, becomes

$$\mathbf{R} = \frac{c^2\mathbf{P}}{E} t + \textbf{constant.} \tag{17.75}$$

Since $\mathbf{P}$ and E are conserved,

$$\mathbf{v} = \frac{d\mathbf{R}}{dt} = \frac{c^2\mathbf{P}}{E}, \tag{17.76}$$

which shows that the point whose position vector is $\mathbf{R}$ moves with a constant velocity. Equation (17.76) is analogous to Eq. (12.15) for a free particle. Further discussion of this result is deferred to Section 23.

Problems

1. In what way are the limits of applicability of electrodynamics related to the existence of an elementary particle such as the electron?

2. Write down the differential equations of motion of an electron in an arbitrarily directed pair of constant homogeneous electric and magnetic fields. Express these in coordinate form using any convenient coordinate system. Do not assume $v \ll c$.

3. Electrons are knocked out of a plane metallic surface by γ-rays and are all assumed to have an initial velocity v_0 perpendicular to the surface. If the number ejected per square centimeter per second is assumed to be

$$n(t) = \begin{cases} n_0, & 0 < t < t_0 \\ 0, & t < 0 \quad \text{or} \quad t > t_0, \end{cases}$$

what average charge density $\rho(x, t)$ will be produced? Let x be the distance from the surface and assume all the electrons returning to the surface are captured. What average electric field $\mathscr{E}(x, t)$ will result?

4. Solve problem 3 with the plane metallic surface replaced by a spherical one.

5. Find the solution of problem 3 subject to the following changes: (a) the electrons are all emitted at $t = 0$ and (b) the electrons have a range of initial velocities. Assume the number ejected per square centimeter per unit velocity range is

$$n(v) = \begin{cases} n_0/v_0, & 0 < v < v_0 \\ 0, & v > v_0. \end{cases}$$

6. Solve problem 5 when the plane surface is replaced by a spherical surface.

7. Why is the field of a system of particles in electrodynamics no longer just a mathematical convenience? Give an example to show the necessity of specifying not only the position and velocities of the particles but their fields as well.

8. Show that Eqs. (15.9) and (15.10) are valid.

9. Show that for a general Lorentz transformation the only invariants of the field are those given by Eqs. (15.9) and (15.10).

10. The dual of a tensor $A_{\alpha\beta}$ is given by $\tilde{A}_{\alpha\beta} = \epsilon_{\alpha\beta\gamma\delta}A_{\gamma\delta}$, where $\epsilon_{\alpha\beta\gamma\delta}$ is defined as in Eq. (15.10). Show that one of the invariants of the field under Lorentz transformation is

$$\frac{i}{8} F_{\alpha\beta}\tilde{F}_{\alpha\beta} = \mathscr{E} \cdot \mathscr{H}$$

and write out the components of $\tilde{F}_{\alpha\beta}$ in terms of the components of $\mathscr{E}$ and $\mathscr{H}$.

11. Express the transformation of $\mathscr{E}$ and $\mathscr{H}$ that apply for a general Lorentz transformation.

12. Show that the characteristic equation of the tensor $F_{\alpha\beta}$ is a quadratic in λ with coefficients expressible in terms of the invariants in Eqs. (15.9) and (15.10). Find the eigenvalues of $F_{\alpha\beta}$.

13. Show that Eq. (16.2) is equivalent to Eqs. (16.1).

14. What is *gauge invariance*? To what form does a gauge transformation reduce for problems in electrostatics?

15. Will the method for producing a uniform magnetic field described in Example 5, p. 148, succeed experimentally?

16. Are there any processes in physics, such as positron-electron annihilation, that would destroy the validity of the continuity equation for charge?

17. Show that any antisymmetric tensor F satisfying Eq. (16.9) must be of the form

$$F_{\alpha\beta} = \frac{\partial \phi_\beta}{\partial x_\alpha} - \frac{\partial \phi_\alpha}{\partial x_\beta},$$

where ϕ_α is some 4-vector.

18. The dual of a tensor $A_{\alpha\beta}$ is given by

$$\tilde{A}_{\alpha\beta} = \epsilon_{\alpha\beta\gamma\delta} A_{\gamma\delta},$$

where $\epsilon_{\alpha\beta\gamma\delta}$ is defined as in Eq. (15.10). Show that Eq. (16.9) can then be expressed in the form

$$\frac{\partial \tilde{F}_{\alpha\beta}}{\partial x_\beta} = 0.$$

19. Show that $\mathscr{E}^2 - \mathscr{H}^2$ and $\mathscr{E} \cdot \mathscr{H}$ are the only independent invariants of the electromagnetic field.

20. Show that the addition of the second term in Eq. (16.12) to F does not alter the equations of motion of the field.

21. Show that Eq. (16.9) and the requirement that $F_{\alpha\beta}$ be an antisymmetric tensor are completely equivalent to Eq. (16.2).

22. How is the momentum density of an electromagnetic field related to the Poynting vector?

23. What is the most important property of the energy-momentum tensor $T_{\alpha\beta}$? Why?

24. What form does $T_{\alpha\beta}$ take in a coordinate system in which $\mathscr{E}$ and $\mathscr{H}$ assume their principal values?

25. Show that the momentum density $\mathbf{G}$ of Eq. (17.24) can be replaced by

$$\mathbf{G} = \frac{1}{4\pi c}(\mathscr{E} \times \mathscr{H} + \mathscr{E} \cdot \nabla \mathbf{A})$$

without changing the expression for the total momentum $\mathbf{P}$ inside an arbitrary closed volume.

26. Prove Eq. (17.38) by expressing the field tensor $F_{\alpha\beta}$ in terms of the potential 4-vector φ_α.

27. Determine the energy-momentum tensor for a plane wave propagating in the direction of the positive Z-axis.

28. Calculate, using Maxwell's stress tensor, the force exerted by a given applied electromagnetic field on a particle of charge q that is uniformly distributed over the volume of a sphere of radius a. Assume the particle is momentarily at rest and integrate the stress tensor over the surface of the particle.

29. Determine the components of the momentum 4-vector for the particle in problem 28 by integrating the energy-momentum tensor over all space.

Chapter VI

Radiation

18. Plane Waves

In the special case when we have only the electromagnetic field, that is, when there are no charges present, there arise certain interesting relationships and simplifications. To begin with the Maxwell–Lorentz equations are in this case

$$\frac{1}{c}\frac{\partial \mathscr{H}}{\partial t} = -\nabla \times \mathscr{E}, \qquad \nabla \cdot \mathscr{H} = 0; \tag{18.1}$$

and

$$\frac{1}{c}\frac{\partial \mathscr{E}}{\partial t} = \nabla \times \mathscr{H}, \qquad \nabla \cdot \mathscr{E} = 0. \tag{18.2}$$

Example 1. Let us investigate solutions of these equations for which $\mathscr{E}$ and $\mathscr{H}$ are independent of x and y. When differential operators are applied to such $\mathscr{E}$ and $\mathscr{H}$ they satisfy the following operator equations:

$$\frac{\partial}{\partial x} = \frac{\partial}{\partial y} = 0, \qquad \nabla = \mathbf{k}\frac{\partial}{\partial z},$$

where $\mathbf{k}$ is a unit vector in the positive Z direction. Equations (18.1) and (18.2) then become

$$\frac{1}{c}\frac{\partial \mathscr{H}}{\partial t} = -\mathbf{k} \times \frac{\partial \mathscr{E}}{\partial z}, \qquad \mathbf{k} \cdot \frac{\partial \mathscr{H}}{\partial z} = 0; \tag{18.3}$$

and

$$\frac{1}{c}\frac{\partial \mathscr{E}}{\partial t} = \mathbf{k} \times \frac{\partial \mathscr{H}}{\partial z}, \qquad \mathbf{k} \cdot \frac{\partial \mathscr{E}}{\partial z} = 0. \tag{18.4}$$

These show that

$$\frac{\partial \mathscr{H}_z}{\partial t} = \frac{\partial}{\partial t}(\mathbf{k} \cdot \mathscr{H}) = \mathbf{k} \cdot \frac{\partial \mathscr{H}}{\partial t} = -c\mathbf{k} \cdot \mathbf{k} \times \frac{\partial \mathscr{E}}{\partial z} = 0,$$

$$\frac{\partial \mathscr{H}_z}{\partial z} = \mathbf{k} \cdot \frac{\partial \mathscr{H}}{\partial z} = 0,$$

and similarly

$$\frac{\partial \mathscr{E}_z}{\partial t} = \frac{\partial \mathscr{E}_z}{\partial z} = 0,$$

or that $\mathscr{H}_z$ and $\mathscr{E}_z$ are constants. $\mathscr{E}_z$ and $\mathscr{H}_z$ represent then the Z components of the constant electric and magnetic fields that may be present.

The variable part is thus perpendicular to the direction of $\mathbf{k}$. For this part

$$\mathbf{k} \cdot \mathscr{E} = 0 \qquad \text{and} \qquad \mathbf{k} \cdot \mathscr{H} = 0. \tag{18.5}$$

Using Eqs. (18.3), (18.4), and (18.5),

$$\frac{1}{c^2}\frac{\partial^2 \mathscr{E}}{\partial t^2} = \frac{1}{c}\frac{\partial}{\partial t}\left(\mathbf{k} \times \frac{\partial \mathscr{H}}{\partial z}\right) = \mathbf{k} \times \frac{\partial}{\partial z}\left(\frac{1}{c}\frac{\partial \mathscr{H}}{\partial t}\right) = -\mathbf{k} \times \frac{\partial}{\partial z}\left(\mathbf{k} \times \frac{\partial \mathscr{E}}{\partial z}\right)$$

$$= -\mathbf{k} \times \left(\mathbf{k} \times \frac{\partial^2 \mathscr{E}}{\partial z^2}\right) = \frac{\partial^2 \mathscr{E}}{\partial z^2} - \mathbf{k}\frac{\partial^2}{\partial z^2}(\mathbf{k} \cdot \mathscr{E}) = \frac{\partial^2 \mathscr{E}}{\partial z^2}$$

or

$$\frac{\partial^2 \mathscr{E}}{\partial z^2} - \frac{1}{c^2}\frac{\partial^2 \mathscr{E}}{\partial t^2} = \left(\frac{\partial}{\partial z} + \frac{1}{c}\frac{\partial}{\partial t}\right)\left(\frac{\partial}{\partial z} - \frac{1}{c}\frac{\partial}{\partial t}\right)\mathscr{E} = 0. \tag{18.6}$$

Putting

$$\xi = z + ct \qquad \text{and} \qquad \eta = z - ct, \tag{18.7}$$

so that

$$z = \tfrac{1}{2}(\xi + \eta) \qquad \text{and} \qquad ct = \tfrac{1}{2}(\xi - \eta),$$

we deduce that

$$\frac{\partial}{\partial \xi} = \frac{\partial z}{\partial \xi}\frac{\partial}{\partial z} + \frac{\partial (ct)}{\partial \xi}\frac{\partial}{\partial (ct)} = \frac{1}{2}\left(\frac{\partial}{\partial z} + \frac{1}{c}\frac{\partial}{\partial t}\right) \tag{18.8}$$

and

$$\frac{\partial}{\partial \eta} = \frac{1}{2}\left(\frac{\partial}{\partial z} - \frac{1}{c}\frac{\partial}{\partial t}\right). \tag{18.9}$$

Equation (18.6) then becomes

$$\frac{\partial^2 \mathscr{E}}{\partial \xi \, \partial \eta} = 0. \tag{18.10}$$

The general solution of this equation is obviously

$$\mathscr{E} = \mathbf{f}_1(\xi) + \mathbf{f}_2(\eta) = \mathbf{f}_1(z + ct) + \mathbf{f}_2(z - ct), \tag{18.11}$$

where $\mathbf{f}_1$ and $\mathbf{f}_2$ are arbitrary vector functions with only the restriction of Eq. (18.5), namely,

$$\mathbf{k} \cdot (\mathbf{f}_1 + \mathbf{f}_2) = \mathbf{k} \cdot \mathscr{E} = 0. \tag{18.12}$$

From Eqs. (18.3) and (18.11) it now follows, since

$$c \frac{\partial \mathbf{f}_1}{\partial z} = \frac{\partial \mathbf{f}_1}{\partial t} \qquad \text{and} \qquad c \frac{\partial \mathbf{f}_2}{\partial z} = - \frac{\partial \mathbf{f}_2}{\partial t},$$

that

$$\frac{\partial \mathscr{H}}{\partial t} = -c\mathbf{k} \times \frac{\partial}{\partial z}(\mathbf{f}_1 + \mathbf{f}_2) = -\mathbf{k} \times \left(\frac{\partial \mathbf{f}_1}{\partial t} - \frac{\partial \mathbf{f}_2}{\partial t} \right)$$

$$= - \frac{\partial}{\partial t} [\mathbf{k} \times (\mathbf{f}_1 - \mathbf{f}_2)],$$

and therefore

$$\mathscr{H} = -\mathbf{k} \times (\mathbf{f}_1 - \mathbf{f}_2) + \mathbf{g}(z). \tag{18.13}$$

Substituting this into the first of Eqs. (18.4), we have

$$\frac{1}{c} \frac{\partial \mathscr{E}}{\partial t} = -\mathbf{k} \times \left[\mathbf{k} \times \left(\frac{\partial \mathbf{f}_1}{\partial z} - \frac{\partial \mathbf{f}_2}{\partial z} \right) \right] + \mathbf{k} \times \frac{d\mathbf{g}}{dz},$$

or

$$\frac{\partial \mathscr{E}}{\partial t} = -\mathbf{k} \times \left[\mathbf{k} \times \left(\frac{\partial \mathbf{f}_1}{\partial t} + \frac{\partial \mathbf{f}_2}{\partial t} \right) \right] - c \frac{d\mathbf{g}}{dz} \times \mathbf{k}$$

$$= -\mathbf{k} \cdot \left(\frac{\partial \mathbf{f}_1}{\partial t} + \frac{\partial \mathbf{f}_2}{\partial t} \right) \mathbf{k} + \frac{\partial}{\partial t}(\mathbf{f}_1 + \mathbf{f}_2) - c \frac{d}{dz}(\mathbf{g} \times \mathbf{k}). \tag{18.14}$$

The first term of the right-hand member vanishes by Eq. (18.12) and the second term is $\partial \mathscr{E} / \partial t$ by Eq. (18.11); hence

$$\frac{d}{dz}(\mathbf{k} \times \mathbf{g}) = 0 \qquad \text{or} \qquad \mathbf{k} \times \mathbf{g} = \textbf{constant}. \tag{18.15}$$

By Eqs. (18.5) and (18.13)

$$\mathbf{k} \cdot \mathscr{H} = -\mathbf{k} \cdot \mathbf{k} \times (\mathbf{f}_1 - \mathbf{f}_2) + \mathbf{k} \cdot \mathbf{g} = \mathbf{k} \cdot \mathbf{g} = 0, \tag{18.16}$$

so that

$$\mathbf{k} \times (\mathbf{g} \times \mathbf{k}) = \mathbf{k} \cdot \mathbf{k}\,\mathbf{g} - \mathbf{k} \cdot \mathbf{g}\,\mathbf{k} = \mathbf{g}. \tag{18.17}$$

Substituting into Eq. (18.13), we may write

$$\begin{aligned} \mathscr{H} &= -\mathbf{k} \times (\mathbf{f}_1 - \mathbf{f}_2) + \mathbf{k} \times (\mathbf{g} \times \mathbf{k}) \\ &= -\mathbf{k} \times [(\mathbf{f}_1 - \tfrac{1}{2}\mathbf{g} \times \mathbf{k}) - (\mathbf{f}_2 + \tfrac{1}{2}\mathbf{g} \times \mathbf{k})] \\ &= -\mathbf{k} \times (\mathbf{f}_1' - \mathbf{f}_2'). \end{aligned} \tag{18.18}$$

The transformation from $\mathbf{f}_1$ and $\mathbf{f}_2$ to $\mathbf{f}_1' = \mathbf{f}_1 - \frac{1}{2}\mathbf{g} \times \mathbf{k}$ and $\mathbf{f}_2' = \mathbf{f}_2 + \frac{1}{2}\mathbf{g} \times \mathbf{k}$ does not alter the form of the expression for $\mathscr{E}$; thus

$$\mathscr{E} = \mathbf{f}_1 + \mathbf{f}_2 = \mathbf{f}_1 - \tfrac{1}{2}\mathbf{g} \times \mathbf{k} + \mathbf{f}_2 + \tfrac{1}{2}\mathbf{g} \times \mathbf{k} = \mathbf{f}_1' + \mathbf{f}_2', \tag{18.19}$$

and since $\mathbf{g} \times \mathbf{k} =$ **constant**, $\mathbf{f}_1'$ and $\mathbf{f}_2'$ are again functions of $z + ct$ and $z - ct$, respectively. This shows that $\mathbf{g}$ can be included in the first term of Eq. (18.13); that is, if $\mathbf{f}_1$ and $\mathbf{f}_2$ are properly chosen, $\mathbf{g} = 0$. Therefore we may take

$$\begin{aligned} \mathscr{E} &= \mathbf{f}_1'(z + ct) + \mathbf{f}_2'(z - ct), \\ \mathscr{H} &= -\mathbf{k} \times [\mathbf{f}_1'(z + ct) - \mathbf{f}_2'(z - ct)], \end{aligned}$$

where $\mathbf{f}_1'$ and $\mathbf{f}_2'$ are arbitrary save for the fact that

$$\mathbf{k} \cdot \mathscr{E} = \mathbf{k} \cdot (\mathbf{f}_1' + \mathbf{f}_2') = 0.$$

Let us express each of the vectors $\mathbf{f}_1'$ and $\mathbf{f}_2'$ as the sum of two vectors—one perpendicular and the other parallel to $\mathbf{k}$. The component of $\mathbf{f}_1'$ parallel to $\mathbf{k}$ is $\mathbf{k} \cdot \mathbf{f}_1'\,\mathbf{k}$; therefore

$$\mathbf{f}_1' = \mathbf{f}_1 + \mathbf{k} \cdot \mathbf{f}_1'\mathbf{k},$$

and likewise

$$\mathbf{f}_2' = \mathbf{f}_2 + \mathbf{k} \cdot \mathbf{f}_2'\mathbf{k},$$

where $\mathbf{f}_1$ and $\mathbf{f}_2$ are now the parts of $\mathbf{f}_1'$ and $\mathbf{f}_2'$ perpendicular to $\mathbf{k}$.

Substituting these expressions for $\mathbf{f}_1'$ and $\mathbf{f}_2'$ in the equations for $\mathscr{E}$ and $\mathscr{H}$, we may write

$$\mathscr{E} = \mathbf{f}_1 + \mathbf{f}_2 + \mathbf{k} \cdot (\mathbf{f}_1' + \mathbf{f}_2')\mathbf{k}$$

and

$$\mathscr{H} = -\mathbf{k} \times [\mathbf{f}_1 - \mathbf{f}_2 + \mathbf{k} \cdot (\mathbf{f}_1' - \mathbf{f}_2')\mathbf{k}].$$

On making use of the condition $\mathbf{k} \cdot (\mathbf{f}_1' + \mathbf{f}_2') = 0$ and the fact that $\mathbf{k} \times \mathbf{k} = 0$, we have finally

$$\mathscr{E} = \mathbf{f}_1(z + ct) + \mathbf{f}_2(z - ct), \tag{18.20}$$

$$\mathscr{H} = -\mathbf{k} \times [\mathbf{f}_1(z + ct) - \mathbf{f}_2(z - ct)], \tag{18.21}$$

where $\mathbf{f}_1$ and $\mathbf{f}_2$ are each perpendicular to $\mathbf{k}$.

From these equations we get

$$\begin{aligned}\mathscr{E} \cdot \mathscr{H} &= -(\mathbf{f}_1 + \mathbf{f}_2) \cdot \mathbf{k} \times (\mathbf{f}_1 - \mathbf{f}_2) \\ &= -\mathbf{f}_1 \cdot \mathbf{k} \times \mathbf{f}_1 - \mathbf{f}_2 \cdot \mathbf{k} \times \mathbf{f}_1 + \mathbf{f}_1 \cdot \mathbf{k} \times \mathbf{f}_2 + \mathbf{f}_2 \cdot \mathbf{k} \times \mathbf{f}_2 \\ &= 2\mathbf{f}_1 \cdot \mathbf{k} \times \mathbf{f}_2 \\ &= 0, \qquad \text{when either } \mathbf{f}_1 \text{ or } \mathbf{f}_2 \text{ is zero.}\end{aligned}$$

Therefore, if $\mathscr{E} = \mathbf{f}_1$ or $\mathscr{E} = \mathbf{f}_2$

$$\mathscr{E} \cdot \mathscr{H} = 0 \qquad \text{or} \qquad \mathscr{E} \perp \mathscr{H}. \tag{18.22}$$

For the cross product we get

$$\begin{aligned}\mathscr{E} \times \mathscr{H} &= -(\mathbf{f}_1 + \mathbf{f}_2) \times [\mathbf{k} \times (\mathbf{f}_1 - \mathbf{f}_2)] \\ &= -(\mathbf{f}_1 + \mathbf{f}_2) \cdot (\mathbf{f}_1 - \mathbf{f}_2)\mathbf{k} + (\mathbf{f}_1 + \mathbf{f}_2) \cdot \mathbf{k}\, (\mathbf{f}_1 - \mathbf{f}_2),\end{aligned}$$

which with the help of Eq. (18.12) reduces to

$$\mathscr{E} \times \mathscr{H} = -(f_1^2 - f_2^2)\mathbf{k}. \tag{18.23}$$

Suppose we divide the fields $\mathscr{E}$ and $\mathscr{H}$ into

$$\mathscr{E} = \mathscr{E}_1 + \mathscr{E}_2, \qquad \mathscr{H} = \mathscr{H}_1 + \mathscr{H}_2,$$

where

$$\begin{aligned}\mathscr{E}_1 &= \mathbf{f}_1(z + ct), & \mathscr{E}_2 &= \mathbf{f}_2(z - ct), \\ \mathscr{H}_1 &= -\mathbf{k} \times \mathbf{f}_1(z + ct), & \mathscr{H}_2 &= \mathbf{k} \times \mathbf{f}_2(z - ct).\end{aligned} \tag{18.24}$$

Consider first the field given by $\mathscr{E}_2$ and $\mathscr{H}_2$. The above equations show that, for any $\mathbf{f}_2$, the field vectors $\mathscr{E}_2$ and $\mathscr{H}_2$ are constant on a surface

$$z - ct = b = \text{constant}. \tag{18.25}$$

This is the equation of a plane perpendicular to the Z-axis moving with the speed c in the positive Z direction. The plane is initially at a distance b from the origin. Obviously there exists such a surface for each value of b. Therefore Eq. (18.25) represents a plane wave moving in the direction of $\mathbf{k}$, the Poynting vector being given by

$$\gamma_2 = \frac{c}{4\pi}(\mathscr{E}_2 \times \mathscr{H}_2) = \frac{c}{4\pi}\, \mathbf{f}_2 \times (\mathbf{k} \times \mathbf{f}_2) = \frac{c}{4\pi}\, \mathbf{f}_2^2\mathbf{k}. \tag{18.26}$$

A similar discussion shows that $\mathbf{f}_1(z + ct)$ represents a wave moving in the direction of $-\mathbf{k}$ and having the Poynting vector

$$\gamma_1 = -\frac{c}{4\pi} f_1^2\mathbf{k}.$$

The general solution is then a superposition of the two waves traveling in opposite directions. The Poynting vector for the general solution is (see Eq. (18.23))

$$\Upsilon = \frac{c}{4\pi}(\mathscr{E} \times \mathscr{H}) = \frac{-c}{4\pi}(f_1^2 - f_2^2)\mathbf{k} = \Upsilon_1 + \Upsilon_2,$$

showing that the Poynting vector relating to the flow of energy across an element of surface for the general solution is just the sum of the Poynting vectors representing the individual energy transfers of the component waves. For each of these waves, as can be seen from Eqs. (18.18) and (18.19),

$$\mathscr{H}^2 = (\pm\mathbf{k} \times \mathscr{E})^2 = \mathscr{E}^2 - (\mathbf{k} \cdot \mathscr{E})^2 = \mathscr{E}^2,$$

so that

$$\mathscr{E} = \mathscr{H}. \tag{18.27}$$

We can eliminate either $\mathscr{E}$ or $\mathscr{H}$ from Eqs. (18.1) and (18.2). Thus

$$\frac{1}{c^2}\frac{\partial^2 \mathscr{H}}{\partial t^2} = -\frac{1}{c}\frac{\partial}{\partial t}(\nabla \times \mathscr{E}) = -\frac{1}{c}\nabla \times \left(\frac{\partial \mathscr{E}}{\partial t}\right) = -\nabla \times (\nabla \times \mathscr{H}),$$

which by Eqs. (A3.37) and (18.1) becomes

$$\frac{1}{c^2}\frac{\partial^2 \mathscr{H}}{\partial t^2} = -(\nabla\nabla \cdot \mathscr{H} - \nabla^2\mathscr{H}) = \nabla^2\mathscr{H};$$

hence

$$\nabla^2\mathscr{H} - \frac{1}{c^2}\frac{\partial^2 \mathscr{H}}{\partial t^2} = 0 \quad \text{or} \quad \frac{\partial^2 \mathscr{H}}{\partial x_\alpha^2} = 0. \tag{18.28}$$

Similarly, we have

$$\nabla^2\mathscr{E} - \frac{1}{c^2}\frac{\partial^2 \mathscr{E}}{\partial t^2} = 0 \quad \text{or} \quad \frac{\partial^2 \mathscr{E}}{\partial x_\alpha^2} = 0. \tag{18.29}$$

Equations (18.1) and (18.2) can be written in the invariant form (see Eqs. (16.9) and (16.22)):

$$\frac{\partial F_{\alpha\beta}}{\partial x_\gamma} + \frac{\partial F_{\beta\gamma}}{\partial x_\alpha} + \frac{\partial F_{\gamma\alpha}}{\partial x_\beta} = 0 \tag{18.30}$$

and

$$\frac{\partial F_{\alpha\beta}}{\partial x_\beta} = 0. \tag{18.31}$$

From Eq. (18.30) we get

$$\frac{\partial^2 F_{\alpha\beta}}{\partial x_\gamma^2} = -\frac{\partial}{\partial x_\gamma}\left(\frac{\partial F_{\beta\gamma}}{\partial x_\alpha} + \frac{\partial F_{\gamma\alpha}}{\partial x_\beta}\right) = -\frac{\partial}{\partial x_\alpha}\left(\frac{\partial F_{\beta\gamma}}{\partial x_\gamma}\right) + \frac{\partial}{\partial x_\beta}\left(\frac{\partial F_{\alpha\gamma}}{\partial x_\gamma}\right)$$

or, using Eq. (18.31),

$$\frac{\partial^2 F_{\alpha\beta}}{\partial x_\gamma^2} = 0. \tag{18.32}$$

For the potentials we have by Eqs. (18.31) and (14.23)

$$\frac{\partial F_{\alpha\beta}}{\partial x_\beta} = \frac{\partial}{\partial x_\beta}\left(\frac{\partial \varphi_\beta}{\partial x_\alpha} - \frac{\partial \varphi_\alpha}{\partial x_\beta}\right) = 0$$

or

$$\frac{\partial^2 \varphi_\alpha}{\partial x_\beta^2} = \frac{\partial}{\partial x_\alpha}\left(\frac{\partial \varphi_\beta}{\partial x_\beta}\right).$$

If we assume, as was previously shown possible in Section 16, that

$$\frac{\partial \varphi_\beta}{\partial x_\beta} = 0, \tag{18.33}$$

then

$$\frac{\partial^2 \varphi_\alpha}{\partial x_\beta^2} = 0. \tag{18.34}$$

Thus, if the condition (18.33) is assumed, each component of all the field quantities ($F_{\alpha\beta}$ of Eq. (18.32) and φ_α of Eq. (18.34)) satisfies the equation

$$\frac{\partial^2 f}{\partial x_\alpha^2} = 0 \tag{18.35}$$

called the *wave equation* or the *d'Alembert equation.* The operator

$$\frac{\partial^2}{\partial x_\alpha^2} = \nabla^2 - \frac{1}{c^2}\frac{\partial^2}{\partial t^2} \equiv \square$$

is knows as the *d'Alembertian.*

The wave equation (Eq. 18.35), being a linear homogeneous equation with constant coefficients, must possess solutions of the form

$$f = a e^{\pm i k_\beta x_\beta}. \tag{18.36}$$

Substitution into Eq. (18.35) gives

$$\frac{\partial^2 f}{\partial x_\alpha^2} = a \frac{\partial}{\partial x_\alpha} \frac{\partial}{\partial x_\alpha} e^{\pm i k_\beta x_\beta} = \pm ia \frac{\partial}{\partial x_\alpha} e^{\pm i k_\beta x_\beta} \frac{\partial}{\partial x_\alpha} (k_\gamma x_\gamma)$$

$$= \pm ia k_\alpha \frac{\partial}{\partial x_\alpha} e^{\pm i k_\beta x_\beta} = -a k_\alpha^2 e^{\pm i k_\beta x_\beta} = 0;$$

therefore

$$k_\alpha^2 = k_1^2 + k_2^2 + k_3^2 + k_4^2 = 0$$

or

$$k_4 = \pm i|\mathbf{k}| = \pm ik, \tag{18.37}$$

where $\mathbf{k} = (k_1, k_2, k_3)$. With this notation, and an arbitrary $\mathbf{k}$,

$$k_\beta x_\beta = \mathbf{k} \cdot \mathbf{r} + (\pm ik)ict = \mathbf{k} \cdot \mathbf{r} \mp ckt. \tag{18.38}$$

If, using the upper sign, we substitute this in Eq. (18.36), we have

$$f = ae^{\pm i(\mathbf{k}\cdot\mathbf{r} - ckt)}. \tag{18.39}$$

The other two solutions, obtained by using the lower sign in Eq. (18.38), correspond merely to another choice of $\mathbf{k}$. Thus, if in Eq. (18.39) $\mathbf{k}$ is replaced by $-\mathbf{k}$, we have, since $|\mathbf{k}| = |-\mathbf{k}|$,

$$f = ae^{\pm i(-\mathbf{k}\cdot\mathbf{r} - ckt)} = ae^{\mp i(\mathbf{k}\cdot\mathbf{r} + ckt)},$$

which are these other solutions. Thus two fundamental solutions are

$$f_1 = ae^{i(\mathbf{k}\cdot\mathbf{r} - ckt)} \quad \text{and} \quad f_2 = be^{-i(\mathbf{k}\cdot\mathbf{r} - ckt)}; \tag{18.40}$$

and we can get both fundamental solutions from

$$k_\alpha = (\mathbf{k}, ik). \tag{18.41}$$

If in the expressions (18.40)

$$\mathbf{k} \cdot \mathbf{r} - ckt = \psi = \text{constant},$$

then f_1 and f_2 are constant. The termini of all $\mathbf{r}$ satisfying this equation for any fixed t determine a surface. Further this equation shows us that the components in the direction $\mathbf{k}$ of all $\mathbf{r}$'s drawn to the surface, for each fixed time t, are equal. Thus the surface is a plane perpendicular to $\mathbf{k}$ and at a distance $ct + \psi/k$ from the origin; hence the plane is moving with the speed c. The solutions given in Eq. (18.40) thus represent plane waves moving in the $\mathbf{k}$ direction with the speed c.

Since f_1 and f_2 are periodic they acquire the same values also on each plane, called a *wave front*, of the set

$$\boldsymbol{k}\cdot\mathbf{r} - ckt = \psi + 2\pi n, \qquad n = -\infty, \ldots, -2, -1, 0, 1, 2, \ldots, \infty.$$

The change $\Delta\mathbf{r}$ in position vector $\mathbf{r}$ required to shift to a neighboring wave front ($n \to n + 1$) is thus seen to satisfy the condition

$$\boldsymbol{k}\cdot\Delta\mathbf{r} = 2\pi.$$

If $\Delta\mathbf{r}$ is in the direction of $\boldsymbol{k}$, i.e., perpendicular to the wave front, then its magnitude is called the *wavelength* of the wave and is given by

$$\lambda = |\Delta\mathbf{r}| = \frac{2\pi}{k}.$$

Likewise, if $\mathbf{r}$ is held fixed and t is increased by T, then one shifts from one wave front to another ($n \to n - 1$) if

$$T = \frac{2\pi}{ck}.$$

T is called the *period* of the wave and its reciprocal

$$\nu = c/\lambda$$

is the *frequency* of the wave.

EXAMPLE 2. In a given coordinate system a plane wave is described by the vector $\boldsymbol{k}$; that is, its direction of propagation is that of $\boldsymbol{k}$ and wavelength $\lambda = 2\pi/k$. What is its wavelength in a coordinate system that is moving with speed v in the positive X direction relatively to the original coordinate system?

Since $k_\alpha = (\boldsymbol{k}, ik)$ is a four-vector, $\boldsymbol{k}$ and k/c transform as $\mathbf{r}$ and t, respectively. Therefore (see last equation of Eqs. (11.8))

$$\frac{k'}{c} = \frac{k/c - vk_x/c^2}{\sqrt{1 - v^2/c^2}}, \tag{18.42}$$

where the primed coordinate system is moving with the speed v along the positive X-axis of the unprimed system.

In the unprimed system let the direction of propagation of the wave make an angle θ with the X-axis; that is, let $\cos\theta = k_x/k$. Dividing Eq.

(18.42) by k/c, we obtain

$$\frac{k'}{k} = \frac{1 - vk_x/ck}{\sqrt{1 - v^2/c^2}} = \frac{1 - (v/c)\cos\theta}{\sqrt{1 - v^2/c^2}}$$

and, since $k = 2\pi/\lambda$ and $k' = 2\pi/\lambda'$,

$$\lambda = \left(\frac{1 - (v/c)\cos\theta}{\sqrt{1 - v^2/c^2}}\right)\lambda'. \tag{18.43}$$

If the primed coordinate system is attached to the source of the wave, λ' may be called the *proper* wavelength of the wave, while λ is the wavelength of the wave as measured by an observer in the unprimed system. This change in wavelength, and hence also in the frequency, is referred to as the *Doppler shift*.

In particular, when the source and the observer are on the X-axis, $\cos\theta = k_x/k = \pm 1$. The upper sign is used when the source and the observer approach each other and the lower when they recede from each other. Thus

$$\lambda = \lambda'\sqrt{\frac{1 \mp v/c}{1 \pm v/c}}, \tag{18.44}$$

or approximately, for small v/c,

$$\lambda \simeq \lambda'(1 \mp v/c). \tag{18.45}$$

Any linear combination of the solutions of Eq. (18.35) given in Eq. (18.40) is again a solution; and, since there is a solution for each value of k, we can take as a solution

$$f = \sum_{s=1}^{n} a_s e^{i(k_s\cdot r - ck_s t)} + \sum_{s=1}^{n} b_s e^{-i(k_s\cdot r - ck_s t)}, \tag{18.46}$$

where the k_s's are arbitrary vectors. In the special case $n = 1$, the solution f is called a *monochromatic wave*. A more general solution than that given in Eq. (18.46) is

$$f = \left(\frac{1}{2\pi}\right)^{3/2}\int [f(k)e^{i(k\cdot r - ckt)} + f^{+}(k)e^{-i(k\cdot r - ckt)}]\, d^3k, \tag{18.47}$$

where $f(k)$ and $f^{+}(k)$ are two arbitrary functions, $d^3k = dk_1\, dk_2\, dk_3$, and the integration is performed over all values of k.

We shall now show that Eq. (18.47) gives a general solution of Eq. (18.35) in the usual mathematical sense. Since Eq. (18.35) is a second-order differential equation, a solution is completely determined if we know

$f(\mathbf{r}, t)$ and $\partial f(\mathbf{r}, t)/\partial t$ for any one moment of time, say $t = 0$. Therefore, in order to prove that our solution is general, it is only necessary to prove that for arbitrary given

$$f(\mathbf{r}, 0) \equiv f_0 \quad \text{and} \quad \left(\frac{\partial f(r, t)}{\partial t}\right)_{t=0} \equiv \frac{\partial f_0}{\partial t}$$

a solution of the form given by Eq. (18.47) exists. Thus we wish to show that, with f_0 and $\partial f^0/\partial t$ given arbitrarily, an $f(\boldsymbol{k})$ and an $f^+(\boldsymbol{k})$ exist which satisfy both Eq. (18.47) and the time derivative of that equation at $t = 0$; that is, which satisfy

$$f_0 = \left(\frac{1}{2\pi}\right)^{3/2} \int [f(\boldsymbol{k})e^{i\boldsymbol{k}\cdot\mathbf{r}} + f^+(\boldsymbol{k})e^{-i\boldsymbol{k}\cdot\mathbf{r}}]\, d^3\boldsymbol{k}$$

and

$$\frac{\partial f_0}{\partial t} = -ic\left(\frac{1}{2\pi}\right)^{3/2} \int [f(\boldsymbol{k})e^{i\boldsymbol{k}\cdot\mathbf{r}} - f^+(\boldsymbol{k})e^{-i\boldsymbol{k}\cdot\mathbf{r}}]\boldsymbol{k}\, d^3\boldsymbol{k}.$$

Changing the variable of integration in the second term of each of these integrals from $\boldsymbol{k}$ to $-\boldsymbol{k}$, we obtain

$$f_0 = \left(\frac{1}{2\pi}\right)^{3/2} \int [f(\boldsymbol{k}) + f^+(-\boldsymbol{k})]e^{i\boldsymbol{k}\cdot\mathbf{r}}\, d^3\boldsymbol{k} \tag{18.48}$$

and

$$\frac{\partial f_0}{\partial t} = -ic\left(\frac{1}{2\pi}\right)^{3/2} \int [f(\boldsymbol{k}) - f^+(-\boldsymbol{k})]\boldsymbol{k}e^{i\boldsymbol{k}\cdot\mathbf{r}}\, d^3\boldsymbol{k}, \tag{18.49}$$

since the range of integration covers all values of $\boldsymbol{k}$. By the inverse Fourier transformation (see Eqs. (A8.94) and (A8.95)), these become

$$f(\boldsymbol{k}) + f^+(-\boldsymbol{k}) = \left(\frac{1}{2\pi}\right)^{3/2} \int f_0 e^{-i\boldsymbol{k}\cdot\mathbf{r}}\, dV \equiv A(\boldsymbol{k}) \tag{18.50}$$

and

$$f(\boldsymbol{k}) - f^+(-\boldsymbol{k}) = \frac{i}{c\boldsymbol{k}}\left(\frac{1}{2\pi}\right)^{3/2} \int \frac{\partial f_0}{\partial t} e^{-i\boldsymbol{k}\cdot\mathbf{r}}\, dV \equiv B(\boldsymbol{k}), \tag{18.51}$$

the functions $A(\boldsymbol{k})$ and $B(\boldsymbol{k})$ standing for the integrals indicated and the integration being performed over all space. Therefore

$$f(\boldsymbol{k}) = \tfrac{1}{2}[A(\boldsymbol{k}) + B(\boldsymbol{k})] \tag{18.52}$$

and

$$f^+(\boldsymbol{k}) = \tfrac{1}{2}[A(-\boldsymbol{k}) - B(-\boldsymbol{k})], \tag{18.53}$$

which are the required functions.

If f_0 and $\partial f_0/\partial t$ are real,

$$A^*(\boldsymbol{k}) = \left(\frac{1}{2\pi}\right)^{3/2} \int f_0 e^{i\boldsymbol{k}\cdot\mathbf{r}}\, dV = A(-\boldsymbol{k})$$

and

$$B^*(\boldsymbol{k}) = \frac{-1}{c\boldsymbol{k}} \left(\frac{1}{2\pi}\right)^{3/2} \int \frac{\partial f_0}{\partial t} e^{i\boldsymbol{k}\cdot\mathbf{r}}\, dV = -B(-\boldsymbol{k}),$$

where as usual the asterisk indicates the complex conjugate of the quantity starred, and therefore

$$f^+(\boldsymbol{k}) = \tfrac{1}{2}[A^*(\boldsymbol{k}) + B^*(\boldsymbol{k})] = f^*(\boldsymbol{k}). \tag{18.54}$$

If we substitute this into Eq. (18.47) and differentiate with respect to x, we have

$$\frac{\partial f}{\partial x} = \left(\frac{1}{2\pi}\right)^{3/2} \int \left[f(\boldsymbol{k}) i \frac{\partial(\boldsymbol{k}\cdot\mathbf{r})}{\partial x} e^{i(\boldsymbol{k}\cdot\mathbf{r} - c\boldsymbol{k}t)} - f^*(\boldsymbol{k}) i \frac{\partial(\boldsymbol{k}\cdot\mathbf{r})}{\partial x} e^{-i(\boldsymbol{k}\cdot\mathbf{r} - c\boldsymbol{k}t)} \right] d^3\boldsymbol{k}$$

$$= \left(\frac{1}{2\pi}\right)^{3/2} \int [i\boldsymbol{k}_x f(\boldsymbol{k}) e^{i(\boldsymbol{k}\cdot\mathbf{r} - c\boldsymbol{k}t)} - i\boldsymbol{k}_x f^*(\boldsymbol{k}) e^{-i(\boldsymbol{k}\cdot\mathbf{r} - c\boldsymbol{k}t)}]\, d^3\boldsymbol{k}.$$

Thus, using similar equations for $\partial f/\partial y$ and $\partial f/\partial z$, we have

$$\nabla f = \left(\frac{1}{2\pi}\right)^{3/2} \int [i\boldsymbol{k} f(\boldsymbol{k}) e^{i(\boldsymbol{k}\cdot\mathbf{r} - c\boldsymbol{k}t)} - i\boldsymbol{k} f^*(\boldsymbol{k}) e^{-i(\boldsymbol{k}\cdot\mathbf{r} - c\boldsymbol{k}t)}]\, d^3\boldsymbol{k}, \tag{18.55}$$

and so the effect of the operator ∇ is to replace $f(\boldsymbol{k})$ by $i\boldsymbol{k}f(\boldsymbol{k})$ and $f^*(\boldsymbol{k})$ by $(i\boldsymbol{k}f(\boldsymbol{k}))^*$. Hence, we say that the operator ∇ operating on f corresponds to multiplication of $f(\boldsymbol{k})$ by $i\boldsymbol{k}$. Likewise we obtain an operator for $\partial/\partial t$; thus,

$$\nabla \sim i\boldsymbol{k} \qquad \text{and} \qquad \frac{\partial}{\partial t} \sim ic\boldsymbol{k}. \tag{18.56}$$

Applying Eqs. (18.47) and (18.54) to each component of the electric and magnetic fields we have

$$\mathscr{E}(\mathbf{r}, t) = \left(\frac{1}{2\pi}\right)^{3/2} \int [\mathscr{E}(\boldsymbol{k}) e^{i(\boldsymbol{k}\cdot\mathbf{r} - c\boldsymbol{k}t)} + \mathscr{E}^*(\boldsymbol{k}) e^{-i(\boldsymbol{k}\cdot\mathbf{r} - c\boldsymbol{k}t)}]\, d^3\boldsymbol{k} \tag{18.57}$$

and

$$\mathscr{H}(\mathbf{r}, t) = \left(\frac{1}{2\pi}\right)^{3/2} \int [\mathscr{H}(\boldsymbol{k}) e^{i(\boldsymbol{k}\cdot\mathbf{r} - c\boldsymbol{k}t)} + \mathscr{H}^*(\boldsymbol{k}) e^{-i(\boldsymbol{k}\cdot\mathbf{r} - c\boldsymbol{k}t)}]\, d^3\boldsymbol{k}. \tag{18.58}$$

$\mathscr{E}(\ell)$ and $\mathscr{H}(\ell)$ are, however, neither independent nor arbitrary, but are subject to certain restrictions corresponding to the Maxwell–Lorentz equations for $\mathscr{E}$ and $\mathscr{H}$. By Eqs. (18.56) and (18.57)

$$\nabla \cdot \mathscr{E} = \left(\frac{1}{2\pi}\right)^{3/2} \int [i\ell \cdot \mathscr{E}(\ell) e^{i(\ell \cdot \mathbf{r} - c\ell t)} - i\ell \cdot \mathscr{E}^*(\ell) e^{-i(\ell \cdot \mathbf{r} - c\ell t)}] \, d^3\ell. \quad (18.59)$$

This is of the form of Eq. (18.47) where

$$\nabla \cdot \mathscr{E} \sim f(\mathbf{r}) \quad \text{and} \quad i\ell \cdot \mathscr{E}(\ell) \sim f(\ell). \quad (18.60)$$

Now Eqs. (18.52) and (18.53), using definitions given in Eqs. (18.50) and (18.51), show us that if $f_0 = \partial f_0/\partial t = 0$, then $f(\ell) = f^*(\ell) = 0$. This requires by the correspondences given in (18.60) that, since $\nabla \cdot \mathscr{E} = 0$ for all t and hence $(\nabla \cdot \mathscr{E})_0 = \partial(\nabla \cdot \mathscr{E})_0/\partial t = 0$,

$$\ell \cdot \mathscr{E}(\ell) = 0. \quad (18.61)$$

Similarly we obtain, corresponding to the equation $(1/c)\, \partial\mathscr{H}/\partial t + \nabla \times \mathscr{E} = 0$,

$$\ell\mathscr{H}(\ell) = \ell \times \mathscr{E}(\ell); \quad (18.62)$$

corresponding to $\nabla \cdot \mathscr{H} = 0$,

$$\ell \cdot \mathscr{H}(\ell) = 0; \quad (18.63)$$

and finally, corresponding to $(1/c)\partial\mathscr{E}/\partial t - \nabla \times \mathscr{H} = 0$,

$$\ell\mathscr{E}(\ell) = -\ell \times \mathscr{H}(\ell). \quad (18.64)$$

The conditions imposed by Eqs. (18.63) and (18.64), however, can be derived from the other two. Thus, multiplying Eq. (18.62) by $\ell\,\cdot$, we have

$$\ell\ell \cdot \mathscr{H}(\ell) = \ell \cdot \ell \times \mathscr{E}(\ell) = 0$$

or Eq. (18.63). To get Eq. (18.64), we multiply Eq. (18.62) by $\ell\, \times$ and use Eq. (18.61), obtaining

$$\ell\,\ell \times \mathscr{H}(\ell) = \ell \times (\ell \times \mathscr{E}(\ell)) = \ell \cdot \mathscr{E}(\ell)\ell - \ell \cdot \ell\mathscr{E}(\ell) = -\ell^2\mathscr{E}(\ell),$$

which is Eq. (18.64) multiplied by ℓ. Hence we see that the Maxwell–Lorentz equations are satisfied by Eqs. (18.57) and (18.58) provided $\mathscr{E}(\ell)$ and $\mathscr{H}(\ell)$ satisfy Eqs. (18.61) and (18.62); these conditions are equivalent to requiring that $\mathscr{E}(\ell)$ and $\mathscr{H}(\ell)$ be equal and that $\mathscr{E}(\ell)$, $\mathscr{H}(\ell)$, and ℓ should form a right-handed orthogonal system of vectors.

We shall now show that $\mathscr{E}(\mathbf{r}, t)$ and $\mathscr{H}(\mathbf{r}, t)$ are completely determined by Eqs. (18.57) and (18.58) when the initial values $\mathscr{E}_0 = \mathscr{E}(\mathbf{r}, 0)$ and

$\mathscr{H}_0 = \mathscr{H}(\mathbf{r}, 0)$ are arbitrarily assigned. Applying the method used in deriving Eq. (18.50) from Eq. (18.47) to each of the component equations of the vector equation (18.57), we obtain in place of Eq. (18.50) the equation

$$\mathscr{E}(\boldsymbol{k}) + \mathscr{E}^*(-\boldsymbol{k}) = \left(\frac{1}{2\pi}\right)^{3/2} \int \mathscr{E}_0 e^{-i\boldsymbol{k}\cdot\mathbf{r}}\, dV \equiv \mathbf{P}(\boldsymbol{k}). \tag{18.65}$$

Similarly, starting with Eq. (18.58), we have

$$\mathscr{H}(\boldsymbol{k}) + \mathscr{H}^*(-\boldsymbol{k}) = \left(\frac{1}{2\pi}\right)^{3/2} \int \mathscr{H}_0 e^{-i\boldsymbol{k}\cdot\mathbf{r}}\, dV \equiv \mathbf{Q}(\boldsymbol{k}). \tag{18.66}$$

Multiplying Eq. (18.66) by $-(\boldsymbol{k}/k)\times$ and using Eq. (18.64) and its complex conjugate, we have

$$-\frac{\boldsymbol{k}}{k}\times \mathscr{H}(\boldsymbol{k}) - \frac{\boldsymbol{k}}{k}\times \mathscr{H}^*(-\boldsymbol{k}) = \mathscr{E}(\boldsymbol{k}) - \mathscr{E}^*(-\boldsymbol{k}) = -\frac{\boldsymbol{k}}{k}\times \mathbf{Q}(\boldsymbol{k}). \tag{18.67}$$

From this equation and Eq. (18.65) we find

$$\mathscr{E}(\boldsymbol{k}) = \frac{1}{2}\left[\mathbf{P}(\boldsymbol{k}) - \frac{\boldsymbol{k}}{k}\times \mathbf{Q}(\boldsymbol{k})\right] \tag{18.68}$$

together with

$$\mathscr{E}^*(-\boldsymbol{k}) = \frac{1}{2}\left[\mathbf{P}(\boldsymbol{k}) + \frac{\boldsymbol{k}}{k}\times \mathbf{Q}(\boldsymbol{k})\right]$$

or

$$\mathscr{E}^*(\boldsymbol{k}) = \frac{1}{2}\left[\mathbf{P}(-\boldsymbol{k}) - \frac{\boldsymbol{k}}{k}\times \mathbf{Q}(-\boldsymbol{k})\right]. \tag{18.69}$$

This is, of course, the complex conjugate of Eq. (18.68), since $\mathbf{P}^*(\boldsymbol{k}) = \mathbf{P}(-\boldsymbol{k})$ and $\mathbf{Q}^*(\boldsymbol{k}) = \mathbf{Q}(-\boldsymbol{k})$, as is evident from Eqs. (18.65) and (18.66). $\mathscr{H}(\boldsymbol{k})$ is now given by Eqs. (18.62) and (18.68).

EXAMPLE 3. If the field at some time $t = 0$ is independent of x and y and is described by the given functions $\mathscr{E}_0(z)$ and $\mathscr{H}_0(z)$, what is the subsequent behavior of the field; that is, what are the functions $\mathscr{E}(\mathbf{r}, t)$ and $\mathscr{H}(\mathbf{r}, t)$?

The given functions must of course satisfy the equations

$$\nabla \cdot \mathscr{E}_0 = 0 \tag{18.70}$$

and

$$\nabla \cdot \mathscr{H}_0 = 0. \tag{18.71}$$

On the other hand, the Fourier transform of each of the components of Eq. (18.65) gives (see Eqs. (A8.94) and (A8.95))

$$\mathscr{E}_0(z) = \left(\frac{1}{2\pi}\right)^{3/2} \int \mathbf{P}(\boldsymbol{k}) e^{i\boldsymbol{k}\cdot\mathbf{r}}\, d^3\boldsymbol{k}; \tag{18.72}$$

so that Eq. (18.70) becomes

$$\nabla \cdot \mathscr{E}_0(z) = \left(\frac{1}{2\pi}\right)^{3/2} \int i\boldsymbol{k} \cdot \mathbf{P}(\boldsymbol{k}) e^{i\boldsymbol{k}\cdot\mathbf{r}}\, d^3\boldsymbol{k} = 0, \tag{18.73}$$

which by the Fourier transformation requires that

$$\boldsymbol{k} \cdot \mathbf{P}(\boldsymbol{k}) = 0. \tag{18.74}$$

Further, since $\mathscr{E}_0(z)$ is independent of x and y,

$$\frac{\partial \mathscr{E}_0(z)}{\partial x} = \left(\frac{1}{2\pi}\right)^{3/2} \int i k_x \mathbf{P}(\boldsymbol{k}) e^{i\boldsymbol{k}\cdot\mathbf{r}}\, d^3\boldsymbol{k} = 0 \tag{18.75}$$

and

$$\frac{\partial \mathscr{E}_0(z)}{\partial y} = \left(\frac{1}{2\pi}\right)^{3/2} \int i k_y \mathbf{P}(\boldsymbol{k}) e^{i\boldsymbol{k}\cdot\mathbf{r}}\, d^3\boldsymbol{k} = 0, \tag{18.76}$$

which by the Fourier transformation leads to

$$k_x \mathbf{P}(\boldsymbol{k}) = k_y \mathbf{P}(\boldsymbol{k}) = 0. \tag{18.77}$$

Thus $\mathbf{P}(\boldsymbol{k}) = 0$, unless $k_x = k_y = 0$. If $\mathscr{E}_0(z)$ of Eq. (18.72) is not identically zero, $\mathbf{P}(\boldsymbol{k})$ must become infinite for $k_x = 0$ and for $k_y = 0$. Assuming that the given $\mathscr{E}_0(z)$ is finite, this requires that $\mathbf{P}(\boldsymbol{k})$ should contain $\delta(k_x)$ and $\delta(k_y)$ as factors (see Eq. (A8.19) and the discussion of the delta function following that equation). Thus we have

$$\mathbf{P}(\boldsymbol{k}) = \delta(k_x)\, \delta(k_y)\, \mathbf{P}_0(k_z), \tag{18.78}$$

where $\mathbf{P}_0(\mathbf{k}_z)$ is some vector. The functional dependence of $\mathbf{P}_0$ upon k_x and k_y need not be included, for the right-hand member of Eq. (18.78) is different from zero only for $k_x = k_y = 0$.

By Eq. (A8.20) we have

$$\int f(x)\delta(x - \lambda)\, dx = f(\lambda).$$

Putting $\lambda = 0$ and $f(x) = xF(x)$, this becomes

$$\int x\delta(x)F(x)\, dx = 0$$

if $F(0) \neq \infty$. In our case we must put $x = k_x$, so that

$$\int k_x \delta(k_x) F(k_x)\, dk_x = 0$$

or, more generally,

$$\int k_x \delta(k_x) F(\mathbf{k})\, d^3k = \iint dk_y\, dk_z k_x \delta(k_x) F(k_x, k_y, k_z)\, dk_x = 0.$$

Thus, the combination $k_x\, \delta(k_x)$ gives zero whenever it occurs in an integrand with a factor that does not become infinite for $k_x = 0$, and may therefore be treated as zero. It then follows that $\mathbf{P}(\mathbf{k})$, as given by Eq. (18.78), will automatically satisfy Eqs. (18.75) and (18.76).

Equation (18.74) now becomes, by virtue of Eqs. (18.77) and (18.78),

$$\begin{aligned} \mathbf{k} \cdot \mathbf{P}(\mathbf{k}) &= k_x P_x(\mathbf{k}) + k_y P_y(\mathbf{k}) + k_z P_z(\mathbf{k}) \\ &= k_z P_z(\mathbf{k}) = k_z\, \delta(k_x)\, \delta(k_y) P_{0z}(k_z) = 0, \end{aligned} \tag{18.79}$$

as a condition on $P_{0z}(k_z)$. This means that $P_{0z}(k_z) = 0$, unless $k_z = 0$, or that $P_{0z}(k_z)$ must be of the form

$$P_{0z}(k_z) = A\delta(k_z). \tag{18.80}$$

There being no increased generality in making $A = A(k_z)$, since the only value of k_z for which $P_{0z}(k_z)$ does not vanish is $k_z = 0$, we may assume A to be constant.

Similar discussion, but starting with Eq. (18.66) and making use of Eqs. (18.71) and the fact that $\mathscr{H}_0 = \mathscr{H}_0(z)$, gives

$$\mathbf{Q}(\mathbf{k}) = \delta(k_x)\delta(k_y)\mathbf{Q}_0(k_z) \tag{18.81}$$

and

$$Q_{0z} = B\delta(k_z), \tag{18.82}$$

where B is some constant.

We now have $\mathscr{E}(\mathbf{r}, t)$ given by Eq. (18.57), with $\mathscr{E}(\mathbf{k})$ given by Eq. (18.68). Therefore

$$\begin{aligned} \mathscr{E}(\mathbf{r}, t) &= \frac{1}{2}\left(\frac{1}{2\pi}\right)^{3/2} \\ &\times \int \left\{ \left[\mathbf{P}(\mathbf{k}) - \frac{\mathbf{k}}{k} \times \mathbf{Q}(\mathbf{k}) \right] e^{i(\mathbf{k}\cdot\mathbf{r} - ckt)} + \text{complex conjugate} \right\} d^3k. \end{aligned} \tag{18.83}$$

For the Z component we have to use

$$\mathscr{E}_z(\mathbf{k}) = \frac{1}{2}\left[\mathbf{P}(\mathbf{k}) - \frac{\mathbf{k}}{k} \times \mathbf{Q}(\mathbf{k})\right]_z = \frac{1}{2}\left[P_z(\mathbf{k}) - \frac{k_x}{k} Q_y(\mathbf{k}) + \frac{k_y}{k} Q_x(\mathbf{k})\right].$$

By Eq. (18.81) the second term of the right-hand member contains $k_x\delta(k_x)$ and the third $k_y\delta(k_y)$; therefore

$$\mathscr{E}_z(k) = \tfrac{1}{2}P_z(k) = \tfrac{1}{2}\delta(k_x)\delta(k_y)\delta(k_z)A = \tfrac{1}{2}A\delta(k). \tag{18.84}$$

Since for any $F(k)$

$$\int F(k)\delta(k)\, dk = F(0),$$

we have

$$\mathscr{E}_z(\mathbf{r}, t) = \frac{1}{2}\left(\frac{1}{2\pi}\right)^{3/2}\int \{A\delta(k)e^{i(k\cdot\mathbf{r}-ckt)} + \text{complex conjugate}\}\, d^3k$$

$$= \frac{1}{2}\left(\frac{1}{2\pi}\right)^{3/2}(A + A^*) = \text{constant}. \tag{18.85}$$

Similarly

$$\mathscr{H}_z(\mathbf{r}, t) = \text{constant}. \tag{18.86}$$

Again, using Eq. (18.83),

$$\frac{\partial\mathscr{E}(\mathbf{r}, t)}{\partial x} = \frac{1}{2}\left(\frac{1}{2\pi}\right)^{3/2}$$

$$\times \int\left[ik_x\mathbf{P}(k) - \frac{k}{k}\times \mathbf{Q}(k)e^{i(k\cdot\mathbf{r}-ckt)} + \text{complex conjugate}\right] d^3k = 0, \tag{18.87}$$

since $\mathbf{P}(k)$ and $\mathbf{Q}(k)$ both contain $\delta(k_x)$, which is here multiplied by k_x. Likewise,

$$\frac{\partial\mathscr{E}(\mathbf{r}, t)}{\partial x} = \frac{\partial\mathscr{H}(\mathbf{r}, t)}{\partial x} = \frac{\partial\mathscr{H}(\mathbf{r}, t)}{\partial y} = 0. \tag{18.88}$$

Equations (18.85), (18.86), (18.87), and (18.88) show that the Z components of the field are constant, and the remaining components depend only on z. The behavior of the field reduces therefore to that discussed in Example 1.

If we take the special case of Eq. (18.46) in which $n = 1$, take k in the positive Z direction, and let $a_1 = a$ and $b_1 = a^+$, we obtain as a particular solution of Eq. (18.35)

$$f = ae^{ik(z-ct)} + a^+e^{-ik(z-ct)}, \tag{18.89}$$

which corresponds to a plane monochromatic wave advancing along the Z-axis. If this is to be real and hence equal to its own complex conjugate, a^+ must be equal to a^*. Thus, for example, the real electric vector of such a monochromatic wave will be given by

$$\mathscr{E} = \mathbf{a}e^{i\mathscr{k}(z-ct)} + \mathbf{a}^*e^{-i\mathscr{k}(z-ct)}. \tag{18.90}$$

Since $\mathbf{a}$ is a complex vector, its magnitude will in general be complex; thus

$$\mathbf{a}\cdot\mathbf{a} = \text{complex scalar} = \rho e^{2i\alpha}, \qquad 0 \leq \alpha \leq \pi, \tag{18.91}$$

with suitable choice of the real quantities ρ and α. Let us define a vector $\mathbf{b}$ by the equation

$$\mathbf{b} = \mathbf{a}e^{i\alpha}; \tag{18.92}$$

then

$$\mathbf{b}\cdot\mathbf{b} = \mathbf{a}\cdot\mathbf{a}\, e^{-2i\alpha} = \rho, \tag{18.93}$$

and by Eqs. (18.90) and (18.92)

$$\begin{aligned}\mathscr{E} &= \mathbf{b}e^{i\mathscr{k}(z-ct)+i\alpha} + \mathbf{b}^*e^{-i\mathscr{k}(z-ct)-i\alpha}\\ &= \mathbf{b}e^{i\varphi} + \mathbf{b}^*e^{-i\varphi},\end{aligned} \tag{18.94}$$

where

$$\varphi = \mathscr{k}(z-ct) + \alpha. \tag{18.95}$$

Expanding the exponentials $e^{i\varphi}$ and $e^{-i\varphi}$ in Eq. (18.94) in terms of cos φ and sin φ, we obtain

$$\begin{aligned}\mathscr{E} &= (\mathbf{b} + \mathbf{b}^*)\cos\varphi + i(\mathbf{b} - \mathbf{b}^*)\sin\varphi\\ &= \mathbf{A}\cos\varphi + \mathbf{B}\sin\varphi,\end{aligned} \tag{18.96}$$

where

$$\mathbf{A} = \mathbf{b} + \mathbf{b}^* \qquad \text{and} \qquad \mathbf{B} = i(\mathbf{b} - \mathbf{b}^*). \tag{18.97}$$

Both $\mathbf{A}$ and $\mathbf{B}$ are real vectors, since they are equal to their own complex conjugates. Moreover,

$$\mathbf{A}\cdot\mathbf{B} = i(\mathbf{b}\cdot\mathbf{b} - \mathbf{b}^*\cdot\mathbf{b}^*) = i(\mathbf{b}\cdot\mathbf{b} - (\mathbf{b}\cdot\mathbf{b})^*);$$

so that, since by Eq. (18.93) $\mathbf{b}\cdot\mathbf{b}$ is real,

$$\mathbf{A}\cdot\mathbf{B} = 0. \tag{18.98}$$

Choosing the axes so that the positive X-axis is in the direction of $\mathbf{A}$, the positive Y direction is paralled or antiparallel to $\mathbf{B}$. Then from Eq. (18.96) the components of $\mathscr{E}$ are

$$\mathscr{E}_x = A\cos\varphi \qquad \text{and} \qquad \mathscr{E}_y = \pm B\sin\varphi. \tag{18.99}$$

From this

$$\frac{\mathscr{E}_x^{\,2}}{A^2} + \frac{\mathscr{E}_y^{\,2}}{B^2} = 1, \tag{18.100}$$

showing that the terminus of the electric vector describes an ellipse. Therefore, a monochromatic plane wave is said to be in general *elliptically polarized*. When $B = A$ the ellipse, of course, becomes a circle and the polarization is said to be *circular*. If in addition $\mathbf{B}$ is in the positive Y direction for a right-handed coordinate system, then Eqs. (18.99) become

$$\mathscr{E}_x = A\cos\varphi \qquad \text{and} \qquad \mathscr{E}_y = A\sin\varphi, \tag{18.101}$$

which, since φ decreases linearly with t, represent the projection on the X- and Y-axes of uniform circular motion of the terminus of $\mathscr{E}$. When looked at in the direction opposite to the direction of propagation, the circular motion is clockwise, and the wave is said to be *right-circularly polarized*. For $\mathbf{B}$ antiparallel to the positive Y direction, $\mathscr{E}_y = -A\sin\varphi$, and the direction of rotation is reversed. The polarization is then described as *left-circular polarization*.

If $\mathbf{B} = 0$, then $\mathscr{E}_y = 0$ and the electric vector at any point has only an X component; hence its terminus oscillates on a line parallel to the X-axis passing through the point. This wave may be described as *polarized in the X direction*. Such a wave may also be said to be *linearly polarized* and to have a line of polarization along the $\mathscr{E}$ vector. In optics, unfortunately, the term *plane polarized* is used, and the *plane of polarization* is taken to be the plane determined by $\mathscr{H}$ and $\mathscr{k}$, a plane perpendicular to $\mathscr{E}$.

19. Retarded and Advanced Potentials

In this section we shall find a general solution of Eqs. (16.60) and (16.61), i.e., of the equations

$$\nabla^2\varphi - \frac{1}{c^2}\frac{\partial^2\varphi}{\partial t^2} = -4\pi\rho \tag{19.1}$$

and

$$\nabla^2\mathbf{A} - \frac{1}{c^2}\frac{\partial^2\mathbf{A}}{\partial t^2} = -4\pi\rho\frac{\mathbf{v}}{c} \tag{19.2}$$

subject to the condition

$$\nabla\cdot\mathbf{A} + \frac{1}{c}\frac{\partial\varphi}{\partial t} = 0. \tag{19.3}$$

Consider first Eq. (19.1). We shall make use of the linear character of this equation in the following way: If ρ is broken up into a number of parts, say $\rho = \sum_i p_i u_i(\mathbf{r}, t)$, and for each i one finds a φ_i satisfying the equation

$$\nabla^2\varphi_i - \frac{1}{c^2}\frac{\partial^2\varphi_i}{\partial t^2} = -4\pi u_i,$$

then $\varphi = \sum_i p_i\varphi_i$ will be a solution of Eq. (19.1). The way in which ρ is broken up is of course completely arbitrary, except that p_i must be independent of the coordinates x, y, z, and the time t.

Instead of the sum we can also use an integral. By using the properties of the delta function (see Eqs. (7.26) or (A8.20)), we may write

$$\rho(x, y, z, t) = \int \delta(\mathbf{r} - \mathbf{r}')\,dV' \int_{-\infty}^{\infty} \delta(t - \tau)\rho(\xi, \eta, \zeta, \tau)\,d\tau,$$

where $\mathbf{r} = (x, y, z)$, $\mathbf{r}' = (\xi, \eta, \zeta)$, $dV' = d\xi\,d\eta\,d\zeta$, and the integration is taken over the whole of r'-space. This is analogous to the equation

$$\rho = \sum_i p_i u_i$$

if one makes the following correspondences:

$$i \sim (\xi, \eta, \zeta, \tau)$$
$$u_i \sim \delta(\mathbf{r} - \mathbf{r}')\delta(t - \tau)$$
$$p_i \sim \rho(\xi, \eta, \zeta, \tau)\,d\tau\,dV'.$$

We shall look for a solution* of

$$\nabla^2\varphi_i - \frac{1}{c^2}\frac{\partial^2\varphi_i}{\partial t^2} = -4\pi\delta(\mathbf{r} - \mathbf{r}')\delta(t - \tau), \tag{19.4}$$

expecting then from our analogy that

$$= \int dV' \int_{-\infty}^{\infty} \varphi_i \rho(\xi, \eta, \zeta, \tau)\,d\tau \tag{19.5}$$

will be a solution of the original equation (19.1).

The right-hand side of Eq. (19.4) is zero except for $\mathbf{r} = \mathbf{r}'$ and $t = \tau$ and thus corresponds to a charge existing instantaneously at $t = \tau$ at the point $\mathbf{r} = \mathbf{r}'$. No preferential direction appears in this equation, and so we would expect the existence of a solution spherically symmetrical about the

* Such a solution is called a Green's function.

point $\mathbf{r}'$. If we put

$$\mathbf{R} = \mathbf{r} - \mathbf{r}', \tag{19.6}$$

the spherically symmetrical solution we are looking for should be a function of R and t; hence, using Eq. (A3.50),

$$\frac{1}{R^2}\frac{\partial}{\partial R}\left(R^2\frac{\partial \varphi_i}{\partial R}\right) - \frac{1}{c^2}\frac{\partial^2 \varphi_i}{\partial t^2} = 0,$$

except at $R = |\mathbf{r} - \mathbf{r}'| = [(x - \xi)^2 + (y - \eta)^2 + (z - \zeta)^2]^{1/2} = 0$.

Since this equation can also be written

$$\frac{1}{R}\frac{\partial^2}{\partial R^2}(R\varphi_i) - \frac{1}{c^2}\frac{\partial^2 \varphi_i}{\partial t^2} = 0,$$

we see that, by letting $\Phi \equiv R\varphi_i$, it may be written

$$\frac{\partial^2 \Phi}{\partial R^2} - \frac{1}{c^2}\frac{\partial^2 \Phi}{\partial t^2} = 0. \tag{19.7}$$

This is of the same form as Eq. (18.6), when we consider only one of the components of $\mathscr{E}$, and has as its general solution (see Eq. (18.11))

$$\Phi = f_1\left(t - \frac{R}{c}\right) + f_2\left(t + \frac{R}{c}\right), \tag{19.8}$$

where f_1 and f_2 are arbitrary functions. This can also be immediately checked by its substitution in Eq. (19.7).

Let us first investigate the case $f_2 \equiv 0$, that is, the case

$$\varphi_i = \frac{\Phi}{R} = \frac{1}{R} f_1\left(t - \frac{R}{c}\right). \tag{19.9}$$

In order to test whether φ_i satisfies Eq. (19.4), we first use Eq. (A3.40) to get

$$\nabla^2 \varphi_i = \frac{1}{R}\nabla^2 f_1 + f_1 \nabla^2\left(\frac{1}{R}\right) + 2\boldsymbol{\nabla}\left(\frac{1}{R}\right)\cdot \boldsymbol{\nabla} f_1,$$

and evaluate the various expressions on the right-hand side. By Eq. (A3.50)

$$\nabla^2 f_1 = \frac{1}{R^2}\frac{\partial}{\partial R}\left(R^2 \frac{\partial f_1}{\partial R}\right)$$

$$= \frac{1}{R^2}\frac{\partial}{\partial R}\left(-\frac{R^2}{c} f_1'\right) = -\frac{2}{Rc} f_1' + \frac{1}{c^2} f_1'',$$

where primes after f_1 designate differentiation with respect to $t - R/c$. By a slight modification of Eq. (8.7) we have

$$\nabla^2\left(\frac{1}{R}\right) = -4\pi\delta(\mathbf{R}) = -4\pi\delta(\mathbf{r} - \mathbf{r}'),$$

and by Eq. (A3.43)

$$\nabla\left(\frac{1}{R}\right) = -\frac{R}{R^3}$$

and

$$\nabla f_1 = \frac{\mathbf{R}}{Rc} f_1';$$

therefore

$$\nabla^2\varphi_i = -\frac{2}{R^2c} f_1' + \frac{1}{c^2R} f_1'' - 4\pi f_1\delta(\mathbf{r} - \mathbf{r}') + \frac{2}{R^2c} f_1'$$

$$= \frac{1}{c^2R} f_1'' - 4\pi f_1\delta(\mathbf{r} - \mathbf{r}').$$

Next we subtract

$$\frac{1}{c^2}\frac{\partial^2\varphi_i}{\partial t^2} = \frac{1}{Rc^2} f_1''$$

from the above to obtain the equation

$$\nabla^2\varphi_i - \frac{1}{c^2}\frac{\partial^2\varphi_i}{\partial t^2} = -4\pi f_1\delta(\mathbf{r} - \mathbf{r}'). \tag{19.10}$$

This must reduce to the right-hand side of Eq. (19.4); therefore for $R = 0$, $f_1 = \delta(t - \tau)$. On the other hand, $f_1 = f_1(t - R/c)$; hence $f_1 = \delta(t - \tau - R/c)$ and

$$\varphi_i = \frac{1}{R}\delta\left(t - \tau - \frac{R}{c}\right). \tag{19.11}$$

The right-hand side of Eq. (19.10) is then $-4\pi\delta(\mathbf{r} - \mathbf{r}')\delta(t - \tau - R/c)$, which, though different in form, is exactly equal to the right-hand side of Eq. (19.4).

Thus, from Eqs. (19.5), (19.9), and (19.11), we determine that

$$\varphi(\mathbf{r}, t) = \int dV' \int_{-\infty}^{\infty} \frac{\rho(\xi, \eta, \zeta, \tau)\, \delta(t - \tau - R/c)}{R} d\tau \tag{19.12}$$

should be a solution of Eq. (19.1). That it is in fact a solution of Eq. (19.1) is easily verified. We have

$$\nabla^2\varphi - \frac{1}{c^2}\frac{\partial^2\varphi}{\partial t^2} = \int dV' \int_{-\infty}^{\infty} \rho(\mathbf{r}', \tau)\left(\nabla^2 - \frac{1}{c^2}\frac{\partial^2}{\partial t^2}\right)\frac{\delta(t - \tau - R/c)}{R}\, d\tau$$

$$= \int dV' \int_{-\infty}^{\infty} \rho(\mathbf{r}', \tau)\left(\nabla^2\varphi_i - \frac{1}{c^2}\frac{\partial^2\varphi_i}{\partial t^2}\right) d\tau$$

$$= -4\pi \int dV' \int_{-\infty}^{\infty} \rho(\mathbf{r}', \tau)\, \delta\left(t - \tau - \frac{R}{c}\right)\delta(\mathbf{r} - \mathbf{r}')\, d\tau$$

by Eqs. (19.10) and (19.11). Integration on the right-hand side with respect to V', due to the presence of the factor $\delta(\mathbf{r} - \mathbf{r}')$, replaces $\mathbf{r}'$ by $\mathbf{r}$ in the rest of the integrand; then

$$\nabla^2\varphi - \frac{1}{c^2}\frac{\partial^2\varphi}{\partial t^2} = -4\pi \int_{-\infty}^{\infty} \rho(\mathbf{r}, \tau)\, \delta(t - \tau)\, d\tau = -4\pi\rho(\mathbf{r}, t),$$

which is Eq. (19.1). By analogy we can at once write that

$$\mathbf{A} = \frac{1}{c}\int dV' \int_{-\infty}^{\infty} \frac{\rho(\xi, \eta, \zeta, \tau)\mathbf{v}(\xi, \eta, \zeta, \tau)\, \delta(t - \tau - R/c)}{R}\, d\tau \tag{19.13}$$

is a solution of Eq. (19.2).

If in Eqs. (19.12) and (19.13) we perform the indicated integration with respect to τ, we obtain

$$\varphi(\mathbf{r}, t) = \int \frac{\rho(\mathbf{r}', t - R/c)}{R}\, dV' = \int \frac{[\rho]}{R}\, dV' \tag{19.14a}$$

and

$$\mathbf{A}(\mathbf{r}, t) = \frac{1}{c}\int \frac{\rho(\mathbf{r}', t - R/c)\mathbf{v}(\mathbf{r}', t - R/c)}{R}\, dV' = \frac{1}{c}\int \frac{[\rho\mathbf{v}]}{R}\, dV', \tag{19.14b}$$

where the brackets are used to indicate that the value of the expression enclosed is to be taken for each point $\mathbf{r}'$ at the time $t - |\mathbf{r} - \mathbf{r}'|/c$, i.e., at a time R/c earlier than t. This is interpreted as allowing for the time of propagation of interaction from the point $\mathbf{r}'$ at which it arises to the point $\mathbf{r}$ at which the potential is being computed. For this reason the potentials given by Eq. (19.14) are called *retarded potentials.*

Had we used f_2 of Eq. (19.8), instead of f_1, we would have come out with *advanced potentials* given by

$$\varphi(\mathbf{r}, t) = \int \frac{\rho(\mathbf{r}', t + R/c)}{R}\, dV' \tag{19.15a}$$

and

$$\mathbf{A}(\mathbf{r}, t) = \frac{1}{c}\int \frac{\rho(\mathbf{r}', t + R/c)\mathbf{v}(\mathbf{r}', t + R/c)}{R}\, dV', \tag{19.15b}$$

which, of course, also satisfy Eqs. (19.1) and (19.2).

Both the retarded and the advanced potentials satisfy also Eq. (19.3). We can verify this as follows:

$$\begin{aligned}
\frac{1}{c}\frac{\partial \varphi}{\partial t} &= \frac{1}{c}\frac{\partial}{\partial t}\iint \frac{\rho(\mathbf{r}', \tau)\,\delta(t - \tau \pm R/c)}{R}\, d\tau\, dV' \\
&= \frac{1}{c}\iint \frac{\rho}{R}\frac{\partial}{\partial t}\,\delta\left(t - \tau \pm \frac{R}{c}\right) d\tau\, dV' \\
&= -\frac{1}{c}\iint \frac{\rho}{R}\frac{\partial}{\partial \tau}\,\delta\left(t - \tau \pm \frac{R}{c}\right) d\tau\, dV' \\
&= -\frac{1}{c}\int \left[\frac{\rho}{R}\,\delta\left(t - \tau \pm \frac{R}{c}\right)\right]_{\tau=-\infty}^{\tau=\infty} dV' \\
&\quad + \frac{1}{c}\iint \frac{1}{R}\left(\frac{\partial \rho}{\partial \tau}\right)\delta\left(t - \tau \pm \frac{R}{c}\right) d\tau\, dV'.
\end{aligned}$$

The first integral is zero because its integrand is everywhere zero. This follows from the fact that the argument of the delta function can not be zero for given $\mathbf{r}$, $\mathbf{r}'$, and t and an infinite τ. Upon performing the integration in the second integral with respect to τ, we get, using Eq. (A8.20),

$$\frac{1}{c}\frac{\partial \varphi}{\partial t} = \frac{1}{c}\int \frac{1}{R}\left(\frac{\partial \rho}{\partial \tau}\right)_{\tau = t \pm R/c} dV'.$$

In order to evaluate the left-hand side of Eq. (19.3) we need to evaluate the terms of the form

$$\frac{\partial A_x}{\partial x} = \frac{1}{c}\iint \rho v_x \frac{\partial}{\partial x}\left(\frac{\delta(t - \tau \pm R/c)}{R}\right) d\tau\, dV'.$$

Since only R contains x, and since $\partial R/\partial x = -\partial R/\partial \xi$, we have

$$\begin{aligned}
\frac{\partial A_x}{\partial x} &= -\frac{1}{c}\iint \rho v_x \frac{\partial}{\partial \xi}\left(\frac{\delta(t - \tau \pm R/c)}{R}\right) d\tau\, dV' \\
&= -\frac{1}{c}\iiint_{-\infty}^{\infty} \left[\frac{\rho v_x}{R}\,\delta\left(t - \tau \pm \frac{R}{c}\right)\right]_{\xi=-\infty}^{\xi=\infty} d\tau\, d\eta\, d\zeta \\
&\quad + \frac{1}{c}\iint \frac{\partial(\rho v_x)}{\partial \xi}\,\frac{\delta(t - \tau \pm R/c)}{R}\, d\tau\, dV'.
\end{aligned}$$

The first integral is again zero. Performing the integration with respect to τ in the second, we get

$$\frac{\partial A_x}{\partial x} = \frac{1}{c}\int \left(\frac{\partial(\rho v_x)}{\partial \xi}\right)_{\tau=t\pm R/c} \frac{dV'}{R}.$$

Combining this with similar expressions for $\partial A_y/\partial y$ and $\partial A_z/\partial z$, we have

$$\nabla \cdot \mathbf{A} + \frac{1}{c}\frac{\partial \varphi}{\partial t} = \int \left(\frac{\partial \rho}{\partial \tau} + \frac{\partial(\rho v_x)}{\partial \xi} + \frac{\partial(\rho v_y)}{\partial \eta} + \frac{\partial(\rho v_z)}{\partial \zeta}\right)_{\tau=t\pm R/c} \frac{dV'}{cR}.$$

Since the continuity equation, Eq. (16.32), holds at any point $\mathbf{r}' = (\xi, \eta, \zeta)$ at any time τ, however it is related to t, the integrand in the above equation vanishes and we have, as required, that

$$\nabla \cdot \mathbf{A} + \frac{1}{c}\frac{\partial \varphi}{\partial t} = 0.$$

Any two solutions of a linear inhomogeneous differential equation, such as Eq. (19.1), must differ only by a solution of the corresponding homogeneous equation. Thus, if φ_1 and φ_2 both satisfy Eq. (19.1), then

$$\nabla^2(\varphi_2 - \varphi_1) - \frac{1}{c^2}\frac{\partial^2(\varphi_2 - \varphi_1)}{\partial t^2}$$

$$= \left(\nabla^2\varphi_2 - \frac{1}{c^2}\frac{\partial^2\varphi_2}{\partial t^2}\right) - \left(\nabla^2\varphi_1 - \frac{1}{c^2}\frac{\partial^2\varphi_1}{\partial t^2}\right) = (-4\pi\rho) - (-4\pi\rho) = 0.$$

Therefore a general solution of such an equation will consist of a particular solution *plus* a general solution of the homogeneous wave equation

$$\nabla^2\psi - \frac{1}{c^2}\frac{\partial^2\psi}{\partial t^2} = 0. \tag{19.16}$$

The same argument applies, of course, to Eq. (19.2).

Thus, we have as general solutions of Eq. (19.1)

$$\varphi(\mathbf{r}, t) = \int \frac{\rho(\mathbf{r}', t - R/c)}{R}\, dV' + \psi_{\text{ret}}(\mathbf{r}, t) \tag{19.17}$$

and

$$\varphi(\mathbf{r}, t) = \int \frac{\rho(\mathbf{r}', t + R/c)}{R}\, dV' + \psi_{\text{adv}}(\mathbf{r}, t), \tag{19.18}$$

with analogous equations holding for $\mathbf{A}(\mathbf{r}, t)$.

Each of complementary solutions $\psi_{\text{ret}}(\mathbf{r}, t)$ and $\psi_{\text{adv}}(\mathbf{r}, t)$ satisfies Eq. (19.16) and is chosen so that $\varphi(\mathbf{r}, t)$ satisfies the initial and boundary conditions imposed by the physical problem being considered. For a correctly specified problem, either of the equations, (19.17) or (19.18), may be used and will lead to the same $\varphi(\mathbf{r}, t)$, but $\psi_{\text{ret}}(\mathbf{r}, t)$ will, of course, in general be different from $\psi_{\text{adv}}(\mathbf{r}, t)$. The determination of the particular solutions, given by the integrals in Eqs. (19.17) and (19.18), is illustrated by the following two examples.

Example 1. Find the retarded and advanced potentials due to a moving particle of charge e.

Let the position of the particle at the time t be given by

$$\mathbf{r}_p(t) = (x_p(t), y_p(t), z_p(t)); \tag{19.19}$$

then the charge density is given by

$$\rho(\mathbf{r}, t) = e\delta(\mathbf{r} - \mathbf{r}_p(t)). \tag{19.20}$$

From Eq. (19.12) the scalar potential is then

$$\varphi(\mathbf{r}, t) = \int dV' \int_{-\infty}^{\infty} \frac{e\delta(\mathbf{r} - \mathbf{r}_p(\tau))\, \delta(t - \tau \pm |\mathbf{r} - \mathbf{r}'|/c)}{|\mathbf{r} - \mathbf{r}'|}\, d\tau,$$

where dV' is a volume element of the $\mathbf{r}'$-space and where the upper and lower signs are to be used for the retarded and the advanced potentials, respectively. Integrating over the $\mathbf{r}'$-space, we get

$$\varphi(\mathbf{r}, t) = e \int_{-\infty}^{\infty} \frac{\delta(t - \tau \mp R/c)}{R}\, d\tau, \tag{19.21}$$

where now

$$\mathbf{R} = \mathbf{R}(\mathbf{r}, \tau) = \mathbf{r} - \mathbf{r}_p(\tau)$$

and

$$R = |\mathbf{R}| = [(x - x_p)^2 + (y - y_p)^2 + (z - z_p)^2]^{1/2}.$$

To perform the next integration, we must take account of the fact that now R is a function of τ. To do this, let $\tau \pm R/c = t'$, and let us change the variable of integration from τ to t'. We have

$$dt' = d\tau \pm \frac{1}{c}\frac{\partial R}{\partial \tau}\, d\tau = \left(1 \pm \frac{1}{c}\frac{\partial R}{\partial \tau}\right) d\tau; \tag{19.22}$$

so that

$$\varphi(\mathbf{r}, t) = e\int_{-\infty}^{\infty} \frac{\delta(t - t')\,dt'}{R\left(1 \pm \dfrac{1}{c}\dfrac{\partial R}{\partial \tau}\right)} = \frac{e}{\left[R\left(1 \pm \dfrac{1}{c}\dfrac{\partial R}{\partial \tau}\right)\right]_{t'=t}}. \tag{19.23}$$

Now

$$\frac{\partial R}{\partial \tau} = \frac{\partial R}{\partial x_p}\frac{dx_p}{d\tau} + \frac{\partial R}{\partial y_p}\frac{dy_p}{d\tau} + \frac{\partial R}{\partial z_p}\frac{dz_p}{d\tau}$$

$$= -\frac{1}{R}\left[(x - x_p)\frac{dx_p}{d\tau} + (y - y_p)\frac{dy_p}{d\tau} + (z - z_p)\frac{dz_p}{d\tau}\right],$$

and, since $\mathbf{R} = \mathbf{r} - \mathbf{r}_p$ and

$$\mathbf{v} = \left(\frac{dx_p}{d\tau}, \frac{dy_p}{d\tau}, \frac{dz_p}{d\tau}\right) = \text{velocity of the particle,}$$

this reduces to

$$\frac{\partial R}{\partial \tau} = -\frac{\mathbf{R}\cdot\mathbf{v}}{R}. \tag{19.24}$$

Using this result and the definition of t', we may rewrite Eq. (19.23) as

$$\varphi(\mathbf{r}, t) = \frac{e}{[R(1 \mp \mathbf{R}\cdot\mathbf{v}/cR)]_{\tau=t\mp R/c}}. \tag{19.25}$$

The corresponding vector potential is, analogously,

$$\mathbf{A}(\mathbf{r}, t) = \frac{e}{c}\left[\frac{\mathbf{v}}{R(1 \mp \mathbf{R}\cdot\mathbf{v}/cR)}\right]_{\tau=t\mp R/c} \tag{19.26}$$

These formulas are the well-known results due to Liénard. The expression $1 - \mathbf{R}\cdot\mathbf{v}/cR$ is called the *Doppler factor*.

For a particle at rest the Doppler factor is equal to 1. In this case $\mathbf{A}(\mathbf{r}, t) = 0$, and

$$\varphi(\mathbf{r}, t) = \left[\frac{e}{R}\right]_{\tau=t\mp R/c};$$

but, since the particle is at rest, $\mathbf{r}_p = \textbf{constant}$ and therefore R is independent of τ. It follows that $\varphi(\mathbf{r}, t)$ is independent of t, and is given by

$$\varphi(\mathbf{r}, t) = \frac{e}{R} = \frac{e}{|\mathbf{r} - \mathbf{r}_p|}, \tag{19.27}$$

the usual Coulomb potential. Both retarded and advanced potentials are thus seen to reduce to the same thing in the case of a particle at rest.

EXAMPLE 2. Find the retarded and the advanced potentials for a particle moving with a uniform velocity **v** along the X-axis.

In this case we may assume

$$x_p(\tau) = v\tau, \qquad y_p = z_p = 0,$$

so that

$$R = [(x - v\tau)^2 + y^2 + z^2]^{1/2}. \tag{19.28}$$

In Eqs. (19.25) and (19.26) R is taken at

$$\tau = t \mp \frac{R}{c}. \tag{19.29}$$

We must solve Eqs. (19.28) and (19.29) simultaneously in order to determine R for arbitrary **r** and t. We have

$$R^2 = (x - v\tau)^2 + y^2 + z^2 = (x - vt \pm Rv/c)^2 + y^2 + z^2,$$

or, solving for R,

$$R = \frac{\pm(x - vt)v/c + [(x - vt)^2 + (1 - v^2/c^2)(y^2 + z^2)]^{1/2}}{1 - v^2/c^2}, \tag{19.30}$$

where the sign of the square root is determined by the fact that for $v = 0$, $R = +(x^2 + y^2 + z^2)^{1/2}$.

The X component of R is

$$R_x = (\mathbf{r} - \mathbf{r}_p)_x = x - x_p = x - v\tau = x - vt \pm \frac{Rv}{c},$$

so that

$$\mathbf{R} \cdot \mathbf{v} = R_x v = v\left(x - vt \pm R\frac{v}{c}\right)$$

and

$$\begin{aligned} R\left(1 \mp \frac{\mathbf{R} \cdot \mathbf{v}}{cR}\right) &= R \mp \frac{v}{c}\left(x - vt \mp R\frac{v}{c}\right) \\ &= R\left(1 - \frac{v^2}{c^2}\right) \mp \frac{v}{c}(x - vt) \\ &= \left[(x - vt)^2 + \left(1 - \frac{v^2}{c^2}\right)(y^2 + z^2)\right]^{1/2}, \end{aligned} \tag{19.31}$$

where in the last step we have made use of Eq. (19.30). Substitution of this result in Eqs. (19.25) and (19.26) gives

$$\varphi(\mathbf{r}, t) = \frac{e}{[(x - vt)^2 + (1 - v^2/c^2)(y^2 + z^2)]^{1/2}} \tag{19.32a}$$

and

$$\mathbf{A}(\mathbf{r}, t) = \frac{e\mathbf{v}/c}{[(x - vt)^2 + (1 - v^2/c^2)(y^2 + z^2)]^{1/2}}. \qquad (19.32b)$$

These potentials lead to the values of $\mathscr{E}$ and $\mathscr{H}$ found in Section 16, Eqs. (16.50) and (16.55). It is to be noted that both retarded and advanced potentials are the same. This is to be explained by the symmetry properties of the world-line with respect to past and future, in case of uniform motion.

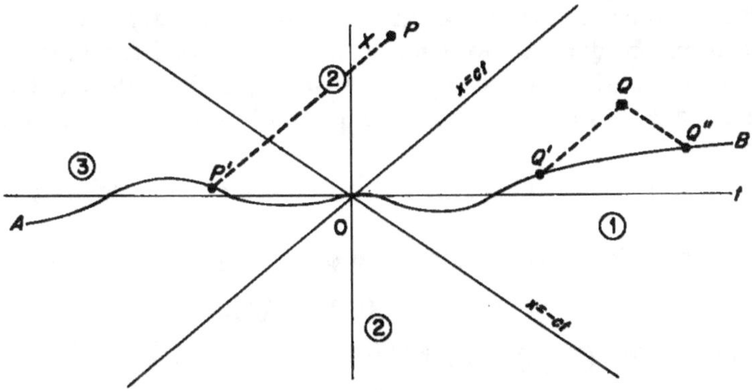

FIG. 19.1. World-line of a particle and the regions of space-time: (1) absolute future, (2) events having a spacelike interval from the origin, and (3) absolute past.

We shall now consider certain properties of the field produced by a moving charge, which will further illustrate the use of retarded and advanced potentials.

For this purpose let us divide space–time into three regions (see Figure 19.1) in the following way: Choose the origin at some point of the world-line, and for this point let $t = 0$. Then let region 1 be that part of space–time for which $t > 0$ and $r < ct$, i.e., the region of absolute future for the particle when it is at the origin; region 2 that portion for which $r > |ct|$; and region 3 the region of absolute past for the particle, i.e., $t < 0$, $r < |ct| = -ct$.

In the figure, as usual, only one of the space coordinates is plotted, namely x; hence one must use Figure 19.1 only for the purpose of making rather general deductions that are independent of the two-dimensional character of the figure.

Let us suppose that the potentials and their time derivatives are given for $t = 0$. These must, of course, be such as to satisfy Eq. (19.3). Equations (19.1) and (19.2) will now enable us to determine φ and $\mathbf{A}$ for all time, if the world-line is known. To simplify our discussion let us restrict ourselves to a consideration of the scalar potential φ. As we have seen, the potential $\varphi(\mathbf{r}, t)$ may be written

$$\varphi(\mathbf{r}, t) = \varphi_{\text{ret}}(\mathbf{r}, t) + \psi_{\text{ret}}(\mathbf{r}, t) = \varphi_{\text{adv}}(\mathbf{r}, t) + \psi_{\text{adv}}(\mathbf{r}, t), \quad (19.33)$$

where $\psi_{\text{ret}}(\mathbf{r}, t)$ and $\psi_{\text{adv}}(\mathbf{r}, t)$ are the complementary functions. The latter must be chosen so as to satisfy the initial and the boundary conditions. In our case the initial conditions, i.e., $\varphi(\mathbf{r}, t)$ and $\partial\varphi(\mathbf{r}, t)/\partial t$ at $t = 0$, are sufficient to permit determination of the complementary functions.

This may be done, for example, as follows: Since ψ (either retarded or advanced) satisfies the homogeneous wave equation (19.16), it may be represented (as seen in Section 18, Eq. (18.47), using Eq. (18.54)) as

$$\psi(\mathbf{r}, t) = \left(\frac{1}{2\pi}\right)^{3/2} \int \{\psi(\boldsymbol{k})e^{i(\boldsymbol{k}\cdot\mathbf{r} - ckt)} + \text{complex conjugate}\}\, d^3\boldsymbol{k}; \quad (19.34)$$

therefore (as in Eq. (18.52), but with ψ replacing f),

$$\psi(\boldsymbol{k}) = \tfrac{1}{2}[A(\boldsymbol{k}) + B(\boldsymbol{k})], \quad (19.35)$$

where (see Eqs. (18.50) and (18.51))

$$A(\boldsymbol{k}) \equiv \left(\frac{1}{2\pi}\right)^{3/2} \int \psi(\mathbf{r}, 0)e^{-i\boldsymbol{k}\cdot\mathbf{r}}\, dV \quad (19.36a)$$

and

$$B(\boldsymbol{k}) \equiv \frac{i}{ck}\left(\frac{1}{2\pi}\right)^{3/2} \int \frac{\partial\psi}{\partial t}(\mathbf{r}, 0)e^{-i\boldsymbol{k}\cdot\mathbf{r}}\, dV. \quad (19.36b)$$

Thus $\psi(\boldsymbol{k})$, and therefore $\psi(\mathbf{r}, t)$, will be determined as soon as we know $\psi(\mathbf{r}, 0)$ and $\partial\psi(\mathbf{r}, 0)/\partial t$. But from Eq. (19.33) we have

$$\psi_{\text{ret}}(\mathbf{r}, 0) = \varphi(\mathbf{r}, 0) - \varphi_{\text{ret}}(\mathbf{r}, 0) \quad (19.37a)$$

and

$$\frac{\partial\psi_{\text{ret}}(\mathbf{r}, 0)}{\partial t} = \frac{\partial\varphi}{\partial t}(\mathbf{r}, 0) - \frac{\partial\varphi_{\text{ret}}(\mathbf{r}, 0)}{\partial t}, \quad (19.37b)$$

where in each equation the first term of the right-hand member is given and the second term may be calculated from Eq. (19.25) as soon as the world-line is given. Similar equations hold, of course, for the advanced potentials.

Consider now any point P and let its coordinates be $(\mathbf{r}, t)$ lying in region 2. The retarded potential at this point will, according to Eq. (19.25), depend upon the distance $R = |\mathbf{r} - \mathbf{r}_p|$ and the velocity v of the particle at the point $\mathbf{r}_p$. This latter point is the location of the particle at the time

$\tau = t - R/c$. Geometrically the last equation is represented on our simplified diagram by the line PP' drawn from P parallel to the line $x = ct$ of the diagram. The point of intersection of this line with the world-line AOB, is the point P', whose location in three dimensions is given by $\mathbf{r}_p$. It must be remembered, of course, that R is not the difference between the ordinates of P and P' on the diagram, but the distance between the corresponding points $\mathbf{r}$ and $\mathbf{r}_p$ in three dimensions. The important thing here is the fact that the retarded potentials in region 2 will depend only upon points like P' lying on the portion AO of the world-line, and not in any way upon the portion OB, in the future from O. In particular this will apply to $\varphi_{\text{ret}}(\mathbf{r}, 0)$ and $\partial\varphi_{\text{ret}}(\mathbf{r}, 0)/\partial t$; consequently, by Eqs. (19.35), (19.36), and (19.37), $\psi_{\text{ret}}(\boldsymbol{k})$ will also depend only on the portion AO of the world-line. Thus, by Eq. (19.34),

$$\psi_{\text{ret}}(\mathbf{r}, t) \text{ is everywhere independent of } OB. \tag{19.38}$$

Similarly, we can obtain the result

$$\psi_{\text{adv}}(\mathbf{r}, t) \text{ is everywhere independent of } AO. \tag{19.39}$$

Consequently, since $\varphi_{\text{ret}}(\mathbf{r}, t)$ in region 2 is determined by the portion AO of the world-line, Eqs. (19.33) and (19.38) require that $\varphi(\mathbf{r}, t)$ in region 2 be independent of OB. A similar argument, moreover, could be carried out with the advanced potential, showing that $\varphi(\mathbf{r}, t)$ in region 2 is also independent of AO. We conclude therefore that

$$\varphi(\mathbf{r}, t) \text{ in region 2 is independent of the world-line,} \tag{19.40}$$

being thus completely determined by the given φ and $\partial\varphi/\partial t$ at $t = 0$.

Consider now the situation in region 1. With each point Q, coordinates $(\mathbf{r}, t)$, in this region there may be associated two points Q' and Q''. Q' bears the same relation to Q as P' does to P; while Q'' represents the position of the particle at the time $t + R/c$, where R is the distance between Q and Q'' in three dimensions. Thus the behavior of the particle at Q'' determines the advanced potential at the point Q. In Figure 19.1, Q'' is the intersection of the world-line of the particle and the line parallel to the line $z = -ct$ passing through Q. Hence $\varphi_{\text{adv}}(\mathbf{r}, t)$ in region 1 depends only on the portion OB of the world-line. This together with the result (19.39) demands that

$$\varphi(\mathbf{r}, t) \text{ in region 1 be independent of } AO. \tag{19.41}$$

Further, Q' also lies on OB, so that in region 1 $\varphi_{\text{ret}}(\mathbf{r}, t)$ is independent of AO. Combining this with the result (19.41), we see that in region 1

$\psi_{\text{ret}}(\mathbf{r}, t) = \varphi(\mathbf{r}, t) - \varphi_{\text{ret}}(\mathbf{r}, t)$ is independent of AO. This and the result (19.38) give

$$\psi_{\text{ret}}(\mathbf{r}, t) \text{ in region 1 is independent of the world-line.} \tag{19.42}$$

A similar argument will hold for region 3, but with advanced potential replacing retarded. We may summarize our results thus: *In region 1 the complementary function $\psi_{\text{ret}}(\mathbf{r}, t)$, in region 2 the total potential $\varphi(\mathbf{r}, t)$, and in region 3 the complementary function $\psi_{\text{adv}}(\mathbf{r}, t)$ are independent of the shape of the world-line, being completely determined by the data given for* $t = 0$.

Thus, for example, if we wish to investigate the effect upon the field in region 3 of various possible past behaviors of the particle, and if in addition we know the field at $t = 0$, the use of the advanced potential would be more convenient, for the complementary function ψ_{adv} would be the same for all such behaviors.

The field produced by a particle. Let us now consider the following question: "What is the field produced by a charge moving in some given way?" Referring to the situation we have been discussing, the field arbitrarily given at the time $t = 0$ cannot be considered as being produced exclusively by the particle, but must be regarded as being complicated by the presence in most cases of some *external* field.

We would like to separate this external field from the field due to the particle and write

$$\varphi(\mathbf{r}, t) = \varphi_{\text{particle}}(\mathbf{r}, t) + \varphi_{\text{external}}(\mathbf{r}, t). \tag{19.43}$$

If we use merely the fact that $\varphi_{\text{particle}}(\mathbf{r}, t)$ must depend only upon the world-line of the particle and satisfy Eq. (19.1), the separation into two terms is not uniquely defined. Thus, if we find one separation, say

$$\varphi(\mathbf{r}, t) = \varphi'_{\text{particle}}(\mathbf{r}, t) + \varphi'_{\text{external}}(\mathbf{r}, t),$$

then

$$\varphi_{\text{particle}} = \varphi'_{\text{particle}} + \Phi_0 \quad \text{and} \quad \varphi_{\text{external}} = \varphi'_{\text{external}} - \Phi_0,$$

where Φ_0 satisfies Eq. (19.16) and is independent of the shape of the world-line, will be also a possible separation satisfying the conditions stated above.

We therefore make two assumptions that together serve to define the phrase *field produced by a particle.*

Assumption 1. At any time $t = t_0$ the potentials due to a particle and their derivatives are independent of the behavior of the particle subsequent *to t_0, depending exclusively upon the portion of the particle's world-line corresponding to $t \leq t_0$.* This is merely an explicit statement of the usual idea that the *effect* cannot occur *before* the *cause*.

Assumption 2. The potentials at the time t_0 due to a particle that has been at rest for all $t \leq t_0$ are the Coulomb potentials, a possible choice of which is: $\mathbf{A} = 0$, $\varphi = e/R$.

Combining Eq. (19.43) with Eq. (19.33), we have

$$\varphi_{\text{particle}} = \varphi - \varphi_{\text{external}} = (\varphi_{\text{ret}} + \psi_{\text{ret}}) - \varphi_{\text{external}}$$
$$= \varphi_{\text{ret}} + \Phi, \tag{19.44}$$

where

$$\Phi = \psi_{\text{ret}} - \varphi_{\text{external}}.$$

Equation (19.44) shows that Φ is equal to $\varphi_{\text{particle}} - \varphi_{\text{ret}}$, both of which depend only upon the world-line of the particle; therefore, Φ depends only on the world-line, not on the given field at $t = 0$. Furthermore, since both $\varphi_{\text{particle}}$ and φ_{ret} satisfy Eq. (19.1), their difference Φ must satisfy the wave equation (19.16). Thus Φ will be determined everywhere and for all times if for any moment of time $t = t_0$ we know $\Phi(\mathbf{r}, t_0)$ and $\partial\Phi(\mathbf{r}, t_0)/\partial t$.

Take, for example $t_0 = 0$. Referring to Figure 19.1, we see that $\varphi_{\text{ret}}(\mathbf{r}, 0)$ and $\partial\varphi_{\text{ret}}(\mathbf{r}, 0)/\partial t$ will depend only upon the portion AO of the world-line. This, by Assumption 1, will also be true of $\varphi_{\text{particle}}(\mathbf{r}, 0)$ and $\partial\varphi_{\text{particle}}(\mathbf{r}, 0)/\partial t$; therefore, by Eq. (19.44), $\Phi(\mathbf{r}, 0)$ and $\partial\Phi(\mathbf{r}, 0)/\partial t$ will depend only upon the portion AO of the world-line.

A similar argument may be applied to any time t_0 to show that $\Phi(\mathbf{r}, t_0)$ and $\partial\Phi(\mathbf{r}, t_0)/\partial t$, and therefore $\Phi(\mathbf{r}, t)$ for all $\mathbf{r}$ and t, are determined by the portion of the world-line prior to t_0.

Thus Φ is independent of the shape of any finite portion of the world-line; for, if this portion extends from $t = t_1$ to $t = t_2 > t_1$, we need only to take $t_0 < t_1$. Moreover, since Φ is independent of the shape of the world-line in the region $t_1 < t < \infty$ for all t_1, we can conclude, by letting $t_1 \to -\infty$, that Φ is independent of the shape of the world-line altogether. It is thus the same for all world-lines.

On the other hand, if the particle has been at rest for all $t \leq t_0$, Assumption 2 gives $\varphi_{\text{particle}} = \varphi_{\text{coulomb}}$; but, by Example 1 above, in this case $\varphi_{\text{ret}} = \varphi_{\text{coulomb}}$. Thus, as seen from this special case, $\Phi = 0$, and Eq. (19.44) becomes

$$\varphi_{\text{particle}}(\mathbf{r}, t) = \varphi_{\text{ret}}(\mathbf{r}, t). \tag{19.45}$$

A similar formula cannot be derived for the advanced potential, as this would necessitate the denial of Assumption 1.

The above discussion depends upon the implicit assumption that the phrase *field produced by a particle* has a physical meaning. Since, however, the total field cannot be separated experimentally into two terms, $\varphi_{\text{particle}} + \varphi_{\text{external}}$, the concept of the field produced by a particle appears highly artificial.

EXAMPLE 3. Find the electric and magnetic fields produced by a moving particle of charge e.

In accordance with the above we must use the retarded potentials. It is convenient to start with Eq. (19.21) and its analog for the vector potential; thus, making use of Eq. (A8.86),

$$\varphi(\mathbf{r}, t) = e\int_{-\infty}^{\infty} \frac{\delta(t-\tau-R/c)}{R}\,d\tau = \frac{e}{2\pi}\iint_{-\infty}^{\infty} e^{i\mu(t-\tau-R/c)}\frac{d\mu\,d\tau}{R} \tag{19.46}$$

and

$$\mathbf{A}(\mathbf{r}, t) = \frac{e}{2\pi c}\iint_{-\infty}^{\infty} e^{i\mu(t-\tau-R/c)}\frac{\mathbf{v}\,d\mu\,d\tau}{R}. \tag{19.47}$$

Here as before $\mathbf{R} = \mathbf{r} - \mathbf{r}_p(\tau)$ as in Eq. (19.21).

From Eqs. (19.47) and (A3.43) we have

$$\nabla\varphi(\mathbf{r}, t) = \frac{e}{2\pi}\iint_{-\infty}^{\infty} e^{i\mu(t-\tau)}\,\nabla\left(\frac{e^{-i\mu R/c}}{R}\right)d\mu\,d\tau$$

$$= \frac{e}{2\pi}\iint_{-\infty}^{\infty} e^{i\mu(t-\tau)}\,\nabla R\,\frac{\partial}{\partial R}\left(\frac{e^{-i\mu R/c}}{R}\right)d\mu\,d\tau.$$

If we let $\mathbf{u} = \nabla R = \mathbf{R}/R$, then

$$\nabla\varphi = -\frac{e}{2\pi}\iint_{-\infty}^{\infty}\left(\frac{1}{R^2} + \frac{i\mu}{cR}\right)\mathbf{u}e^{i\mu(t-\tau-R/c)}\,d\mu\,d\tau$$

$$= \frac{-e}{2\pi}\iint_{-\infty}^{\infty}\mathbf{u}e^{i\mu(t-\tau-R/c)}\frac{d\mu\,d\tau}{R^2} - \frac{e}{2\pi c}\frac{\partial}{\partial t}\iint_{-\infty}^{\infty}\mathbf{u}e^{i\mu(t-\tau-R/c)}\frac{d\mu\,d\tau}{R}$$

and similarly

$$\frac{\partial \mathbf{A}}{\partial t} = \frac{e}{2\pi} \frac{\partial}{\partial t} \iint_{-\infty}^{\infty} \mathbf{v} e^{i\mu(t-\tau-R/c)} \frac{d\mu\, d\tau}{cR}.$$

Thus

$$\mathscr{E} = -\nabla\varphi - \frac{1}{c}\frac{\partial \mathbf{A}}{\partial t} = \frac{e}{2\pi} \iint_{-\infty}^{\infty} \mathbf{u} e^{i\mu(t-\tau-R/c)} \frac{d\mu\, d\tau}{R^2}$$

$$+ \frac{e}{2\pi c}\frac{\partial}{\partial t} \iint_{-\infty}^{\infty} \frac{\mathbf{u} - \mathbf{v}/c}{R} e^{i\mu(t-\tau-R/c)}\, d\mu\, d\tau. \quad (19.48)$$

Analogously

$$\mathscr{H} = \nabla \times \mathbf{A} = \frac{e}{2\pi c} \iint_{-\infty}^{\infty} e^{i\mu(t-\tau)}\, \nabla \times \left(\frac{e^{-i\mu R/c}\mathbf{v}}{R}\right) d\mu\, d\tau$$

$$= -\frac{e}{2\pi c} \iint_{-\infty}^{\infty} e^{i\mu(t-\tau)}\, \mathbf{v} \times \nabla\left(\frac{e^{-i\mu R/c}}{R}\right) d\mu\, d\tau$$

$$= \frac{e}{2\pi c} \iint_{-\infty}^{\infty} \frac{\mathbf{v} \times \mathbf{u}}{R^2} e^{i\mu(t-\tau-R/c)}\, d\mu\, d\tau$$

$$+ \frac{e}{2\pi c^2}\frac{\partial}{\partial t} \iint_{-\infty}^{\infty} \frac{\mathbf{v} \times \mathbf{u}}{R} e^{i\mu(t-\tau-R/c)}\, d\mu\, d\tau. \quad (19.49)$$

Integration with respect to μ gives

$$\mathscr{E} = e\int_{-\infty}^{\infty} \frac{\mathbf{u}}{R^2}\,\delta\left(t - \tau - \frac{R}{c}\right) d\tau + \frac{e}{c}\frac{\partial}{\partial t}\int_{-\infty}^{\infty} \frac{\mathbf{u} - \mathbf{v}/c}{R}\,\delta\left(t - \tau - \frac{R}{c}\right) d\tau \quad (19.50)$$

and

$$\mathscr{H} = \frac{e}{c}\int_{-\infty}^{\infty} \frac{\mathbf{v} \times \mathbf{u}}{R^2}\,\delta\left(t - \tau - \frac{R}{c}\right) d\tau + \frac{e}{c^2}\frac{\partial}{\partial t}\int_{-\infty}^{\infty} \frac{\mathbf{v} \times \mathbf{u}}{R}\,\delta\left(t - \tau - \frac{R}{c}\right) d\tau. \quad (19.51)$$

Integration with respect to τ is performed as for Eq. (19.21). This gives

$$\mathscr{E} = \left[\frac{e\mathbf{u}}{KR^2}\right]_{\tau=t-R/c} + \frac{\partial}{\partial t}\left[\frac{e(\mathbf{u} - \mathbf{v}/c)}{cKR}\right]_{\tau=t-R/c} \quad (19.52)$$

and

$$\mathscr{H} = \left[\frac{e\mathbf{v}\times\mathbf{u}}{cKR^2}\right]_{\tau=t-R/c} + \frac{\partial}{\partial t}\left[\frac{e\mathbf{v}\times\mathbf{u}}{c^2KR}\right]_{\tau=t-R/c}, \tag{19.53}$$

where

$$K = 1 - \frac{\mathbf{R}\cdot\mathbf{v}}{cR} = 1 - \frac{\mathbf{u}\cdot\mathbf{v}}{c} = 1 + \frac{1}{c}\frac{\partial R}{\partial \tau} \tag{19.54}$$

is the *Doppler factor.*

Equations (19.52) and (19.53) require the evaluation of expressions of the form

$$\frac{\partial}{\partial t}[(\cdots)]_{\tau=t-R/c} = \frac{\partial}{\partial t}[F(\mathbf{r},\mathbf{v},\tau)]_{\tau=t-R/c} = \frac{\partial}{\partial t}F(\mathbf{r},\mathbf{v},\tau(\mathbf{r},t))$$

$$= \left[\frac{\partial F}{\partial \tau}\frac{\partial \tau}{\partial t}\right]_{\tau=t-R/c},$$

where $\tau(\mathbf{r}, t)$ expresses the dependence of τ on $\mathbf{r}$ and t that is imposed by the equation

$$\tau = t - \frac{R}{c} = t - \frac{|\mathbf{r}-\mathbf{r}_p(\tau)|}{c}.$$

If we differentiate this last equation with respect to t, we find that

$$\frac{\partial \tau}{\partial t} = 1 - \frac{1}{c}\frac{\partial R}{\partial t} = 1 - \frac{1}{c}\frac{\partial R}{\partial \tau}\frac{\partial \tau}{\partial t}$$

or

$$\frac{\partial \tau}{\partial t} = \frac{1}{\left(1 + \dfrac{1}{c}\dfrac{\partial R}{\partial \tau}\right)} = \frac{1}{K}.$$

We thus conclude that

$$\frac{\partial}{\partial t}[(\cdots)]_{\tau=t-R/c} = \left[\frac{1}{K}\frac{\partial}{\partial \tau}(\cdots)\right]_{\tau=t-R/c}.$$

Using this formula, one may replace Eqs. (19.52) and (19.53) by

$$\mathscr{E}(\mathbf{r},t) = e\left[\frac{\mathbf{u}}{KR^2} + \frac{1}{cK}\frac{\partial}{\partial \tau}\left(\frac{\mathbf{R}}{KR^2}\right) - \frac{1}{c^2K}\frac{\partial}{\partial \tau}\left(\frac{\mathbf{v}}{KR}\right)\right]_{\tau=t-R/c}$$

and

$$\mathscr{H}(\mathbf{r},t) = e\left[\frac{\mathbf{v}\times\mathbf{u}}{cKR^2} + \frac{1}{c^2K}\frac{\partial}{\partial \tau}\left(\frac{\mathbf{v}\times\mathbf{u}}{KR}\right)\right]_{\tau=t-R/c}.$$

After considerable algebra using the formulas

$$\frac{\partial \mathbf{R}}{\partial \tau} = -\frac{\partial \mathbf{r}_p}{\partial \tau} = -\mathbf{v}$$

$$\frac{\partial R}{\partial \tau} = \frac{\partial R}{\partial \mathbf{r}_p} \cdot \frac{\partial \mathbf{r}_p}{\partial \tau} = -\frac{\mathbf{R} \cdot \mathbf{v}}{R}$$

$$\frac{\partial K}{\partial \tau} = \frac{\partial}{\partial \tau}\left(1 - \frac{\mathbf{R} \cdot \mathbf{v}}{cR}\right) = \frac{v^2 - \mathbf{R} \cdot \dot{\mathbf{v}}}{cR} + \frac{\mathbf{R} \cdot \mathbf{v}}{cR^2}\frac{\partial R}{\partial \tau} = \frac{v^2 - \mathbf{R} \cdot \dot{\mathbf{v}} - (\mathbf{u} \cdot \mathbf{v})^2}{cR}$$

one obtains the formulas

$$\mathscr{E}(\mathbf{r}, t) = e\left[\frac{(\mathbf{u} - \mathbf{v}/c)(\dot{\mathbf{v}} \cdot \mathbf{R} + c^2 - v^2)}{c^2K^3R^2} - \frac{\dot{\mathbf{v}}}{c^2K^2R}\right]_{\tau=t-R/c} \tag{19.55}$$

and

$$\mathscr{H}(\mathbf{r}. t) = e\left[\frac{(\mathbf{v} \times \mathbf{u})(\dot{\mathbf{v}} \cdot \mathbf{R} + c^2 - v^2)}{c^3K^3R^2} + \frac{\dot{\mathbf{v}} \times \mathbf{u}}{c^2K^2R}\right]_{\tau=t-R/c}, \tag{19.56}$$

where $\mathbf{v} = d\mathbf{r}_p/d\tau$, $\dot{\mathbf{v}} = d\mathbf{v}/d\tau$, $\mathbf{R} = \mathbf{r} - \mathbf{r}_p$, $\mathbf{u} = \mathbf{R}/R$, and $K = 1 - (\mathbf{u} \cdot \mathbf{v})/c$.

EXAMPLE 4. Calculate the radiation from a moving charge.

Consider first the case when $\dot{\mathbf{v}} = 0$; then, by Eqs. (19.55) and (19.56),

$$\mathscr{E} = e\left[\frac{(\mathbf{u} - \mathbf{v}/c)(1 - v^2/c^2)}{K^3R^2}\right] \tag{19.57}$$

and

$$\mathscr{H} = e\left[\frac{\mathbf{v} \times \mathbf{u}\,(1 - v^2/c^2)}{cK^3R^2}\right]. \tag{19.58}$$

Here, as in the remainder of this section, all quantities dependent on τ are taken at $\tau = t - R/c$. From Eqs. (19.57) and (19.58) it follows that Poynting's vector is proportional to $1/R^4$, and its integral over a surface at large distance approaches zero as $1/R^2$. The energy thus remains in the finite portion of space around the charges, that is, there is no radiation.

If $\dot{\mathbf{v}} \neq 0$, the terms not containing $\dot{\mathbf{v}}$ become negligible at large distances. Thus, for large R the fields approach

$$\mathscr{E} = e\left[\frac{(\mathbf{u} - \mathbf{v}/c)\,\dot{\mathbf{v}} \cdot \mathbf{u} - K\dot{\mathbf{v}}}{c^2K^3R}\right], \tag{19.59}$$

$$\mathscr{H} = e\left[\frac{(\mathbf{v} \times \mathbf{u})\,\dot{\mathbf{v}} \cdot \mathbf{u} + cK\,\dot{\mathbf{v}} \times \mathbf{u}}{c^3K^3R}\right]. \tag{19.60}$$

We note that in this case

$$\mathbf{u} \times \mathscr{E} = e\left[\mathbf{u} \times \frac{(\mathbf{u} - \mathbf{v}/c)\dot{\mathbf{v}} \cdot \mathbf{u} - K\dot{\mathbf{v}}}{c^2K^3R}\right]$$
$$= e\left[\frac{-(\mathbf{u} \times \mathbf{v}/c)\,\dot{\mathbf{v}} \cdot \mathbf{u} - K\,\mathbf{u} \times \dot{\mathbf{v}}}{c^2K^3R}\right],$$

so that

$$\mathbf{u} \times \mathscr{E} = \mathscr{H}. \tag{19.61}$$

Thus $\mathscr{H}$ is perpendicular to $\mathbf{u}$ and to $\mathscr{E}$. Also

$$\mathbf{u} \cdot \mathscr{E} = e\left[\mathbf{u} \cdot \frac{(\mathbf{u} - \mathbf{v}/c)\,\dot{\mathbf{v}} \cdot \mathbf{u} - K\dot{\mathbf{v}}}{c^2K^3R}\right]$$
$$= e\left[\frac{(1 - \mathbf{u} \cdot \mathbf{v}/c)\,\dot{\mathbf{v}} \cdot \mathbf{u} - K\,\mathbf{u} \cdot \dot{\mathbf{v}}}{c^2K^3R}\right] = 0, \tag{19.62}$$

and hence $\mathbf{u}$ is perpendicular to $\mathscr{E}$. Since $\mathbf{u}$ is a unit vector, Eqs. (19.61) and (19.62) mean that $\mathscr{E} = \mathscr{H}$, and that $\mathscr{E}$, $\mathscr{H}$, and $\mathbf{u}$ form a right-handed set of orthogonal vectors. Poynting's vector is therefore given by

$$\Upsilon = \frac{c}{4\pi}\,\mathscr{E} \times \mathscr{H} = \frac{c}{4\pi}\,\mathscr{E}^2\mathbf{u}. \tag{19.63}$$

From Eq. (19.59)

$$\mathscr{E}^2 = \frac{e^2}{c^4K^6R^2}\left\{(\dot{\mathbf{v}} \cdot \mathbf{u})^2\left(\mathbf{u} - \frac{\mathbf{v}}{c}\right)^2 + K^2\dot{\mathbf{v}}^2 - 2K\dot{\mathbf{v}} \cdot \left(\mathbf{u} - \frac{\mathbf{v}}{c}\right)\dot{\mathbf{v}} \cdot \mathbf{u}\right\},$$

which, remembering Eq. (19.54), becomes

$$\mathscr{E}^2 = \frac{e^2}{c^4K^6R^2}\left\{(\dot{\mathbf{v}} \cdot \mathbf{u})^2\left(1 + \frac{v^2}{c^2} - \frac{2\mathbf{u} \cdot \mathbf{v}}{c}\right)\right.$$
$$\left. + K^2\dot{v}^2 - 2\left(1 - \frac{\mathbf{u} \cdot \mathbf{v}}{c}\right)\left(\mathbf{u} \cdot \dot{\mathbf{v}} - \frac{\mathbf{v} \cdot \dot{\mathbf{v}}}{c}\right)\dot{\mathbf{v}} \cdot \mathbf{u}\right\}$$
$$= \frac{e^2}{c^4K^6R^2}\left\{K^2\dot{v}^2 - \left(1 - \frac{v^2}{c^2}\right)(\dot{\mathbf{v}} \cdot \mathbf{u})^2 + \frac{2\mathbf{u} \cdot \dot{\mathbf{v}}\,\mathbf{v} \cdot \dot{\mathbf{v}}}{c}\left(1 - \frac{\mathbf{u} \cdot \mathbf{v}}{c}\right)\right\}. \tag{19.64}$$

Hence

$$\Upsilon = \frac{e^2}{4\pi c^3K^6R^2}\left\{K^2\dot{v}^2 - \left(1 - \frac{v^2}{c^2}\right)(\mathbf{u} \cdot \dot{\mathbf{v}})^2 + \frac{2K\mathbf{u} \cdot \dot{\mathbf{v}}\,\mathbf{v} \cdot \dot{\mathbf{v}}}{c}\right\}\mathbf{u}. \tag{19.65}$$

If we choose the coordinate system so that at the time $\tau = t - R/c$ the particle is momentarily at rest at the origin; then $\mathbf{v} = 0$, $K = 1$, and $R = r$. In that case Eq. (19.65) becomes

$$\Upsilon = \frac{e^2}{4\pi c^3 r^2}\{\dot{v}^2 - (\mathbf{u}\cdot\dot{\mathbf{v}})^2\}\mathbf{u} = \frac{e^2(\mathbf{u}\times\dot{\mathbf{v}})^2\mathbf{u}}{4\pi c^3 r^2}. \tag{19.66}$$

If we integrate this vector over the surface of a sphere of very large radius $r = a$, then $\mathbf{u} = \mathbf{r}/r$ is parallel to $d\mathbf{S}$ and hence

$$\int \Upsilon\cdot d\mathbf{S} = \frac{e^2}{4\pi c^3}\int \frac{(\mathbf{u}\times\dot{\mathbf{v}})^2}{a^2}\,\mathbf{u}\cdot d\mathbf{S} = \frac{e^2}{4\pi c^3 a^2}\int (\mathbf{u}\times\dot{\mathbf{v}})^2\,dS$$

$$= \frac{e^2}{4\pi c^3}\int (\mathbf{u}\times\dot{\mathbf{v}})^2\,d\omega, \tag{19.67}$$

where $d\omega = dS/a^2$ is a differential element of the solid angle about the origin. If θ is the angle between $\mathbf{u}$ and $\dot{\mathbf{v}}$, Eq. (19.67) indicates that $e^2\dot{v}^2\sin^2\theta/4\pi c^3$ is the radiation energy per unit of time per unit solid angle emitted at the time $\tau = t - a/c$ in a direction forming an angle θ with the acceleration $\dot{\mathbf{v}}$. Introducing a polar coordinate system with the polar axis along $\dot{\mathbf{v}}$, we have

$$d\omega = \sin\theta\,d\theta\,d\varphi.$$

The total radiation passing through the surface per unit of time is therefore

$$\int \Upsilon\cdot d\mathbf{S} = \frac{e^2}{4\pi c^3}\int (\mathbf{u}\times\dot{\mathbf{v}})^2\,d\omega = \frac{e^2\dot{v}^2}{4\pi c^3}\int_0^{\pi}\sin^3\theta\,d\theta\int_0^{2\pi}d\varphi$$

$$= \frac{2e^2\dot{v}^2}{3c^3}. \tag{19.68}$$

20. Spherical Waves

The electromagnetic radiation from a system of charges contained in a finite region of space R, when observed at large distances from the system, will be shown to take the form of spherical waves. We will also show that these waves at sufficiently large distances depend on the changes with time of the lowest-order nonzero electric and magnetic moments. The electric field was related to the electric moments in Section 9 for the electrostatic case of stationary charges. This treatment must now be generalized to describe the electromagnetic field produced by a system of moving charges.

We shall start with Eqs. (19.48) and (19.49), which for a system of particles can be written ($\mathbf{R}_s$, $\mathbf{u}_s$, and $\mathbf{v}_s$ are functions of τ but not of t)

$$\mathscr{E}(\mathbf{r}, t) = \sum_s \frac{e_s}{2\pi} \iint_{-\infty}^{\infty} \left(\frac{\mathbf{u}_s}{R_s^2} + \frac{(\mathbf{u}_s - \mathbf{v}_s/c)}{cR_s} \frac{\partial}{\partial t} \right) e^{i\mu(t-\tau-R_s/c)}\, d\mu\, d\tau \quad (20.1)$$

and

$$\mathscr{H}(\mathbf{r}, t) = \sum_s \frac{e_s}{2\pi c} \iint_{-\infty}^{\infty} \frac{\mathbf{v}_s \times \mathbf{u}_s}{R_s} \left(\frac{1}{R_s} + \frac{1}{c} \frac{\partial}{\partial t} \right) e^{i\mu(t-\tau-R_s/c)}\, d\mu\, d\tau. \quad (20.2)$$

Here $\mathbf{r}_s$ is the position vector of the sth particle, and $\mathbf{R}_s = \mathbf{r} - \mathbf{r}_s$, $R_s = |\mathbf{r} - \mathbf{r}_s|$, and $\mathbf{u}_s = \mathbf{R}_s/R_s$. Performing the indicated differentiation, we obtain

$$\mathscr{E} = \sum_s \frac{e_s}{2\pi} \iint_{-\infty}^{\infty} \frac{\mathbf{u}_s}{R_s^2} + \frac{i\mu(\mathbf{u}_s - \mathbf{v}_s/c)}{cR_s} e^{i\mu(t-\tau-R_s/c)}\, d\mu\, d\tau, \quad (20.3)$$

$$\mathscr{H} = \sum_s \frac{e_s}{2\pi c} \iint_{-\infty}^{\infty} \frac{\mathbf{v}_s \times \mathbf{u}_s}{R_s} \left(\frac{1}{R_s} + \frac{i\mu}{c} \right) e^{i\mu(t-\tau-R_s/c)}\, d\mu\, d\tau. \quad (20.4)$$

Let the origin of our coordinate system be chosen at some suitable point C, called the center of the system, within the region R occupied by the particles The field is then to be determined only at those remote points $\mathbf{r}$ satisfying the requirement that $|\mathbf{r}| \gg |\mathbf{r}_s|$ for all position vectors $\mathbf{r}_s$ of the system of charges e_s. For such $\mathbf{r}$, since then $\mathbf{R}_s \simeq \mathbf{r}$, the first terms in the integrands of Eqs. (20.3) and (20.4) drop off as $1/r^2$ and contribute to $\mathscr{E}$ and $\mathscr{H}$ terms having this asymptotic behavior. Since the corresponding Poynting vector drops off as $1/r^4$, these terms do not contribute to the radiation field and will therefore be dropped. $\mathbf{R}_s$ will then enter in the integrand only through $F(R_s) = e^{-i\mu R_s/c}/R_s$ and through $\mathbf{u}_s$. For large r, R_s will approach r, and $F(R_s)$ will behave approximately as $e^{-i\mu r/c}/r$. We therefore let

$$F(R_s) = \frac{1}{r} e^{-i\mu r/c} f,$$

which defines f. Again letting $\mathbf{r}/r = \mathbf{u}$, we have

$$f = F(R_s) r e^{i\mu r/c} = \frac{r}{R_s} e^{-i(\mu/c)(R_s - r)} = \frac{r}{|\mathbf{r} - \mathbf{r}_s|} e^{-i(\mu/c)(|\mathbf{r}-\mathbf{r}_s|-r)}$$

$$= \frac{1}{|\mathbf{u} - \mathbf{r}_s/r|} e^{-i(\mu/c)(r(|\mathbf{u}-\mathbf{r}_s/r|-1)}$$

This is a function of $\mathbf{r}_s/r$ and $\mathbf{r}$ and can be expanded in powers of $\mathbf{r}_s/r$.

The term independent of $\mathbf{r}_s/r$, the only one that does not vanish as $\mathbf{r}_s/r \to 0$, is

$$\lim_{r_s/r\to 0} f = \lim_{r\to\infty} \frac{1}{|\mathbf{u} - \mathbf{r}_s/r|} \exp\left[-i\frac{\mu}{c}\frac{(|\mathbf{u} - \mathbf{r}_s/r| - 1)r_s}{|\mathbf{r}_s/r|}\right].$$

Since

$$\lim_{r\to\infty}\left|\mathbf{u} - \frac{\mathbf{r}_s}{r}\right| = |\mathbf{u}| = 1$$

and

$$\begin{aligned}\lim_{r\to\infty} r\left(\left|\mathbf{u} - \frac{\mathbf{r}_s}{r}\right| - 1\right) &= \lim_{r\to\infty}\left\{r\left[\left(\mathbf{u} - \frac{\mathbf{r}_s}{r}\right)\cdot\left(\mathbf{u} - \frac{\mathbf{r}_s}{r}\right)\right]^{1/2} - r\right\} \\ &= \lim_{r\to\infty}\left[r\left(1 - \frac{2}{r}\mathbf{u}\cdot\mathbf{r}_s + \frac{r_s^2}{r^2}\right)^{1/2} - r\right] \\ &= \lim_{r\to\infty}\left[r\left(1 - \frac{\mathbf{u}\cdot\mathbf{r}_s}{r}\right) - r\right] = -\mathbf{u}\cdot\mathbf{r}_s,\end{aligned}$$

we conclude that

$$f = e^{i(\mu/c)\mathbf{u}\cdot\mathbf{r}_s} + \text{terms vanishing as } \mathbf{r}\to\infty;$$

hence for large r

$$F(R_s) = \frac{1}{r}e^{-i(\mu/c)(r - \mathbf{u}\cdot\mathbf{r}_s)},$$

the other terms being negligible. Similarly, we find

$$\begin{aligned}\mathbf{u}_s = \frac{\mathbf{R}_s}{R_s} = \frac{\mathbf{r} - \mathbf{r}_s}{|\mathbf{r} - \mathbf{r}_s|} &= \frac{\mathbf{u} - \mathbf{r}_s/r}{|\mathbf{u} - \mathbf{r}_s/r|} \\ &= \mathbf{u} + \text{terms vanishing as } r\to\infty,\end{aligned}$$

and thus, for large r, Eqs. (20.3) and (20.4) become

$$\mathscr{E}(\mathbf{r}, t) = \frac{1}{r}\sum_s \frac{e_s}{2\pi c}\iint_{-\infty}^{\infty} i\mu\left(\mathbf{u} - \frac{\mathbf{v}_s}{c}\right)e^{i\mu(t-\tau-r/c)+i\mu\mathbf{u}\cdot\mathbf{r}_s/c}\,d\mu\,d\tau, \quad (20.5)$$

$$\mathscr{H}(\mathbf{r}, t) = \frac{1}{r}\sum_s \frac{e_s}{2\pi c^2}\iint_{-\infty}^{\infty} i\mu\,\mathbf{v}_s\times\mathbf{u}\,e^{i\mu(t-\tau-r/c)+i\mu\mathbf{u}\cdot\mathbf{r}_s/c}\,d\mu\,d\tau. \quad (20.6)$$

We shall now expand $e^{i\mu\mathbf{u}\cdot\mathbf{r}_s/c}$ in powers of the exponent and consider separately each power of $\mathbf{u}\cdot\mathbf{r}_s$. We note, however, that each factor $\mathbf{u}\cdot\mathbf{r}_s$ will always occur multiplied by $i\mu$. The effect of this will be, as we shall see, to convert $\mathbf{r}_s$ into $\dot{\mathbf{r}}_s = \mathbf{v}_s$; therefore, when collecting terms of the same power of $\mu\mathbf{r}_s$, we regard $\mathbf{v}_s$, occurring in Eq. (20.5), as being of the first power in $\mu\mathbf{r}_s$. Since

$$e^{i\mu\mathbf{u}\cdot\mathbf{r}_s/c} = 1 + \frac{i\mu\,\mathbf{u}\cdot\mathbf{r}_s}{c} - \frac{\mu^2(\mathbf{u}\cdot\mathbf{r}_s)^2}{2c^2} - \frac{i\mu^3(\mathbf{u}\cdot\mathbf{r}_s)^3}{6c^3} + \cdots,$$

we have for the terms of zeroth power in $\mu\mathbf{r}_s$

$$\mathscr{E}_0 = \frac{1}{2\pi c}\sum_s e_s \iint_{-\infty}^{\infty} \frac{i\mu\mathbf{u}}{r} e^{i\mu(t-\tau-r/c)}\,d\mu\,d\tau \tag{20.7}$$

and

$$\mathscr{H}_0 = 0. \tag{20.8}$$

Moreover, by letting $Q = \sum_s e_s$ and using Eqs. (A8.86) and (A8.21), we find that

$$\begin{aligned}\mathscr{E}_0 &= \frac{1}{2\pi c}\int_{-\infty}^{\infty}\frac{Q\mathbf{u}}{r}\left[\frac{\partial}{\partial t}\int_{-\infty}^{\infty} e^{i\mu(t-\tau-r/c)}\,d\mu\right]d\tau \\ &= \frac{1}{c}\int_{-\infty}^{\infty}\frac{Q\mathbf{u}}{r}\,\delta'\left(t-\tau-\frac{r}{c}\right)d\tau = \left[\frac{\dot{Q}\mathbf{u}}{cr}\right]_{\tau=t-r/c} = 0,\end{aligned} \tag{20.9}$$

because the total charge Q, being conserved, does not vary with time.

For radiation fields corresponding to terms of the first power in $\mu\mathbf{r}_s$ we have

$$\mathscr{E}_1 = \frac{1}{r}\sum_s \frac{e_s}{2\pi c^2}\iint_{-\infty}^{\infty} i\mu e^{i\mu(t-\tau-r/c)}(i\mu\,\mathbf{u}\cdot\mathbf{r}_s\,\mathbf{u} - \mathbf{v}_s)\,d\mu\,d\tau$$

and

$$\mathscr{H}_1 = \frac{1}{r}\sum_s \frac{e_s}{2\pi c^2}\iint_{-\infty}^{\infty} i\mu\,\mathbf{v}_s\times\mathbf{u}\,e^{i\mu(t-\tau-r/c)}\,d\mu\,d\tau.$$

Now $\sum e_s\mathbf{u}\cdot\mathbf{r}_s = \mathbf{u}\cdot\sum e_s\mathbf{r}_s = \mathbf{u}\cdot\mathscr{p}$ and $\sum e_s\mathbf{v}_s = \dot{\mathscr{p}}$ where $\mathscr{p}$ is the electric dipole moment introduced in Section 9; hence

$$\begin{aligned}\mathscr{E}_1 &= \frac{1}{2\pi c^2}\frac{1}{r}\int_{-\infty}^{\infty} d\tau\,\mathbf{u}\cdot\mathscr{p}\,\mathbf{u}\,\frac{\partial^2}{\partial t^2}\int_{-\infty}^{\infty} e^{i\mu(t-\tau-r/c)}\,d\mu \\ &\quad - \frac{1}{2\pi c^2}\frac{1}{r}\int_{-\infty}^{\infty} d\tau\,\dot{\mathscr{p}}\,\frac{\partial}{\partial t}\int_{-\infty}^{\infty} e^{i\mu(t-\tau-r/c)}\,d\mu \\ &= \left[\frac{\mathbf{u}\mathbf{u}\cdot\ddot{\mathscr{p}} - \ddot{\mathscr{p}}}{c^2 r}\right]_{\tau=t-r/c} = \left[\frac{(\ddot{\mathscr{p}}\times\mathbf{u})\times\mathbf{u}}{c^2 r}\right]_{\tau=t-r/c}\end{aligned} \tag{20.10}$$

and

$$\mathscr{H}_1 = \frac{1}{2\pi c^2} \iint_{-\infty}^{\infty} i\mu \not{p} \times \mathbf{u}\, e^{i\mu(t-\tau-r/c)}\, d\mu\, d\tau = \left[\frac{\ddot{\not{p}} \times \mathbf{u}}{c^2 r}\right]_{\tau=t-r/c}. \tag{20.11}$$

We note that

$$\mathbf{u} \cdot \mathscr{E}_1 = 0 \qquad \text{and} \qquad \mathbf{u} \times \mathscr{E}_1 = \mathscr{H}_1, \tag{20.12}$$

so that $\mathscr{E}_1$, $\mathscr{H}_1$, and $\mathbf{u}$ form a right-handed system. Also, since $\mathbf{u}$ is a unit vector, we have $\mathscr{H}_1 = \mathscr{E}_1$.

Poynting's vector to this order of approximation is

$$\boldsymbol{\gamma}_1 = \frac{c}{4\pi} \mathscr{E}_1 \times \mathscr{H}_1 = \frac{c}{4\pi} \mathscr{H}_1^2 \mathbf{u} = \frac{\mathbf{u}}{4\pi c^3 r^2} [(\ddot{\not{p}} \times \mathbf{u})^2]_{\tau=t-r/c}$$

$$= \mathbf{u}\left[\frac{\ddot{\not{p}}^2 \sin^2\theta}{4\pi c^3 r^2}\right]_{\tau=t-r/c}, \tag{20.13}$$

where θ is the angle between $\ddot{\not{p}}$ and $\mathbf{u}$. The total radiation corresponding to $\boldsymbol{\gamma}_1$ is, integrating over a sphere of radius r,

$$\int \boldsymbol{\gamma}_1 \cdot d\mathbf{S} = \left[\frac{\ddot{\not{p}}^2}{4\pi c^3} \int_0^{\pi} d\theta \sin^3\theta \int_0^{2\pi} d\varphi\right]_{\tau=t-r/c}$$

$$= \left[\frac{\ddot{\not{p}}^2}{2c^3} \int_0^{\pi} \sin^3\theta\, d\theta\right]_{\tau=t-r/c} = \left[\frac{2\ddot{\not{p}}^2}{3c^3}\right]_{\tau=t-r/c} \tag{20.14}$$

in agreement with Eq. (19.68). This is called the *dipole radiation*. We have seen in Section 9, Eq. (9.6), that when the origin of the coordinate system is shifted, so that $\mathbf{r}_s \to \mathbf{r}_s' = \mathbf{r}_s + \mathbf{a}$, the electric dipole moment changes from $\not{p}$ to

$$\not{p}' = \not{p} + Q\mathbf{a}. \tag{20.15}$$

Since charge is conserved, one knows that $\dot{Q} = 0$ and thus that

$$\dot{\not{p}}' = \dot{\not{p}} + \dot{Q}\mathbf{a} = \dot{\not{p}}, \qquad \text{and} \qquad \ddot{\not{p}}' = \ddot{\not{p}}. \tag{20.16}$$

The dipole radiation is therefore unaltered by such a shift of the origin.

Corresponding to terms of the second power in $\mu \mathbf{r}_s$ we have

$$\mathscr{E}_2 = \frac{1}{r} \sum_s \frac{e_s}{2\pi c} \iint_{-\infty}^{\infty} i\mu e^{i\mu(t-\tau-r/c)} \left[\frac{1}{2}\left(\frac{i\mu\, \mathbf{u} \cdot \mathbf{r}_s}{c}\right)^2 \mathbf{u} - \frac{i\mu\, \mathbf{u} \cdot \mathbf{r}_s}{c} \frac{\mathbf{v}_s}{c}\right] d\mu\, d\tau \tag{20.17}$$

and

$$\mathscr{H}_2 = \frac{1}{r}\sum_s \frac{e_s}{2\pi c^2}\iint\limits_{-\infty}^{\infty} i\mu\mathbf{v}_s \times \mathbf{u}\,\frac{i\mu\,\mathbf{u}\cdot\mathbf{r}_s}{c}\,e^{i\mu(t-\tau-r/c)}\,d\mu\,d\tau. \tag{20.18}$$

Equation (20.17) can be written

$$\begin{aligned}\mathscr{E}_2 &= \frac{1}{4\pi c^3 r}\sum_s e_s\int_{-\infty}^{\infty} d\tau\,(\mathbf{u}\cdot\mathbf{r}_s)^2\mathbf{u}\,\frac{\partial^3}{\partial t^3}\int_{-\infty}^{\infty} e^{i\mu(t-\tau-r/c)}\,d\mu \\ &\quad - \frac{1}{2\pi c^3 r}\sum_s e_s\int_{-\infty}^{\infty} d\tau\,\mathbf{u}\cdot\mathbf{r}_s\,\mathbf{v}_s\,\frac{\partial^2}{\partial t^2}\int_{-\infty}^{\infty} e^{i\mu(t-\tau-r/c)}\,d\mu \\ &= \left[\frac{\partial^2}{\partial t^2}\sum_s \frac{e_s}{c^3 r}(\mathbf{u}\cdot\mathbf{r}_s\,\mathbf{u}\cdot\mathbf{v}_s\,\mathbf{u} - \mathbf{u}\cdot\mathbf{r}_s\,\mathbf{v}_s)\right]_{\tau=t-r/c}.\end{aligned} \tag{20.19}$$

On the other hand, by Eq. (9.24), one has

$$\mathbf{Q} = \sum_s e_s(3\mathbf{r}_s\mathbf{r}_s - \mathbf{r}_s\cdot\mathbf{r}_s\,\mathbf{I}); \tag{20.20}$$

and, by Eq. (16.43),

$$\mathbf{m} = \frac{1}{2}\sum_s e_s\,\mathbf{r}_s \times \frac{\mathbf{v}_s}{c}, \tag{20.21}$$

where $\mathbf{Q}$ and $\mathbf{m}$ are the quadrupole and the magnetic moments, respectively. From Eqs. (20.20) and (20.21)* one has

$$\begin{aligned} c\,\mathbf{u}\times\mathbf{m} &= \tfrac{1}{2}\sum_s e_s(\mathbf{u}\cdot\mathbf{v}_s\mathbf{r}_s - \mathbf{u}\cdot\mathbf{r}_s\mathbf{v}_s), \\ \mathbf{u}\cdot\dot{\mathbf{Q}} &= \sum_s e_s(3\mathbf{u}\cdot\mathbf{r}_s\mathbf{v}_s + 3\mathbf{u}\cdot\mathbf{v}_s\mathbf{r}_s - 2\mathbf{r}_s\cdot\mathbf{v}_s\mathbf{u}), \\ \mathbf{u}\cdot\dot{\mathbf{Q}}\cdot\mathbf{u} &= 2\sum_s e_s(3\mathbf{u}\cdot\mathbf{r}_s\mathbf{u}\cdot\mathbf{v}_s - \mathbf{r}_s\cdot\mathbf{v}_s), \\ \mathbf{u}\cdot\dot{\mathbf{Q}}\times\mathbf{u} &= 3\sum_s e_s(\mathbf{u}\cdot\mathbf{r}_s\mathbf{v}_s\times\mathbf{u} + \mathbf{u}\cdot\mathbf{v}_s\mathbf{r}_s\times\mathbf{u}) \end{aligned} \tag{20.22}$$

and

$$\begin{aligned}\mathbf{u}\times[(\mathbf{u}\cdot\dot{\mathbf{Q}})\times\mathbf{u}] &= \mathbf{u}\cdot\dot{\mathbf{Q}} - \mathbf{u}\cdot\dot{\mathbf{Q}}\cdot\mathbf{u}\,\mathbf{u} \\ &= 3\sum_s e_s(\mathbf{u}\cdot\mathbf{r}_s\mathbf{v}_s + \mathbf{u}\cdot\mathbf{v}_s\mathbf{r}_s - 2\mathbf{u}\cdot\mathbf{r}_s\mathbf{u}\cdot\mathbf{v}_s\mathbf{u});\end{aligned} \tag{20.23}$$

hence, using Eqs. (20.19), (20.22), and (20.23),

$$\mathscr{E}_2 = \left[\frac{\partial^2}{\partial\tau^2}\frac{1}{c^3 r}\left\{c\,\mathbf{u}\times\mathbf{m} - \frac{1}{6}\mathbf{u}\times[(\mathbf{u}\cdot\dot{\mathbf{Q}})\times\mathbf{u}]\right\}\right]_{\tau=t-r/c}$$

* See Appendix A7 for a treatment of dyadics.

Since, as is easily shown, $(\mathbf{u} \cdot \dot{\mathbf{Q}}) \times \mathbf{u} = \mathbf{u} \cdot (\dot{\mathbf{Q}} \times \mathbf{u}) = \mathbf{u} \cdot \dot{\mathbf{Q}} \times \mathbf{u}$, we can write

$$\mathscr{E}_2 = \left[\frac{\mathbf{u} \times (c\ddot{\mathbf{m}} - \frac{1}{6}\mathbf{u} \cdot \dddot{\mathbf{Q}} \times \mathbf{u})}{c^3 r}\right]_{\tau=t-r/c}. \tag{20.24}$$

Similarly, from Eq. (20.18), one has

$$\mathscr{H}_2 = \frac{1}{2\pi c^3 r} \sum_s e_s \int_{-\infty}^{\infty} d\tau\, \mathbf{u} \cdot \mathbf{r}_s \mathbf{v}_s \times \mathbf{u} \frac{\partial^2}{\partial t^2} \int_{-\infty}^{\infty} e^{i\mu(t-\tau-r/c)}\, d\mu. \tag{20.25}$$

Since, as is easy to verify,

$$\sum_s e_s \mathbf{u} \cdot \mathbf{r}_s \mathbf{v}_s \times \mathbf{u} = (c\mathbf{m} \times \mathbf{u}) \times \mathbf{u} + \tfrac{1}{6}\mathbf{u} \cdot \dot{\mathbf{Q}} \times \mathbf{u} = (c\mathbf{m} \times \mathbf{u} + \tfrac{1}{6}\mathbf{u} \cdot \dot{\mathbf{Q}}) \times \mathbf{u},$$

Eq. (20.25) reduces to

$$\mathscr{H}_2 = \left[\frac{(c\ddot{\mathbf{m}} \times \mathbf{u} + \frac{1}{6}\mathbf{u} \cdot \dddot{\mathbf{Q}}) \times \mathbf{u}}{c^3 r}\right]_{\tau=t-r/c}. \tag{20.26}$$

Again

$$\mathbf{u} \cdot \mathscr{E}_2 = 0 \tag{20.27}$$

and

$$\mathbf{u} \times \mathscr{E}_2 = \frac{\mathbf{u}\,\mathbf{u} \cdot (c\ddot{\mathbf{m}} - \frac{1}{6}\mathbf{u} \cdot \dddot{\mathbf{Q}} \times \mathbf{u}) - (c\ddot{\mathbf{m}} - \frac{1}{6}\mathbf{u} \cdot \dddot{\mathbf{Q}} \times \mathbf{u})}{c^3 r} = \mathscr{H}_2, \tag{20.28}$$

because from the fourth equation of Eqs. (20.22) the triple scalar product $\mathbf{u} \cdot (\mathbf{u} \cdot \dddot{\mathbf{Q}}) \times \mathbf{u}$ is zero. Thus, $\mathscr{E}_2$, $\mathscr{H}_2$, and $\mathbf{u}$ again form a right-handed system of orthogonal vectors.

The radiated energy in this case is seen by Eq. (A2.27) to be

$$\begin{aligned}\int \boldsymbol{\Upsilon}_2 \cdot d\mathbf{S} &= \frac{c}{4\pi} \int \mathscr{E}_2 \times \mathscr{H}_2 \cdot d\mathbf{S} = \frac{c}{4\pi} \int \mathscr{E}_2^2\, dS \\ &= \frac{1}{4\pi c^5} \int_0^{\pi} \int_0^{2\pi} \Big[c^2 \ddot{m}^2 - \frac{c}{3}\ddot{\mathbf{m}} \cdot (\mathbf{u} \cdot \dddot{\mathbf{Q}} \times \mathbf{u}) + \tfrac{1}{36}(\mathbf{u} \cdot \dddot{\mathbf{Q}} \times \mathbf{u})^2 \\ &\quad - c^2(\mathbf{u} \cdot \ddot{\mathbf{m}})^2 \Big]_{\tau=t-r/c} \sin\theta\, d\theta\, d\varphi. \end{aligned} \tag{20.29}$$

The first and last terms can be evaluated by choosing the polar axis in the direction of $\ddot{\mathbf{m}}$; then $\mathbf{u} \cdot \ddot{\mathbf{m}} = \ddot{m} \cos\theta$ and these terms become

$$\begin{aligned}&\frac{1}{2c^3} \int_0^{\pi} [\ddot{m}^2 - (\mathbf{u} \cdot \ddot{\mathbf{m}})^2]_{\tau=t-r/c} \sin\theta\, d\theta \\ &\quad = \left[\frac{\ddot{m}}{2c^3} \int_0^{\pi} (\sin\theta - \cos^2\theta \sin\theta)\, d\theta\right]_{\tau=t-r/c} = \left[\frac{2\ddot{m}}{3c^3}\right]_{\tau=t-r/c}. \end{aligned} \tag{20.30}$$

Since $\dddot{Q}_{jk}$ is real and symmetric, we know that we can always choose a coordinate system in which $\dddot{Q}$ is diagonal. For this coordinate system

$$\mathbf{u} \cdot \dddot{\mathbf{Q}} = u_1 \dddot{Q}_{11} \mathbf{i} + u_2 \dddot{Q}_{22} \mathbf{j} + u_3 \dddot{Q}_{33} \mathbf{k},$$

where **i**, **j**, **k** are the unit vectors along the axes, and hence by Eq. (A2.21)

$$\ddot{\mathbf{m}} \cdot (\mathbf{u} \cdot \dddot{\mathbf{Q}} \times \mathbf{u}) = \ddot{\mathbf{m}} \cdot (\mathbf{u} \cdot \dddot{\mathbf{Q}}) \times \mathbf{u} = \begin{vmatrix} \ddot{m}_1 & \ddot{m}_2 & \ddot{m}_3 \\ u_1 \dddot{Q}_{11} & u_2 \dddot{Q}_{22} & u_3 \dddot{Q}_{33} \\ u_1 & u_2 & u_3 \end{vmatrix}$$

$$= \ddot{m}_1 u_2 u_3 (\dddot{Q}_{22} - \dddot{Q}_{33}) + \ddot{m}_2 u_3 u_1 (\dddot{Q}_{33} - \dddot{Q}_{11}) + \ddot{m}_3 u_1 u_2 (\dddot{Q}_{11} - \dddot{Q}_{22}). \quad (20.31)$$

Introducing a polar coordinate system with the polar axis along the Z-axis and letting θ and φ be the direction angles of **u**, we have

$$\begin{gathered} u_1 = \sin\theta\cos\varphi, \qquad u_2 = \sin\theta\sin\varphi, \qquad u_3 = \cos\theta, \\ u_2 u_3 = \sin\theta\cos\theta\sin\varphi, \\ u_3 u_1 = \sin\theta\cos\theta\cos\varphi, \\ u_1 u_2 = \sin^2\theta\sin\varphi\cos\varphi. \end{gathered} \quad (20.32)$$

The second integral in Eq. (20.29) thus vanishes by virtue of Eqs. (20.31) and (20.32) and the fact that

$$\int_0^{2\pi} \sin\varphi \, d\varphi = \int_0^{2\pi} \cos\varphi \, d\varphi = \int_0^{2\pi} \sin\varphi\cos\varphi \, d\varphi = 0.$$

The third integral in Eq. (20.29) has as its integrand $(\mathbf{u} \cdot \dddot{\mathbf{Q}} \times \mathbf{u})^2$, which by Eq. (A2.27) becomes

$$(\mathbf{u} \cdot \dddot{\mathbf{Q}} \times \mathbf{u})^2 = [(\mathbf{u} \cdot \dddot{\mathbf{Q}}) \times \mathbf{u}]^2 = (\mathbf{u} \cdot \dddot{\mathbf{Q}})^2 - (\mathbf{u} \cdot \dddot{\mathbf{Q}} \cdot \mathbf{u})^2. \quad (20.33)$$

Since

$$(\mathbf{u} \cdot \dddot{\mathbf{Q}})^2 = u_1^2 \dddot{Q}_{11}^2 + u_2^2 \dddot{Q}_{22}^2 + u_3^2 \dddot{Q}_{33}^2 \quad (20.34)$$

and

$$(\mathbf{u} \cdot \dddot{\mathbf{Q}} \cdot \mathbf{u}) = u_1^2 \dddot{Q}_{11} + u_2^2 \dddot{Q}_{22} + u_3^2 \dddot{Q}_{33},$$

we have, therefore, by Eq. (20.33)

$$\begin{aligned} (\mathbf{u} \cdot \dddot{\mathbf{Q}} \times \mathbf{u})^2 = u_1^2 \dddot{Q}_{11}^2 + u_2^2 \dddot{Q}_{22}^2 + u_3^2 \dddot{Q}_{33}^2 - [u_1^4 \dddot{Q}_{11}^2 \\ + u_2^4 \dddot{Q}_{22}^2 + u_3^4 \dddot{Q}_{33}^2 + 2u_1^2 u_2^2 \dddot{Q}_{11} \dddot{Q}_{22} \\ + 2u_2^2 u_3^2 \dddot{Q}_{22} \dddot{Q}_{33} + 2u_3^2 u_1^2 \dddot{Q}_{33} \dddot{Q}_{11}]. \end{aligned} \quad (20.35)$$

The third integral in Eq. (20.29) thus involves the following integrals:

$$\iint u_1^2 \sin\theta\, d\theta\, d\varphi = \iint u_2^2 \sin\theta\, d\theta\, d\varphi = \iint u_3^2 \sin\theta\, d\theta\, d\varphi = \frac{4\pi}{3},$$

$$\iint u_1^4 \sin\theta\, d\theta\, d\varphi = \iint u_2^4 \sin\theta\, d\theta\, d\varphi = \iint u_3^4 \sin\theta\, d\theta\, d\varphi = \frac{4\pi}{5}, \tag{20.36}$$

$$\iint u_1^2 u_2^2 \sin\theta\, d\theta\, d\varphi = \iint u_2^2 u_3^2 \sin\theta\, d\theta\, d\varphi = \iint u_3^2 u_1^2 \sin\theta\, d\theta\, d\varphi = \frac{4\pi}{15},$$

where in each case the limits are 0 to π for θ and 0 to 2π for φ. The third integral in Eq. (20.29), by virtue of Eqs. (20.35) and (20.36), reduces to

$$\frac{1}{4\pi c^5}\int_0^\pi \int_0^{2\pi} [\tfrac{1}{36}(\mathbf{u}\cdot \dddot{\mathbf{Q}} \times \mathbf{u})^2]_{\tau=t-r/c} \sin\theta\, d\theta\, d\varphi$$
$$= \frac{1}{144\pi c^5}\left\{\frac{4\pi}{3}(\dddot{Q}_{11}^2 + \dddot{Q}_{22}^2 + \dddot{Q}_{33}^2) - \frac{4\pi}{5}(\dddot{Q}_{11}^2 + \dddot{Q}_{22}^2 + \dddot{Q}_{33}^2)\right.$$
$$\left. - \frac{8\pi}{15}(\dddot{Q}_{11}\dddot{Q}_{22} + \dddot{Q}_{22}\dddot{Q}_{33} + \dddot{Q}_{33}\dddot{Q}_{11})\right\},$$

and thus Eq. (20.29) becomes

$$\int \boldsymbol{\Upsilon}_2 \cdot d\mathbf{S} = \left[\frac{2\dddot{m}^2}{3c^3} + \frac{1}{270c^5}\right.$$
$$\left.\times (\dddot{Q}_{11}^2 + \dddot{Q}_{22}^2 + \dddot{Q}_{33}^2 - \dddot{Q}_{11}\dddot{Q}_{22} - \dddot{Q}_{22}\dddot{Q}_{33} - \dddot{Q}_{33}\dddot{Q}_{11})\right]_{\tau=t-r/c}. \tag{20.37}$$

The expression in parentheses can be written, using the summation convention, in the form

$$\tfrac{1}{2}\{3(\dddot{Q}_{11}^2 + \dddot{Q}_{22}^2 + \dddot{Q}_{33}^2) - (\dddot{Q}_{11} + \dddot{Q}_{22} + \dddot{Q}_{33})(\dddot{Q}_{11} + \dddot{Q}_{22} + \dddot{Q}_{33})\}$$
$$= \tfrac{1}{2}\{3\dddot{Q}_{jk}\dddot{Q}_{jk} - \dddot{Q}_{jj}\dddot{Q}_{kk}\},$$

because, $\dot{Q}_{jk}$ being diagonal, $\dddot{Q}_{jk}\dddot{Q}_{jk} - (\dddot{Q}_{11}^2 + \dddot{Q}_{22}^2 + \dddot{Q}_{33}^2)$ consists only of vanishing terms. Furthermore, using Eq. (20.20),

$$Q_{jj} = \sum_s e_s \sum_j (3x_{sj}x_{sj} - \delta_{jj}\sum_k x_{sk}x_{sk})$$
$$= \sum_s e_s(3\mathbf{r}_s\cdot\mathbf{r}_s - 3\mathbf{r}_s\cdot\mathbf{r}_s) = 0;$$

therefore, finally,

$$\int \boldsymbol{\Upsilon}_2 \cdot d\mathbf{S} = \frac{2\dddot{m}^2}{3c^3} + \frac{(\dddot{Q}_{jk})^2}{180c^5}. \tag{20.38}$$

This is in an invariant form and is therefore valid for all coordinate systems. The first term may be called the *magnetic-moment radiation* and the second the *quadrupole radiation.*

EXAMPLE 1. Find the first-order and second-order radiation from a particle of charge e that is moving with simple harmonic motion, $\mathbf{r} = \mathbf{r}_0 \sin 2\pi\nu t$, about the origin.

Since $\not{p} = e\mathbf{r}$ and $\ddot{\mathbf{r}} = -(2\pi\nu)^2\mathbf{r}$, we have $\ddot{\not{p}} = -4\pi^2\nu^2(e\mathbf{r}_0)\sin 2\pi\nu t$ and thus the average dipole radiation is

$$\text{aver}\int \mathbf{Y}_1 \cdot d\mathbf{S} = \frac{32\pi^4\nu^4(er_0)^2}{3c^3}\overline{\sin^2 2\pi\nu\left(t - \frac{r}{c}\right)} = \frac{16\pi^4\nu^4(er_0)^2}{3c^3}, \quad (20.39)$$

where the bar indicates a time average.

To obtain the second-order radiation we note first from Eq. (20.21) that

$$\mathbf{m} = \frac{1}{2}\, e\mathbf{r} \times \frac{\mathbf{v}}{c}$$

$$\dot{\mathbf{m}} = \frac{1}{2}\, e\left(\mathbf{v} \times \frac{\mathbf{v}}{c} + \mathbf{r} \times \ddot{\mathbf{r}}\right) = -\frac{1}{2}\, e(2\pi\nu)^2\mathbf{r} \times \mathbf{r} = 0.$$

Thus, the magnetic moment is constant, and there is no radiation due to it. By differentiating Eq. (20.20) and substituting for $\ddot{\mathbf{r}}$ we find that

$$\dddot{\mathbf{Q}} = -(4\pi\nu)^2\dot{\mathbf{Q}}, \quad (20.40)$$

where

$$\dot{\mathbf{Q}} = e(3\mathbf{r}\mathbf{v} + 3\mathbf{v}\mathbf{r} - 2\mathbf{r} \cdot \mathbf{v}\mathbf{I}); \quad (20.41)$$

therefore

$$\int \mathbf{Y}_2 \cdot d\mathbf{S} = \left[\frac{64\pi^4\nu^4(\dot{Q}_{jk})^2}{45c^5}\right]_{\tau=t-r/c}. \quad (20.42)$$

It can be readily shown that for a single particle

$$(\dot{Q}_{jk})^2 = 6e^2[3r^2v^2 - (\mathbf{r} \cdot \mathbf{v})^2], \quad (20.43)$$

which for our case reduces to

$$\begin{aligned}(\dot{Q}_{jk})^2 &= 12e^2r_0^4(2\pi\nu)^2 \cos^2 2\pi\nu t \sin^2 2\pi\nu t \\ &= 12\pi^2\nu^2e^2r_0^4 \sin^2 4\pi\nu t. \quad (20.44)\end{aligned}$$

Substituting this in Eq. (20.42) and taking the time average, we have

$$\text{aver}\int \boldsymbol{\Upsilon}_2 \cdot d\mathbf{S} = \frac{256\pi^6\nu^6e^2r_0^4}{15c^5}\,\overline{\sin^2 4\pi\nu\left(t - \frac{r}{c}\right)}$$

$$= \frac{128\pi^6\nu^6e^2r_0^4}{15c^5}. \tag{20.45}$$

By Eqs. (20.39) and (20.45) the ratio of the quadrupole radiation to the dipole radiation is given by

$$\frac{8\pi^2\nu^2r_0^2}{5c^2} = \frac{4}{5}\frac{\overline{v^2}}{c^2}, \tag{20.46}$$

where $\overline{v^2}$ is the average value of the square of the velocity.

EXAMPLE 2. Suppose the motion of a particle can be resolved into a series

$$\mathbf{r}_p(t) = \sum_n \mathbf{a}_n \cos(2\pi\nu_n t + \delta_n), \tag{20.47}$$

where ν_n and δ_n are arbitrary, and $\mathbf{a}_n$ the corresponding vector amplitudes. Find the corresponding dipole radiation.

We have at once

$$\ddot{p} = e\ddot{\mathbf{r}}_p = -(2\pi)^2 e \sum_n \nu_n^2 \mathbf{a}_n \cos(2\pi\nu_n t + \delta_n),$$

so that, using Eq. (20.14),

$$\int \boldsymbol{\Upsilon}_1 \cdot d\mathbf{S} = \frac{2}{3c^3}(2\pi)^4 e^2 \sum_{n,m} \nu_n^2\nu_m^2 \mathbf{a}_n \cdot \mathbf{a}_m$$
$$\times\, [\cos(2\pi\nu_n\tau + \delta_n)\cos(2\pi\nu_m\tau + \delta_m)]_{\tau=t-r/c}. \tag{20.48}$$

In averaging over a long time T, we have

$$\lim_{T\to\infty}\frac{1}{T}\int_0^T \cos(2\pi\nu_n\tau + \delta_n)\cos(2\pi\nu_m\tau + \delta_m)\,d\tau = 0, \qquad \text{if} \quad n \neq m,$$

and

$$\lim_{T\to\infty}\frac{1}{T}\int_0^T \cos^2(2\pi\nu_n\tau + \delta_n)\,d\tau = \frac{1}{2};$$

hence

$$\text{aver}\int \boldsymbol{\Upsilon}_1 \cdot d\mathbf{S} = \frac{(2\pi)^4 e^2}{3c^3}\sum_n \nu_n^4 a_n^2. \tag{20.49}$$

Thus, each frequency ν_n contributes, on the average,

$$\frac{(2\pi\nu_n)^4 e^2}{3c^3}\, a_n^2 \tag{20.50}$$

per unit of time to the energy radiated as dipole radiation.

EXAMPLE 3. Find the dipole radiation from a particle that moves away from the origin and returns in such a way that its position vector $r_p(t)$ can be resolved into a Fourier integral

$$\mathbf{r}_p(t) = \int_{-\infty}^{\infty} \mathbf{a}(\nu) e^{2\pi i\nu t}\, d\nu. \tag{20.51}$$

Differentiating each side of Eq. (20.51) with respect to time, we find that

$$\ddot{p} = e\ddot{\mathbf{r}}_p = -(2\pi)^2 e \int_{-\infty}^{\infty} \nu^2 \mathbf{a}(\nu) e^{2\pi i\nu t}\, d\nu, \tag{20.52}$$

and thus by Eq. (20.14) that

$$\int \Upsilon_1 \cdot d\mathbf{S} = \frac{2}{3c^3}(2\pi)^4 e^2 \left[\iint_{-\infty}^{\infty} \nu^2 \mu^2 \mathbf{a}(\nu) \cdot \mathbf{a}(\mu) e^{2\pi i(\nu+\mu)\tau}\, d\nu\, d\mu \right]_{\tau = t - r/c}. \tag{20.53}$$

The time average of the right-hand side can readily be shown to vanish. Thus one cannot, as in Example 2, find an *average radiation per unit of time* for each frequency (that isn't identically zero) or per unit frequency range. This arises from the fact that the particle motion must be expandable in a Fourier integral.* This requires that the particle be essentially at the origin for all but a finite interval of time, i.e., $\mathbf{r}_p(t) \to 0$ as $|t| \to \infty$. The average radiation over all time is therefore necessarily zero.

One can, however, calculate the *total* energy radiated by finding

$$\int_{-\infty}^{\infty} dt \int \Upsilon_1 \cdot d\mathbf{S} = \frac{2(2\pi)^4 e^2}{3c^3} \int_{-\infty}^{\infty} dt \iint_{-\infty}^{\infty} \nu^2 \mu^2 \mathbf{a}(\nu) \cdot \mathbf{a}(\mu) e^{2\pi i(\nu+\mu)(t - r/c)}\, d\nu\, d\mu.$$

By shifting the variable of integration in the time integration from t to

* We disregard here the possibility that $a(\nu)$ might involve singular functions, such as the Dirac delta function.

$\tau = t - r/c$ this becomes

$$\int_{-\infty}^{\infty} dt \int \Upsilon_1 \cdot d\mathbf{S} = \frac{2(2\pi)^5 e^2}{3c^3} \iint_{-\infty}^{\infty} \nu^2 \mu^2 \mathbf{a}(\nu) \cdot \mathbf{a}(\mu) \left[\frac{1}{2\pi} \int_{-\infty}^{\infty} e^{2\pi i(\nu+\mu)\tau} \, d\tau \right] d\nu \, d\mu$$

$$= \frac{2(2\pi)^5 e^2}{3c^3} \iint_{-\infty}^{\infty} \nu^2 \mu^2 \mathbf{a}(\nu) \cdot \mathbf{a}(\mu) \delta(\nu + \mu) \, d\nu \, d\mu$$

$$= \frac{2(2\pi)^5 e^2}{3c^3} \int_{-\infty}^{\infty} \nu^4 \mathbf{a}(\nu) \cdot \mathbf{a}(-\nu) \, d\nu. \tag{20.54}$$

The integrand is an even function of ν; therefore,

$$\int_{-\infty}^{\infty} d\tau \int \Upsilon_1 \cdot d\mathbf{S} = \frac{4(2\pi)^5 e^2}{3c^3} \int_0^{\infty} \nu^4 \mathbf{a}(\nu) \cdot \mathbf{a}(-\nu) \, d\nu. \tag{20.55}$$

Thus, in this case, the *total* radiation per unit frequency range is

$$\frac{128\pi^5 e^2 \nu^4}{3} \mathbf{a}(\nu) \cdot \mathbf{a}(-\nu). \tag{20.56}$$

EXAMPLE 4. Calculate the radiation due to a pair of similar particles moving uniformly in a circle of radius a and remaining 180° apart.

Taking the origin at the center of the circle, we have

$$\not{p} = e(\mathbf{r}_1 + \mathbf{r}_2) = e(\mathbf{r}_1 - \mathbf{r}_1) = 0, \tag{20.57}$$

since now

$$\mathbf{r}_2 = -\mathbf{r}_1 .$$

Also, since

$$\ddot{\mathbf{r}}_s = -(2\pi\nu)^2 \mathbf{r}_s \qquad \text{and} \qquad \mathbf{m} = \frac{1}{2} \sum_s e_s \mathbf{r}_s \times \frac{\mathbf{v}_s}{c},$$

we find that

$$\dot{\mathbf{m}} = \frac{1}{2} e \sum_s \left(\mathbf{v}_s \times \frac{\mathbf{v}_s}{c} + \mathbf{r}_s \times \frac{\ddot{\mathbf{r}}_s}{c} \right) = -\frac{1}{2} e \sum_s (2\pi\nu)^2 \, \mathbf{r}_s \times \mathbf{r}_s = 0. \tag{20.58}$$

The quadrupole moment radiation is given by a generalization to two or more particles of Eqs. (20.41) and (20.42). Corresponding to Eq. (20.41), we have

$$\dot{Q}_{jk} = \sum_s e_s (3x_j^s \dot{x}_k^s + 3\dot{x}_j^s x_k^s - 2x_i^s \dot{x}_i^s \delta_{jk}), \tag{20.59}$$

where x_j^s are the components of $\mathbf{r}_s$. Thus, we have

$$
\begin{aligned}
(\dot{Q}_{jk})^2 &= \sum_{ss'} e_s e_{s'} (3x_j^s \dot{x}_k^s + 3\dot{x}_j^s x_k^s - 2x_i^s \dot{x}_i^s \delta_{jk})(3x_j^{s'} \dot{x}_k^{s'} + 3\dot{x}_j^{s'} x_k^{s'} - 2x_\ell^{s'} \dot{x}_\ell^{s'} \delta_{jk}) \\
&= 6 \sum_{ss'} e_s e_{s'} (3x_j^s x_j^{s'} \dot{x}_k^s \dot{x}_k^{s'} + 3x_j^s \dot{x}_j^{s'} x_k^{s'} \dot{x}_k^s - 2x_j^s \dot{x}_j^s x_k^{s'} \dot{x}_k^{s'}) \\
&= 6 \sum_{ss'} e_s e_{s'} (3\mathbf{r}_s \cdot \mathbf{r}_{s'}\, \mathbf{v}_s \cdot \mathbf{v}_{s'} + 3\mathbf{r}_s \cdot \mathbf{v}_{s'}\, \mathbf{r}_{s'} \cdot \mathbf{v}_s - 2\, \mathbf{r}_s \cdot \mathbf{v}_s\, \mathbf{r}_{s'} \cdot \mathbf{v}_{s'}).
\end{aligned}
\tag{20.60}
$$

In our case $\mathbf{r}_s \cdot \mathbf{v}_s = \mathbf{r}_{s'} \cdot \mathbf{v}_{s'} = \mathbf{r}_s \cdot \mathbf{v}_{s'} = \mathbf{r}_{s'} \cdot \mathbf{v}_s = 0$, and $e_1 = e_2 = e$; hence

$$
(\dot{Q}_{jk})^2 = 18e^2 \sum_{ss'} \mathbf{r}_s \cdot \mathbf{r}_{s'}\, \mathbf{v}_s \cdot \mathbf{v}_{s'}. \tag{20.61}
$$

Since s and s' are either 1 or 2, and $\mathbf{r}_2 = -\mathbf{r}_1$

$$
\mathbf{r}_s \cdot \mathbf{r}_{s'} = \begin{Bmatrix} a^2, & \text{if} \quad s = s', \\ -a^2, & \text{if} \quad s \neq s', \end{Bmatrix} = a^2(2\delta_{ss'} - 1), \tag{20.62}
$$

and, similarly,

$$
\mathbf{v}_s \cdot \mathbf{v}_{s'} = v^2(2\delta_{ss'} - 1). \tag{20.63}
$$

Substituting these in Eq. (20.60), we have

$$
(\dot{Q}_{jk})^2 = 18e^2a^2v^2 \sum_{ss'} (4\delta_{ss'}^2 - 4\delta_{ss'} + 1),
$$

and, since $\delta_{ss'} = \delta_{ss'}^2$ and $v = 2\pi\nu a$, this becomes

$$
(\dot{Q}_{jk})^2 = 18e^2a^2v^2 \sum_{ss'} 1 = 72e^2a^2v^2 = 72e^2(2\pi\nu)^2a^4.
$$

Thus, by Eq. (20.42), the radiation per unit of time is

$$
\frac{(4\pi\nu)^6 e^2 a^4}{10c^5} \quad \text{or} \quad \frac{32e^2c}{5a^2}\left(\frac{v}{c}\right)^6. \tag{20.64}
$$

21. Standing Waves—Radiation in a Box

The general solution of the Maxwell–Lorentz equations in the absence of charges, as we have seen in Section 18, Eqs. (18.57), (18.58), (18.61), and (18.62), is

$$\mathscr{E}(\mathbf{r}, t) = \left(\frac{1}{2\pi}\right)^{3/2} \int [\mathscr{E}(\boldsymbol{k})e^{i(\boldsymbol{k}\cdot\mathbf{r}-c k t)} + \mathscr{E}^*(\boldsymbol{k})e^{-i(\boldsymbol{k}\cdot\mathbf{r}-c k t)}]\, d^3\boldsymbol{k}, \tag{21.1}$$

in which $d^3\boldsymbol{k} = dk_x\, dk_y\, dk_z$ and

$$\boldsymbol{k} \cdot \mathscr{E}(\boldsymbol{k}) = 0. \tag{21.2}$$

A similar equation holds for $\mathscr{H}$ with

$$\mathscr{H}(\boldsymbol{k}) = \frac{\boldsymbol{k}}{k} \times \mathscr{E}(\boldsymbol{k}). \tag{21.3}$$

Thus both $\mathscr{E}(\mathbf{r}, t)$ and $\mathscr{H}(\mathbf{r}, t)$ are determined as soon as we know $\mathscr{E}(\boldsymbol{k})$.

Electric field in a box. Suppose now that the electromagnetic field is enclosed in a rectangular box with *perfectly reflecting walls.* For our purpose we shall define a perfectly reflecting wall as a surface at which the tangential component of $\mathscr{E}(\mathbf{r}, t)$ is zero for all t. Suppose that the box is located in the first quadrant so that three of its edges coincide with the coordinate axes. Let the lengths of these edges be L_x, L_y, and L_z, respectively. Further, we assume that there is no radiation outside the box, so that the tangential component of $\mathscr{E}(\mathbf{r}, t)$ is zero along the whole of each of the six planes $x = 0$, $y = 0$, $z = 0$, $x = L_x$, $y = L_y$, $z = L_z$.

We shall now successively impose these boundary conditions on the $\mathscr{E}(\mathbf{r}, t)$ given by Eq. (21.1), starting with the condition that the tangential component is zero for $x = 0$. This means that

$$\mathscr{E}_y(\mathbf{r}, t) = 0 \quad \text{and} \quad \mathscr{E}_z(\mathbf{r}, t) = 0, \quad \text{for} \quad x = 0. \tag{21.4}$$

The first of these two conditions is

$$\mathscr{E}_y(0, y, z, t) = \left(\frac{1}{2\pi}\right)^{3/2} \int [\mathscr{E}_y(\boldsymbol{k})e^{i(yk_y+zk_z-ckt)} + \mathscr{E}_y^*(\boldsymbol{k})e^{-i(yk_y+zk_z-ckt)}]\, d^3\boldsymbol{k} = 0; \tag{21.5}$$

or, changing in the second term the variable of integration from $\boldsymbol{k}$ to $-\boldsymbol{k}$, which does not alter $k = |\boldsymbol{k}|$,

$$\iint_{-\infty}^{\infty} e^{i(yk_y+zk_z)}\, dk_y\, dk_z \int_{-\infty}^{\infty} [\mathscr{E}_y(\boldsymbol{k})e^{-ickt} + \mathscr{E}_y^*(-\boldsymbol{k})e^{ickt}]\, dk_x = 0. \tag{21.6}$$

Since Eq. (21.6) is to be valid for all y and z, it can be shown that this equation requires that

$$\int_{-\infty}^{\infty} [\mathscr{E}_y(\mathscr{k})e^{-ic\mathscr{k}t} + \mathscr{E}_y^*(-\mathscr{k})e^{ic\mathscr{k}t}]\, d\mathscr{k}_x = 0. \tag{21.7}$$

Multiplying the left-hand side of Eq. (21.7) by $(c/2\pi)e^{ic\mathscr{k}'t}\,dt$, where $\mathscr{k}' = (\mathscr{k}_x', \mathscr{k}_y, \mathscr{k}_z)$, integrating from $t = -\infty$ to $t = \infty$, and noting that

$$\frac{1}{2\pi}\int_{-\infty}^{\infty} e^{i(\mathscr{k}'\pm\mathscr{k})ct}\, d(ct) = \delta(\mathscr{k}' \pm \mathscr{k}) = \delta(\mathscr{k} \pm \mathscr{k}'),$$

we obtain the equation

$$\int_{-\infty}^{\infty} [\mathscr{E}_y(\mathscr{k})\delta(\mathscr{k} - \mathscr{k}') + \mathscr{E}_y^*(-\mathscr{k})\delta(\mathscr{k} + \mathscr{k}')]\, d\mathscr{k}_x = 0. \tag{21.8}$$

This equation must hold for all $\mathscr{k}' = (\mathscr{k}_x', \mathscr{k}_y, \mathscr{k}_z)$, and in particular for $\mathscr{k}' = 0$, i.e., $\mathscr{k}_x' = \mathscr{k}_y = \mathscr{k}_z = 0$. Equation (21.8) then gives

$$\int_{-\infty}^{\infty} [\mathscr{E}_y(\mathscr{k}_x, 0, 0) + \mathscr{E}_y^*(-\mathscr{k}_x, 0, 0)]\delta(\mathscr{k})\, d\mathscr{k}_x = 0$$

or, on integration, since now $\mathscr{k} = |\mathscr{k}_x|$,

$$\mathscr{E}_y(0) + E_y^*(0) = 0. \tag{21.9}$$

This shows that the Y component of the integrand in Eq. (21.1) is zero for $\mathscr{k} = 0$, i.e., for $\nu = 0$ (since $\mathscr{k} = 2\pi/\lambda = 2\pi\nu/c$). Thus, there is no electrostatic field in the Y direction; and since the Y direction is not essentially different from the X and the Z directions, all components of the electrostatic field are zero.

For $\mathscr{k}' \neq 0$, and hence $\mathscr{k} + \mathscr{k} \neq 0$, the second term in Eq. (21.8) vanishes, since then $\delta(\mathscr{k} + \mathscr{k}') \equiv 0$. Thus, for $\mathscr{k}' \neq 0$, Eq. (21.8) becomes

$$\int_{-\infty}^{\infty} \mathscr{E}_y(\mathscr{k}_x, \mathscr{k}_y, \mathscr{k}_z)\delta(\mathscr{k} - \mathscr{k}')\, d\mathscr{k}_x = 0.$$

There are in general two values of $\mathscr{k}_x$, namely $\pm\mathscr{k}_x'$, for which $\mathscr{k} = \mathscr{k}'$; therefore this equation is equivalent to

$$\mathscr{E}_y(\mathscr{k}_x', \mathscr{k}_y, \mathscr{k}_z) + \mathscr{E}_y(-\mathscr{k}_x', \mathscr{k}_y, \mathscr{k}_z) = 0.$$

Since $\mathscr{k}_x'$ is arbitrary, we can drop the primes, obtaining

$$\mathscr{E}_y(\mathscr{k}_x, \mathscr{k}_y, \mathscr{k}_z) = -\mathscr{E}_y(-\mathscr{k}_x, \mathscr{k}_y, \mathscr{k}_z) \tag{21.10}$$

or, in other words, that $\mathscr{E}_y(\mathscr{k})$ is an odd function of $\mathscr{k}_x$.

This condition is also sufficient to satisfy Eq. (21.7), as can be seen by dividing the range of integration with respect to k_x in Eq. (21.7) into the range $-\infty$ to 0 and the range 0 to ∞, which by Eq. (21.10) mutually cancel. Equation (21.10) includes also Eq. (21.9), since for $k = 0$ it gives $\mathscr{E}_y(0) = \mathscr{E}_y^*(0) = 0$. Although this is a greater restriction on $\mathscr{E}_y(0)$ than Eq. (21.9), $\mathscr{E}(\mathbf{r}, t)$ is not further restricted. This can be seen from the fact that the imaginary part of $\mathscr{E}(0)$ does not alter the integrand in Eq. (21.1).

We obtain an exactly similar result for $\mathscr{E}_z$, namely,

$$\mathscr{E}_z(k_x, k_y, k_z) = -\mathscr{E}_z(-k_x, k_y, k_z) \tag{21.11}$$

by using the second of the conditions in Eqs. (21.4).

In addition, by using Eq. (21.2) we have

$$k \cdot \mathscr{E}(k) = k_x\mathscr{E}_x(k_x, k_y, k_z) + k_y\mathscr{E}_y(k_x, k_y, k_z) + k_z\mathscr{E}_z(k_x, k_y, k_z) = 0$$

and, changing k_x to $-k_x$,

$$-k_x\mathscr{E}_x(-k_x, k_y, k_z) + k_y\mathscr{E}_y(-k_x, k_y, k_z) + k_z\mathscr{E}_z(-k_x, k_y, k_z) = 0.$$

Adding the last two equations, and taking into account Eqs. (21.10) and (21.11), we have

$$k_x[\mathscr{E}_x(k_x, k_y, k_z) - \mathscr{E}_x(-k_x, k_y, k_z)] = 0,$$

so that

$$\mathscr{E}_x(k_x, k_y, k_z) = \mathscr{E}_x(-k_x, k_y, k_z). \tag{21.12}$$

Equations (21.10) to (21.12) show that for every wave corresponding to a particular $k = (k_x, k_y, k_z)$ and $\mathscr{E}(k)$ there must occur, as the result of the existence of the wall at $x = 0$, another wave with $k' = (-k_x, k_y, k_z)$ and

$$\mathscr{E}(k') = (\mathscr{E}_x(k), -\mathscr{E}_y(k), -\mathscr{E}_z(k)). \tag{21.13}$$

One of the two waves, that which is propagated toward the wall, is called the *incident wave;* the other is termed the *reflected wave.*

If k and $\mathscr{E}(k)$ satisfy Eq. (21.2), so will k' and $\mathscr{E}(k')$, since

$$k' \cdot \mathscr{E}(k') = -k_x\mathscr{E}_x(k) - k_y\mathscr{E}_y(k) - k_z\mathscr{E}_z(k) = -k \cdot \mathscr{E}(k). \tag{21.14}$$

Let $\mathbf{i}$ be, as usual, the unit vector in the X direction; then by Eq. (A2.21)

$$(\mathbf{i} \times k) \cdot k' = \begin{vmatrix} 1 & 0 & 0 \\ k_x & k_y & k_z \\ -k_x & k_y & k_z \end{vmatrix} = 0 \tag{21.15}$$

and

$$\left|\mathbf{i}\cdot\frac{\mathbf{k}'}{k'}\right| = \left|\frac{-k_x}{k}\right| = \frac{k_x}{\sqrt{k_x^2+k_y^2+k_z^2}} = \left|\mathbf{i}\cdot\frac{\mathbf{k}}{k}\right|. \tag{21.16}$$

Equations (21.15) and (21.16) show that the plane determined by the directions of propagation of the two waves contains the normal to the wall, and this normal makes equal angles with the two directions of propagation. These are the usual laws of reflection.

There is, however, another way of exhibiting the meaning of Eqs. (21.10) to (21.12). Taking for example $\mathscr{E}_x$, we have

$$\begin{aligned}\mathscr{E}_x(\mathbf{r}, t) &= \left(\frac{1}{2\pi}\right)^{3/2}\int [\mathscr{E}_x(\mathbf{k})e^{i(\mathbf{k}\cdot\mathbf{r}-ckt)} + \text{complex conjugate}]\, d^3k \\ &= \left(\frac{1}{2\pi}\right)^{3/2}\iint_{-\infty}^{\infty} dk_y\, dk_z\Bigg[\int_{-\infty}^{0}\mathscr{E}_x(k_x, k_y, k_z)e^{i(xk_x+yk_y+zk_z-ckt)}\, dk_x \\ &\quad + \int_0^{\infty}\mathscr{E}_x(k_x, k_y, k_z)e^{i(xk_x+yk_y+zk_z-ckt)}\, dk_x \\ &\quad + \text{complex conjugate}\Bigg];\end{aligned}$$

or, changing the variable of integration in $\int_{-\infty}^{0}\cdots dk_x$ from k_x to $-k_x$ and using Eq. (21.12),

$$\begin{aligned}\mathscr{E}_x(\mathbf{r}, t) &= \left(\frac{1}{2\pi}\right)^{3/2}\iint_{-\infty}^{\infty} dk_y\, dk_z\int_0^{\infty}\{\mathscr{E}_x(k_x, k_y, k_z)[e^{ixk_x} + e^{-ixk_x}]e^{i(yk_y+zk_z-ckt)} \\ &\quad + \text{complex conjugate}\}\, dk_x \\ &= 2\left(\frac{1}{2\pi}\right)^{3/2}\int_{-\infty}^{\infty} dk_y\, dk_z\int_0^{\infty}\cos(xk_x)[\mathscr{E}_x(\mathbf{k})e^{i(yk_y+zk_z-ckt)} \\ &\quad + \text{complex conjugate}]\, dk_x \\ &= 2\int_0^{\infty} f_x(k_x; y, z, t)\cos(xk_x)\, dk_x,\end{aligned} \tag{21.17}$$

where

$$f_x(k_x; y, z, t) = \left(\frac{1}{2\pi}\right)^{3/2}\iint_{-\infty}^{\infty}[\mathscr{E}_x(\mathbf{k})e^{i(yk_y+zk_z-ckt)} + \text{complex conjugate}]\, dk_y\, dk_z.$$

We see that $\mathcal{E}_x(\mathbf{r}, t)$ may be considered as consisting of real waves of the type $f_x \cos(x k_x)$. At any given instant of time this expression varies sinusoidally with x (y and z being fixed). In addition the points of maxima and minima, as well as the points for which the expression vanishes, are not changed with time. Such a wave is said to be a *standing wave*. This particular wave can be said to be a standing wave with respect to x.

For $\mathcal{E}_y(\mathbf{r}, t)$ and $\mathcal{E}_z(\mathbf{r}, t)$ we obtain by analogous procedure

$$\mathcal{E}_y(\mathbf{r}, t) = 2\int_0^\infty g_y(y, z, t)\sin(x k_x)\, dk_x \tag{21.18}$$

and

$$\mathcal{E}_z(\mathbf{r}, t) = 2\int_0^\infty g_z(y, z, t)\sin(x k_x)\, d k_x, \tag{21.19}$$

where g_y and g_z are two components of

$$\mathbf{g} = i\left(\frac{1}{2\pi}\right)^{3/2} \iint_{-\infty}^{\infty} [e^{i(yk_y+zk_z-ckt)}\, \boldsymbol{\mathcal{E}}(\boldsymbol{k}) - \text{complex conjugate}]\, dk_y\, dk_z. \tag{21.20}$$

Thus $\mathcal{E}_y(\mathbf{r}, t)$ and $\mathcal{E}_z(\mathbf{r}, t)$ are also composed of standing waves with respect to x, but now $\sin(x k_x)$ enters instead of $\cos(x k_x)$, which is to be expected, since $\mathcal{E}_y$ and $\mathcal{E}_z$ must vanish for $x = 0$.

In a similar way, imposing upon $\boldsymbol{\mathcal{E}}(\mathbf{r}, t)$ the conditions at $y = 0$ and $z = 0$, as given by Eqs. (21.17) to (21.19), we obtain finally

$$\begin{aligned}
\mathcal{E}_x(\mathbf{r}, t) &= 8\left(\frac{1}{2\pi}\right)^{3/2} \int F_x \cos(x k_x)\sin(y k_y)\sin(z k_z)\, d^3 k, \\
\mathcal{E}_y(\mathbf{r}, t) &= 8\left(\frac{1}{2\pi}\right)^{3/2} \int F_y \sin(x k_x)\cos(y k_y)\sin(z k_z)\, d^3 k, \\
\mathcal{E}_z(\mathbf{r}, t) &= 8\left(\frac{1}{2\pi}\right)^{3/2} \int F_z \sin(x k_x)\sin(y k_y)\cos(z k_z)\, d^3 k,
\end{aligned} \tag{21.21}$$

where F_x, F_y, F_z are the components of

$$\mathbf{F}(\boldsymbol{k}, t) = -[\boldsymbol{\mathcal{E}}(\boldsymbol{k})e^{-ickt} + \boldsymbol{\mathcal{E}}^*(\boldsymbol{k})e^{ickt}]. \tag{21.22}$$

The integration is over that portion of $\boldsymbol{k}$-space for which each component of $\boldsymbol{k}$ is positive. For this reason $\boldsymbol{\mathcal{E}}(\boldsymbol{k})$ in Eq. (21.22) and $\mathbf{F}(\boldsymbol{k}, t)$ in Eqs. (21.21) and (21.22) need have no special properties with respect to changes in sign of the components of $\boldsymbol{k}$. Thus the conditions along the planes $x = 0$

$y = 0$, and $z = 0$ reduce completely to Eqs. (21.21) and (21.22) with an arbitrary $\mathscr{E}(\boldsymbol{k})$. It is evident that $\mathscr{E}(\mathbf{r}, t)$ consists of standing waves with respect to all three variables.

We next consider the condition that the tangential component of $\mathscr{E}(\mathbf{r}, t)$ must also vanish for $x = L_x$, $y = L_y$, and $z = L_z$. Thus, for example, $\mathscr{E}_y(\mathbf{r}, t)$ must vanish for $x = L_x$. This, upon substitution into Eq. (21.1), gives

$$\mathscr{E}_y(L_x, y, z, t) = \left(\frac{1}{2\pi}\right)^{3/2} \int [\mathscr{E}_y(k) e^{i(L_x k_x + y k_y + z k_z - ckt)} + \text{complex conjugate}]\, d^3k = 0. \qquad (21.23)$$

We can treat this equation exactly as we did Eq. (21.5). The analog of Eq. (21.6) is now

$$\iint_{-\infty}^{\infty} e^{i(yk_y + zk_z)}\, dk_y\, dk_z \int_{-\infty}^{\infty} [\mathscr{E}_y(k) e^{-ickt} + \mathscr{E}_y^*(-k) e^{ickt}] e^{iL_x k_x}\, dk_x = 0,$$

and therefore, as for Eq. (21.7), we must require that

$$\int_{-\infty}^{\infty} [\mathscr{E}_y(k) e^{-ickt} + \mathscr{E}_y^*(-k) e^{ickt}] e^{iL_x k_x}\, dk_x = 0.$$

Moreover, since we know by Eq. (21.10) that $\mathscr{E}_y(k)$ is an odd function of k_x, we can also write this equation, remembering that $e^{iL_x k_x} = \cos(L_x k_x) + i \sin(L_x k_x)$, in the form

$$\int_{-\infty}^{\infty} [\mathscr{E}_y(k) e^{-ickt} + \mathscr{E}_y^*(-k) e^{ickt}] \sin(L_x k_x)\, dk_x = 0.$$

Analogously to Eq. (21.8) we now have

$$\int_{-\infty}^{\infty} [\mathscr{E}_y(k)\delta(k - k') + \mathscr{E}_y^*(-k)\delta(k + k')] \sin(L_x k_x)\, dk_x = 0.$$

For $\mathbf{k}' \neq 0$, we obtain, analogously to Eq. (21.10), the condition

$$\mathscr{E}_y(k_x, k_y, k_z)\sin(L_x k_x) + \mathscr{E}_y(-k_x, k_y, k_z)\sin(-L_x k_x) = 0.$$

With the help of Eq. (21.10) this becomes

$$\sin(L_x k_x)\, \mathscr{E}_y(k) = 0; \qquad (21.24)$$

thus $\mathscr{E}_y(k) = 0$, unless $L_x k_x = \ell\pi$, where ℓ is an integer.

Similarly, since $\mathscr{E}_y(\mathbf{r}, t)$ must also vanish for $z = L_z$, $\mathscr{E}_y(k) = 0$, unless $L_z k_z = n\pi$, where n is an integer. Combining these with analogous conditions on $\mathscr{E}_x(k)$ and $\mathscr{E}_z(k)$, we have

$$\mathscr{E}_x(k) = 0, \qquad \text{unless} \quad k_y = \frac{m\pi}{L_y} \quad \text{and} \quad k_z = \frac{n\pi}{L_z}; \tag{21.25}$$

$$\mathscr{E}_y(k) = 0, \qquad \text{unless} \quad k_z = \frac{n\pi}{L_z} \quad \text{and} \quad k_x = \frac{\ell\pi}{L}; \tag{21.26}$$

$$\mathscr{E}_z(k) = 0, \qquad \text{unless} \quad k_x = \frac{\ell\pi}{L_x} \quad \text{and} \quad k_y = \frac{m\pi}{L_y}. \tag{21.27}$$

Consider next the expression

$$k_x \mathscr{E}_x(k) + k_y \mathscr{E}_y(k) + k_z \mathscr{E}_z(k),$$

which by Eq. (21.2) must be zero. If one term fails to vanish, there must be at least one other nonzero term. Suppose that $\mathscr{E}(k) \neq 0$; then one of the components, say $\mathscr{E}_x(k)$, is not zero, and hence by Eq. (21.25)

$$k_y = \frac{m\pi}{L_y} \quad \text{and} \quad k_z = \frac{n\pi}{L_z}.$$

If $k_x = 0$, we can write

$$k_x = \frac{\ell\pi}{L_x} \quad \text{where} \quad \ell = 0.$$

On the other hand, if $k_x \neq 0$, then $k_x \mathscr{E}_x(k)$ does not vanish, and therefore either $\mathscr{E}_y(k)$ or $\mathscr{E}_z(k)$ is not zero; in either case (see Eqs. (21.26) and (21.27))

$$k_x = \frac{\ell\pi}{L_x}.$$

We thus conclude generally that in order to have $\mathscr{E}(k) \neq 0$, we must have

$$k_x = \frac{\ell\pi}{L_x}, \qquad k_y = \frac{m\pi}{L_y}, \qquad k_z = \frac{n\pi}{L_z}, \tag{21.28}$$

where ℓ, m, n are integers.

Since $\mathscr{E}(k)$ is zero except at the discrete values of k given by Eq. (20.28), it must behave as a delta function at these values, otherwise the integral in Eq. (21.1) would vanish. Since $\mathbf{F}(k, t)$ of Eq. (21.22) is linear in $\mathscr{E}(k)$, the integrals of Eqs. (21.21) will reduce to sums over the values of k for which $\mathscr{E}(\mathscr{k})$ does not vanish, namely for

$$\mathscr{k} = \mathscr{k}_{\ell,m,n} = \left(\frac{\ell\pi}{L_x}, \frac{m\pi}{L_y}, \frac{n\pi}{L_z}\right), \qquad \ell, m, n = 0, 1, 2 \ldots. \tag{21.29}$$

Thus, we shall have

$$\begin{aligned} \mathscr{E}_x(\mathbf{r}, t) &= \sqrt{\frac{8}{V}} \sum f_x(\mathscr{k}, t)\cos(x\mathscr{k}_x)\sin(y\mathscr{k}_y)\sin(z\mathscr{k}_z), \\ \mathscr{E}_y(\mathbf{r}, t) &= \sqrt{\frac{8}{V}} \sum f_y(\mathscr{k}, t)\sin(x\mathscr{k}_x)\cos(y\mathscr{k}_y)\sin(z\mathscr{k}_z), \\ \mathscr{E}_z(\mathbf{r}, t) &= \sqrt{\frac{8}{V}} \sum f_z(\mathscr{k}, t)\sin(x\mathscr{k}_x)\sin(y\mathscr{k}_y)\cos(z\mathscr{k}_z), \end{aligned} \tag{21.30}$$

where the f's are components of some real vector $\mathbf{f}(\mathscr{k}, t)$, and $\sqrt{8/V}$ is a normalizing factor introduced for convenience. V is the volume of the box, and the summations are over all $\mathscr{k}$ satisfying Eq. (21.29).

Comparison of Eqs. (21.21) and (21.30) shows that the latter can be obtained from the former by putting

$$\mathbf{F}(\mathscr{k}, t) = \left(\frac{\pi^3}{V}\right)^{1/2} \mathbf{f}(\mathscr{k}, t) \sum_{\ell,m,n=0}^{\infty} \delta(\mathscr{k} - \mathscr{k}_{\ell,m,n}), \tag{21.31}$$

where $\mathscr{k}_{\ell,m,n}$, is given by Eq. (21.29). Comparing this result with Eq. (21.22), we see that $\mathbf{f}(\mathscr{k}, t)$ must be of the form

$$\mathbf{f}(\mathscr{k}, t) = \mathbf{a}(\mathscr{k})e^{-ic\mathscr{k}t} + \mathbf{a}^*(\mathscr{k})e^{ic\mathscr{k}t}, \tag{21.32}$$

where $\mathbf{a}(\mathscr{k})$ is some vector function of $\mathscr{k}$ and

$$\mathscr{E}(\mathscr{k}) = -\left(\frac{\pi^3}{V}\right)^{1/2} \mathbf{a}(\mathscr{k}) \sum_{\ell,m,n=0}^{\infty} \delta(\mathscr{k} - \mathscr{k}_{\ell,m,n}). \tag{21.33}$$

Insofar as $\mathbf{a}(\mathscr{k})$ may be regarded as arbitrary, the dependence of $\mathbf{f}(\mathscr{k}, t)$ on time is expressed by regarding it as a general real vector solution of the differential equation

$$\frac{\partial^2\mathbf{f}(\mathscr{k}, t)}{\partial t^2} + c^2\mathscr{k}^2\mathbf{f}(\mathscr{k}, t) = 0. \tag{21.34}$$

This equation, if used with Eqs. (21.30), shows that $\mathscr{E}(\mathbf{r}, t)$ satisfies the wave equation

$$\nabla^2\mathscr{E}(\mathbf{r}, t) - \frac{1}{c^2}\frac{\partial^2\mathscr{E}(\mathbf{r}, t)}{\partial t^2} = 0. \tag{21.35}$$

Furthermore, $\mathscr{E}(\mathbf{r}, t)$ as given by Eqs. (21.30) and (21.32) satisfies all the boundary conditions regardless of the magnitude of $\mathbf{a}(\boldsymbol{k})$. The condition $\boldsymbol{k} \cdot \mathscr{E}(\boldsymbol{k}) = 0$, however, requires that $\boldsymbol{k} \cdot \mathbf{f}(\boldsymbol{k}, t) = 0$ and hence that

$$\boldsymbol{k} \cdot \mathbf{a}(\boldsymbol{k}) = 0. \tag{21.36}$$

It is clear from the above that we may disregard Eqs. (21.31) and (21.33) and consider Eqs. (21.29), (21.30), (21.32), and (21.36) as fully specifying the nature of the electric field that may exist in a box with perfectly reflecting walls. The method of derivation shows that it is a general solution.

Obtaining the magnetic field. By Eq. (21.3) the Fourier components $\mathscr{H}(\boldsymbol{k})$ of the magnetic field are related to the $\mathscr{E}(\boldsymbol{k})$ as follows:

$$\mathscr{H}(\boldsymbol{k}) = \frac{\boldsymbol{k}}{k} \times \mathscr{E}(\boldsymbol{k}).$$

In particular, we have

$$\mathscr{H}_x(\boldsymbol{k}) = \frac{1}{k}[k_y \mathscr{E}_z(\boldsymbol{k}) - k_z \mathscr{E}_y(\boldsymbol{k})].$$

Using the fact that $\mathscr{E}_y(\boldsymbol{k})$ and $\mathscr{E}_z(\boldsymbol{k})$ are odd functions of k_x, we have

$$\begin{aligned}\mathscr{H}_x(-k_x, k_y, k_z) &= \frac{1}{k}[k_y \mathscr{E}_z(-k_x, k_y, k_z) - k_z \mathscr{E}_y(-k_x, k_y, k_z)] \\ &= -\frac{1}{k}[k_y \mathscr{E}_z(k_x, k_y, k_z) - k_z \mathscr{E}_y(k_x, k_y, k_z)];\end{aligned}$$

therefore,

$$\mathscr{H}_x(-k_x, k_y, k_z) = -\mathscr{H}_x(k_x, k_y, k_z). \tag{21.37}$$

Analogously, using equations similar to Eqs. (21.10) and (21.12), one has

$$\begin{aligned}\mathscr{H}_x(k_x, -k_y, k_z) &= \frac{1}{k}[(-k_y)\mathscr{E}_z(k_x, -k_y, k_z) - k_z \mathscr{E}_y(k_x, -k_y, k_z)] \\ &= \frac{1}{k}[k_y \mathscr{E}_z(k_x, k_y, k_z) - k_z \mathscr{E}_y(k_x, k_y, k_z)] \\ &= \mathscr{H}_x(k_x, k_y, k_z),\end{aligned} \tag{21.38}$$

and similarly,

$$\mathscr{H}_x(k_x, k_y, -k_z) = \mathscr{H}_x(k_x, k_y, k_z). \tag{21.39}$$

It follows at once from Eqs. (21.3) and (21.28) that $\mathscr{H}(\boldsymbol{k}) = 0$ except at the discrete values $\boldsymbol{k} = \boldsymbol{k}_{\ell,m,n}$, where the $\boldsymbol{k}_{\ell,m,n}$ are given by Eq. (21.29). We also know that $\mathscr{H}(\mathbf{r}, t)$ satisfies an equation of the same form as Eq. (21.1) (see Eq. (18.58)). These facts and the symmetry properties of $\mathscr{H}(\boldsymbol{k})$, which are described by Eqs. (21.37), (21.38), (21.39), and similar conditions on the other components of $\mathscr{H}(\boldsymbol{k})$ lead to the equations

$$\begin{aligned}
\mathscr{H}_x &= \sqrt{\frac{8}{V}} \sum g_x(\boldsymbol{k}, t)\sin(xk_x)\cos(yk_y)\cos(zk_z), \\
\mathscr{H}_y &= \sqrt{\frac{8}{V}} \sum g_y(\boldsymbol{k}, t)\cos(xk_x)\sin(yk_y)\cos(zk_z), \\
\mathscr{H}_z &= \sqrt{\frac{8}{V}} \sum g_z(\boldsymbol{k}, t)\cos(xk_x)\cos(yk_y)\sin(zk_z),
\end{aligned} \tag{21.40}$$

where again the summation is over all $\boldsymbol{k}$ satisfying Eq. (21.29) and

$$\mathbf{g}(\boldsymbol{k}, t) = -\frac{i\boldsymbol{k}}{k} \times [\mathbf{a}(\boldsymbol{k})e^{-ickt} - \mathbf{a}^*(\boldsymbol{k})e^{ickt}]. \tag{21.41}$$

This equation can be written in the form

$$\mathbf{g}(\boldsymbol{k}, t) = \mathbf{b}(\boldsymbol{k})e^{-ickt} + \mathbf{b}^*(\boldsymbol{k})e^{ickt} \tag{21.42}$$

with $\mathbf{b}(\boldsymbol{k}) = -i\boldsymbol{k} \times \mathbf{a}(\boldsymbol{k})/k$. By comparing Eqs. (21.32) and (21.41) we note also that

$$\mathbf{g}(\boldsymbol{k}, t) = \frac{\boldsymbol{k}}{ck^2} \times \frac{\partial}{\partial t} \mathbf{f}(\boldsymbol{k}, t). \tag{21.43}$$

The combinations of functions of x, y, and z occurring in Eqs. (21.30) and (21.40) are each a complete orthogonal set for functions of $\mathbf{r}$ over the volume of the box; i.e., for $0 \leq x \leq L_x$, $0 \leq y \leq L_y$, and $0 \leq z \leq L_z$ (see Appendix A8). Thus, for example, putting

$$\varphi_{\boldsymbol{k}}(\mathbf{r}) = \sqrt{\frac{8}{V}} \cos(xk_x)\sin(yk_y)\sin(zk_z), \tag{21.44}$$

we have

$$\int_V \varphi_{\boldsymbol{k}}^*(\mathbf{r})\varphi_{\boldsymbol{k}'}(\mathbf{r})\, dV = \delta_{\boldsymbol{k}\boldsymbol{k}'}. \tag{21.45}$$

It is for this reason that the normalizing factor $\sqrt{8/V}$ was introduced in Eqs. (21.30).

Expressing the fields in terms of the potentials. The field given by Eqs. (21.30) and (21.40) can also be obtained from potentials of the form

$$\begin{aligned}
A_x(\mathbf{r}, t) &= \sqrt{\frac{8}{V}} \sum A_x(\boldsymbol{k}, t)\cos(x k_x)\sin(y k_y)\sin(z k_z), \\
A_y(\mathbf{r}, t) &= \sqrt{\frac{8}{V}} \sum A_y(\boldsymbol{k}, t)\sin(x k_x)\cos(y k_y)\sin(z k_z), \\
A_z(\mathbf{r}, t) &= \sqrt{\frac{8}{V}} \sum A_z(\boldsymbol{k}, t)\sin(x k_x)\sin(y k_y)\cos(z k_z), \\
\varphi(\mathbf{r}, t) &= \sqrt{\frac{8}{V}} \sum \varphi(\boldsymbol{k}, t)\sin(x k_x)\sin(y k_y)\sin(z k_z),
\end{aligned} \tag{21.46}$$

summation being again over all $\boldsymbol{k}$ satisfying Eq. (21.29), by means of basic equations, Eqs. (14.15) and (14.16). If one sets

$$\begin{aligned}
\mathbf{A}(\boldsymbol{k}, t) &= \boldsymbol{\lambda}(\boldsymbol{k})e^{-ickt} + \boldsymbol{\lambda}^*(\boldsymbol{k})e^{ickt}, \\
\varphi(\boldsymbol{k}, t) &= \mu(\boldsymbol{k})e^{-ickt} + \mu^*(\boldsymbol{k})e^{ickt},
\end{aligned} \tag{21.47}$$

then Eq. (14.15),

$$\mathscr{E}(\mathbf{r}, t) = -\boldsymbol{\nabla}\varphi(\mathbf{r}, t) - \frac{1}{c}\frac{\partial}{\partial t}\mathbf{A}(\mathbf{r}, t),$$

requires that

$$\mathbf{a}(\boldsymbol{k}) = -\boldsymbol{k}\mu(\boldsymbol{k}) + ik\boldsymbol{\lambda}(\boldsymbol{k}), \tag{21.48}$$

$\mathbf{a}(\boldsymbol{k})$ being the vector occurring in Eqs. (21.32) and (21.36). On the other hand, Eq. (14.16),

$$\mathscr{H}(\mathbf{r}, t) = \boldsymbol{\nabla} \times \mathbf{A}(\mathbf{r}, t),$$

can be shown to be satisfied if

$$\boldsymbol{k} \times \left[\boldsymbol{\lambda}(\boldsymbol{k}) + \frac{i}{k}\mathbf{a}(\boldsymbol{k})\right] = 0,$$

which is consistent with Eq. (21.48).

The Lorentz condition

$$\boldsymbol{\nabla} \cdot \mathbf{A} + \frac{1}{c}\frac{\partial \varphi}{\partial t} = 0$$

can easily be shown to require that

$$\boldsymbol{k} \cdot \boldsymbol{\lambda}(\boldsymbol{k}) = -ik\mu(\boldsymbol{k}). \tag{21.49}$$

Thus, by Eq. (21.47), $\mathbf{A}(k)$ has in general a *longitudinal component*, a component in the direction of k.

As a special choice of potentials one may take $\varphi(\mathbf{r}, t) \equiv 0$; then $\varphi(k, t) = 0$ and by Eq. (21.47) $\mu(k) = 0$. For this case, by Eq. (21.48),

$$\mathbf{a}(k) = ik\boldsymbol{\lambda}(k) \tag{21.50}$$

and thus by Eq. (21.49)

$$k \cdot \mathbf{a}(k) = ik\, k \cdot \boldsymbol{\lambda}(k) = k^2\mu(k) = 0.$$

This requires that the longitudinal component of $\mathbf{A}(k)$ vanish.

Number of modes per unit frequency range. It is often important in applications to know the number of standing waves that can exist in the box, with frequencies lying between ν and $\nu + d\nu$. This is a matter of counting the number of choices of the integers ℓ, m, n that lead to $k_{\ell,m,n}$ corresponding to this frequency interval. Now, the frequency is given in terms of k by $\nu = ck/2\pi$ (since $k = 2\pi\nu/c$); that is, by

$$\nu = \frac{c}{2\pi}(k_x^2 + k_y^2 + k_z^2)^{1/2}. \tag{21.51}$$

Thus, the values of ℓ, m, n that give frequencies between ν and $\nu + d\nu$ give values of k between $(2\pi/c)\nu$ and $(2\pi/c)(\nu + d\nu)$. If we plot values of k_x, k_y, k_z satisfying Eq. (21.29) along three Cartesian axes, we obtain a rectangular lattice of points in what we shall call k-space. The distance of each such point from the origin is the value of k. We need first to determine the number of lattice points lying in the first octant between the sphere of radius $k = (2\pi/c)\nu$ and one of radius $k = (2\pi/c)(\nu + d\nu)$.

If planes are passed parallel to the coordinate planes, half-way between adjacent lattice points, they will form boxes with the sides π/L_x, π/L_y, and π/L_z (i.e., of volume π^3/V) and each such box will contain one lattice point at its center. Thus the lattice points have the density V/π^3 per unit volume of k-space and are uniformly distributed in the positive octant. Now, the volume of k-space between the two spheres just referred to is

$$4\pi k^2\, dk = 4\pi\left(\frac{2\pi\nu}{c}\right)^2 \frac{2\pi}{c}\, d\nu = \frac{32\pi^4\nu^2}{c^3}\, d\nu.$$

The volume in the positive octant is $4\pi^4\nu^2\, d\nu/c^3$. Thus the number of lattice points in question is approximately

$$\frac{V}{\pi^3}\frac{4\pi^4\nu^2\,d\nu}{c^3} = \frac{4\pi\nu^2 V\,d\nu}{c^3}. \tag{21.52}$$

It is usual to count two standing waves for each frequency, since there are two independent polarizations possible for each $\boldsymbol{k}$. This corresponds also to the fact that the three components of $\mathbf{a}(\boldsymbol{k})$ of Eq. (21.32) must satisfy a single equation, namely, Eq. (21.36). We conclude, therefore, that the number of standing waves with frequencies between ν and $\nu + d\nu$ is approximately

$$\frac{8\pi\nu^2}{c^3}\,V\,d\nu. \tag{21.53}$$

The standing waves considered above may be assigned amplitudes and phases independently and thus they represent the *modes of the cavity* formed by the perfectly conducting walls of the box. The density of modes per unit volume per unit frequency range is thus given by

$$\rho_{\text{modes}} = \frac{8\pi\nu^2}{c^3}. \tag{21.54}$$

One does not usually consider the number of running waves, because they are not independent. Each of the standing waves is made up generally of eight tightly coupled running waves having propagation vectors corresponding to the eight possible choices of sign for the vector $(\pm k_x, \pm k_y, \pm k_z)$. If, of course, one of the components k_x, k_y, k_z is zero, the number of running waves in a group reduces to four. On the other hand, it can be seen from Eq. (21.30) that, if two of these components are zero, the electric field $\mathscr{E}(\mathbf{r}, t)$ must vanish; thus this possibility can be ignored.

We see that the number of waves for a given frequency range is independent of the shape of the rectangular box and is proportional to its volume. Weyl showed that this number is completely independent of the shape of the box,* for ν sufficiently large, and hence that Eq. (21.54) holds generally.

Periodic boundary conditions. We shall now show that the same density of modes as given by Eq. (21.54) results if we assume, instead of perfectly conducting walls, *periodic boundary conditions*. These conditions

* H. Weyl, *Crelles J.*, **144**, 163 (1912).

can be expressed by assuming that the whole of space is divided up into boxes having dimensions L_x, L_y, and L_z and that the conditions in all the boxes are to be exactly the same at all times.

Let us analyze the field in terms of the four-potential $\varphi_\alpha(\mathbf{r}, t)$, and let us consider the discrete set of vectors (which we shall call *lattice vectors*)

$$\mathbf{r}_{\ell,m,n} = (\ell L_x, mL_y, nL_z), \tag{21.55}$$

where ℓ, m, and n are integers. Then a displacement from any point $\mathbf{r}$ within one box to a point $\mathbf{r} + \mathbf{r}_{\ell,m,n}$ will bring us to a corresponding point of another box. Our conditions may therefore be written

$$\varphi_\alpha(\mathbf{r} + \mathbf{r}_{\ell,m,n}, t) = \varphi_\alpha(\mathbf{r}, t). \tag{21.56}$$

Using for $\varphi_\alpha(\mathbf{r}, t)$ the general real solution of the wave equation given by Eqs. (18.47) and (18.54), we have

$$\varphi_\alpha(\mathbf{r}, t) = \left(\frac{1}{2\pi}\right)^{3/2} \int [\varphi_\alpha(\boldsymbol{k}) e^{i(\boldsymbol{k}\cdot\mathbf{r} - ckt)} + \text{complex conjugate}]\, d^3\boldsymbol{k}. \tag{21.57}$$

Combining this equation with Eq. (21.56), we find that

$$\int [\varphi_\alpha(\boldsymbol{k})(1 - e^{i\boldsymbol{k}\cdot\mathbf{r}_{\ell,m,n}}) e^{i(\boldsymbol{k}\cdot\mathbf{r} - ckt)} + \text{complex conjugate}]\, d^3\boldsymbol{k} = 0. \tag{21.58}$$

Since this must hold for all t, we must have

$$\varphi_\alpha(\boldsymbol{k})(1 - e^{i\boldsymbol{k}\cdot\mathbf{r}_{\ell,m,n}}) = 0. \tag{21.59}$$

This means that $\varphi_\alpha(\boldsymbol{k})$ must vanish everywhere except at those $\boldsymbol{k}$'s satisfying the condition

$$\boldsymbol{k} \cdot \mathbf{r}_{\ell,m,n} = 2\pi s_{\ell,m,n}, \qquad s_{\ell,m,n} = \text{integer}, \tag{21.60}$$

for all ℓ, m, and n.

If we take $\ell = 1$ and $m = n = 0$, then from Eq. (21.55)

$$\mathbf{r}_{1,0,0} = (L_x, 0, 0)$$

and Eq. (21.60) becomes

$$L_x k_x = 2\pi p,$$

where $p \equiv s_{1,0,0}$ is an integer. Similar conditions, of course, apply to k_y and k_z and we see that $\varphi_\alpha(\boldsymbol{k})$ must vanish unless $\boldsymbol{k}$ is equal to one of the *reciprocal lattice vectors*

$$\boldsymbol{k}_{p,q,r} \equiv \left(\frac{2\pi p}{L_x}, \frac{2\pi q}{L_y}, \frac{2\pi r}{L_z}\right), \tag{21.61}$$

where p, q, and r are any three integers.

If one replaces k by $-k$ in the complex conjugate term in Eq. (21.57), one can factor out $e^{ik\cdot r}$ in the integrand. Then, since $\varphi_\alpha(k)$ vanishes except at the k's given in Eq. (21.61), the integral in Eq. (21.57) becomes a sum and one can write

$$\varphi_\alpha(\mathbf{r}, t) = \sqrt{\frac{1}{V}} \sum f_\alpha(k, t) e^{ik\cdot \mathbf{r}}, \tag{21.62}$$

where summation is over all the reciprocal lattice vectors of Eq. (21.61). The normalization factor $1/\sqrt{V}$ is chosen so that the function

$$\psi(\mathbf{r}) \equiv \sqrt{\frac{1}{V}}\, e^{ik\cdot \mathbf{r}}, \tag{21.63}$$

where k is any of the reciprocal lattice vectors, is normalized so that

$$\int_V \psi_k^*(\mathbf{r}) \psi_{k'}(\mathbf{r})\, dV = \delta_{kk'} \equiv \delta_{\ell\ell'} \delta_{mm'} \delta_{nn'}, \tag{21.64}$$

where V is the volume of one box.

The expression for $\varphi_\alpha(\mathbf{r}, t)$ given by Eqs. (21.61) and (21.62) obviously satisfies the periodicity condition (21.56). If it is substituited in the wave equation

$$\nabla^2 \varphi_\alpha(\mathbf{r}, t) - \frac{1}{c^2} \frac{\partial^2 \varphi_\alpha(\mathbf{r}, t)}{\partial t^2} = 0,$$

one obtains the equation

$$\frac{\partial^2 f_\alpha(k, t)}{\partial t^2} + c^2 k^2 f_\alpha(k, t) = 0. \tag{21.65}$$

A general solution of this equation is

$$f_\alpha(k, t) = a_\alpha(k) e^{-ickt} + b_\alpha(k) e^{ickt},$$

where $a_\alpha(k)$ and $b_\alpha(k)$ are arbitrary functions of k.

If we put

$$\varphi_\alpha(\mathbf{r}, t) = (\mathbf{A}(\mathbf{r}, t), i\varphi(\mathbf{r}, t)), \tag{21.66}$$

$$f_\alpha(k, t) = (\mathbf{f}(k, t), if_0(k, t)), \tag{21.67}$$

$$a_\alpha(k) = (\mathbf{a}(k), ia_0(k)), \tag{21.68}$$

then, by Eq. (21.62),

$$\mathbf{A}(\mathbf{r}, t) = \sqrt{\frac{1}{V}} \sum \mathbf{f}(k, t) e^{ik\cdot \mathbf{r}} \tag{21.69}$$

and

$$\varphi(\mathbf{r}, t) = \sqrt{\frac{1}{V}} \sum f_0(k, t) e^{ik\cdot \mathbf{r}}. \tag{21.70}$$

In order for $\mathbf{A}(\mathbf{r}, t)$ and $\varphi(\mathbf{r}, t)$ to be real, we must have

$$\mathbf{f}^*(-\boldsymbol{k}, t) = \mathbf{f}(\boldsymbol{k}, t)$$

and

$$f_0^*(-\boldsymbol{k}, t) = f_0(\boldsymbol{k}, t);$$

therefore

$$\mathbf{f}(\boldsymbol{k}, t) = \mathbf{a}(\boldsymbol{k})e^{-ic\boldsymbol{k}t} + \mathbf{a}^*(-\boldsymbol{k})e^{ic\boldsymbol{k}t} \tag{21.71}$$

and

$$f_0(\boldsymbol{k}, t) = a_0(\boldsymbol{k})e^{-ic\boldsymbol{k}t} + a_0^*(-\boldsymbol{k})e^{ic\boldsymbol{k}t}, \tag{21.72}$$

$\mathbf{a}(\boldsymbol{k})$ and $a_0(\boldsymbol{k})$ being, respectively, arbitrary vector and scalar functions of $\boldsymbol{k}$.

The Lorentz condition

$$\frac{\partial \varphi_\alpha(\mathbf{r}, t)}{\partial x_\alpha} = \nabla \cdot \mathbf{A}(\mathbf{r}, t) + \frac{1}{c}\frac{\partial \varphi(\mathbf{r}, t)}{\partial t} = 0, \tag{21.73}$$

which $\varphi_\alpha(\mathbf{r}, t)$ must satisfy, now becomes

$$\boldsymbol{k} \cdot \mathbf{a}(\boldsymbol{k}) = \boldsymbol{k} a_0(\boldsymbol{k}). \tag{21.74}$$

The running waves corresponding to the propagation vectors given in Eq. (21.61) can be assigned arbitrary amplitudes and phases and still meet the boundary conditions. Thus they can be looked upon as the modes of the electromagnetic field for these periodic boundary conditions. The number of modes corresponding to the first octant of $\boldsymbol{k}$-space (ℓ, m, n all positive) is just one-eighth of that given in (21.53) because the steps between the allowed values of the components of $\boldsymbol{k}$ are twice as large as before (see Eq. (21.29)). In the present case, however, the components ℓ, m, n are not restricted to being positive and the total number of modes is just eight times the number in the first octant. We thus end up with the same density of modes as that given by Eq. (21.54) for a box with perfectly conducting walls.

22. Geometrical Optics and Its Limitations

In this book we are interested in geometrical optics only as a limiting case of the theory of electromagnetic waves. Recognition of the wave nature of light is due primarily to Fresnel and Huygens, and is based on the phenomenon of *light interference*, the occurrence under certain conditions of bright and dark regions, where geometrical optics would predict merely one or the other. The assumption of the wave nature of light makes possible the interpretation of such phenomena, since waves may reinforce or annihilate each other, depending upon their phase relation.

The idea that light is an electromagnetic wave is due to Maxwell. It is based primarily on three facts: the phenomena of *polarization*, which evidences the transverse character of the wave, mentioned in Section 18; the possibility of propagation of light in a vacuum; and the equality between the experimentally determined speed of light in vacuum and the speed of propagation of interaction c—the latter having been determined by purely electromagnetic experiments.

Once the identification had been made, it became possible to show that conclusions of the electromagnetic theory of waves, some of which we have treated in the last four sections, apply to light; visible light corresponds to waves with wavelength between, approximately, 4×10^{-5} and 7×10^{-5} cm.

In the last fifty years, the difficulties of accounting for the experimental facts connected with interaction of light with matter, and the process of emission of light by matter, has led to a generalized theory, quantum electrodynamics, in which light possesses both particle and wave nature, but neither completely.

One of the early difficulties of the wave theory of light was to account for the facts of geometrical optics. In this section we shall show how the facts of geometrical optics can be derived from the wave theory. We shall see that geometrical optics is an approximation to the wave theory and is valid only when certain restrictions are fulfilled.

We have seen in Section 18, Eq. (18.35), that the Cartesian components of all electrodynamic quantities, in the absence of matter, satisfy the wave equation

$$\nabla^2 f - \frac{1}{c^2}\frac{\partial^2 f}{\partial t^2} = 0.$$

We shall consider, however, the more general wave equation

$$\nabla^2 f - \frac{1}{v^2}\frac{\partial^2 f}{\partial t^2} = 0, \tag{22.1}$$

where v is a function of the coordinates (x, y, z) and perhaps also of some parameter ν that may enter into the form of the solution f. Thus, for example, if we assume a solution of the form*

$$f = g(x, y, z)e^{-2\pi i\nu t}, \tag{22.2}$$

Eq. (22.1) becomes

$$\nabla^2 g + \frac{(2\pi\nu)^2}{v^2} g = 0. \tag{22.3}$$

As stated earlier, v may contain ν as a parameter. Since only v^2 enters into Eq. (22.1), we may assume v to be positive.

Equations of this type occur when one considers light passing through a material medium, and takes as f any Cartesian component of $\mathscr{E}$ and $\mathscr{H}$ averaged over volumes containing a large number of particles, but small in comparison with all dimensions of importance in the problem being considered. The wave equation for empty space previously considered is, of course, a special case of Eq. (22.1), obtained by setting v = constant = c.

We shall assume†

$$f = A(x, y, z)e^{i\varphi(x,y,z,t)}, \tag{22.4}$$

where A and φ are real and $A > 0$. We call φ the *phase* of the wave thus represented. This is, of course, more general than Eq. (22.2), which may be obtained from Eq. (22.4) by putting

$$\varphi(x, y, z, t) = \theta(x, y, z) - 2\pi\nu t \tag{22.5}$$

and letting

$$g = Ae^{i\theta}. \tag{22.6}$$

If we have a solution of Eq. (22.1) of the form given in Eq. (22.4) we can consider the real (or imaginary) part of it as representing a scalar field quantity or a Cartesian component of a field vector. Note that the real and the imaginary parts of f are then themselves solutions of Eq. (22.1).

The *intensity* of the light is a measure of the energy flux and is thus proportional to the magnitude of the Poynting vector. The latter is quadratic in the field quantities and hence the intensity is proportional to $|f|^2 = A^2$.

At any given time $t = t_1$ the equation

$$\varphi(\mathbf{r}, t_1) = C,$$

* Some authors use a positive sign in the exponent. A useful convention that is often used to keep things straight is to use $j = -i$. The exponential in Eq. (22.2) becomes then $e^{j\omega t}$, where $\omega = 2\pi\nu$, in keeping with the practice in electrical engineering.

† Omission of t as a variable in A makes the intensity of light constant at each point in space.

where C is a constant, defines a *surface of constant phase.* Two such surfaces corresponding to $C = C_1$ and $C = C_1 + 2\pi$, and hence to the same value of $e^{i\varphi}$, are usually said to be a *wavelength* λ apart. This definition of a wavelength is exact only if this distance is a constant; however, we may make use of it to obtain a more general definition.

Keeping t constant, we have

$$d\varphi = d\mathbf{r} \cdot \nabla\varphi,$$

as is shown in Appendix A3 (see Eq. (A3.59)). If $d\mathbf{r}$ is tangent to a surface of constant phase, $d\varphi = 0$ and hence $\nabla\varphi$ must be perpendicular to the surface. Thus, for a displacement $d\mathbf{r}$ along a normal to the surface

$$d\varphi = |\nabla\varphi|\, dr. \tag{22.7}$$

If $|\nabla\varphi|$ is constant, the wavelength λ as defined above must be such that, for $dr = \lambda$, $d\varphi = 2\pi$, i.e.,

$$2\pi = |\nabla\varphi|\lambda. \tag{22.8}$$

We now note that this equation provides us with a natural definition of λ even when $|\nabla\varphi|$ is a function of $\mathbf{r}$ and t. Of course, for this general case the wavelength λ is also a function of $\mathbf{r}$ and t.

Similarly, at any given point $\mathbf{r} =$ **constant** we have

$$d\varphi = \frac{\partial\varphi}{\partial t}\, dt,$$

which enables us to define the period T of the wave by the equation

$$2\pi = -\frac{\partial\varphi}{\partial t}\, T. \tag{22.9}$$

Let $-\partial\varphi/\partial t \equiv \omega$; then

$$\omega = \frac{2\pi}{T} = 2\pi\nu, \tag{22.10}$$

where ν is the frequency. This is in agreement with the special choice of φ given in Eq. (22.5). The minus sign in Eq. (22.9) secures this agreement, and is introduced for future convenience (see Eq. (22.25)).

Let us now suppose that A varies so slowly, or that λ is so small, that

$$\lambda^2|\nabla^2 A| \ll A. \tag{22.11}$$

With this assumption it is possible to find approximate solutions of the wave equation corresponding to geometrical optics.

We have, using Eqs. (22.4), (A3.18), and (A3.40),

$$\nabla^2 f = A\nabla^2 e^{i\varphi} + 2\nabla A \cdot \nabla e^{i\varphi} + e^{i\varphi}\nabla^2 A, \qquad \nabla e^{i\varphi} = ie^{i\varphi}\nabla\varphi,$$

and

$$\nabla^2 e^{i\varphi} = ie^{i\varphi}\nabla^2\varphi - e^{i\varphi}(\nabla\varphi)^2,$$

so that Eq. (22.1) becomes

$$iA\,\nabla^2\varphi - A(\nabla\varphi)^2 + 2i\,\nabla A \cdot \nabla\varphi + \nabla^2 A + \frac{A}{v^2}\left(\frac{\partial\varphi}{\partial t}\right)^2 - \frac{Ai}{v^2}\frac{\partial^2\varphi}{\partial t^2} = 0. \tag{22.12}$$

Separating the real and the imaginary parts, we obtain

$$A(\nabla\varphi)^2 - \frac{A}{v^2}\left(\frac{\partial\varphi}{\partial t}\right)^2 - \nabla^2 A = 0 \tag{22.13}$$

and

$$A\,\nabla^2\varphi + 2\,\nabla A \cdot \nabla\varphi - \frac{A}{v^2}\frac{\partial^2\varphi}{\partial t^2} = 0. \tag{22.14}$$

Since $(\nabla\varphi)^2 = 4\pi^2/\lambda^2$, we obtain, using inequality (22.11),

$$|\nabla^2 A| \ll \frac{A}{\lambda^2} = \frac{A(\nabla\varphi)^2}{4\pi^2} < A(\nabla\varphi)^2.$$

The last term of Eq. (22.13) may therefore be neglected in comparison with the first, and we have

$$(\nabla\varphi)^2 - \frac{1}{v^2}\left(\frac{\partial\varphi}{\partial t}\right)^2 = 0. \tag{22.15}$$

Rays as paths of imaginary particles. If we put

$$W = \alpha\varphi, \tag{22.16}$$

where α is some constant, Eq. (22.15) may be written as

$$(\nabla W)^2 - \frac{1}{v^2}\left(\frac{\partial W}{\partial t}\right)^2 = 0, \tag{22.17}$$

and may be regarded as the Hamilton–Jacobi partial differential equation for an imaginary particle.

Since (see Eqs. (2.19) and (2.22))

$$\mathbf{p} = \nabla W \qquad \text{and} \qquad H = -\partial W/\partial t, \tag{22.18}$$

Eq. (22.17) is equivalent to

$$H^2 = v^2p^2$$

or, choosing a positive sign,

$$H = v(x, y, z)\,p. \tag{22.19}$$

The canonical equations of motion (see Eqs. (2.17)) are therefore

$$\dot{\mathbf{r}} = \mathbf{v} = \frac{\partial H}{\partial \mathbf{p}} = \frac{v\mathbf{p}}{p} \tag{22.20}$$

and

$$\dot{\mathbf{p}} = -\frac{\partial H}{\partial \mathbf{r}} = -p\,\nabla v. \tag{22.21}$$

Equation (22.20) shows that the velocity $\mathbf{v}$ of the imaginary particle is numerically equal to $v(x, y, z)$ and is directed along $\mathbf{p}$.

Every solution of Eq. (22.17) is a function of $\mathbf{r}$ and t and hence

$$\mathbf{p} = \nabla W = \mathbf{p}(\mathbf{r}, t). \tag{22.22}$$

This defines at each point the direction vector

$$\mathbf{u}(\mathbf{r}, t) = \frac{\mathbf{v}}{v} = \frac{\mathbf{p}}{p}. \tag{22.23}$$

A solution of Eq. (22.17) therefore defines a vector field

$$\mathbf{u} = \frac{\nabla W}{|\nabla W|} = \frac{\nabla \varphi}{|\nabla \varphi|}, \tag{22.24}$$

which gives at each point and moment of time the direction of a possible trajectory of a particle. If each such trajectory is regarded as a *ray* of light, a solution of Eq. (22.17) corresponds to a bundle of rays. We shall see that the bundles of rays so defined obey all the laws of geometrical optics.

Newton held that light consists of particles traveling along rays. His interpretation of the nature of light would correspond to taking Eq. (22.17) as fundamental. We have seen, however, that it is only an approximation, depending upon the assumption (22.11) for its validity, and therefore failing in those cases where this assumption is invalid.

If we introduce the *propagation vector* $\boldsymbol{k}$ by letting

$$\boldsymbol{k} = \nabla \varphi \qquad \text{and} \qquad \omega = -\partial \varphi / \partial t, \tag{22.25}$$

then the energy and momentum of the imaginary light particle would be*

$$\mathbf{p} = \nabla W = \alpha \mathbf{k} \quad \text{and} \quad H = -\partial W/\partial t = \alpha\omega. \tag{22.26}$$

Rectilinear propagation and the laws of reflection and refraction. Although the laws of geometrical optics are often expressed in terms of the behavior of individual rays, geometrical optics actually operates with light beams that are bundles of rays, all of which are normal to some surface. We shall call such a bundle of rays a *normal bundle*. In obtaining geometrical optics as a limiting case of the wave motion we find that it is only for such bundles that geometrical optics is valid. Thus, let $d\mathbf{r}$ represent any displacement along a surface $\varphi =$ constant; then $d\varphi = \nabla\varphi \cdot d\mathbf{r} = 0$, so that

$$\mathbf{u} \cdot d\mathbf{r} = \frac{\nabla\varphi \cdot d\mathbf{r}}{|\nabla\varphi|} = \frac{d\varphi}{|\nabla\varphi|} = 0. \tag{22.27}$$

It follows that $\mathbf{u}$ is perpendicular to $d\mathbf{r}$ and is hence normal to the surface $\varphi =$ constant. The bundle is thus a normal bundle. As typical examples of such bundles we may take radial rays and parallel rays.

In geometrical optics the rays for each frequency are considered separately and one assumes that Eq. (22.5) holds for the individual frequency, or color, components in the light. If we make this assumption, then for each component

$$\frac{\partial\varphi}{\partial t} = -\omega = \text{constant}, \tag{22.28}$$

and thus

$$\frac{\partial}{\partial t}\mathbf{k} = \frac{\partial}{\partial t}\nabla\varphi = \nabla\frac{\partial\varphi}{\partial t} = -\nabla\omega = 0. \tag{22.29}$$

Also, from Eq. (22.15) and (22.25) we have

$$k = \left|\frac{1}{v}\frac{\partial\varphi}{\partial t}\right| = \frac{\omega}{v}. \tag{22.30}$$

Let us calculate the curvature of a ray. This can be defined as the rate of change of the direction per unit length of the ray, say $d\theta/ds$. Now the change of direction $d\theta$ is simply $|d\mathbf{u}|$, since $\mathbf{u}$ is along the ray and has unit

* These equations are made use of in quantum mechanics, in which corresponding to each particle of momentum p and energy H there is associated a wave. The k and ω of this wave satisfy Eq. (22.26) with $\alpha = \hbar = h/2\pi$, where h is Planck's constant.

magnitude. Thus,

$$\text{curvature} = \left|\frac{d\mathbf{u}}{ds}\right| = \left|\frac{d\mathbf{r}\cdot\nabla\mathbf{u}}{ds}\right| = |\mathbf{u}\cdot\nabla\mathbf{u}|, \tag{22.31}$$

since $d\mathbf{r}/ds = \mathbf{u}$.

Another expression for the curvature may be obtained by noting that $\nabla\mathbf{u}\cdot\mathbf{u} = \nabla 1 = 0$ and expanding $\nabla\,\mathbf{u}\cdot\mathbf{u}$ by formula (A3.28); thus

$$\nabla\mathbf{u}\cdot\mathbf{u} = 2\,\mathbf{u}\cdot\nabla\mathbf{u} + 2\,\mathbf{u}\times(\nabla\times\mathbf{u}) = 0; \tag{22.32}$$

so that

$$\text{curvature} = |\mathbf{u}\cdot\nabla\mathbf{u}| = |\mathbf{u}\times(\nabla\times\mathbf{u})|. \tag{22.33}$$

By use of Eqs. (22.24) and (22.25), $\mathbf{u} = \boldsymbol{k}/k$, and therefore

$$\nabla\times\mathbf{u} = \nabla\times\frac{\boldsymbol{k}}{k} = \frac{1}{k}\nabla\times\boldsymbol{k} - \boldsymbol{k}\times\nabla\frac{1}{k} = -\boldsymbol{k}\times\nabla\frac{1}{k}, \tag{22.34}$$

since $\nabla\times\boldsymbol{k} = \nabla\times\nabla\varphi = 0$. Using these results, we have

$$\mathbf{u}\cdot\nabla\times\mathbf{u} = -\frac{\boldsymbol{k}}{k}\cdot\boldsymbol{k}\times\nabla\frac{1}{k}. \tag{22.35}$$

This shows that $\mathbf{u}$ is perpendicular to $\nabla\times\mathbf{u}$ and hence that

$$\text{curvature} = |\mathbf{u}\times\nabla\times\mathbf{u}| = |\nabla\times\mathbf{u}|. \tag{22.36}$$

Thus, the necessary and sufficient condition for the bundle of rays to be rectilinear is that

$$\nabla\times\mathbf{u} = 0. \tag{22.37}$$

This occurs, for example, when the medium is homogeneous, i.e., when v = constant; for then Eq. (22.30) gives k = constant, and hence, by Eq. (22.34), $\nabla\times\mathbf{u} = 0$. Rectilinear propagation of light in a homogeneous medium is one of the fundamental laws of geometrical optics.

We shall now investigate the behavior of a ray of light at the junction of two media. For this purpose let us suppose that a surface S separates two media in which v is equal to v and v', respectively, each a constant. Equation (22.30) shows that there must be a discontinuity in k at the surface, and therefore in $\boldsymbol{k}$. Let $\boldsymbol{k}$ and $\boldsymbol{k}'$ be the corresponding values of the propagation vector on the two sides of the surface.

Consider the value of $\oint \boldsymbol{k}\cdot d\mathbf{r}$ taken around an infinitesimal circuit C (see Figure 22.1) composed of parts parallel and perpendicular to the surface, and such that the perpendicular parts are negligibly small in

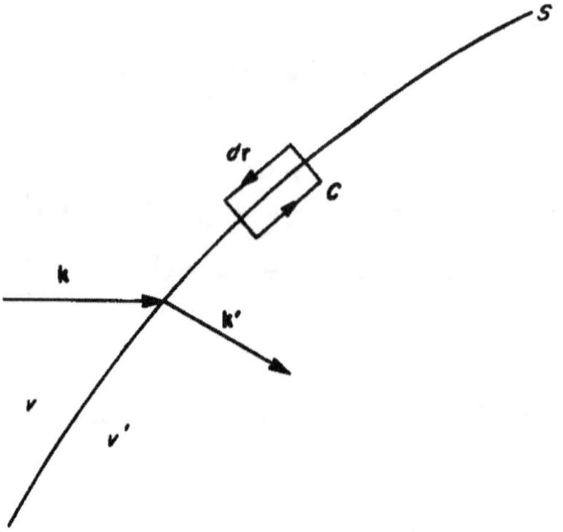

FIG. 22.1. Line integral $\oint \mathbf{k} \cdot d\mathbf{r}$ taken around a contour C at the interface between two media.

comparison with the parallel parts. The integral is then essentially

$$\oint \boldsymbol{k} \cdot d\mathbf{r} = (\boldsymbol{k} - \boldsymbol{k}') \cdot \delta\mathbf{r},$$

where $\delta\mathbf{r}$ is a vector representing the length and direction of one of the long sides of the path C. On the other hand, by Eq. (22.25)

$$\oint \boldsymbol{k} \cdot d\mathbf{r} = \oint \nabla\varphi \cdot d\mathbf{r} = \oint d\varphi = 0; \tag{22.38}$$

therefore

$$(\boldsymbol{k} - \boldsymbol{k}') \cdot \delta\mathbf{r} = 0 \tag{22.39}$$

for any $\delta\mathbf{r}$ parallel to the surface. One concludes therefore that $\boldsymbol{k} - \boldsymbol{k}'$ must be normal to the surface. This means that

$$\mathbf{N} \times (\boldsymbol{k} - \boldsymbol{k}') = 0 \tag{22.40}$$

or

$$\boldsymbol{k} - \boldsymbol{k}' = f\mathbf{N}, \tag{22.41}$$

where $\mathbf{N}$ is a unit vector normal to the surface, and f some scalar function of the coordinates.

For a reflection $\boldsymbol{k}'$ occurs on the same side of the surface as $\boldsymbol{k}$. This means that $\boldsymbol{k}$ is now a two-valued function of position. Equation (22.38) is of course still true, but in the integration we must change from $\boldsymbol{k}$ to $\boldsymbol{k}'$, or vice versa, every time the path touches the surface.

Consider the path $PAP'BP$ in Figure 22.2, where the points A and B lie on the reflecting surface S. Applying Eq. (22.38) to this path, we obtain

$$\int_{PA} \boldsymbol{k} \cdot d\mathbf{r} + \int_{AP'} \boldsymbol{k}' \cdot d\mathbf{r} + \int_{P'B} \boldsymbol{k}' \cdot d\mathbf{r} + \int_{BP} \boldsymbol{k} \cdot d\mathbf{r} = 0.$$

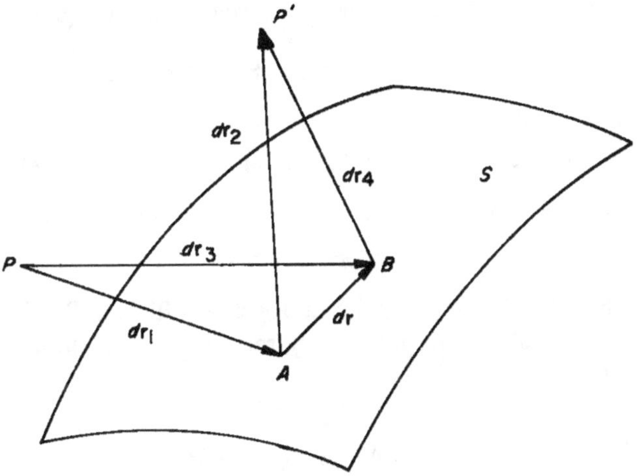

FIG. 22.2. Virtual paths of rays at a reflecting surface as they relate to the law of reflection.

If the vectors from P to A, A to P', P to B, B to P', and A to B, which are designated $\delta\mathbf{r}_1$, $\delta\mathbf{r}_2$, $\delta\mathbf{r}_3$, $\delta\mathbf{r}_4$, and $\delta\mathbf{r}$, respectively, are all infinitesimal, this equation becomes

$$\boldsymbol{k} \cdot \delta\mathbf{r}_1 + \boldsymbol{k}' \cdot \delta\mathbf{r}_2 - \boldsymbol{k}' \cdot \delta\mathbf{r}_4 - \boldsymbol{k} \cdot \delta\mathbf{r}_3 = 0.$$

From the figure

$$\delta\mathbf{r}_3 = \delta\mathbf{r}_1 + \delta\mathbf{r}$$

and

$$\delta\mathbf{r}_2 = \delta\mathbf{r}_4 + \delta\mathbf{r};$$

therefore, we have

$$\boldsymbol{k} \cdot \delta\mathbf{r}_1 + \boldsymbol{k}' \cdot (\delta\mathbf{r}_4 + \delta\mathbf{r}) - \boldsymbol{k}' \cdot \delta\mathbf{r}_4 - \boldsymbol{k} \cdot (\delta\mathbf{r}_1 + \delta\mathbf{r}) = 0$$

or

$$(\boldsymbol{k} - \boldsymbol{k}') \cdot \delta\mathbf{r} = 0. \quad (22.42)$$

Since $\delta \mathbf{r}$ is along the surface, this equation is the same as Eq. (22.39). Thus Eqs. (22.40) and (22.41) apply to reflection as well as refraction.

From Eq. (22.41) it is obvious that N is in the plane of $\boldsymbol{k}$ and $\boldsymbol{k}'$; that is, the incident ray, the normal to the surface, and the refracted or reflected rays are in the same plane.

For reflection $v = v'$, $k = k'$, but $\boldsymbol{k} \neq \boldsymbol{k}'$, since $\boldsymbol{k}$ is directed into the surface, while $\boldsymbol{k}'$ is directed out of the surface. Multiplying Eq. (22.40) by $\mathbf{N} \times$ and remembering that $N^2 = 1$, we obtain the equation

$$\mathbf{N} \cdot (\boldsymbol{k} - \boldsymbol{k}')\mathbf{N} + \boldsymbol{k}' = \boldsymbol{k}. \tag{22.43}$$

Squaring both sides, we have

$$[\mathbf{N} \cdot (\boldsymbol{k} - \boldsymbol{k}')]^2 + 2\boldsymbol{k}' \cdot \mathbf{N}\,\mathbf{N} \cdot (\boldsymbol{k} - \boldsymbol{k}') + k'^2 = k^2,$$

and, since $k' = k$, this reduces to $(\mathbf{N} \cdot \boldsymbol{k})^2 - (\mathbf{N} \cdot \boldsymbol{k}')^2 = 0$, or to

$$\mathbf{N} \cdot \boldsymbol{k}' = \pm \mathbf{N} \cdot \boldsymbol{k}. \tag{22.44}$$

Thus, either $\mathbf{N} \cdot (\boldsymbol{k} - \boldsymbol{k}') = 0$, which together with Eq. (22.43) would give $\boldsymbol{k} = \boldsymbol{k}'$ and would correspond to refraction for the special case when $v = v'$, or

$$\mathbf{N} \cdot \boldsymbol{k}' = -\mathbf{N} \cdot \boldsymbol{k}. \tag{22.45}$$

This substituted into Eq. (22.43) gives for reflection

$$\boldsymbol{k}' = \boldsymbol{k} - 2\boldsymbol{k} \cdot \mathbf{N}\,\mathbf{N}. \tag{22.46}$$

From (22.45) and the fact $k = k'$ we can obtain a more familiar statement of the *law of reflection*, namely

$$\left| \mathbf{N} \cdot \frac{\boldsymbol{k}}{k} \right| = \left| \mathbf{N} \cdot \frac{\boldsymbol{k}'}{k'} \right| \tag{22.47}$$

or, stated in words: The angle of incidence is equal to the angle of reflection.

The corresponding law for refraction is obtained from Eq. (22.40), namely

$$|\mathbf{N} \times \boldsymbol{k}| = |\mathbf{N} \times \boldsymbol{k}'|$$

or

$$k \sin \theta = k' \sin \theta', \tag{22.48}$$

where θ and θ' are the angles of incidence and refraction, respectively.

The law of reflection ($\theta' = \theta$) would also follow from this equation by noting that then $k' = k$.

If we introduce *indices of refraction n* and n' for the two media defined by

$$n = \frac{c}{v} \quad \text{and} \quad n' = \frac{c}{v'}, \tag{22.49}$$

then by Eq. (22.30)

$$\frac{k}{k'} = \frac{v'}{v} = \frac{n}{n'},$$

and Eq. (22.48) becomes

$$n \sin \theta = n' \sin \theta'. \tag{22.50}$$

This is the form of the *law of refraction* most commonly used.

Note that we have obtained the laws of reflection and refraction by applying the geometrical optics approximation to the wave equation, Eq. (22.1).

Fermat's principle. Since

$$\mathbf{k} = k\mathbf{u} = \frac{\omega\mathbf{u}}{v} = \frac{\omega}{c} n\mathbf{u}, \tag{22.51}$$

Eq. (22.38) can be replaced by the statement that

$$E(P, P') = \int_{P'}^{P} n\mathbf{u} \cdot d\mathbf{r} \tag{22.52}$$

is independent of the path.

In geometrical optics the function E plays an important role and is called the *eikonal.* Since the integral in Eq. (22.51) is independent of the path, by fixing P' we make $E(P, P')$ a function of only the upper limit P. Thus for a change $d\mathbf{r}$ in the upper limit we have

$$dE = \nabla E \cdot d\mathbf{r} = n\mathbf{u} \cdot d\mathbf{r}$$

and hence $\nabla E = n\mathbf{u}$ or

$$(\nabla E)^2 = n^2, \tag{22.53}$$

which is a fundamental equation of geometrical optics.

Equation (22.53) can of course be used when n is a function of position and forms a convenient starting point for geometrical optics in a non-homogeneous medium.

To justify this approach we first show that, when we are dealing with a given frequency component $\omega = 2\pi\nu$, Eq. (22.53) is equivalent to Eq. (22.15). For this case the phase is

$$\varphi(\mathbf{r}, t) = \theta(\mathbf{r}) - \omega t,$$

in agreement with Eq. (22.5), and we let

$$E = \frac{c\,\theta(\mathbf{r})}{\omega}. \tag{22.54}$$

One then observes, using Eq. (22.28), that

$$\nabla E = \frac{c}{\omega}\nabla\theta(\mathbf{r}) = \frac{c}{\omega}\nabla\varphi(\mathbf{r}, t),$$

$$n = \frac{c}{v} = \frac{-c}{\omega}\left(\frac{1}{v}\frac{\partial\varphi}{\partial t}\right),$$

and thus that Eqs. (22.15) and (22.53) are indeed equivalent.

From Eq. (22.52)

$$\int_{P'AP} n\mathbf{u}\cdot d\mathbf{r} = \int_{P'BP} n\mathbf{u}\cdot d\mathbf{r},$$

where the letters A and B designate distinct paths taken between P' and P. Now let the path designated by A be along the actual ray so that $d\mathbf{r} = \mathbf{u}\,dr$ and the path B be sufficient near to the path A that $\mathbf{u}\cdot d\mathbf{r} = dr\cos\theta$ is always positive, then

$$\int_{P'AP} n\,dr = \int_{P'BP} n\cos\theta\,dr \leq \int_{P'BP} n\,dr. \tag{22.55}$$

This condition can be expressed by saying that for sufficiently close points P' and P, so that $\cos\theta$ is positive for all neighboring paths, the actual path A yields the minimum value of the *optical path length*

$$\int_{P'}^{P} n\,dr.$$

This result may be expressed in the variational form

$$\delta\int_{P'}^{P} n\,dr = 0 \tag{22.56}$$

and is known as *Fermat's principle*.

Intensity variation in geometrical optics. Returning now to Eq. (22.14), we multiply it by A and obtain*

$$A^2\nabla^2\varphi + \nabla A^2 \cdot \nabla\varphi - \frac{A^2}{v^2}\frac{\partial^2\varphi}{\partial t^2} = 0.$$

Putting $\nabla\varphi = \boldsymbol{k}$ and $-\partial\varphi/\partial t = \omega$, this becomes

$$A^2\nabla \cdot \boldsymbol{k} + \boldsymbol{k} \cdot \nabla A^2 + \frac{A^2}{v^2}\frac{\partial\omega}{\partial t} = 0,$$

and thus, since A is assumed independent of t,

$$\nabla \cdot A^2\boldsymbol{k} + \frac{\partial}{\partial t}\frac{A^2\omega}{v^2} = 0. \tag{22.57}$$

If we let

$$\rho = \alpha\frac{A^2\omega}{v^2} = \frac{\alpha A^2 n^2\omega}{c^2}, \tag{22.58}$$

then

$$\rho\mathbf{v} = \frac{\alpha A^2\omega}{v}\frac{\mathbf{v}}{v} = A^2\boldsymbol{k}, \tag{22.59}$$

by Eq. (22.23), (22.30), and (22.51). Equation (22.57) can thus be written

$$\nabla \cdot \rho\mathbf{v} + \frac{\partial\rho}{\partial t} = 0. \tag{22.60}$$

This is the equation of conservation of energy, provided that, by the proper choice of α, ρ is made equal to the energy density. Since the energy of a single imaginary light particle was found to be $\alpha\omega$ (see Eq. (22.58)), the density of light particles in the beam, by Eq. (22.58), must be A^2n^2/c^2. On the other hand, by Eq. (22.59), the magnitude of $\alpha A^2\boldsymbol{k} = \rho\mathbf{v}$ is the energy passing through a unit area perpendicular to $\mathbf{v}$ per unit time, i.e., the *intensity of illumination*.

When ω is a constant, Eq. (22.57) becomes

$$\nabla \cdot A^2\boldsymbol{k} = 0, \tag{22.61}$$

where, as before,

$$\boldsymbol{k} = \nabla\varphi = \nabla\theta. \tag{22.62}$$

This is an equation determining A^2. Equation (22.61) may thus be

* Since this equation is generally disregarded, it is often stated that geometrical optics holds only when, in addition to inequality (22.11), $\nabla \cdot \boldsymbol{k}$ and ∇A^2 are small. The discussion that follows shows that this is not the case.

considered as a supplement to geometrical optics, giving the intensity of illumination $\alpha A^2 \boldsymbol{k}$ at each point of the bundle of rays. We shall now show that Eq. (22.61) can always be solved.

A general solution of the equation $\nabla \cdot \mathbf{P} = 0$ is $\mathbf{P} = \nabla u \times \nabla v$, where u and v are arbitrary (see Theorem 8, Appendix A3); therefore the solution of Eq. (22.61) must be of the form

$$A^2 \boldsymbol{k} = \nabla u \times \nabla v. \tag{22.63}$$

With u and v arbitrary, however, $\nabla u \times \nabla v$ will not, in general, be in the direction of $\boldsymbol{k}$. To insure that $\nabla u \times \nabla v$ is properly directed it is necessary and sufficient that

$$\boldsymbol{k} \cdot \nabla u = \boldsymbol{k} \cdot \nabla v = 0. \tag{22.64}$$

That these conditions are necessary can be seen if we multiply Eq. (22.63) by $\cdot \nabla u$ and $\cdot \nabla v$, successively. We note also that Eqs. (22.64) are sufficient to insure that $\boldsymbol{k}$ is either parallel or antiparallel to $\nabla u \times \nabla v$, because with these conditions

$$\boldsymbol{k} \times (\nabla u \times \nabla v) = \boldsymbol{k} \cdot \nabla v \nabla u - \boldsymbol{k} \cdot \nabla u \nabla v = 0.$$

Equations (22.64) have an infinite number of solutions, and it would be generally difficult to find the particular u and v that would satisfy the boundary conditions that would ordinarily be imposed on A^2. We may however proceed as follows:

Let α and β be any two independent solutions of Eqs. (22.64); that is,

$$\boldsymbol{k} \cdot \nabla \alpha = \boldsymbol{k} \cdot \nabla \beta = 0. \tag{22.65}$$

In addition, let the labeling be such that $\nabla \alpha \times \nabla \beta$, rather than $\nabla \beta \times \nabla \alpha$, is parallel to $\boldsymbol{k}$. Having found α and β, we may use α, β, and θ as a set of curvilinear coordinates, Eqs. (22.65) together with the fact that $\nabla \theta = \boldsymbol{k}$ insuring their independence. Now let $u = u(\alpha, \beta, \theta)$ and $v = v(\alpha, \beta, \theta)$ be any two independent functions satisfying Eqs. (22.64). We must then have

$$\boldsymbol{k} \cdot \nabla u = \boldsymbol{k} \cdot \left(\frac{\partial u}{\partial \alpha} \nabla \alpha + \frac{\partial u}{\partial \beta} \nabla \beta + \frac{\partial u}{\partial \theta} \nabla \theta \right) = 0$$

and

$$\boldsymbol{k} \cdot \nabla v = \boldsymbol{k} \cdot \left(\frac{\partial v}{\partial \alpha} \nabla \alpha + \frac{\partial v}{\partial \beta} \nabla \beta + \frac{\partial v}{\partial \theta} \nabla \theta \right) = 0.$$

With the help of Eqs. (22.65) and the equation $\nabla \theta = \boldsymbol{k}$, we see that these conditions on u and v reduce to

$$\frac{\partial u}{\partial \theta} = \frac{\partial v}{\partial \theta} = 0.$$

Thus, u and v are functions of α and β only. In other words, any solution of one of the equations in Eqs. (22.64) is necessarily a function of any two given independent solutions, such as α and β. We now have

$$\begin{aligned}\nabla u \times \nabla v &= \left(\frac{\partial u}{\partial \alpha}\nabla\alpha + \frac{\partial u}{\partial \beta}\nabla\beta\right) \times \left(\frac{\partial v}{\partial \alpha}\nabla\alpha + \frac{\partial v}{\partial \beta}\nabla\beta\right)\\ &= \left(\frac{\partial u}{\partial \alpha}\frac{\partial v}{\partial \beta} - \frac{\partial u}{\partial \beta}\frac{\partial v}{\partial \alpha}\right)\nabla\alpha \times \nabla\beta = f(\alpha, \beta)\nabla\alpha \times \nabla\beta,\end{aligned}$$

so that Eq. (22.63) becomes

$$A^2\mathbf{k} = f(\alpha, \beta)\nabla\alpha \times \nabla\beta. \tag{22.66}$$

The arbitrariness of u and v of Eq. (22.63) has thus been replaced by the arbitrariness of the function $f(\alpha, \beta)$. Since $\nabla\alpha \times \nabla\beta$ is already in the direction of $\mathbf{k}$ and since A^2 is positive, we must require that $f(\alpha, \beta)$ be nonnegative everywhere, or we could replace it by $|f(\alpha, \beta)|$. That $f(\alpha, \beta)$ is otherwise arbitrary can be seen by evaluating $\nabla \cdot A^2\mathbf{k}$; thus by Eqs. (A3.23) and (A3.26)

$$\begin{aligned}\nabla \cdot A^2\mathbf{k} &= \nabla \cdot [f(\alpha, \beta)\nabla\alpha \times \nabla\beta] = \nabla\alpha \times \nabla\beta \cdot \nabla f + f\nabla \cdot (\nabla\alpha \times \nabla\beta)\\ &= (\nabla\alpha \times \nabla\beta) \cdot \left(\frac{\partial f}{\partial \alpha}\nabla\alpha + \frac{\partial f}{\partial \beta}\nabla\beta\right) = 0.\end{aligned}$$

This shows that Eq. (22.61) is satisfied for all $f(\alpha, \beta)$.

Multiplying Eq. (22.66) by $\mathbf{k}\,\cdot$ and dividing by k^2, we obtain

$$A^2 = \frac{f(\alpha, \beta)\mathbf{k} \cdot \nabla\alpha \times \nabla\beta}{k^2} = \frac{wf(\alpha, \beta)}{k}, \tag{22.67}$$

where $w = |\nabla\alpha \times \nabla\beta|$, since $\mathbf{k}$ is in the direction of $\nabla\alpha \times \nabla\beta$. This is the required solution, $f(\alpha, \beta)$ being a positive-valued function that is determined in each case by the data.

To show how this may be done, suppose, as would usually be the case, that A^2 is given for all points of some surface of constant θ, say $\theta = \theta_0$, i.e., along a surface normal to the bundle of rays. The quantities A^2, w, and k are ordinarily functions of α, β, and θ, but on the surface $\theta = \theta_0$ they are functions only of α and β and hence may be expressed as $A_0^2(\alpha, \beta)$,

$w_0(\alpha, \beta)$, and $k_0(\alpha, \beta)$. From Eq. (22.67) we have, therefore,

$$A_0^2(\alpha, \beta) = \frac{w_0(\alpha, \beta) f(\alpha, \beta)}{k_0(\alpha, \beta)}$$

so that

$$f(\alpha, \beta) = \frac{k_0 A_0^2}{w_0}, \tag{22.68}$$

a positive-valued function; and finally, by substitution into Eq. (22.67),

$$A^2 = A_0^2 \frac{k_0 w}{k w_0} \quad \text{or} \quad I = I_0 \frac{w}{w_0}, \tag{22.69}$$

where $I = A^2 k$ is the intensity of illumination at α, β, θ and I_0 that at α, β, θ_0.

Limits of applicability of geometrical optics. We will now investigate the limits of applicability of geometrical optics. We have seen that the theoretical basis for geometrical optics is Eq. (22.15), which results from the wave equation if we assume condition (22.11). We have already observed that Eq. (22.14), which in cases contemplated in geometrical optics reduces to Eq. (22.61), is usually disregarded. Since this equation is to be solved for A using any φ obtained from Eq. (22.15), it can be regarded as supplementing the equations of geometrical optics by giving the intensity of illumination for each point of the beam. Thus, to determine whether or not geometrical optics is applicable in any particular case, one could proceed as follows: first, determine $k = \nabla\varphi$ by solving Eq. (22.15) subject to the conditions imposed upon the beam in the particular case in question; second, determine A by the use of Eq. (22.69), or the more general Eq. (22.67); finally, check to see whether A thus determined satisfies condition (22.11).

An important general case is the variation in intensity at the edge of a beam of light. Let us consider a path (see Figure 22.3) starting from some point P just outside the beam, where $A \simeq 0$ and $\nabla A \simeq 0$, and ending at a point Q just inside the beam, where A is assumed to have approximately the value it has in the main portion of the beam. The path is assumed to be everywhere parallel to ∇A and to have a length a. Since we are interested in finding the maximum rate of transition from dark to light possible in geometrical optics, we shall assume that a is small; hence the curvature of the path will not be important.

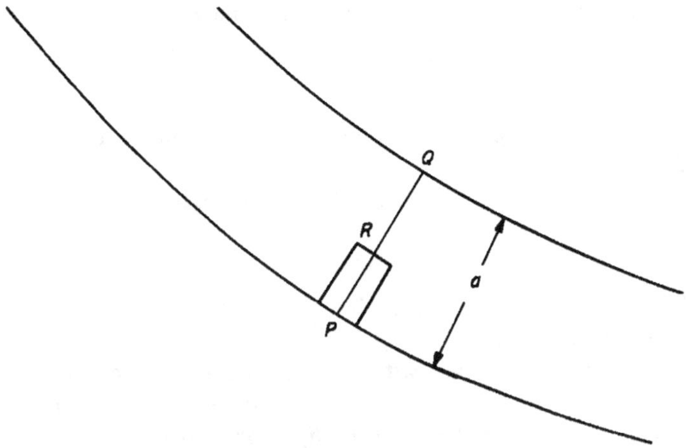

FIG. 22.3. Intensity variation at the edge of a light beam required for geometrical optics to be applicable.

Since the path PQ is very nearly straight, we have

$$|\nabla A|_{\text{average}} = \frac{A(Q) - A(P)}{a} \simeq \frac{A}{a}.$$

Let the largest value of $|\nabla A|$ that occurs at some point R along PQ be labeled $|\nabla A|_{\max}$; moreover, let T be a thin tube surrounding PR generated by lines having, as does PR, everywhere the direction of ∇A. Let the average length and average cross-sectional area of the tube be labeled ℓ and s, respectively.

We have at once

$$|\nabla A|_{\max} \geq |\nabla A|_{\text{aver}} \simeq \frac{A}{a}, \tag{22.70}$$

and from the definition of $|\nabla^2 A|_{\text{aver}}$ and Gauss' theorem (see Eq. (A4.8))

$$|\nabla^2 A|_{\text{aver}} \simeq \frac{1}{\ell s}\int_V |\nabla^2 A|\, dV \geq \frac{1}{\ell s}\left|\int_V \nabla \cdot \nabla A\, dV\right| = \frac{1}{\ell s}\left|\int_\Sigma \nabla A \cdot d\mathbf{S}\right|, \tag{22.71}$$

where Σ includes both the lateral surface of T and its end surfaces. On the lateral surface $\nabla A \cdot d\mathbf{S}$ is zero and on the end at P one has $|\nabla A| \simeq 0$. Thus Eq. (22.71) becomes

$$|\nabla^2 A|_{\text{aver}} \geq \frac{1}{\ell s}\left|\int_{\Sigma_R} \nabla A \cdot d\mathbf{S}\right| \simeq \frac{1}{\ell}|\nabla A|_{\max} \geq \frac{A}{a\ell} \geq \frac{A}{a^2}.$$

where Σ_R is the end surface at R of the tube T, and one concludes that

$$|\nabla^2 A|_{\max} \geq |\nabla^2 A|_{\text{aver}} \geq \frac{A}{a^2}. \tag{22.72}$$

If geometrical optics is to hold, we may combine this inequality with (22.11) to show that

$$\lambda^2|\nabla^2 A|_{\max} \ll A \leq a^2|\nabla^2 A|_{\max}$$

or

$$\lambda \ll a. \tag{22.73}$$

Thus, we see that the extinction of light at the shadow edge cannot be accomplished discontinuously within the geometrical optics approximation, but must take place within a distance a large compared to a wavelength. In particular, any attempt to isolate something approaching a single ray by letting the beam of light pass through an opening whose diameter is small compared to a wavelength, thus making $a \ll \lambda$ and violating the inequality of (22.71), leads to a situation in which geometrical optics and thus the concept of rays is no longer fully applicable. In other words, while geometrical optics seems to predict sharp boundaries between light and shadow, it requires a transition region with a thickness large compared to a wavelength λ to be applicable.

In order for a beam of light to have approximately the characteristics predicted by geometrical optics, namely, sharp boundaries, it is necessary that its diameter D be large compared to the thickness a of the transition region. We thus conclude that for such a beam

$$D \gg a \gg \lambda. \tag{22.74}$$

General restrictions imposed by the wave nature of light. Consider for simplicity the scalar wave equation

$$\nabla^2 f - \frac{1}{c^2}\frac{\partial^2 f}{\partial t^2} = 0 \tag{22.75}$$

relating to the propagation of an electromagnetic wave, such as light, in free space. This is a special case of Eq. (22.1) corresponding to a constant phase velocity $v = c$. A general solution to this equation is given by Eq. (18.47) and can be written

$$f(\mathbf{r}, t) = \text{Re}\left(\int_{k\text{-space}} A(k)e^{i(k\cdot\mathbf{r}-\omega t)}\, d^3k\right), \tag{22.76}$$

where $\omega = ck$ and Re( ) stands for the real part of the quantity.

A plane wave is obtained by taking $A(k) = A_0\delta(k - k_0)$, because then

$$f(\mathbf{r}, t) = \mathrm{Re}(A_0 e^{i(k_0\cdot\mathbf{r}-\omega t)}),$$

which is seen to represent a wave having a definite frequency $\omega = ck_0$ and wavelength $\lambda = 2\pi/k_0$. While it has the property of a beam in that it propagates in a definite direction k_0, it has no restriction to a given region of space.

To restrict $f(\mathbf{r}, t)$ to a beam it is necessary to require that $A(k)$ be appreciably different from zero only over some region K of k-space which we will assume to be centered about $k = k_0$. If only one frequency $\omega = ck$ is permitted, the region K for which $A(k) \neq 0$ is restricted to the sphere $|k| = k_0$ in k-space; otherwise K is a volume of k-space.

Suppose $\mathbf{r}_A$ is outside the beam and $\mathbf{r}_B = \mathbf{r}_A + \mathbf{a}$ is inside the beam; then by Eq. (22.76)

$$f(\mathbf{r}_B, t) = \mathrm{Re}\left(\int_K A(k)e^{ik\cdot\mathbf{a}}e^{i(k\cdot\mathbf{r}_A-\omega t)}\, d^3k\right). \tag{22.77}$$

If the factor $e^{ik\cdot\mathbf{a}}$ were replaced by $e^{i\theta}$, θ equal to a constant, then the integral would reduce to $e^{i\theta}f(\mathbf{r}_A, t) \simeq 0$. It is therefore clear that for at least two k's, say k_1 and k_2, within the region K of k-space we must have $e^{ik_2\cdot\mathbf{a}}$ appreciably different from $e^{ik_1\cdot\mathbf{a}}$. This in turn requires that*

$$(k_2 - k_1)\cdot\mathbf{a} = \Delta k\cdot\mathbf{a} \approx \pi. \tag{22.78}$$

If $\mathbf{a} = a\mathbf{i}$, where $\mathbf{i}$ is a unit vector along the X-axis, then Eq. (22.78) becomes

$$\Delta k_x\, a \approx \pi. \tag{22.79}$$

This equation means that a sharp boundary of thickness a can be achieved only at the expense of having the direction of the beam, nominally given by k_0, uncertain to the extent that the k_x component has a range Δk. This is the basis of the diffraction phenomenon in wave optics.

If the uncertainty Δk_x is known, then the transition from dark to light in the X direction must be over a distance at least as large as the a obtained from Eq. (22.79), namely,

$$a \approx \frac{\pi}{\Delta k_x}.$$

Thus the width Δx of the beam in this direction must satisfy the condition

$$\Delta x \gtrsim \frac{\pi}{\Delta k_x}. \tag{22.80}$$

* The sign $\approx$ is to be read "is of the order of."

EXAMPLE 1. Determine the resolving power of a microscope.

Consider the light rays coming from the object being observed and entering the objective lens L of the microscope shown in Figure 22.4. Since the propagation vector k for a ray in this bundle lies in the solid angle subtended by L at the object, its X component has an uncertainty

$$\Delta k_x \simeq 2k \sin \alpha. \tag{22.81}$$

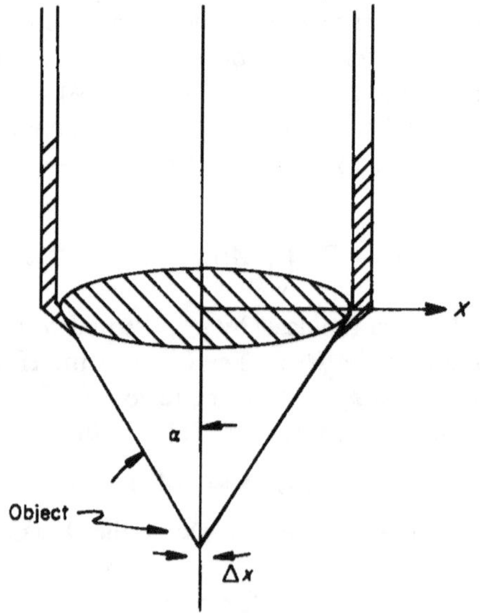

FIG. 22.4. Limitations of the ray theory for analyzing a microscope and its consequent resolving power.

This means that rays cannot be identified as coming from an area that is any smaller in its extension Δx in the X direction than that required by the inequality in (22.80), namely,

$$\Delta x \gtrsim \frac{2\pi}{\Delta k_x} = \frac{2\pi}{2k \sin \alpha} = \frac{\lambda}{2 \sin \alpha}. \tag{22.82}$$

A similar equation holds for Δy.

We interpret the above to mean that one cannot determine the source of the rays within the positional uncertainty Δx given by Eq. (22.82) and hence one cannot resolve two points in the object that are closer than this Δx. The quantity $\lambda/\sin \alpha$ (or sometimes $\kappa\lambda/\sin \alpha$, where $\kappa \approx 1$) is called the *resolving power* of the microscope.

EXAMPLE 2. Determine the resolving power of a telescope.

The light that enters the telescope in Figure 22.5 from the object has an uncertainty in the k_α given again by Eq. (22.81), and therefore Eq. (22.82) also applies. Now, however, since α is very small, we have

$$\sin\alpha \simeq \alpha \simeq \frac{D}{2R},$$

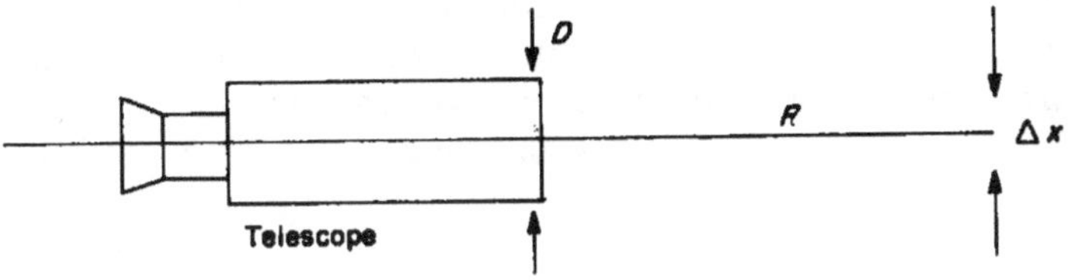

Fig. 22.5. Parameters entering into the resolving power of a telescope.

and thus, by Eq. (22.82),

$$\Delta x \gtrsim \frac{\lambda}{2\alpha} \simeq \frac{\lambda R}{D}$$

is the minimum distance between points in the object that are distinguishable in the telescope.

The angle

$$\Delta\theta \simeq \frac{\Delta x}{R} \simeq \frac{\lambda}{D} \tag{22.83}$$

subtended by Δx at the telescope is seen to be independent of the distance R to the object. This ratio λ/D is referred to as the *resolving power* of a telescope.

23. Particle Properties of Electromagnetic Waves

In the previous section we have seen that with certain approximations light may be considered as consisting of particles traveling along rays and possessing energy and momentum proportional to ω and k, respectively. The constant of proportionality is left arbitrary. In this section we shall consider other particle-like properties of light that do not depend upon the approximations of geometrical optics. We have seen in Section 17 that the energy and momentum of the field is obtained from the four-vector

$$P_\alpha = \int T_{\alpha\beta}\, dS_\beta \tag{23.1}$$

when the integration is over a three-dimensional hyperplane normal to the time direction; in other words, over the instantaneous space.

The integral P_α is, of course, a four-dimensional vector, provided the hypersurface of integration is invariantly defined; but, in attempting to calculate energy and momentum, each observer would normally use his own three-dimensional space as the hypersurface. In particular, two

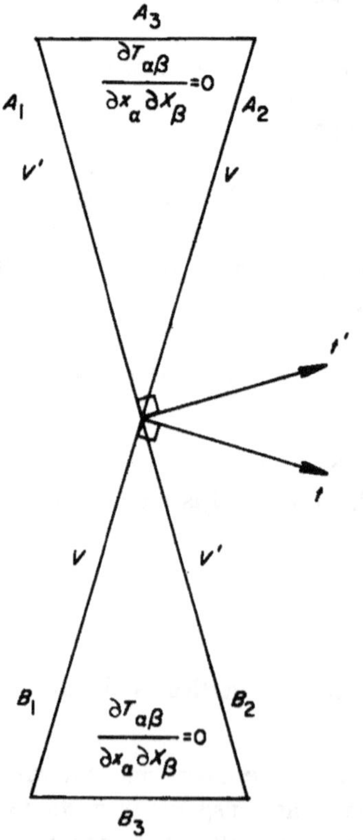

Fig. 23.1. Formation of two closed hypersurfaces A and B by the instantaneous spaces V and V' for two observers. Each is formed by a portion of V, a portion of V', and a portion (A_3 or B_3) of the cylinder at infinity.

observers would obtain

$$P_\alpha = \int T_{\alpha 4}\, dV \tag{23.2}$$

and

$$P'_\alpha = \int T'_{\alpha 4}\, dV' \tag{23.3}$$

as the momentum–energy four-vector. We wish to investigate the relation between these two expressions.

Consider the two closed regions of space–time Ω_1 and Ω_2 of Figure 23.1 formed by the intersection hypersurfaces S, S^*, and Σ. Since

$$\frac{\partial T_{\alpha\beta}}{\partial x_\beta} = 0 \tag{23.4}$$

within both regions, we have by Gauss' theorem

$$\int_{\Omega_1} \frac{\partial T_{\alpha\beta}}{\partial x_\beta}\, d\Omega = -\int_{S_1} T_{\alpha\beta}\, dS_\beta + \int_{S_1^*} T_{\alpha\beta}\, dS_\beta^* + \int_{\Sigma_1} T_{\alpha\beta}\, d\Sigma = 0,$$

$$\int_{\Omega_2} \frac{\partial T_{\alpha\beta}}{\partial x_\beta}\, d\Omega = \int_{S_2} T_{\alpha\beta}\, dS_\beta - \int_{S_2^*} T_{\alpha\beta}\, dS_\beta^* + \int_{\Sigma_2} T_{\alpha\beta}\, d\Sigma = 0,$$

where dS_β and dS_β^* represent elements of the hypersurfaces S and S^* using the convention that the normal is toward the positive time direction. The normal is taken to be outwardly directed on Σ. Subtracting, we have

$$\int_{S^*} T_{\alpha\beta}\, dS_\beta^* - \int_S T_{\alpha\beta}\, dS_\beta = \int_{\Sigma_2} T_{\alpha\beta}\, d\Sigma - \int_{\Sigma_1} T_{\alpha\beta}\, d\Sigma_\beta . \tag{23.5}$$

Moreover, the right-hand side ordinarily goes to zero as R approaches infinity. The latter follows from the fact that the field strengths $\mathscr{E}$ and $\mathscr{H}$ produced by charges and currents within a finite region of space fall off as $1/R^2$; thus $T_{\alpha\beta}$, being quadratic in these field strengths, falls off as $1/R^4$, while the volume of the hypersurfaces Σ_1 and Σ_2 increase as R^3.

We shall, in any case, consider only those cases for which the right-hand side of Eq. (23.5) goes to zero as $R \to \infty$. Then Eq. (23.5) becomes

$$\int_S T_{\alpha\beta}\, dS_\beta = \int_{S^*} T_{\alpha\beta}\, dS_\beta^* ,$$

and hence

$$P_\alpha = \int_S T_{\alpha 4}\, dV = \int_S T_{\alpha\beta}\, dS_\beta = \int_{S^*} T_{\alpha\beta}\, dS^*_\beta, \tag{23.6}$$

where all vectors and tensors are expressed in the unprimed coordinate system. The right-hand side of Eq. (23.6), being the integral of the differential four-vector $T_{\alpha\beta}\, dS^*_\beta$, transforms under a Lorentz transformation as a four-vector. Thus in the primed coordinate system, in which $dS^{*\prime}_\beta = (0, 0, 0, dV')$, it becomes

$$\int_{S^*} T'_{\alpha\beta}\, dS^{*\prime}_\beta = \int_{S^*} T'_{\alpha 4}\, dV' = P'_\alpha,$$

where P'_α was defined in Eq. (23.3). In other words, the energy–momentum four-vectors obtained as in Eqs. (23.2) and (23.3) by having each observer integrate the energy–momentum tensor in his coordinate system over his own instantaneous space are related by the required Lorentz transformations for a four-vector.

The above conclusion rests on the validity of Eq. (23.4) and on the vanishing of the right-hand side of Eq. (23.5) in the limit of $R \to \infty$. When no particles are present and the electromagnetic field drops off rapidly as one leaves some finite region of space both conditions can be met. Such a field is usually called an electromagnetic *wave packet*. With such a wave packet the energy and momentum form a four-vector P_α, analogous to that for a particle, and its components P_x, P_y, P_z, and iE/c transform as x, y, z, and ict under a Lorentz transformation. Thus, written in vector form as in Eq. (11.39), the general Lorentz transformation for the momentum $\mathbf{P}$ and energy E of a wave packet is

$$\mathbf{P}' = \mathbf{P} + \mathbf{v}\left[\frac{(\gamma - 1)\mathbf{v}\cdot\mathbf{P}}{v^2} - \frac{\gamma E}{c^2}\right],$$

$$E' = \gamma(E - \mathbf{v}\cdot\mathbf{P}), \tag{23.7}$$

where, as before,

$$\gamma = \left(1 - \frac{v^2}{c^2}\right)^{-1/2}$$

Interchanging the primed and unprimed quantities and replacing $\mathbf{v}$ by $-\mathbf{v}$, we obtain the inverse transformations

$$\mathbf{P} = \mathbf{P}' + \mathbf{v}\left[\frac{(\gamma - 1)\mathbf{v}\cdot\mathbf{P}'}{v^2} + \frac{\gamma E'}{c^2}\right],$$

$$E = \gamma(E' + \mathbf{v}\cdot\mathbf{P}'). \tag{23.8}$$

In general, it is possible to find a coordinate system for which $\mathbf{P}' = 0$. The corresponding $\mathbf{v}$ may then be called the velocity of the wave packet. Putting $\mathbf{P}' = 0$ and $E' = Mc^2$, which defines the rest mass M, into Eq. (23.8), we obtain

$$\mathbf{P} = \frac{M\mathbf{v}}{\sqrt{1 - v^2/c^2}} \quad \text{and} \quad E = \frac{Mc^2}{\sqrt{1 - v^2/c^2}}, \tag{23.9}$$

which are exactly the same relations as for a particle of mass M.

In addition, we have

$$\mathbf{v} = \frac{c^2\mathbf{P}}{E}, \tag{23.10}$$

which is Eq. (17.76).

The square of the magnitude of the momentum four-vector is an invariant and by Eqs. (23.9) is seen to be given by

$$P^2 - \frac{E^2}{c^2} = -M^2c^2.$$

For $M = 0$ the energy is thus directly proportional to the magnitude of the momentum, being given by

$$E = cP. \tag{23.11}$$

The velocity of the wave packet by Eq. (23.10) is then

$$\mathbf{v} = \frac{c^2\mathbf{P}}{cP} = c\,\frac{\mathbf{P}}{P}, \tag{23.12}$$

and hence its speed is always equal to c. The expressions for the momentum and energy given by Eqs. (23.9) thus become indeterminant.

We can still, however, use the integral for P_α in Eq. (23.2) to determine the energy and momentum, provided of course the integral exists, and can thus write Eq. (23.11) in the form

$$\frac{1}{8\pi}\int(\mathscr{E}^2 + \mathscr{H}^2)\,dV = \frac{1}{4\pi}\left|\int \mathscr{E}\times\mathscr{H}\,dV\right| \tag{23.13}$$

We may at this point make use of the following inequalities:

$$\frac{1}{8\pi}\int(\mathscr{E}^2+\mathscr{H}^2)\,dV=\frac{1}{8\pi}\int[(\mathscr{E}-\mathscr{H})^2+2\mathscr{E}\mathscr{H}]\,dV$$

$$\geq\frac{1}{4\pi}\int\mathscr{E}\mathscr{H}\,dV, \tag{23.14}$$

$$\int\mathscr{E}\mathscr{H}\,dV\geq\int|\mathscr{E}\times\mathscr{H}|\,dV, \tag{23.15}$$

$$\int|\mathscr{E}\times\mathscr{H}|\,dV\geq\left|\int\mathscr{E}\times\mathscr{H}\,dV\right|. \tag{23.16}$$

Thus, in order that Eq. (23.13) hold, it is necessary and sufficient that the equality, instead of the inequality, should be true in each of the Eqs. (23.14), (23.15), and (23.16). The first of these requires that

$$(\mathscr{E}-\mathscr{H})^2=0 \qquad \text{or} \qquad \mathscr{E}=\mathscr{H}, \tag{23.17}$$

the second that

$$\mathscr{E}\mathscr{H}=|\mathscr{E}\times\mathscr{H}| \qquad \text{or} \qquad \mathscr{E}\cdot\mathscr{H}=0, \tag{23.18}$$

and the third that

$$\mathscr{E}\times\mathscr{H}\ \text{has a constant direction.} \tag{23.19}$$

In other words, $\mathscr{E}$ is everywhere perpendicular and equal in magnitude to $\mathscr{H}$, and both are perpendicular to the fixed direction of flow of energy given by Poynting's vector $(c/4\pi)\,\mathscr{E}\times\mathscr{H}$.

It is interesting to consider also the structure of such a wave packet when it is regarded as being composed of plane waves. With the use of the general solution of the electromagnetic equations in the absence of charges, as introduced in Section 18, we have by Eqs. (18.57), (18.58), (18.61), and (18.62)

$$\mathscr{E}(\mathbf{r},t)=\left(\frac{1}{2\pi}\right)^{3/2}\int[\mathscr{E}(\mathscr{k})e^{i(\mathscr{k}\cdot\mathbf{r}-c\mathscr{k}t)}+\mathscr{E}^*(\mathscr{k})e^{-i(\mathscr{k}\cdot\mathbf{r}-c\mathscr{k}t)}]\,d^3\mathscr{k} \tag{23.20}$$

and

$$\mathscr{H}(\mathbf{r},t)=\left(\frac{1}{2\pi}\right)^{3/2}\int\frac{\mathscr{k}}{\mathscr{k}}\times[\mathscr{E}(\mathscr{k})e^{i(\mathscr{k}\cdot\mathbf{r}-c\mathscr{k}t)}+\mathscr{E}^*(\mathscr{k})e^{-i(\mathscr{k}\cdot\mathbf{r}-c\mathscr{k}t)}]\,d^3\mathscr{k}, \tag{23.21}$$

with

$$\mathscr{k}\cdot\mathscr{E}(\mathscr{k})=\mathscr{k}\cdot\mathscr{E}^*(\mathscr{k})=0. \tag{23.22}$$

To calculate the energy and momentum we need a formula for

$\int f(\mathbf{r}, t)g(\mathbf{r}, t)\, dV$, where f and g are functions satisfying the wave equation, expressed in terms of the plane wave amplitudes $f(\mathscr{k})$ and $g(\mathscr{k})$. If $f(\mathbf{r}, t)$ and $g(\mathbf{r}, t)$ are expressed in a form analogous to Eq. (23.20), we obtain by the use of Eq. (A8.102)

$$\begin{aligned}
\int f(\mathbf{r}, t)g(\mathbf{r}, t)\, dV &= \left(\frac{1}{2\pi}\right)^3 \iiint [f(\mathscr{k})e^{i(\mathscr{k}\cdot\mathbf{r}-c\mathscr{k}t)} + f^*(\mathscr{k})e^{-i(\mathscr{k}\cdot\mathbf{r}-c\mathscr{k}t)}] \\
&\qquad \cdot [g(\mathscr{k}')e^{i(\mathscr{k}'\cdot\mathbf{r}-c\mathscr{k}'t)} + g^*(\mathscr{k}')e^{-i(\mathscr{k}'\cdot\mathbf{r}-c\mathscr{k}'t)}]\, d^3\mathscr{k}\, d^3\mathscr{k}'\, dV \\
&= \left(\frac{1}{2\pi}\right)^3 \iint \Big[f(\mathscr{k})g(\mathscr{k}')e^{-ic(\mathscr{k}+\mathscr{k}')t} \int e^{i(\mathscr{k}+\mathscr{k}')\cdot\mathbf{r}}\, dV \\
&\quad + f^*(\mathscr{k})g^*(\mathscr{k}')e^{ic(\mathscr{k}+\mathscr{k}')t} \int e^{-i(\mathscr{k}+\mathscr{k}')\cdot\mathbf{r}}\, dV \\
&\quad + f(\mathscr{k})g^*(\mathscr{k}')e^{-ic(\mathscr{k}-\mathscr{k}')t} \int e^{i(\mathscr{k}-\mathscr{k}')\cdot\mathbf{r}}\, dV \\
&\quad + f^*(\mathscr{k})g(\mathscr{k}')e^{ic(\mathscr{k}-\mathscr{k}')t} \int e^{-i(\mathscr{k}-\mathscr{k}')\cdot\mathbf{r}}\, dV \Big]\, d^3\mathscr{k}\, d^3\mathscr{k}' \\
&= \iint [f(\mathscr{k})g(\mathscr{k}')e^{-ic(\mathscr{k}+\mathscr{k}')t}\delta(\mathscr{k}+\mathscr{k}') \\
&\quad + f^*(\mathscr{k})g^*(\mathscr{k}')e^{ic(\mathscr{k}+\mathscr{k}')t}\delta(\mathscr{k}+\mathscr{k}') \\
&\quad + f(\mathscr{k})g^*(\mathscr{k}')e^{-ic(\mathscr{k}-\mathscr{k}')t}\delta(\mathscr{k}-\mathscr{k}') \\
&\quad + f^*(\mathscr{k})g(\mathscr{k}')e^{ic(\mathscr{k}-\mathscr{k}')t}\delta(\mathscr{k}-\mathscr{k}')]\, d^3\mathscr{k}\, d^3\mathscr{k}'.
\end{aligned}$$

Integrating with respect to $\mathscr{k}'$, we obtain the desired formula:

$$\begin{aligned}
\int fg\, dV = \int [f(\mathscr{k})g(-\mathscr{k})e^{-2ic\mathscr{k}t} &+ f^*(\mathscr{k})g^*(-\mathscr{k})e^{2ic\mathscr{k}t} \\
&+ f(\mathscr{k})g^*(\mathscr{k}) + f^*(\mathscr{k})g(\mathscr{k})]\, d^3\mathscr{k}.
\end{aligned} \tag{23.23}$$

Applying this formula to components of $\mathscr{E}$, we have

$$\begin{aligned}
\int \mathscr{E}^2 f V = \int [\mathscr{E}(\mathscr{k}) \cdot \mathscr{E}(-\mathscr{k})e^{-2ic\mathscr{k}t} \\
+ \mathscr{E}^*(\mathscr{k}) \cdot \mathscr{E}^*(-\mathscr{k})e^{2ic\mathscr{k}t} + 2\mathscr{E}(\mathscr{k}) \cdot \mathscr{E}^*(\mathscr{k})]\, d^3\mathscr{k}.
\end{aligned} \tag{23.24}$$

Likewise applying Eq. (23.23) to compute $\int \mathscr{H}^2\, dV$, where now $f(\mathscr{k})$ and $g(\mathscr{k})$ are components of $\mathscr{H}(\mathscr{k}) = (\mathscr{k}/\mathscr{k}) \times \mathscr{E}(\mathscr{k})$, and making use of Eqs. (23.22) and (A2.27), we find that

$$\begin{aligned}
\int \mathscr{H}^2\, dV = \int [-\mathscr{E}(\mathscr{k}) \cdot \mathscr{E}(-\mathscr{k})e^{-2ic\mathscr{k}t} \\
-\mathscr{E}^*(\mathscr{k}) \cdot \mathscr{E}^*(-\mathscr{k})e^{2ic\mathscr{k}t} + 2\mathscr{E}(\mathscr{k}) \cdot \mathscr{E}^*(\mathscr{k})]\, d^3\mathscr{k};
\end{aligned} \tag{23.25}$$

therefore

$$E = \frac{1}{8\pi}\int(\mathscr{E}^2 + \mathscr{H}^2)\,dV = \frac{1}{2\pi}\int \mathscr{E}(\boldsymbol{k})\cdot\mathscr{E}^*(\boldsymbol{k})\,d^3\boldsymbol{k}. \tag{23.26}$$

Similarly

$$\int \mathscr{E}\times\mathscr{H}\,dV = \int [\mathscr{E}(\boldsymbol{k})\times\mathscr{H}(-\boldsymbol{k})e^{-2ickt} + \mathscr{E}^*(\boldsymbol{k})\times\mathscr{H}^*(-\boldsymbol{k})e^{2ickt} + \mathscr{E}(\boldsymbol{k})\times\mathscr{H}^*(\boldsymbol{k}) + \mathscr{E}^*(\boldsymbol{k})\times\mathscr{H}(\boldsymbol{k})]\,d^3\boldsymbol{k},$$

and by the application of Eqs. (18.62) and (A2.25) this becomes

$$\int \mathscr{E}\times\mathscr{H}\,dV = \int [-\mathscr{E}(\boldsymbol{k})\cdot\mathscr{E}(-\boldsymbol{k})e^{-2ickt} - \mathscr{E}^*(\boldsymbol{k})\cdot\mathscr{E}^*(-\boldsymbol{k})e^{2ickt} + 2\mathscr{E}(\boldsymbol{k})\cdot\mathscr{E}^*(\boldsymbol{k})]\frac{\boldsymbol{k}}{k}\,d^3\boldsymbol{k}. \tag{23.27}$$

The first two terms of the integrand merely change their sign when we replace $\boldsymbol{k}$ by $-\boldsymbol{k}$; therefore they vanish on integration. The momentum vector is thus given by

$$\mathbf{P} = \frac{1}{4\pi c}\int \mathscr{E}\times\mathscr{H}\,dV = \frac{1}{2\pi c}\int \mathscr{E}(\boldsymbol{k})\cdot\mathscr{E}^*(\boldsymbol{k})\frac{\boldsymbol{k}}{k}\,d^3\boldsymbol{k}. \tag{23.28}$$

Equation (23.11), or, if you like, Eq. (23.13), will not hold unless

$$E^2 = \frac{1}{4\pi^2}\iint [\mathscr{E}(\boldsymbol{k})\cdot\mathscr{E}^*(\boldsymbol{k})][\mathscr{E}(\boldsymbol{k}')\cdot\mathscr{E}^*(\boldsymbol{k}')]\,d^3\boldsymbol{k}\,d^3\boldsymbol{k}' \tag{23.29}$$

is equal to

$$c^2P^2 = \frac{1}{4\pi^2}\iint \frac{\boldsymbol{k}}{k}\cdot\frac{\boldsymbol{k}'}{k'}[\mathscr{E}(\boldsymbol{k})\cdot\mathscr{E}^*(\boldsymbol{k})][E(\boldsymbol{k}')\cdot\mathscr{E}^*(\boldsymbol{k}')]\,d^3\boldsymbol{k}\,d^3\boldsymbol{k}'. \tag{23.30}$$

The integrand in Eq. (23.29), being the sum of products of the type $\mathscr{E}_i(\boldsymbol{k})\mathscr{E}_i^*(\boldsymbol{k})\mathscr{E}_j(\boldsymbol{k}')\mathscr{E}_j^*(\boldsymbol{k}')$, is always positive or zero. In Eq. (23.30) the same expression occurs multiplied by $\boldsymbol{k}\cdot\boldsymbol{k}'/kk'$, which in magnitude is always less than 1. The two integrals will therefore not be equal unless the quantity

$$[\mathscr{E}(\boldsymbol{k})\cdot\mathscr{E}^*(\boldsymbol{k})][\mathscr{E}(\boldsymbol{k}')\cdot\mathscr{E}^*(\boldsymbol{k}')]$$

vanishes whenever $\boldsymbol{k}\cdot\boldsymbol{k}' \neq kk'$.

Let $\boldsymbol{k}_0$ be a particular value of $\boldsymbol{k}'$ for which $\mathscr{E}(\boldsymbol{k}_0) \neq 0$. Consider the set of $\boldsymbol{k}$ for which $\mathscr{E}(\boldsymbol{k}) \neq 0$; then we must have

$$\boldsymbol{k}\cdot\boldsymbol{k}_0 = kk_0 \tag{23.31}$$

or

$$\boldsymbol{k} = \xi \boldsymbol{k}_0, \tag{23.32}$$

where $\xi \geq 0$. Thus, all $\boldsymbol{k}$ for which $\boldsymbol{\mathscr{E}}(\boldsymbol{k})$ does not vanish are parallel to some fixed $\boldsymbol{k}_0$, and a wave packet having no mass must be of the form

$$\boldsymbol{\mathscr{E}}(\mathbf{r}, t) = \int_0^\infty [\boldsymbol{\mathscr{E}}(\xi) e^{i\xi(\boldsymbol{k}_0\cdot\mathbf{r} - c k_0 t)} + \boldsymbol{\mathscr{E}}^*(\xi) e^{-i\xi(\boldsymbol{k}_0\cdot\mathbf{r} - c k_0 t)}]\, d\xi, \tag{23.33}$$

i.e., must consist only of waves moving with the speed c in a common direction. The direction of $\boldsymbol{\mathscr{E}}(\xi)$, which represents the polarization of the component plane waves, can still vary arbitrarily with ξ.

Wave packets. We shall next consider the possibility of building up wave packets so small as to correspond to a particle not only with regard to the relationship between its momentum and energy but also in spatial extent. For this purpose, we shall take a general homogeneous and isotropic medium and suppose only that individual waves composing the wave packet satisfy the wave equation (22.1), namely,

$$\nabla^2 f - \frac{1}{v^2}\frac{\partial^2 f}{\partial t^2} = 0, \tag{23.34}$$

where f represents a scalar field or some Cartesian component of a vector field.

Letting

$$f(\mathbf{r}, t; \boldsymbol{k}) = e^{i(\boldsymbol{k}\cdot\mathbf{r} - \omega t)}, \tag{23.35}$$

we have

$$\nabla^2 f(\mathbf{r}, t; \boldsymbol{k}) = -k^2 f(\mathbf{r}, t; \boldsymbol{k})$$

and

$$\frac{\partial^2}{\partial t^2} f(\mathbf{r}, t; \boldsymbol{k}) = -\omega^2 f(\mathbf{r}, t; \boldsymbol{k}),$$

so that $f(\mathbf{r}, t; \boldsymbol{k})$ satisfies Eq. (23.34) provided only that $\omega = kv$. In a *dispersive* medium the phase velocity v will depend on k and we write

$$\omega = kv(k). \tag{23.36}$$

The medium is then still homogeneous and isotropic, because the wave equation (23.34) depends in the same way on x, y, and z, and the phase velocity is independent of the direction of $\boldsymbol{k}$.

A wave packet satisfying Eq. (23.34) is then given by

$$f(\mathbf{r}, t) = \left(\frac{1}{2\pi}\right)^{3/2} \int f(\boldsymbol{k}) e^{i(\boldsymbol{k}\cdot\mathbf{r} - \omega(\boldsymbol{k})t)}\, d^3\boldsymbol{k}. \tag{23.37}$$

Its initial value at $t = 0$ is

$$f(\mathbf{r}, 0) = \left(\frac{1}{2\pi}\right)^{3/2} \int f(\boldsymbol{k}) e^{i\boldsymbol{k}\cdot\mathbf{r}}\, d^3\boldsymbol{k}. \tag{23.38}$$

This is merely the Fourier expansion of an arbitrary function $f(r, 0)$, whose transform is

$$f(\boldsymbol{k}) = \left(\frac{1}{2\pi}\right)^{3/2} \int f(\mathbf{r}, 0) e^{-i\boldsymbol{k}\cdot\mathbf{r}}\, dV. \tag{23.39}$$

We shall suppose that the energy density is proportional to $|f|^2 = ff^*$ and shall define the center of the wave packet at $t = 0$ as the point given by

$$\mathbf{r}_0 = \frac{\int \mathbf{r}\, |f(\mathbf{r}, 0)|^2\, dV}{\int |f(\mathbf{r}, 0)|^2\, dV}. \tag{23.40}$$

This is analogous to obtaining a center of mass when the mass density is $|f|^2$, or to obtaining a weighted average of $\mathbf{r}$ with the weighting function $|f|^2$. Let us suppose, for simplicity, that the origin of the coordinate system is so chosen that $\mathbf{r}_0 = 0$, i.e., that the center of the wave packet is initially at the origin.

As a measure of spread of the wave packet in the X direction we can take

$$\Delta x = 2\left[\frac{\int x^2\, |f(\mathbf{r}, 0)|^2\, dV}{\int |f(\mathbf{r}, 0)|^2\, dV}\right]^{1/2}. \tag{23.41}$$

This is twice the *standard deviation* from the mean (in this case $x = 0$) for a probability density function proportional to $|f(\mathbf{r}, 0)|^2$.

Similarly we can regard $|f(\boldsymbol{k})|^2$ as the weight function for $\boldsymbol{k}$. Then, the average $\boldsymbol{k}$ will be given by

$$\boldsymbol{k}_0 = \frac{\int \boldsymbol{k}\, |f(\boldsymbol{k})|^2\, d^3\boldsymbol{k}}{\int |f(\boldsymbol{k})|^2\, d^3\boldsymbol{k}}. \tag{23.42}$$

Analogous to Eq. (23.41), the measure of the spread of $\boldsymbol{k}_x$ is

$$\Delta \boldsymbol{k}_x = 2\left[\frac{\int (\boldsymbol{k}_x - \boldsymbol{k}_{0x})^2 |f(\boldsymbol{k})|^2 d^3\boldsymbol{k}}{\int |f(\boldsymbol{k})|^2\, d^3\boldsymbol{k}}\right]^{1/2}, \tag{23.43}$$

with similar equations for the other components.

Let us now transform Eqs. (23.41) and (23.43) by putting

$$f(\mathbf{r}, 0) = e^{i\mathbf{k}_0\cdot\mathbf{r}} g(\mathbf{r}). \tag{23.44}$$

Since

$$|f(\mathbf{r}, 0)|^2 = e^{i\mathbf{k}_0\cdot\mathbf{r}} g(\mathbf{r}) e^{-i\mathbf{k}_0\cdot\mathbf{r}} g^*(\mathbf{r}) = |g(\mathbf{r})|^2,$$

Eq. (23.41) becomes

$$\Delta x = 2\left[\frac{\int x^2 |g|^2\, dV}{N}\right]^{1/2}, \tag{23.45}$$

where

$$N = \int |f(\mathbf{r}, 0)|^2\, dV = \int |g|^2\, dV. \tag{23.46}$$

Also

$$\begin{aligned}\frac{\partial g}{\partial x} = \frac{\partial}{\partial x}[e^{-i\mathbf{k}_0\cdot\mathbf{r}} f(\mathbf{r}, 0)] &= \frac{\partial}{\partial x}\left[\left(\frac{1}{2\pi}\right)^{3/2}\int f(\mathbf{k}) e^{i(\mathbf{k}-\mathbf{k}_0)\cdot\mathbf{r}}\, d^3\mathbf{k}\right] \\ &= i\left(\frac{1}{2\pi}\right)^{3/2}\int (k_x - k_{0x}) f(\mathbf{k}) e^{i(\mathbf{k}-\mathbf{k}_0)\cdot\mathbf{r}}\, d^3\mathbf{k},\end{aligned} \tag{23.47}$$

so that

$$\begin{aligned}\int\left|\frac{\partial g}{\partial x}\right|^2 dV &= \int \frac{\partial g}{\partial x}\frac{\partial g^*}{\partial x}\, dV \\ &= \left(\frac{1}{2\pi}\right)^3 \iiint (k_x - k_{0x})(k'_x - k_{0x}) f(\mathbf{k}) f^*(\mathbf{k}') e^{i(\mathbf{k}-\mathbf{k}')\cdot\mathbf{r}}\, d^3\mathbf{k}\, d^3\mathbf{k}'\, dV \\ &= \iint (k_x - k_{0x})(k'_x - k_{0x}) f(\mathbf{k}) f^*(\mathbf{k}')\delta(\mathbf{k} - \mathbf{k}')\, d^3\mathbf{k}\, d^3\mathbf{k}' \\ &= \int (k_x - k_{0x})^2 |f(\mathbf{k})|^2\, d^3\mathbf{k},\end{aligned} \tag{23.48}$$

which is the numerator in Eq. (23.43). For the denominator in that equation we have, using Eq. (23.39),

$$\begin{aligned}\int |f(\mathbf{k})|^2\, d^3\mathbf{k} &= \left(\frac{1}{2\pi}\right)^3 \iiint f(\mathbf{r}, 0) f^*(\mathbf{r}', 0) e^{i(\mathbf{r}'-\mathbf{r})\cdot\mathbf{k}}\, dV\, dV'\, d^3\mathbf{k} \\ &= \iint f(\mathbf{r}, 0) f^*(\mathbf{r}', 0)\delta(\mathbf{r}' - \mathbf{r})\, dV\, dV' = \int |f(\mathbf{r}, 0)|^2\, dV = N.\end{aligned} \tag{23.49}$$

Thus, using Eqs. (23.43), (23.45), (23.48), and (23.49),

$$(\Delta x\, \Delta k_x)^2 = \frac{16}{N^2}\left[\int x^2 |g|^2\, dV\right]\cdot\left[\int\left|\frac{\partial g}{\partial x}\right|^2 dV\right]. \tag{23.50}$$

Now, since the square of the absolute value of any real quantity is either positive or zero, we have

$$\left|\frac{xg}{\alpha} + \frac{\partial g}{\partial x}\right|^2 \geq 0 \tag{23.51}$$

for any positive real number α. This requires that

$$\begin{aligned}\left|\frac{xg}{\alpha} + \frac{\partial g}{\partial x}\right|^2 &= \left(\frac{xg}{\alpha} + \frac{\partial g}{\partial x}\right)\left(\frac{xg^*}{\alpha} + \frac{\partial g^*}{\partial x}\right) \\ &= \frac{x^2\,|g|^2}{\alpha^2} + \left|\frac{\partial g}{\partial x}\right|^2 + \frac{x}{\alpha}\left(g\frac{\partial g^*}{\partial x} + g^*\frac{\partial g}{\partial x}\right) \\ &= \frac{x^2\,|g|^2}{\alpha^2} + \left|\frac{\partial g}{\partial x}\right|^2 + \frac{x}{\alpha}\frac{\partial}{\partial x}|g|^2 \\ &= \frac{x^2\,|g|^2}{\alpha^2} + \left|\frac{\partial g}{\partial x}\right|^2 + \frac{1}{\alpha}\frac{\partial}{\partial x}(x\,|g|^2) - \frac{1}{\alpha}|g|^2 \geq 0\end{aligned}$$

or

$$\left|\frac{\partial g}{\partial x}\right|^2 \geq \frac{1}{\alpha}|g|^2 - \frac{x^2\,|g|^2}{\alpha^2} - \frac{1}{\alpha}\frac{\partial}{\partial x}(x\,|g|^2).$$

Multiplying by $dV = dx\,dy\,dz$ and integrating, we obtain

$$\int\left|\frac{\partial g}{\partial x}\right|^2 dV \geq \frac{1}{\alpha}\int|g|^2\,dV - \frac{1}{\alpha^2}\int x^2\,|g|^2\,dV - \frac{1}{\alpha}\iint[x\,|g|^2]_{x=-\infty}^{x=\infty}\,dy\,dz.$$

If we now require that $\lim_{|x|\to\infty}(x\,|g|^2) = 0$ (otherwise $\int x^2\,|g|^2\,dV$ would not converge), then $[x\,|g|^2]_{x=-\infty}^{x=\infty} = 0$ and the inequality becomes

$$\int\left|\frac{\partial g}{\partial x}\right|^2 dV \geq \frac{N}{\alpha} - \frac{1}{\alpha^2}\int x^2\,|g|^2\,dV. \tag{23.52}$$

If we let $\alpha = (2/N)\int x^2|g|^2\,dV$, which makes the right-hand member a maximum, this inequality can be written in the form

$$4\int x^2\,|g|^2\,dV \cdot \int\left|\frac{\partial g}{\partial x}\right|^2 dV \geq N^2. \tag{23.53}$$

Substituted into Eq. (23.50) this gives the *uncertainty relation*

$$\Delta x\,\Delta k_x \geq 2. \tag{23.54}$$

A similar inequality holds, of course, for the other two directions.

This inequality shows that the smaller the space within which the wave packet is restricted, the greater the range of k it must contain.

Phase and group velocities of a wave packet. Let us now investigate the behavior of the wave packet given by Eq. (23.37). With $\boldsymbol{k}_0$ as defined in Eq. (23.42), let

$$\boldsymbol{k} = \boldsymbol{k}_0 + \boldsymbol{k}'. \tag{23.55}$$

Then, by Taylor's expansion formula,

$$\omega(\boldsymbol{k}) = \omega(\boldsymbol{k}_0) + \boldsymbol{k}' \cdot \left(\frac{\partial \omega}{\partial \boldsymbol{k}}\right)_{\boldsymbol{k}_0} + \text{higher terms.} \tag{23.56}$$

If now Δk_x, Δk_y, and Δk_z are small, so that $f(\boldsymbol{k})$ in Eq. (23.37) will be very small whenever k_x', k_y', and k_z' are appreciably different from k_{x0}, k_{y0}, k_{z0}, respectively, the higher terms may be omitted, and we obtain, approximately,

$$\begin{aligned} f(\mathbf{r}, t) &= \left(\frac{1}{2\pi}\right)^{3/2} \int f(\boldsymbol{k}) e^{i[\boldsymbol{k}\cdot\mathbf{r} - \omega(\boldsymbol{k})t]}\, d^3\boldsymbol{k} \\ &= \left(\frac{1}{2\pi}\right)^{3/2} \int f(\boldsymbol{k}_0 + \boldsymbol{k}') \\ &\quad \times \exp\left(i\left\{(\boldsymbol{k}_0 + \boldsymbol{k}') \cdot \mathbf{r} - \left[\omega(\boldsymbol{k}_0) + \boldsymbol{k}' \cdot \left(\frac{\partial \omega}{\partial \boldsymbol{k}}\right)_{\boldsymbol{k}_0}\right] t\right\}\right) d^3\boldsymbol{k}' \\ &= A(\mathbf{r}, t)\, e^{i[\boldsymbol{k}_0\cdot\mathbf{r} - \omega(\boldsymbol{k}_0)t]}, \end{aligned} \tag{23.57}$$

where

$$A(\mathbf{r}, t) = \left(\frac{1}{2\pi}\right)^{3/2} \int f(\boldsymbol{k}_0 + \boldsymbol{k}') \exp\left(i\left\{\boldsymbol{k}' \cdot \left[\mathbf{r} - \left(\frac{\partial \omega}{\partial \boldsymbol{k}}\right)_{\boldsymbol{k}_0}\right] t\right\}\right) d^3\boldsymbol{k}'. \tag{23.58}$$

If initially $f(\mathbf{r}, t)$ is essentially different from zero only within some element of volume $\Delta x\, \Delta y\, \Delta z$, A must have the same property. On the other hand, Eq. (23.58) shows that A contains $\mathbf{r}$ and t only in the combination

$$\mathbf{r}(t) \equiv \mathbf{r} - \left(\frac{\partial \omega}{\partial \boldsymbol{k}}\right)_{\boldsymbol{k}_0} t, \tag{23.59}$$

and thus A at the time t will have the same value at $\mathbf{r}(t)$ that it initially ($t = 0$) had at $\mathbf{r}$. This shows that the wave packet is moving as a whole with the velocity

$$\mathbf{V} = \left(\frac{\partial \omega}{\partial \boldsymbol{k}}\right)_{\boldsymbol{k}_0} = \left(\frac{\partial \omega}{\partial k} \frac{\partial k}{\partial \boldsymbol{k}}\right)_{k_0} = \frac{\boldsymbol{k}_0}{k_0}\left(\frac{\partial \omega}{\partial k}\right)_{k_0}, \tag{23.60}$$

which is known as the *group velocity* of the waves, in contradistinction to the *phase velocity* **v**. The magnitude of the group velocity is

$$V = |\mathbf{V}| = \frac{\partial \omega}{\partial k}, \tag{23.61}$$

which, since $k = 2\pi/\lambda$ and $\omega = 2\pi\nu$, can also be written

$$V = \frac{\partial \nu}{\partial(1/\lambda)}. \tag{23.62}$$

For a nondispersive medium

$$v(k) = v = \text{constant}$$

and by Eq. (23.36)

$$V = \frac{\partial \omega}{\partial k} = \frac{\partial(kv)}{\partial k} = v.$$

Thus the group velocity is the same as the phase velocity. This is true for an electromagnetic wave packet in free space, because then $v(k) = c$.

We see from the above analysis that if the wave packet is composed only of k near k_0, i.e., $\Delta k = (\Delta k_x, \Delta k_y, \Delta k_z)$ is small, then the packet moves as a whole with group velocity **V**, whose magnitude is given by Eq. (23.61). Except for a nondispersive medium there is also a tendency for the wave packet to spread out in space as time increases. This effect becomes important for Δk large and can be explained as follows:

We may divide the region of the k-space, around k_0, in which $f(k)$ is appreciably different from zero, into a number of smaller regions R_1, R_2, . . . , R_n centered about k_1, k_2, . . . , k_n, respectively. Then the function $f(\mathbf{r}, t)$ representing the wave packet is expressible as

$$f(\mathbf{r}, t) = \sum_{p=1}^{n} f_p(\mathbf{r}, t), \tag{23.63}$$

where the $f_p(\mathbf{r}, t)$ are given by

$$f_p(\mathbf{r}, t) = \left(\frac{1}{2\pi}\right)^{3/2} \int_{R_p} f(k)\, e^{i[k\cdot\mathbf{r} - \omega(k)t]}\, d^3k \tag{23.64}$$

and have themselves the structure of wave packets. The original wave packet can thus be viewed as a composite of wave packets $f_p(\mathbf{r}, t)$, $p = 1$, $2, \ldots, n$, each moving with its own group velocity

$$\mathbf{V}_p = \left(\frac{\partial \omega}{\partial k}\right)_{k_p}.$$

If these are not the same, the individual wave packets will drift apart, and the composite wave packet will spread out. The spreading of the wave packet will be the more rapid the larger is Δk. Combining this result with the uncertainty relation given by the inequality (23.54), we may conclude that in a dispersive medium the smaller the wave packet the more rapidly it will spread. This assumes, of course, that the wave packet is not otherwise changed.

Problems

1. What are plane electromagnetic waves? How are they expressed using the propagation vector **k**?

2. Give expressions for the electric and magnetic fields of a plane wave propagating in the direction of the positive Y-axis. What relations exist between the components of $\mathscr{E}$?

3. In what ways can a plane monochromatic wave be polarized?

4. The expressions for f_1 and f_2 given in Eq. (18.40) are complex valued and therefore cannot directly represent a real wave. What interpretation, or change, is needed to express a real wave? What is the *phase* of the wave at the origin when $t = t_1$? What is the amplitude of the wave?

5. Show that if complex field quantities **E** and **H** are used to represent the real field quantities $\mathscr{E}$ and $\mathscr{H}$, as in Eq. (18.40), and if one defines a *complex Poynting vector*

$$\mathbf{S} = \frac{c}{8\pi}\mathbf{E} \times \mathbf{H}^*,$$

then the real part of S is just the time average of the real Poynting vector

$$\Upsilon = \frac{c}{4\pi}\mathscr{E} \times \mathscr{H}.$$

6. What two assumptions can be used to define the phrase "field produced by a particle"? Why are the retarded potentials used in calculating the field produced by a system of particles?

7. How would the results of Examples 3 and 4, pp. 202 and 205, be modified if advanced potentials were used in place of retarded potentials? What unnatural features do these new results exhibit?

8. Write down an explicit expression for the *Doppler factor* in the spherical coordinates of the point of observation R for a particle at the origin oscillating along the Z-axis with simple harmonic motion. Assume the maximum displacement of the particle is very small compared to R.

9. Derive Eqs. (19.55) and (19.56).

10. Obtain expressions for the dipole and quadrupole moments in terms of a charge density $\rho(r)$.

11. Assuming that the current and potential on a thin half-wave cylindrical antenna is the same as for an open-ended lossless transmission line a half-wavelength long, what functions of position and time describe the potential, current, and charge per unit length? What ratio defines the impedance of the antenna? What electric dipole moment is produced? Are there higher electric moments produced? Is there a magnetic moment?

12. What is the polarization of the field radiated by the antenna in problem 11? How does the Poynting vector at a large distance from the antenna depend on the angle θ between the antenna and the direction of observation?

13. When is the quadrupole radiation important in comparison to the dipole radiation of a spherical wave?

14. What is a *standing wave*? Mathematically describe such a wave in the three space dimensions.

15. Show that the requirement that the tangential component of the electric field $\mathscr{E}$ vanish on the surface of a box leads to standing waves only, while periodic boundary conditions lead to both running and standing waves.

16. How many standing wave modes are there in a cubical box one centimeter on a side whose wavelengths lie between 4000 Å and 4010 Å? Does this answer depend on the shape of the box? Does it depend on whether one uses periodic boundary conditions or the conditions that the tangential components of $\mathscr{E}$ vanish on the surface of the volume?

17. What evidence is there for light being an electromagnetic wave?

18. Why is Eq. (22.17) said to pertain to an imaginary particle? In what way is its trajectory like a ray of light?

19. Define the following: (a) normal bundle of rays, (b) eikonal, (c) Fermat's principle.

20. How can one determine whether geometrical optics is applicable in a particular problem?

21. How sharply can the edge of a shadow be defined? Explain.

22. If a microscope has an objective lens which subtends a solid angle of π steradian relative to the object being observed, what approximately is the minimum distance that can be resolved with this microscope using blue light?

23. Can a wave packet consist of waves moving in widely different directions?

24. Given that the wave packet for the electric field $\mathscr{E}$ is of the form $\mathbf{f}(\mathbf{r}, t)$ one is justified in assuming that the energy density is proportional to $|\mathbf{f}(\mathbf{r}, t)|^2$. Why?

25. What is meant by a dispersive medium and what is the significance of the group velocity $\mathbf{v}_g$ for such a medium?

26. If a wave packet is to represent a particle of zero mass, what restrictions must it satisfy?

27. How far can the analogy, described in Section 23, between the behavior of electromagnetic waves and that of particles be carried? Why can it not be carried further?

Chapter VII

Dynamics of Charged Particles

24. Charged Particles and Field as a Dynamical System

In Section 14 we obtained equations of motion for particles in a given electromagnetic field, the Lorentz equations of motion of Eq. (14.17), by treating particles as test particles; that is, we assumed that particles themselves have charges so small that they produce no appreciable fields of their own. In Section 16 we obtained the field equations, the Maxwell–Lorentz equations. These equations show how the fields produced by charged particles are connected with the charges and velocities of the particles, when these velocities are assumed to be known.

The question nevertheless arises as to how one deals with the dynamics of actual particles, such as electrons, which in many problems cannot be regarded as test particles. In fact, in all cases where the fields produced by the particles, moving in a given field, cannot be neglected, we have neither test particles nor particles with motions given *a priori*. How are such cases to be treated?

The method adopted in classical electrodynamics is to make the further assumption that *the electromagnetic field occurring in the Lorentz equations of motion is the total field*, composed of the given external field and the field produced by the particles, in accordance with the Maxwell–Lorentz equations. In other words, the Lorentz equations of motion and the Maxwell–Lorentz field equations are to be solved as a system of simultaneous equations, giving both the motion of the particles and the changes in the fields produced by them. As was said at the end of Section 14, this corresponds to making a metaphysical assumption as to the behavior of particles in the absence of their own field.

This procedure, although it leads to very serious difficulties, is the only natural one in classical electrodynamics, the difficulties resulting from it being due to the limitations inherent in the theory.

In fact, one can arrive at the classical procedure by considering the Lagrangian of the total system—the particles and the field. The Lagrangian of a single particle without its field is given by Eq. (12.7); therefore, the Lagrangian for the particles alone is given by

$$L_p = -\sum_s m_s c^2 \sqrt{1 - v_s^2/c^2}. \tag{24.1}$$

The Lagrangian of the total field is given by Eq. (17.1),

$$L_f = \frac{1}{8\pi} \int (\mathscr{E}^2 - \mathscr{H}^2)\, dV. \tag{24.2}$$

Finally for the Lagrangian of interaction between this field and the particles we have, using Eq. (16.14) and the fact that $W = \int L\, dt$,

$$L_{pf} = \sum_s \left[\left(\frac{e_s}{c}\right) \mathbf{A}(\mathbf{r}_s, t) \cdot \mathbf{v}_s - e_s \varphi(\mathbf{r}_s, t) \right] = \int s_\alpha \varphi_\alpha(\mathbf{r}, t)\, dV, \tag{24.3}$$

where $\varphi_\alpha(\mathbf{r}, t)$ is defined in Eq. (14.2) and s_α in Eq. (16.18). As we have seen, in classical electrodynamics the Lagrangian of the total system is obtained by putting

$$L = L_{\text{total}} = L_p + L_f + L_{pf}. \tag{24.4}$$

After this is done, the equations of motion and the field equations are obtained by setting

$$\delta \int L\, dt = 0 \tag{24.5}$$

under the assumption that the $\mathbf{r}_s$, the coordinates of the particles, and the $\varphi_\alpha(\mathbf{r}, t)$, the coordinates of the field, are independently varied. As is easily seen, one obtains in this way exactly the Lorentz equations of motion for the particles and the Maxwell–Lorentz equations for the field. The $\mathscr{E}$ and $\mathscr{H}$ entering these equations clearly refer to the total fields. The above procedure, moreover, is based only on Hamilton's principle and basic requirements as to the forms of the Lagrangians L_p, L_f, and L_{pf} that have previously been justified in detail. This procedure is thus not open to amendment except through the introduction of auxiliary assumptions. Since the usual procedure, mentioned earlier, agrees with this basic approach, its difficulties would appear to involve the limitations inherent in the very roots of classical electrodynamics.

We shall now develop the consequences of Eq. (24.4), discussing the difficulties as we come upon them. Let us calculate the total energy and the total momentum of the system. Instead of doing this *ab initio*, we can simplify our work by making use of the results we have previously obtained. We introduce first the following general considerations.

Suppose we have a system consisting of two parts, part A having generalized coordinates q_i^A and part B with generalized coordinates q_i^B. The total Lagrangian will then be of the form

$$L = L_A + L_B + M, \tag{24.6}$$

where M represents the interaction between the two parts.

The total momentum of the system (see Eq. (3.14)) is given by

$$\mathbf{P} = \frac{\partial L}{\partial \dot{q}_i} \frac{\partial q_i}{\partial \mathbf{r}}; \tag{24.7}$$

hence in our case

$$\begin{aligned} \mathbf{P} &= \frac{\partial L}{\partial \dot{q}_i^A} \frac{\partial q_i^A}{\partial \mathbf{r}} + \frac{\partial L}{\partial \dot{q}_i^B} \frac{\partial q_i^B}{\partial \mathbf{r}} \\ &= \frac{\partial (L_A + M)}{\partial \dot{q}_i^A} \frac{\partial q_i^A}{\partial \mathbf{r}} + \frac{\partial (L_B + M)}{\partial \dot{q}_i^B} \frac{\partial q_i^B}{\partial \mathbf{r}}. \end{aligned} \tag{24.8}$$

Since $L_A + M$ is the Lagrangian of part A *in the presence of part B*, and $L_B + M$ of part B *in the presence of part A*, we have

$$\mathbf{P} = \tilde{\mathbf{P}}_A + \tilde{\mathbf{P}}_B, \tag{24.9}$$

where $\tilde{\mathbf{P}}_A$ and $\tilde{\mathbf{P}}_B$ are the respective momenta of the parts A and B taking into account their interaction.

Applying this to our case, we let A correspond to the particles and B to the field; then $\tilde{\mathbf{P}}_A$ is the momentum of the particles in the presence of the field, which by Eq. (14.6) is

$$\tilde{\mathbf{P}}_A = \mathbf{P}_{\text{particles}} = \sum_s \left[\frac{m_s \mathbf{v}_s}{\sqrt{1 - v_s^2/c^2}} + \frac{e_s}{c} \mathbf{A}(\mathbf{r}_s, t) \right]. \tag{24.10}$$

The momentum of the field in the presence of particles may be obtained from Eqs. (17.55) and (17.56), provided we first determine a suitable hyperplane Σ over which the integration in Eq. (17.55) is to be performed. It seems very reasonable and natural to assume that Σ is a spacelike hyperplane and thus that in some coordinate system, which we designate as the prime coordinate system, it is just the instantaneous space corresponding to a fixed time, say $t' = 0$. Expressed somewhat differently, the hyperplane is the set of all points in space–time having primed coordinates of the form $(x', y', z', 0)$.

Since a differential element dS_β of Σ is associated with the normal to Σ and thus has the direction of the t'-axis, it has only a fourth component in the primed coordinate system. The momentum four-vector for the field in the primed coordinate system, according to Eq. (17.55), is then

$$P'_\alpha = \int_\Sigma T'_{\alpha\beta}\, dS'_\beta = \int_{\text{all space}} T'_{\alpha 4}\, dV', \tag{24.11}$$

where by Eq. (17.56)

$$T'_{\alpha 4} = \frac{1}{4\pi i c}\left(F'_{\alpha\gamma}F'_{4\gamma} - \frac{1}{4}\,\delta_{\alpha 4}F'_{\epsilon\gamma}F'_{\epsilon\gamma}\right) + \frac{1}{ic}\,(\delta_{\alpha 4}\varphi'_\gamma s'_\gamma - \varphi'_\alpha s'_4). \tag{24.12}$$

For $\alpha = k \neq 4$, Eq. (24.12) reduces, by Eqs. (14.25), to

$$T'_{k4} = \frac{1}{4\pi c}\,(\mathscr{E}' \times \mathscr{H}')_k - \frac{\rho' A'_k}{c}, \tag{24.13}$$

and thus the momentum of the field in the primed coordinate system is

$$\tilde{\mathbf{P}}'_B = \frac{1}{4\pi c}\int_{\text{all space}} \mathscr{E}' \times \mathscr{H}'\, dV' - \frac{1}{c}\int_{\text{all space}} \rho'\mathbf{A}'\, dV'. \tag{24.14}$$

For a system of particles $\rho = \sum_s e_s\, \delta(\mathbf{r}' - \mathbf{r}'_s)$ and the last integral in Eq. (24.14) becomes $\sum_s e_s\mathbf{A}'(\mathbf{r}'_{s'}, t')$. Thus, by expressing Eq. (24.10) in the primed coordinate system and using Eq. (24.9) we have for the total momentum of the particles and field in the primed coordinate system

$$\mathbf{P}' = \tilde{\mathbf{P}}'_A + \tilde{\mathbf{P}}'_B = \sum_s \frac{m_s \mathbf{v}'_s}{\sqrt{1 - v'^2_s/c^2}} + \frac{1}{4\pi c}\int \mathscr{E}' \times \mathscr{H}\, dV'. \tag{24.15}$$

It is important to note that Eq. (24.15) was derived for a particular coordinate system. If one assumes, as has historically been common practice, that the field contributes a momentum

$$\mathbf{P}_{\text{field}} = \frac{1}{4\pi c}\int \mathscr{E} \times \mathscr{H}\, dV \tag{24.16}$$

to that contributed directly by the bare particles, one finds that $\mathbf{P}_{\text{field}}$ does not transform properly under a Lorentz transformation. This apparent difficulty has led to a great deal of confusion in the fundamental ideas of electrodynamics that has only recently been successfully cleared up.* It even led Poincaré to introduce nonelectromagnetic forces to correct the overall transformation properties. We see, however, from our development that dS_β can be reduced to dV only for the primed coordinate system and that Eq. (24.16) is thus not a valid equation.

Returning now to Eq. (24.6), we have for the total Hamiltonian

$$H = -L + \sum_i p_i^A \dot{q}_i^A + \sum_j p_j^B \dot{q}_j^B, \qquad (24.17$$

and for the Hamiltonians of the two parts (each in the presence of the other)

$$\tilde{H}_A = -L_A - M + \sum_i p_i^A \dot{q}_i^A, \qquad \tilde{H}_B = -L_B - M + \sum_j p_j^B \dot{q}_j^B. \qquad (24.18)$$

We thus observe that

$$H = \tilde{H}_A + \tilde{H}_B + M, \qquad (24.19)$$

which shows that $H \neq \tilde{H}_A + \tilde{H}_B$.

The Hamiltonian for the particles in the primed coordinate system is given by

$$\tilde{H}_A' = H'_{\text{particles}} = \sum_s \left[\frac{m_s c^2}{\sqrt{1 - v_s'^2/c^2}} + e_s \varphi'(\mathbf{r}_s', t')\right] \qquad (24.20)$$

and the interaction Lagrangian, see Eq. (24.3), by

$$M' = L'_{\varphi f} = \int s_\alpha' \varphi_\alpha'(\mathbf{r}', t')\, dV'. \qquad (24.21)$$

The Hamiltonian for the field in this coordinate system is

$$\tilde{H}_B' = H'_{\text{field}} = -icP_4' = -ic\int T_{44}'\, dV',$$

which by Eq. (24.12) becomes

$$\begin{aligned}\tilde{H}_B' &= -\frac{1}{4\pi}\int\left[F_{4\gamma}'F_{4\gamma}' - \frac{1}{4}F_{\epsilon\gamma}'F_{\epsilon\gamma}' + 4\pi(\varphi_\gamma' s_\gamma' - \varphi_4' s_4')\right] dV' \\ &= \frac{1}{8\pi}\int(\mathscr{E}'^2 + \mathscr{H}'^2)\, dV' - \int s_\gamma'\varphi_\gamma'\, dV' + \int s_4'\varphi_4'\, dV'. \qquad (24.22)\end{aligned}$$

* See F. Rohrlich, *Classical Charged Particles*, Addison-Wesley, Reading, Mass., 1965, Chapter 2.

For point particles $\rho = \sum_s e_s\delta(\mathbf{r} - \mathbf{r}_s)$, and we have by Eqs. (14.2) and (16.18)

$$\int s_4'\varphi_4'\, dV' = -\int \rho'\varphi'\, dV' = -\sum_s e_s\varphi'(\mathbf{r}_s', t'). \tag{24.23}$$

It now follows from Eqs. (24.19) to (24.23) that

$$H' = \tilde{H}_A' + \tilde{H}_B' + M'$$
$$= \sum_s \frac{m_s c^2}{\sqrt{1 - v_s'^2/c^2}} + \frac{1}{8\pi}\int (\mathscr{E}'^2 + \mathscr{H}'^2)\, dV'. \tag{24.24}$$

Again, the form of this equation must change if we change to another coordinate system.

Treatment of a single isolated particle. The total momentum of a single isolated particle in the primed coordinate system, by Eq. (24.15), is given by

$$\mathbf{P}' = \frac{m_0\mathbf{v}'}{\sqrt{1 - v'^2/c^2}} + \frac{1}{4\pi c}\int \mathscr{E}' \times \mathscr{H}'\, dV', \tag{24.25}$$

where $\mathbf{v}'$ is the velocity of the particle and m_0 its mass without the field present. If we are not in the primed coordinate system, then, since the two sides of Eq. (24.25) transform differently under a Lorentz transformation, the equation is no longer valid.

Consider a coordinate system in which the particle is at rest, so that $\mathbf{v} = 0$ and $\mathscr{H} = 0$. Since the particle together with its field is at rest, one would expect the total momentum to be zero. Thus in this coordinate system Eq. (24.25) holds, leading us to conclude that this is the primed coordinate system as defined earlier. In other words, the hyperplane Σ over which we integrate in Eq. (24.11) is just the instantaneous space of a coordinate system moving with the particle.

Since the total momentum $\mathbf{P}'$ is zero in the primed coordinate system, the momentum four-vector P_α' is completely determined by the energy $E' = -iP_4/c$ of the particle and field as calculated in the rest frame. Since we are considering an isolated system, $E' = H'$ and thus we have, from Eq. (24.24) and the fact that $\mathbf{v}' = 0$ and $\mathscr{H}' = 0$,

$$E' = m_0c^2 + \frac{1}{8\pi}\int \mathscr{E}'^2\, dV'. \tag{24.26}$$

Unfortunately, the integral in Eq. (24.26) is infinite for a point charge.

Anticipating this difficulty, we take as our model of an electron* at rest a spherically symmetrical charge distribution $\rho(r)$ that vanishes outside a sphere of radius a centered at the origin of the primed coordinate system. From a consideration of the spherical symmetry we know that

$$\mathcal{E} = f(r)\frac{\mathbf{r}}{r}. \tag{24.27}$$

Moreover, by integrating both sides of the Maxwell equation

$$\nabla \cdot \mathcal{E} = 4\pi\rho(r)$$

over the volume V of a sphere of radius r, we have

$$q(r) \equiv \int_V \rho \, dV = \frac{1}{4\pi}\int_V \nabla \cdot \mathcal{E} \, dV = \frac{1}{4\pi}\int_{\text{surface}} \mathcal{E} \cdot d\mathbf{S} = r^2 f(r),$$

and thus from Eq. (24.27)

$$\mathcal{E} = \frac{q(r)\mathbf{r}}{r^3}. \tag{24.28}$$

For such a spherically symmetrical model of an electron Eq. (24.26) reduces to

$$\begin{aligned} E' &= m_0c^2 + \frac{1}{2}\int_0^a r^2\mathcal{E}^2 \, dr + \frac{1}{2}\int_a^\infty r^2\mathcal{E}^2 \, dr \\ &= m_0c^2 + \frac{1}{2}\int_0^a \frac{q^2(r)}{r^2} \, dr + \frac{e^2}{2}\int_a^\infty \frac{dr}{r^2}, \end{aligned} \tag{24.29}$$

where we have used the fact that $q(r) = e$ when $r > a$. The middle term on the right represents the energy of the field inside the electron and is necessarily positive. It can be made zero by assuming $q(r) = 0$ for $r < a$ and thus that all the charge is on the surface of the particle. Leaving out this term, we have that

$$E' \geq m_0c^2 + \frac{e^2}{2a}. \tag{24.30}$$

This equation shows at once that the energy becomes infinite as the radius a of the electron approaches zero.

* Because of the importance of the electron in practical applications, we shall often refer in the remaining sections of this book to this particular charged particle, instead of using throughout the general term "charged particle." We shall, however, continue to use e as the charge of a particle; thus for an electron e is negative.

Since the energy and momentum of the field transforms properly under a Lorentz transformation, we can associate an *electromagnetic mass* m_e with the electromagnetic field. We then have

$$E' = (m_0 + m_e)c^2 = mc^2, \tag{24.31}$$

where m is the total mass of the particle and m_0 its *bare mass* (mass without the field). Comparing Eqs. (24.29) and (24.31), we have

$$m_e \geq \frac{e^2}{2ac^2}, \tag{24.32}$$

which is seen to be infinite for a point electron. Taking all the mass to be electromagnetic ($m = m_e$), we see from this equation that

$$a \geq \frac{e^2}{2mc^2} = \text{classical electron radius.} \tag{24.33}$$

The classical electron radius is expressed in terms of fundamental constants and has a magnitude of 1.4089×10^{-13} cm. This is small compared to the Compton wavelength for an electron, $\lambda\!\!\!^{-} = \hbar/mc = 3.8612 \times 10^{-11}$ cm, and thus the phenomena that might be related to such a finite radius for the electron can be expected to be masked by quantum-mechanical effects.

25. Motion of an Electron in a Given External Field

Following our plan of treatment of a charged particle, we shall investigate the dynamic behavior of a charge distribution ρ, restricted to a small spherical volume of radius a. We shall endeavor to see to what extent this behavior is independent of the charge distribution ρ, so long as this distribution is spherically symmetrical, within such a sphere.

It will be shown that there are several very serious difficulties with an extended electron, which are largely removed by going to Dirac's theory of the point electron; moreover, this point-electron theory is in keeping with the mass renormalization treatment of quantum electrodynamics.

Field produced by an extended charge. To calculate the field produced by this charge distribution, at points within the distribution, we shall make use of Eqs. (19.12) and (19.13), namely,

$$\varphi(\mathbf{r}, t) = dV' \int_{-\infty}^{\infty} \frac{\rho(\xi, \eta, \zeta, \tau)\delta(t - \tau - R/c)\, d\tau}{R} \tag{25.1}$$

and

$$\mathbf{A}(\mathbf{r}, t) = \frac{1}{c} \int dV' \int_{-\infty}^{\infty} \frac{\rho(\xi, \eta, \zeta, \tau)\mathbf{v}(\xi, \eta, \zeta, \tau)\delta(t - \tau - R/c)\, d\tau}{R}, \tag{25.2}$$

where

$$dV' = d\xi\, d\eta\, d\zeta,$$

$$R = [(x - \xi)^2 + (y - \eta)^2 + (z - \zeta)^2]^{1/2} = |\mathbf{r} - \mathbf{r}'|, \tag{25.3}$$

$$\mathbf{r} = (x, y, z), \qquad \text{and} \qquad \mathbf{r}' = (\xi, \eta, \zeta).$$

Introducing from (A8.91) the relation

$$\frac{1}{2\pi} \int_{-\infty}^{\infty} e^{i\mu(t-\tau-R/c)}\, d\mu = \delta(t - \tau - R/c) \tag{25.4}$$

into Eqs. (25.1) and (25.2), we have

$$\varphi(\mathbf{r}, t) = \frac{1}{2\pi} \int dV' \iint_{-\infty}^{\infty} \frac{\rho(\mathbf{r}', \tau)}{R}\, e^{i\mu(t-\tau-R/c)}\, d\mu\, d\tau \tag{25.5}$$

and

$$\mathbf{A}(\mathbf{r}, t) = \frac{1}{2\pi c} \int dV' \iint_{-\infty}^{\infty} \frac{\rho(\mathbf{r}', \tau)\mathbf{v}(\mathbf{r}', \tau)}{R}\, e^{i\mu(t-\tau-R/c)}\, d\mu\, d\tau. \tag{25.6}$$

Consider the total charge as being composed of elementary charges $de' = \rho(\xi, \eta, \zeta, \tau)\, dV'$, their position being given by the vector $\mathbf{r}'$. Each such charge may be identified by its position $\mathbf{r}_0'$, at some fixed time; and its motion may be specified by giving its coordinates as function of τ; thus

$$\mathbf{r}' = \mathbf{r}'(\mathbf{r}_0', \tau). \tag{25.7}$$

We shall suppose that $\mathbf{r}_0'$ is the position of de' at the time τ_0; then in integration with respect to e' one can take*

$$\rho(\mathbf{r}', \tau)\, dV' = de' = \rho(\mathbf{r}_0')\, d\xi_0\, d\eta_0\, d\zeta_0 = \rho(\mathbf{r}_0')\, dV_0'. \tag{25.8}$$

* This assumes, of course, continuity in the behavior of elements of charge that are near one another.

Equations (25.5) and (25.6) then become

$$\varphi(\mathbf{r}, t) = \frac{1}{2\pi} \int de' \iint_{-\infty}^{\infty} e^{i\mu(t-\tau-R/c)} \frac{d\mu\, d\tau}{R} \tag{25.9}$$

and

$$\mathbf{A}(\mathbf{r}, t) = \frac{1}{2\pi c} \int de' \iint_{-\infty}^{\infty} e^{i\mu(t-\tau-R/c)} \frac{\mathbf{v}\, d\mu\, d\tau}{R}, \tag{25.10}$$

where $\mathbf{R} = \mathbf{r} - \mathbf{r}'$ and $\mathbf{v} = \mathbf{v}(\mathbf{r}_0', \tau)$ are function of $\mathbf{r}_0'$ and τ.

By following the same procedure used to derive Eq. (19.48) from Eqs. (19.46) and (19.47), we see that

$$\mathscr{E}(\mathbf{r}, t) = \frac{1}{2\pi} \int de' \iint_{-\infty}^{\infty} \left[\frac{\mathbf{R}}{R^3} + \left(\frac{\mathbf{R}}{R^2} - \frac{\mathbf{v}/c}{R}\right) \frac{1}{c} \frac{\partial}{\partial t}\right] e^{i\mu(t-\tau-R/c)}\, d\mu\, d\tau. \tag{25.11}$$

We expand the exponential occurring in the integrand in powers of R/c in the following way:

$$\begin{aligned} e^{i\mu(t-\tau-R/c)} &= e^{i\mu(t-\tau)} \sum_{n=0}^{\infty} \frac{(-i\mu R/c)^n}{n!} \\ &= \sum_{n=0}^{\infty} \frac{(-R/c)^n}{n!} \frac{\partial^n}{\partial t^n} e^{i\mu(t-\tau)}. \end{aligned} \tag{25.12}$$

Equation (25.11) then becomes

$$\begin{aligned} \mathscr{E}(\mathbf{r}, t) &= \frac{1}{2\pi} \int de' \sum_{n=0}^{\infty} \frac{(-1)^n}{n!} \int_{-\infty}^{\infty} \left(\frac{R}{c}\right)^n \left[\frac{\mathbf{R}}{R^3} \frac{\partial^n}{\partial t^n} \int_{-\infty}^{\infty} e^{i\mu(t-\tau)}\, d\mu \right. \\ &\quad \left. + \frac{1}{c}\left(\frac{\mathbf{R}}{R^2} - \frac{\mathbf{v}/c}{R}\right) \frac{\partial^{n+1}}{\partial t^{n+1}} \int_{-\infty}^{\infty} e^{i\mu(t-\tau)}\, d\mu\right] d\tau \\ &= \int de' \sum_{n=0}^{\infty} \frac{(-1)^n}{n!} \int_{-\infty}^{\infty} \left(\frac{R}{c}\right)^n \left[\frac{\mathbf{R}}{R^3} \frac{\partial^n}{\partial t^n} \delta(t-\tau) \right. \\ &\quad \left. + \frac{1}{c}\left(\frac{\mathbf{R}}{R^2} - \frac{\mathbf{v}/c}{R}\right) \frac{\partial^{n+1}}{\partial t^{n+1}} \delta(t-\tau)\right] d\tau. \end{aligned}$$

Since $\mathbf{R}$ does not contain the time t explicitly, we have

$$\mathscr{E}(\mathbf{r}, t) = \int de' \sum_{n=0}^{\infty} \frac{1}{n!}\left(-\frac{1}{c}\right)^n \left[\frac{\partial^n}{\partial t^n}\int_{-\infty}^{\infty} R^{n-3}\mathbf{R}\delta(t-\tau)\,d\tau + \frac{1}{c}\frac{\partial^{n+1}}{\partial t^{n+1}}\int_{-\infty}^{\infty}\left(R^{n-2}\mathbf{R} - R^{n-1}\frac{\mathbf{v}}{c}\right)\delta(t-\tau)\,d\tau\right]. \quad (25.13)$$

Integrating with respect to τ replaces the variable τ in the integral by t, so that

$$\mathscr{E}(\mathbf{r}, t) = \int de' \sum_{n=0}^{\infty} \frac{1}{n!}\left(-\frac{1}{c}\right)^n \left[\frac{d^n}{dt^n}(R^{n-3}\mathbf{R}) + \frac{1}{c}\frac{d^{n+1}}{dt^{n+1}}\left(R^{n-2}\mathbf{R} - R^{n-1}\frac{\mathbf{v}}{c}\right)\right], \quad (25.14)$$

where

$$\mathbf{R} = \mathbf{r} - \mathbf{r}'(\mathbf{r}_0', t). \quad (25.15)$$

Now $\mathbf{r}$ and $\mathbf{r}_0'$ are to be treated as parameters in our calculations, which explains the use in Eq. (25.14) of the total derivatives.

Since

$$\sum_{n=0}^{\infty} \frac{1}{n!}\left(-\frac{1}{c}\right)^n \frac{d^n(R^{n-3}\mathbf{R})}{dt^n} = R^{-3}\mathbf{R} - \frac{1}{c}\frac{d(R^{-2}\mathbf{R})}{dt} + \sum_{m=0}^{\infty}\frac{1}{(m+2)!}\left(-\frac{1}{c}\right)^{m+2}\frac{d^{m+2}(R^{m-1}\mathbf{R})}{dt^{m+2}} \quad (25.16)$$

and

$$\sum_{n=0}^{\infty} \frac{1}{n!}\left(-\frac{1}{c}\right)^n \frac{1}{c}\frac{d^{n+1}(R^{n-2}\mathbf{R})}{dt^{n+1}} = \frac{1}{c}\frac{d(R^{-2}\mathbf{R})}{dt} - \sum_{q=0}^{\infty}\frac{1}{(q+1)!}\left(-\frac{1}{c}\right)^{q+2}\frac{d^{q+2}(R^{q-1}\mathbf{R})}{dt^{q+2}}, \quad (25.17)$$

where we put $m = n - 2$ and $q = n - 1$, Eq. (25.14) can be written

$$\mathscr{E}(\mathbf{r}, t) = \int de' \left[R^{-3}\mathbf{R} - \sum_{n=0}^{\infty}\frac{n+1}{(n+2)!}\left(-\frac{1}{c}\right)^{n+2}\frac{d^{n+2}(R^{n-1}\mathbf{R})}{dt^{n+2}} - \sum_{n=0}^{\infty}\frac{1}{n!}\left(-\frac{1}{c}\right)^{n+2}\frac{d^{n+1}(R^{n-1}\mathbf{v})}{dt^{n+1}}\right]. \quad (25.18)$$

Furthermore, by Leibnitz's theorem

$$\frac{d^{n+2}(R^{n-1}\mathbf{R})}{dt^{n+2}} = \sum_{s=0}^{n+2} \frac{(n+2)!}{s!\,(n+2-s)!} \frac{d^{n+2-s}R^{n-1}}{dt^{n+2-s}} \frac{d^s\mathbf{R}}{dt^s};$$

so that, putting $s = k + 1$,

$$\frac{d^{n+2}}{dt^{n+2}}(R^{n-1}\mathbf{R}) = \mathbf{R}\frac{d^{n+2}R^{n-1}}{dt^{n+2}} - \sum_{k=0}^{n+1} \frac{(n+2)!}{(k+1)!\,(n+1-k)!} \frac{d^{n+1-k}R^{n-1}}{dt^{n+1-k}} \frac{d^k\mathbf{v}}{dt^k}, \tag{25.19}$$

since $d\mathbf{R}/dt = -d\mathbf{r}'/dt = -\mathbf{v}$. Likewise

$$\frac{d^{n+1}(R^{n-1}\mathbf{v})}{dt^{n+1}} = \sum_{k=0}^{n+1} \frac{(n+1)!}{k!\,(n+1-k)!} \frac{d^{n+1-k}R^{n-1}}{dt^{n+1-k}} \frac{d^k\mathbf{v}}{dt^k}; \tag{25.20}$$

therefore,

$$\mathscr{E}(\mathbf{r}, t) = \int\Bigg[R^{-3}\mathbf{R} - \mathbf{R}\sum_{n=0}^{\infty} \frac{(n+1)}{(n+2)!}\left(-\frac{1}{c}\right)^{n+2} \frac{d^{n+2}R^{n-1}}{dt^{n+2}} - \sum_{n=0}^{\infty}\sum_{k=1}^{n+1} \frac{(n+1)k}{(n+1-k)!\,(k+1)!}\left(-\frac{1}{c}\right)^{n+2} \mathbf{v}^{(k)} \frac{d^{n+1-k}R^{n-1}}{dt^{n+1-k}}\Bigg]\,de', \tag{25.21}$$

where $\mathbf{v}^{(k)}$ is the kth derivative of $\mathbf{v}$.

For simplicity of calculation we shall put $\mathbf{v} = 0$; that is, we choose a coordinate system in which the particle is at rest and therefore has a spherically symmetrical charge distribution. We can afterwards obtain the result for any other coordinate system by performing a Lorentz transformation. Possibility of such a choice of the coordinate system implies, however, that the whole charge distribution moves together as a rigid body.*

Now

$$\frac{dR}{dt} = \frac{d}{dt}(\mathbf{R}\cdot\mathbf{R})^{1/2} = \frac{\mathbf{R}}{R}\cdot\frac{d\mathbf{R}}{dt} = \frac{-\mathbf{R}\cdot\mathbf{v}}{R};$$

so that

$$\frac{dR^m}{dt} = mR^{m-1}\frac{dR}{dt} = -mR^{m-2}\mathbf{R}\cdot\mathbf{v},$$

* A more extensive investigation by G. A. Schott, in his *Electromagnetic Radiation* (Cambridge Univ. Press, London, 1912) shows that (for a spherically symmetrical charge distribution) the relative motion of parts does not affect the result.

and similarly

$$\frac{d^2R^m}{dt^2} = m(m-2)R^{m-4}(\mathbf{R}\cdot\mathbf{v})^2 + mR^{m-2}(v^2 - \mathbf{R}\cdot\dot{\mathbf{v}}).$$

Proceeding in this way to d^6R^m/dt^6 and then putting $\mathbf{v} = 0$ in each expression, we obtain

$$\frac{dR^m}{dt} = 0, \tag{25.22}$$

$$\frac{d^2R^m}{dt^2} = -mR^{m-2}\mathbf{R}\cdot\dot{\mathbf{v}}, \tag{25.23}$$

$$\frac{d^3R^m}{dt^3} = -mR^{m-2}\mathbf{R}\cdot\ddot{\mathbf{v}}, \tag{25.24}$$

$$\frac{d^4R^m}{dt^4} = 3m(m-2)R^{m-4}(\mathbf{R}\cdot\dot{\mathbf{v}})^2 + mR^{m-2}(3\dot{v}^2 - \mathbf{R}\cdot\mathbf{v}^{(3)}), \tag{25.25}$$

$$\frac{d^5R^m}{dt^5} = 10m(m-2)R^{m-4}(\mathbf{R}\cdot\dot{\mathbf{v}})(\mathbf{R}\cdot\ddot{\mathbf{v}}) + mR^{m-2}(10\dot{\mathbf{v}}\cdot\ddot{\mathbf{v}} - \mathbf{R}\cdot\mathbf{v}^{(4)}), \tag{25.26}$$

$$\begin{aligned}\frac{d^6R^m}{dt^6} &= -15m(m-2)(m-4)R^{m-6}(\mathbf{R}\cdot\dot{\mathbf{v}})^3 \\ &\quad + 5m(m-2)R^{m-4}R^{m-6}[2(\mathbf{R}\cdot\ddot{\mathbf{v}})^2 - 3\mathbf{R}\cdot\dot{\mathbf{v}}(3\dot{v}^2 - \mathbf{R}\cdot\mathbf{v}^{(3)})] \\ &\quad + mR^{m-2}(10\ddot{v}^2 + 15\dot{\mathbf{v}}\cdot\mathbf{v}^{(3)} - \mathbf{R}\cdot\mathbf{v}^{(5)}).\end{aligned} \tag{25.27}$$

It can now be seen that the lowest power of R occurring in d^sR^m/dt^s, when $\mathbf{v}$ is put equal to zero, is greater than or equal to $m - \frac{1}{2}s$. We wish to obtain the expansion of $\mathscr{E}(\mathbf{r}, t)$ up to the first power of R. Considering the first summation in Eq. (25.21), since it has R as a factor, we need only those terms for which $m - \frac{1}{2}s = (n-1) - \frac{1}{2}(n+2) < 0$, or $n < 4$, and $n + 2 < 6$. Thus, in this summation the sixth derivative is the highest needed. Similarly, in the double summation we need only terms for which $m - \frac{1}{2}s = (n-1) - \frac{1}{2}(n+1-k) < 1$, or $n + k < 5$, and $n + 1 - k = (n+k) + 1 - 2k < 5 + 1 - 2k < 6$. Therefore Eqs. (25.22) to (25.27) will provide us with all the derivatives needed.

Further, since $n + k < 5$, and in Eq. (25.21) $k < n + 1$, we need retain only the terms in the double sum for which $k - 1 < n < 5 - k$. As a consequence we may require that $k < 3$. If we change the order of the summation in the double sum of Eq. (25.21) and apply the restrictions

$k < 3$ and $n < 5 - k$, that equation can be written

$$\mathscr{E}(\mathbf{r}, t) = \int \left\{ \frac{\mathbf{R}}{R^3} - \frac{\mathbf{R}}{c^2}\left[\frac{1}{2!}\frac{d^2R^{-1}}{dt^2} - \frac{2}{3!\,c}\frac{d^3R^0}{dt^3} + \frac{3}{4!\,c^2}\frac{d^4R}{dt^4} - \frac{4}{5!\,c^3}\frac{d^5R^2}{dt^5} + \frac{5}{6!\,c^4}\frac{d^6R^3}{dt^6} - \cdots\right] - \sum_{n=0}^{4}\frac{n+1}{n!\,2!\,c^2}\left(-\frac{1}{c}\right)^n \dot{\mathbf{v}}\frac{d^nR^{n-1}}{dt^n} - \sum_{n=1}^{3}\frac{2(n+1)}{(n-1)!\,3!\,c^2}\left(-\frac{1}{c}\right)^n \ddot{\mathbf{v}}\frac{d^{n-1}R^{n-1}}{dt^{n-1}} - \frac{3\cdot 3}{4!\,c^4}\dddot{\mathbf{v}}R - \cdots \right\} de', \tag{25.28}$$

where a row of dots indicates terms containing R to a power higher than the first. Substituting from Eqs. (25.22) to (25.27) and collecting terms, we finally obtain

$$\mathscr{E}(\mathbf{r}, t) = \int \left\{ \frac{\mathbf{u}}{R^2} - \frac{1}{2c^2R}[\mathbf{u}\,\mathbf{u}\cdot\dot{\mathbf{v}} + \dot{\mathbf{v}}] + \frac{1}{c^3}\left[\frac{2}{3}\ddot{\mathbf{v}} + \frac{3}{4c}\dot{\mathbf{v}}\,\mathbf{u}\cdot\dot{\mathbf{v}} - \frac{3}{8c}\mathbf{u}(\dot{v}^2 - (\mathbf{u}\cdot\dot{\mathbf{v}})^2)\right] - \frac{R}{c^4}\left[\frac{3}{8}\dddot{\mathbf{v}} - \frac{1}{8}\mathbf{u}\,\mathbf{u}\cdot\dddot{\mathbf{v}} - \frac{2}{3c}(\mathbf{u}\,\dot{\mathbf{v}}\cdot\ddot{\mathbf{v}} - \dot{\mathbf{v}}\,\mathbf{u}\cdot\ddot{\mathbf{v}} - 2\ddot{\mathbf{v}}\,\mathbf{u}\cdot\dot{\mathbf{v}}) + \frac{5}{16c^2}(3(\dot{\mathbf{v}} - \mathbf{u}\,\mathbf{u}\cdot\dot{\mathbf{v}})\dot{\mathbf{v}}^2 + 3(\mathbf{u}\cdot\dot{\mathbf{v}})^2\dot{\mathbf{v}} + \mathbf{u}(\mathbf{u}\cdot\dot{\mathbf{v}})^3)\right] + \cdots \right\} de', \tag{25.29}$$

where $\mathbf{u} = \mathbf{R}/R$.

The terms in the last brackets as well as the higher terms are hopelessly dependent upon the charge distribution; therefore, unless these terms can be neglected, one is unable to interpret the consequences of Eq. (25.29) consistently. On the other hand, if

$$\frac{a}{c}\dot{v} \ll c, \qquad \frac{a}{c}\ddot{v} \ll \dot{v}, \qquad \frac{a}{c}\dddot{v} \ll \ddot{v}, \qquad \text{etc.}, \tag{25.30}$$

the terms containing first and higher powers of R can be neglected in comparison with the preceding terms, since $R < 2a$. Note that R is the distance between points within the sphere at only slightly different times.

Self-force for a charged particle at rest. The force that the field produced by the particle exerts upon the particle itself, since $\mathbf{v}(\mathbf{r}, t) = 0$, is given by

$$\mathbf{F}_{\text{self}} = \int \rho(\mathbf{r}, t)\left[\mathscr{E}(\mathbf{r}, t) + \frac{1}{c}\mathbf{v}(\mathbf{r}, t) \times \mathscr{H}(\mathbf{r}, t)\right] dV = \int \mathscr{E}(\mathbf{r}, t)\, de, \quad (25.31)$$

where, analogous to our treatment of Eqs. (25.5) and (25.6), we put $\rho(\mathbf{r}, t)\, dV = de$. Thus, using Eq. (25.29) and neglecting the higher-order terms indicated above, we have

$$\mathbf{F}_{\text{self}} = \iint \left\{ \frac{\mathbf{u}}{R^2} - \frac{1}{2c^2R}[\mathbf{u}\,\mathbf{u}\cdot\dot{\mathbf{v}} + \dot{\mathbf{v}}] + \frac{1}{c^3}\left[\frac{2}{3}\ddot{\mathbf{v}} + \frac{3}{4c}\dot{\mathbf{v}}\,\mathbf{u}\cdot\dot{\mathbf{v}} - \frac{3}{8c}\mathbf{u}(\dot{v}^2 - (\mathbf{u}\cdot\dot{\mathbf{v}})^2)\right]\right\} de\, de'. \quad (25.32)$$

This result was obtained by calculating the force due to the element de' acting upon the element de, with $\mathbf{R} = \mathbf{r} - \mathbf{r}'$. Obviously the opposite procedure should give the same result. Interchanging de and de' merely reverses the direction of $\mathbf{u}$; thus we can also write

$$\mathbf{F}_{\text{self}} = \iint \left\{ -\frac{\mathbf{u}}{R^2} - \frac{1}{2c^2R}[\mathbf{u}\,\mathbf{u}\cdot\dot{\mathbf{v}} + \dot{\mathbf{v}}] + \frac{1}{c^3}\left[\frac{2}{3}\ddot{\mathbf{v}} - \frac{3}{4c}\dot{\mathbf{v}}\,\mathbf{u}\cdot\dot{\mathbf{v}} + \frac{3}{8c}\mathbf{u}(\dot{v}^2 - (\mathbf{u}\cdot\dot{\mathbf{v}})^2)\right]\right\} de\, de'.$$

Adding this equation to the previous one and dividing by 2, we obtain for the self-force

$$\mathbf{F}_{\text{self}} = -\frac{1}{2c^2}\iint (\mathbf{u}\mathbf{u}\cdot\dot{\mathbf{v}} + \dot{\mathbf{v}})\frac{de\, de'}{R} + \frac{2e^2}{3c^3}\ddot{\mathbf{v}}, \quad (25.33)$$

where we have used the fact that

$$\iint \frac{2\ddot{\mathbf{v}}}{3c^2}\, de\, de' = \frac{2\ddot{\mathbf{v}}}{3c^2}\int de \int de' = \frac{2e^2}{3c^3}\ddot{\mathbf{v}}. \quad (25.34)$$

The cancelling of the other terms means that they would be found to vanish if we actually performed the integration, due to the symmetry of the charge distribution.

We shall now consider the integral

$$\mathbf{J} = \iint \frac{\mathbf{u}\mathbf{u}\cdot\dot{\mathbf{v}}\, de\, de'}{R}. \quad (25.35)$$

If we put $\mathbf{u} = \mathbf{u}_0 + \mathbf{u}_1$, where $\mathbf{u}_0$ is parallel to $\dot{\mathbf{v}}$ and $\mathbf{u}_1$ is perpendicular to $\dot{\mathbf{v}}$, then

$$\mathbf{J} = \iint \frac{\mathbf{u}_0\mathbf{u}_0 \cdot \dot{\mathbf{v}}}{R}\, de\, de' + \iint \frac{\mathbf{u}_1\mathbf{u}_0 \cdot \dot{\mathbf{v}}}{R}\, de\, de'. \tag{25.36}$$

If $\mathbf{R}$ is rotated 180° about an axis parallel to $\dot{\mathbf{v}}$ running through the center of the charge distribution, the vector $\mathbf{u} = \mathbf{u}_0 + \mathbf{u}_1$ becomes $\mathbf{u}_0 - \mathbf{u}_1$. Therefore, in the second integral of Eq. (25.36), for every elementary contribution $\mathbf{u}_1\,\mathbf{u}_0 \cdot \dot{\mathbf{v}}\, de\, de'/R$ there is also an equal and opposite contribution $-\mathbf{u}_1\,\mathbf{u}_0 \cdot \dot{\mathbf{v}}\, de\, de'/R$. This integral is thus zero.

Since $\mathbf{u}_0$ is parallel to $\dot{\mathbf{v}}$, we have $\mathbf{u}_0 \cdot \dot{\mathbf{v}} = u_0\dot{v}$ and

$$\mathbf{u}_0\mathbf{u}_0 \cdot \dot{\mathbf{v}} = \dot{\mathbf{v}}u_0^2 = \dot{\mathbf{v}}(\mathbf{s} \cdot \mathbf{u})^2, \tag{25.37}$$

where $\mathbf{s}$ is a unit vector in the direction of $\dot{\mathbf{v}}$; therefore

$$\mathbf{J} = \dot{\mathbf{v}} \iint \frac{(\mathbf{s} \cdot \mathbf{u})^2\, de\, de'}{R}. \tag{25.38}$$

The structure involved in the integration is a spherically symmetrical charge distribution and a unit vector $\mathbf{s}$. Since all directions in space are indistinguishable, except as they relate to the coordinate system employed, the integral in Eq. (25.38) must be independent of the direction of $\mathbf{s}$. The average of this integral over all possible directions of $\mathbf{s}$ must therefore equal the integral, and hence

$$\iint \frac{(\mathbf{s} \cdot \mathbf{u})^2\, de\, de'}{R} = \frac{1}{4\pi} \int d\omega \iint \frac{(\mathbf{s} \cdot \mathbf{u})^2\, de\, de'}{R} = \frac{1}{4\pi} \iint \frac{de\, de'}{R} \int (\mathbf{s} \cdot \mathbf{u})^2\, d\omega, \tag{25.39}$$

where $d\omega$ is an element of the solid angle within which a particular $\mathbf{s}$ lies. Choosing a polar coordinate system with the polar axis along $\mathbf{u}$, we have

$$\mathbf{s} \cdot \mathbf{u} = \cos\theta \qquad \text{and} \qquad d\omega = \sin\theta\, d\theta\, d\varphi \tag{25.40}$$

and hence that

$$\frac{1}{4\pi} \int (\mathbf{s} \cdot \mathbf{u})^2\, d\omega = \frac{1}{4\pi} \int_0^\pi \cos^2\theta \sin\theta\, d\theta \int_0^{2\pi} d\varphi = \frac{1}{2} \int_0^\pi \cos^2\theta \sin\theta\, d\theta = \frac{1}{3}. \tag{25.41}$$

Thus from Eqs. (25.37) and (25.39) we have

$$\mathbf{J} = \frac{\dot{\mathbf{v}}}{3} \iint \frac{de\, de'}{R} \tag{25.42}$$

and (see Eqs. (25.33) and (25.34))

$$\mathbf{F}_{\text{self}} = -\frac{2\dot{\mathbf{v}}}{3c^2}\iint \frac{de\,de'}{R} + \frac{2e^2}{3c^3}\ddot{\mathbf{v}}. \tag{25.43}$$

We also know that

$$\iint \frac{de\,de'}{R} = 2E_0\,; \tag{25.44}$$

that is, this double integral is just twice the electrostatic energy E_0 of the charge distribution. The factor 2 occurs because the electrostatic energy $de\,de'/R$ of each pair of differential charge elements de and de' occurs in the integration twice. We may, moreover, let

$$m_e' = \frac{2}{3c^2}\iint \frac{de\,de'}{R} = \frac{4}{3}\frac{E_0}{c^2} = \frac{4}{3}m_e, \tag{25.45}$$

where m_e stands for the electromagnetic mass of the electron as defined in Section 24 and m_e' that based on the present treatment of the self-force.* Substituting from Eq. (25.45) into Eq. (25.43), we finally have

$$\mathbf{F}_{\text{self}} = -m_e'\dot{\mathbf{v}} + \frac{2e^2}{3c^3}\ddot{\mathbf{v}}. \tag{25.46}$$

Self-force on a moving particle. We must now perform the necessary transformations to determine the self-force in a coordinate system in which the particle moves with the velocity $\mathbf{v}$. To this end we must recall that the three components of kinetic momentum $\mathbf{P}$ and i/c times the kinetic energy T constitute a four-vector. Since by Eqs. (14.14) and (14.26)

$$\frac{d\mathbf{P}}{dt} = \mathbf{F} = \textbf{Lorentz force} \qquad \text{and} \qquad \frac{dT}{dt} = e\,\mathbf{v}\cdot\boldsymbol{\mathscr{E}},$$

we can write that

$$\left(d\mathbf{P}, \frac{i}{c}\,dT\right) = \left(\mathbf{F}\,dt, \frac{ie}{c}\,\mathbf{v}\cdot\boldsymbol{\mathscr{E}}\,dt\right), \tag{25.47}$$

which expresses the equality of two differential four-vectors. The total force acting on the bare particle is $\mathbf{F} + \mathbf{F}_{\text{self}}$, where $\mathbf{F}$ is the Lorentz force

* The occurrence of the factor $\frac{4}{3}$ in the definition of the electromagnetic mass m_e' that arises in the treatment of the self-force of an extended electron is a serious fault in the model that will be discussed later in this section.

and $\mathbf{F}_{\text{self}}$ the self-force. Thus, if $\mathbf{P}$ is the momentum of the bare particle, one has

$$\frac{d\mathbf{P}}{dt} = \mathbf{F}_{\text{self}} + \mathbf{F}, \tag{25.48}$$

and hence $\mathbf{F}_{\text{self}}\, dt$ should transform as $d\mathbf{P}$ and $\mathbf{F}\, dt$. Now

$$f_\alpha = (\mathbf{f}, f_4) = \left(\frac{\mathbf{F}_{\text{self}}\, dt}{m_e'}, f_4\right) \tag{25.49}$$

is a four-vector if f_4 is properly chosen. An equation connecting the fourth components of the vectors given in Eqs. (25.47) and (25.49) in the same way as the other components are connected by Eq. (25.48),

$$\frac{i}{c}\, dT = m_e' f_4 + \frac{ie}{c} \mathbf{v} \cdot \mathscr{E}\, dt, \tag{25.50}$$

insures a proper choice of f_4.

Let the coordinate system for which $\mathbf{v} = 0$ be designated by primes; then, by Eq. (14.12), $T^2 = c^2P^2 + m^2c^4$ and

$$\frac{dT'}{dt} = \frac{c^2\mathbf{P}' \cdot \dot{\mathbf{P}}'}{T'} = 0,$$

because $\mathbf{P}' = 0$. Since the velocity of the particle is zero in the primed system and $dT' = 0$, Eq. (25.50) requires that $f_4' = 0$. Thus, by Eq. (25.46), the components of f_α in the primed system are

$$\mathbf{f}' = (-\dot{\mathbf{v}}' + b\ddot{\mathbf{v}}')\, dt'$$

and

$$f_4' = 0, \tag{25.51}$$

where $b = 2e^2/3m_e'c^3 = e^2/2m_ec^3$.

Transforming now to the unprimed system by using the generalized Lorentz transformation given by Eqs. (11.46), with $\mathbf{r}$ replaced by $\mathbf{f}$, $\mathbf{r}'$ by $\mathbf{f}'$, and t' by $f_4'/ic = 0$, we have

$$\mathbf{f} = \mathbf{f}' + \frac{\gamma - 1}{v^2} \mathbf{v}\mathbf{v} \cdot \mathbf{f}', \tag{25.52}$$

where $\gamma = (1 - v^2/c^2)^{-1/2}$. Substitution from Eq. (25.51) now gives

$$\mathbf{f} = -\left(\dot{\mathbf{v}}' + \frac{\gamma - 1}{v^2} \mathbf{v}\mathbf{v} \cdot \dot{\mathbf{v}}'\right) dt' + b\left(\ddot{\mathbf{v}}' + \frac{\gamma - 1}{v^2} \mathbf{v}\mathbf{v} \cdot \ddot{\mathbf{v}}'\right) dt'. \tag{25.53}$$

To complete our transformation we must express the primed quantities

$\dot{\mathbf{v}}\,dt' = d\dot{\mathbf{v}}'$ and $\dot{\mathbf{v}}'\,dt' = d\dot{\mathbf{v}}'$ in terms of the corresponding unprimed quantities.

From the general Lorentz transformation of Eqs. (11.46) we have

$$d\mathbf{r}' = d\mathbf{r} + \frac{\gamma - 1}{v^2}\mathbf{v}\mathbf{v}\cdot d\mathbf{r} - \gamma\mathbf{v}\,dt \tag{25.54}$$

and

$$dt' = \gamma\left(dt - \frac{\mathbf{v}\cdot d\mathbf{r}}{c^2}\right). \tag{25.55}$$

Dividing each side of Eq. (25.54) by the corresponding side of Eq. (25.55), we have

$$\mathbf{v}_p' \equiv \frac{d\mathbf{r}'}{dt'} = \frac{\mathbf{v}_p + \dfrac{\gamma - 1}{v^2}\mathbf{v}\mathbf{v}\cdot\mathbf{v}_p - \gamma\mathbf{v}}{\gamma\left(1 - \dfrac{\mathbf{v}\cdot\mathbf{v}_p}{c^2}\right)}, \tag{25.56}$$

where $\mathbf{v}_p$ and $\mathbf{v}_p'$ are the velocities of a particle in the unprimed and the primed coordinate systems, respectively. The purpose of this use of the subscript p is to indicate those $\mathbf{v}$'s that are changing with the time, thereby differentiating them from the constant $\mathbf{v}$ giving the relative velocity of the coordinate systems. After taking differentials the subscript p can be dropped, and thus we set

$$\mathbf{v}_p = \mathbf{v} \quad \text{and} \quad \mathbf{v}_p' = \mathbf{v}' = 0.$$

The differential of Eq. (25.56) is

$$d\mathbf{v}_p' = \frac{\left(1 - \dfrac{\mathbf{v}\cdot\mathbf{v}_p}{c^2}\right)\dot{\mathbf{v}}_p + \left(\dfrac{\mathbf{v}_p}{c^2} - \dfrac{(\gamma - 1)}{\gamma v^2}\mathbf{v}\right)\mathbf{v}\cdot\dot{\mathbf{v}}_p}{\gamma\left(1 - \dfrac{\mathbf{v}\cdot\mathbf{v}_p}{c^2}\right)^2}\,dt. \tag{25.57}$$

For $\mathbf{v}_p = \mathbf{v}$ this simplifies to

$$d\mathbf{v}' = \gamma\left(\dot{\mathbf{v}} + \frac{\gamma - 1}{v^2}\mathbf{v}\mathbf{v}\cdot\dot{\mathbf{v}}\right)dt. \tag{25.58}$$

Dividing each side of Eq. (25.57) by the corresponding side of Eq. (25.55), we obtain

$$\dot{\mathbf{v}}_p' = \frac{\left(1 - \dfrac{\mathbf{v}\cdot\mathbf{v}_p}{c^2}\right)\dot{\mathbf{v}}_p + \left(\dfrac{\mathbf{v}_p}{c^2} - \dfrac{(\gamma - 1)}{\gamma v^2}\mathbf{v}\right)\mathbf{v}\cdot\dot{\mathbf{v}}_p}{\gamma^2\left(1 - \dfrac{\mathbf{v}\cdot\mathbf{v}_p}{c^2}\right)^3}. \tag{25.59}$$

We can take the differential of each side, as before, to obtain

$$d\dot{\mathbf{v}}_p' = \left\{\ddot{\mathbf{v}}_p\left(1 - \frac{\mathbf{v}\cdot\mathbf{v}_p}{c^2}\right)^2 + \frac{3\dot{\mathbf{v}}_p\mathbf{v}\cdot\dot{\mathbf{v}}_p}{c^2}\left(1 - \frac{\dot{\mathbf{v}}\cdot\mathbf{v}_p}{c^2}\right) + \left(\frac{\mathbf{v}_p}{c^2} - \frac{(\gamma-1)\mathbf{v}}{\gamma v^2}\right)\right.$$
$$\times \left.\left[\frac{3(\mathbf{v}\cdot\dot{\mathbf{v}}_p)^2}{c^2} + \mathbf{v}\cdot\ddot{\mathbf{v}}_p\left(1 - \frac{\mathbf{v}\cdot\mathbf{v}_p}{c^2}\right)\right]\right\}\gamma^{-2}\left(1 - \frac{\mathbf{v}\cdot\mathbf{v}_p}{c^2}\right)^{-4} dt, \quad (25.60)$$

which, for $\mathbf{v}_p = \mathbf{v}$, reduces to

$$d\dot{\mathbf{v}}' = \gamma^2\left\{\ddot{\mathbf{v}} + \frac{3\gamma^2\mathbf{v}\cdot\dot{\mathbf{v}}}{c^2}\dot{\mathbf{v}} + \frac{\gamma-1}{v^2}\left[\frac{3\gamma^2(\mathbf{v}\cdot\dot{\mathbf{v}})^2}{c^2} + \mathbf{v}\cdot\ddot{\mathbf{v}}\right]\mathbf{v}\right\} dt. \quad (25.61)$$

Substituting from Eqs. (25.58) and (25.61) into Eq. (25.53), using Eq. (25.49), and noticing that $\gamma\dot{\mathbf{v}} + \gamma^3\mathbf{v}\cdot\dot{\mathbf{v}}\mathbf{v}/c^2 = d(\gamma\mathbf{v})/dt$, we finally obtain

$$\mathbf{F}_{\text{self}} = -\frac{d}{dt}\frac{m_e'v}{\sqrt{1 - v^2/c^2}}$$
$$+ \frac{2e^2}{3c^3}\left\{\ddot{\mathbf{v}} + \frac{3\gamma^2\mathbf{v}\cdot\dot{\mathbf{v}}}{c^2}\dot{\mathbf{v}} + \frac{\gamma^2}{c^2}\left[\mathbf{v}\cdot\ddot{\mathbf{v}} + \frac{3\gamma^2(\mathbf{v}\cdot\dot{\mathbf{v}})^2}{c^2}\right]\mathbf{v}\right\}\gamma^2. \quad (25.62)$$

The equation of motion of the particle of charge e in an external field $\mathscr{E}$ and $\mathscr{H}$ (see Eq. (25.48)) is

$$\frac{d}{dt}\frac{m_0\mathbf{v}}{\sqrt{1 - v^2/c^2}} = e\left(\mathscr{E} + \frac{\mathbf{v}}{c}\times\mathscr{H}\right) + \mathbf{F}_{\text{self}}, \quad (25.63)$$

where m_0 is the mass of the bare particle. Using the expression for $\mathbf{F}_{\text{self}}$ given by Eq. (25.62), this equation can be written

$$\frac{d}{dt}\frac{(m_0 + m_e')\mathbf{v}}{\sqrt{1 - v^2/c^2}} = e\left(\mathscr{E} + \frac{\mathbf{v}}{c}\times\mathscr{H}\right)$$
$$+ \frac{2e^2}{3c^2}\left\{\ddot{\mathbf{v}} + \frac{3\gamma^2\mathbf{v}\cdot\dot{\mathbf{v}}}{c^2}\dot{\mathbf{v}} + \frac{\gamma^2}{c^2}\left[\mathbf{v}\cdot\ddot{\mathbf{v}} + \frac{3\gamma^2(\mathbf{v}\cdot\dot{\mathbf{v}})^2}{c^2}\right]\mathbf{v}\right\}\gamma^2. \quad (25.64)$$

Thus, as in Eq. (24.34) (but with m_e' replacing m_e), m_e' is merely added to the nonelectromagnetic mass m_0. There is no meaningful way of separating the total mass $m_0 + m_e'$ into these two parts, the separation being possible only if the structure of the particle is known. We shall use m to designate the total mass of the particle, which is experimentally determinable. Our equation of motion of the particle therefore becomes

$$\frac{d}{dt}\frac{m\mathbf{v}}{\sqrt{1-v^2/c^2}} = e\left(\mathscr{E} + \frac{\mathbf{v}}{c}\times\mathscr{H}\right)$$
$$+\frac{2e^2}{3c}\left[\frac{\ddot{\mathbf{v}}}{c^2-v^2} + \frac{3\mathbf{v}\cdot\dot{\mathbf{v}}\dot{\mathbf{v}} + \mathbf{v}\cdot\ddot{\mathbf{v}}\mathbf{v}}{(c^2-v^2)^2} + \frac{3(\mathbf{v}\cdot\dot{\mathbf{v}})^2\mathbf{v}}{(c^2-v^2)^3}\right], \tag{25.65}$$

where $\mathscr{E}$ and $\mathscr{H}$ are the external fields.

In deriving this equation we made use of the inequalities (25.30), where a is the radius of the charge distribution representing the electron. We obtained a radius for the electron of the order of $a = e^2/2mc^2$ by assuming that the mass is wholly, or mainly, electromagnetic in nature. For want of a better assumption we shall take a in these inequalities to be this classical electron radius.

Equation (25.65) can also be written in the following covariant form:

$$mc^2\frac{du_\alpha}{ds} = eF_{\alpha\gamma}u_\gamma + \frac{2e^2}{3}\left[\frac{d^2u_\alpha}{ds} - \left(\frac{du_\gamma}{ds}\right)^2 u_\alpha\right], \tag{25.66}$$

where, as in Eq. (11.24),

$$u_\alpha = \frac{dx_\alpha}{ds} = \left(\frac{\mathbf{v}}{c}\gamma,\ i\gamma\right) \tag{25.67}$$

and $F_{\alpha\gamma}$ is the electromagnetic field tensor defined by Eq. (14.25). The space components of this equation give rise to Eq. (25.65) while the fourth component ($\alpha = 4$) becomes by Eq. (25.67)

$$mc^2 i\dot{\gamma}\frac{dt}{ds} = im\frac{\gamma^4\mathbf{v}\cdot\dot{\mathbf{v}}}{c} = ie\frac{\gamma\mathbf{v}}{c}\cdot\mathscr{E} + \frac{2e^2}{3}\left[\frac{i\gamma^5\mathbf{v}\cdot\ddot{\mathbf{v}}}{c^4} + \frac{3i\gamma^7(\mathbf{v}\cdot\dot{\mathbf{v}})^2}{c^6}\right],$$

which simplifies to

$$m\gamma^3 = e\mathbf{v}\cdot\mathscr{E} + \frac{2e^2\gamma^4}{3c^3}\left[\mathbf{v}\cdot\ddot{\mathbf{v}} + \frac{3\gamma^2}{c^2}(\mathbf{v}\cdot\dot{\mathbf{v}})^2\right]. \tag{25.68}$$

This equation, however, is not a new condition but a consequence of Eq. (25.65). We can establish this by using the identity

$$\frac{d}{dt}\frac{m\mathbf{v}}{\sqrt{1-v^2/c^2}} = m\gamma\left(\dot{\mathbf{v}} + \frac{\gamma^2}{c^2}\mathbf{v}\cdot\dot{\mathbf{v}}\,\mathbf{v}\right) \tag{25.69}$$

to convert the left-hand side and then multiplying by $\mathbf{v}\cdot$ on both sides of the equation. The resulting equation

$$m\gamma \mathbf{v}\cdot\dot{\mathbf{v}}\left(1+\frac{\gamma^2v^2}{c^2}\right) = e\mathbf{v}\cdot\mathscr{E}$$
$$+\frac{2e^2}{3c^3}\left[\gamma^2\mathbf{v}\cdot\ddot{\mathbf{v}}\left(1+\frac{\gamma^2v^2}{c^2}\right)+\frac{3\gamma^4(\mathbf{v}\cdot\dot{\mathbf{v}})^2}{c^2}\left(1+\frac{\gamma^2v^2}{c^2}\right)\right]$$

is seen to agree with Eq. (25.68), when one observes that

$$1+\gamma^2\frac{v^2}{c^2} = 1+\gamma^2\left(\frac{v^2}{c^2}-1\right)+\gamma^2 = \gamma^2.$$

To obtain a nonrelativistic approximation to Eq. (25.65), we neglect v^2 in comparison with c^2 and obtain at once

$$m\dot{\mathbf{v}} = e\left(\mathscr{E}+\frac{\mathbf{v}}{c}\times\mathscr{H}\right)+\frac{2e^2}{3c^3}\left[\ddot{\mathbf{v}}+\frac{3\mathbf{v}\cdot\dot{\mathbf{v}}\dot{\mathbf{v}}}{c^2}+\frac{\mathbf{v}\cdot\ddot{\mathbf{v}}\mathbf{v}}{c^2}+\frac{3(\mathbf{v}\cdot\dot{\mathbf{v}})^2\mathbf{v}}{c^4}\right].$$

In this same approximation

$$\left|\frac{\mathbf{v}\cdot\ddot{\mathbf{v}}\mathbf{v}}{c^2}\right| \le \frac{v^2}{c^2}\ddot{v} \ll \ddot{v}$$

and

$$\left|\frac{(\mathbf{v}\cdot\dot{\mathbf{v}})^2\mathbf{v}}{c^4}\right| \le \frac{|\mathbf{v}\cdot\dot{\mathbf{v}}|v^2\dot{v}}{c^4} \ll \left|\frac{\mathbf{v}\cdot\dot{\mathbf{v}}\dot{\mathbf{v}}}{c^2}\right|$$

so that the last two terms in the brackets can be neglected in comparison with the preceding two. Introducing a constant

$$a' = \frac{4}{3}a = \frac{2e^2}{3mc^2}, \tag{25.70}$$

where a is the classical electron radius, we can show that

$$\frac{2e^2}{c^5}|\mathbf{v}\cdot\dot{\mathbf{v}}\dot{\mathbf{v}}| = \frac{3ma'}{c^3}|\mathbf{v}\cdot\dot{\mathbf{v}}\dot{\mathbf{v}}| < \frac{3m}{c^2}|\mathbf{v}|\left|\frac{a'}{c}\dot{\mathbf{v}}\right|\dot{v} \ll \frac{3mv}{c}\dot{v}$$

and thus that the second term in the brackets represents a very small correction to the left-hand member and can be neglected. We therefore obtain

$$m\left(\dot{\mathbf{v}}-\frac{a'}{c}\ddot{\mathbf{v}}\right) = e\left(\mathscr{E}+\frac{\mathbf{v}}{c}\times\mathscr{H}\right) = \textbf{Lorentz force} \tag{25.71}$$

as the nonrelativistic approximation to the equation of motion of a charged particle of charge e and mass m, correct to the first order in v/c.

Now $a' = 2e^2/3mc^2$ is $\frac{4}{3}$ the classical electron radius a defined in Section 24 and has the value 1.879×10^{-13} cm, and thus the coefficient a'/c of $\ddot{\mathbf{v}}$ in Eq. (25.71) represents the very short time (6.26×10^{-24} sec) it takes for light to travel this distance.

Difficulties with Lorentz extended electron model. While it was possible to derive the equation of motion given by Eq. (26.65) of an electron as a finite charge subject to the self-force produced by its own field and while this equation yields results that are, at least to the first order, in agreement with a wide class of experiments, the Lorentz extended electron model has some very serious drawbacks. Some of the objections to the model are the following:

1. The structure-dependent terms in the expression for the self-force were dropped without any real justification. If they are left in, the theory becomes unmanageable because of the complexity of the equations and the impossibility of obtaining from physical measurements a unique structure for the electron.

2. Neither a rigid-sphere model nor one that distorts under forces is wholly satisfactory. A rigid sphere contradicts relativity in that a signal can be propagated across it with infinite velocity. On the other hand, an elastic-sphere model permits motion of the parts of an electron and makes it no longer an elementary object (the material point of Section 2). If we try to live with an electron whose parts can move relative to each other, we are faced with the need to make ad hoc assumptions about the structure that cannot be checked with experiments.

3. The Lorentz model, being a charged distribution of only one sign, is unstable without the introduction of nonelectromagnetic forces, such as were proposed by Poincaré. The introduction of these forces leads again to ad hoc assumptions that cannot be experimentally checked.

4. The electromagnetic mass m_e' determined from the computation of the self-force is $\frac{4}{3}$ the electromagnetic mass m_e obtained in Section 24 by considering the energy and momentum stored in the electromagnetic field. In all early treatments of the Lorentz electron, this troublesome factor $\frac{4}{3}$ occurred in the expression for the momentum stored in the field and led to the belief that the momentum and energy of the electron's field did not transform properly under a Lorentz transformation. First Fermi* in

* E. Fermi, *Physik. Z.*, **23**, 340 (1922); *Atti Accad. Nazl. Lincei*, **31**, 184 and 306 (1922).

1922, then Wilson* in 1936, Kwal† in 1949, and finally Rohrlich‡ in 1960 pointed out the fallacy, when particles are present, of letting the momentum be given by

$$\mathbf{P} = \frac{1}{4\pi c}\int \mathscr{E} \times \mathscr{H}\, dV$$

in all coordinate systems. It is only quite recently that the correct treatment, using a fixed hypersurface for the integrations (see Section 24), has been generally accepted in the literature.

5. The extended electron is not in keeping with most treatments of quantum electrodynamics, which use a point electron. In this more general, or covering, theory the infinities, such as in the mass, have been eliminated (starting with the proposal by Kramers in 1947) by a covariant subtraction process called *renormalization*. A similar idea was introduced even earlier (in 1938) by Dirac in his treatment of a classical point electron. This point electron model is thus in better agreement with quantum electrodynamics.

Dirac's model of a point electron. In 1938 Dirac§ proposed to treat the point electron by considering only the momentum and energy flow across a cylinder enclosing the world-line of the particle. This leads to the equations of motion of the electron. Since the electron is associated with a point, the singularity of the electromagnetic field, one avoids the question of the stability of the model.

We define $F^{\text{in}}_{\mu\nu}$ and $F^{\text{out}}_{\mu\nu}$ in terms of the actual field $F^{\text{act}}_{\mu\nu}$ by the equations

$$F^{\text{act}}_{\mu\nu} = F^{\text{ret}}_{\mu\nu} + F^{\text{in}}_{\mu\nu} \tag{25.72}$$

and

$$F^{\text{act}}_{\mu\nu} = F^{\text{adv}}_{\mu\nu} + F^{\text{out}}_{\mu\nu}, \tag{25.73}$$

where $F^{\text{ret}}_{\mu\nu}$ and $F^{\text{adv}}_{\mu\nu}$ are the retarded and advanced field tensors.

The field radiated by the electron $F^{\text{rad}}_{\mu\nu}$ we define to be $F^{\text{out}}_{\mu\nu} - F^{\text{in}}_{\mu\nu}$ and thus

$$F^{\text{rad}}_{\mu\nu} = F^{\text{out}}_{\mu\nu} - F^{\text{in}}_{\mu\nu} = F^{\text{ret}}_{\mu\nu} - F^{\text{adv}}_{\mu\nu}. \tag{25.74}$$

* W. Wilson, *Proc. Phys. Soc.* (*London*), **48**, 736 (1936).

† B. Kwal, *J. Phys. Radium*, **10**, 103 (1949).

‡ F. Rohrlich, *Am. J. Phys.*, **28**, 639 (1960). For an enlightening history of classical electron models and the troublesome factor $\frac{4}{3}$ see also F. Rohrlich, *Classical Charged Particles*, Addison-Wesley, Reading, Mass., 1965, Chap. 2.

§ P. A. M. Dirac, *Proc. Roy. Soc.* (*London*), **A167**, 148 (1938).

Dirac shows that this definition agrees with the usual definition of the radiated field when the latter has a unique meaning. His definition permits one to speak of the radiation component of the field everywhere, even right near the electron, at all times. One does not, however, need to look upon Eq. (25.74) other than as a definition of a component of the field labelled $F_{\mu\nu}^{\text{rad}}$. Dirac shows $F_{\mu\nu}^{\text{rad}}$ to be free of singularities and to be given in terms of the worldline of the electron by

$$F_{\mu\nu}^{\text{rad}} = \frac{4e}{3}\left(\frac{dz_\mu}{ds}\frac{d^3z_\nu}{ds^3} - \frac{d^3z_\mu}{ds^3}\frac{dz_\nu}{ds}\right), \tag{25.75}$$

where s is the proper time, i.e., the integrated interval along the worldline, and $z_\mu = z_\mu(s)$ are the space–time coordinates at s.

He defines also another component of the field labeled $f_{\mu\nu}$ by the equation

$$f_{\mu\nu} = F_{\mu\nu}^{\text{act}} - \tfrac{1}{2}(F_{\mu\nu}^{\text{ret}} + F_{\mu\nu}^{\text{adv}}) \tag{25.76}$$

or, from Eqs. (25.72) and (25.73),

$$f_{\mu\nu} = \tfrac{1}{2}(F_{\mu\nu}^{\text{in}} + F_{\mu\nu}^{\text{out}}). \tag{25.77}$$

To obtain the equations of motion of the electron we surround the singularity of the field representing the world-line of the electron by a thin tube and calculate the energy and momentum flow across the hypersurface (two space dimensions and one time dimension) forming the walls of the tube. We require that the energy and momentum flow out from a finite length of the tube equal the difference between the energy and momentum at the two end points. This means that this difference must be a perfect differential.

We calculate this energy and momentum flow across the walls of the tube by means of the energy–momentum tensor

$$T_{\alpha\beta} = \frac{1}{4\pi ic}\left(F_{\alpha\gamma}F_{\beta\gamma} - \frac{1}{4}\,\delta_{\alpha\beta}F_{\epsilon\gamma}F_{\epsilon\gamma}\right)$$

of Eq. (17.35). The results obtained from two tubes of different cross section will be the same because the region of space–time lying between them is free of singularities and thus (see Eq. (17.38))

$$\frac{\partial T_{\alpha\beta}}{\partial x_\beta} = 0.$$

This insures, as we have seen in previous sections, that the total flow out of the two tubes will be the same.

Dirac shows that the momentum–energy vector representing the net flow of momentum and energy out of the walls of a tube of spherical cross section is given by

$$\int [\tfrac{1}{2}e^2\epsilon^{-1}\dot{u}_\mu - ef_{\mu\nu}u_\nu]\, ds, \tag{25.78}$$

where ϵ is the radius of the tube and

$$u_\mu = \frac{dz_\mu}{ds} \tag{25.79}$$

is the four-velocity of the particle. Here terms that vanish as $\epsilon \to 0$ have been dropped.

We now use the assumption the momentum–energy four-vector that represents the flow out of the sides of the tube is dependent only on the velocities, accelerations, etc. at the end point. If we take the near end of the tube to be fixed and consider how the flow out depends on s, we have that the integral in (25.78) is equal to some four-vector B_μ that is a function of the four-vectors u_μ, $\dot{u}_\mu$, etc. of the particle at the far end of the tube. We thus set

$$\tfrac{1}{2}e^2\epsilon^{-1}\dot{u}_\mu - ef_{\mu\nu}u_\nu = \dot{B}_\mu. \tag{25.80}$$

From Eq. (11.25) we have

$$u_\mu^2 = -1 \tag{25.81}$$

and thus, by differentiating,

$$u_\mu \dot{u}_\mu = 0 \tag{25.82}$$

and

$$u_\mu \ddot{u}_\mu + \dot{u}_\mu^2 = 0. \tag{25.83}$$

Making use of Eqs. (25.80) and (25.82) and the fact that $f_{\mu\nu}$ is an antisymmetric tensor, one obtains as a requirement on B_μ that

$$u_\mu \dot{B}_\mu = \tfrac{1}{2}e^2\epsilon^{-1}u_\mu\dot{u}_\mu - eu_\mu f_{\mu\nu}u_\nu = 0. \tag{25.84}$$

Dirac makes the assumption at this point that

$$B_\mu = ku_\mu, \tag{25.85}$$

where k is a constant. He justifies this assumption on the basis that all other choices satisfying Eq. (25.84) are much more complicated and one would "hardly expect them to apply to a simple thing like an electron."

Substituting from Eq. (25.85) into Eq. (25.80), we have

$$mc^2\dot{u}_\mu = ef_{\mu\nu}u_\nu \tag{25.86}$$

as our equation of motion for the electron, where

$$mc^2 = \tfrac{1}{2}e^2\epsilon^{-1} - k \tag{25.87}$$

and m plays the role of the mass of the electron. We can take m to be finite and independent of ϵ. This is seen to require that

$$k = \tfrac{1}{2}e^2\epsilon^{-1} - mc^2$$

become infinite as $\epsilon \to 0$.

The equation of motion of Eq. (25.86) is not suitable for practical problems. We need to introduce the incident field $F_{\mu\nu}^{\text{in}}$. From Eqs. (25.74), (25.75), and (25.77)

$$f_{\mu\nu} = F_{\mu\nu}^{\text{in}} + \frac{1}{2}F_{\mu\nu}^{\text{rad}} = F_{\mu\nu}^{\text{in}} - \frac{2e}{3}(\ddot{u}_\mu u_\nu - u_\mu \ddot{u}_\nu) \tag{25.88}$$

and thus, using Eqs. (25.81) and (25.83), Eqs. (25.86) can be written

$$mc^2\dot{u}_\mu = eF_{\mu\nu}u_\nu + \frac{2e^2}{3}(\ddot{u}_\mu - u_\mu \dot{u}_\nu^2). \tag{25.89}$$

This equation is the same as that given by Eq. (25.66) for the Lorentz extended-electron model. While Eq. (25.66) was considered as only approximate, in that higher-order structure-dependent terms were dropped, Eq. (25.89) can be considered to hold exactly within the limits of validity of classical electrodynamics.

The four-vector $-B_\mu$ can be interpreted as the momentum–energy four-vector for the momentum and energy residing within the tube at any value of the proper time s. From Eq. (25.85) and the fact that k does to infinity as ϵ^{-1} as $\epsilon \to 0$ we see that the energy within the tube must be negative and must tend to $-\infty$ as ϵ tends to zero. This negative energy compensates for the large positive Coulomb field just outside the tube. A model would thus be an infinite negative mass at the point electron that compensates for the infinite mass due to the Coulomb field. We need not be too disturbed by the strangeness of this model, since what is needed for a point electron is not so much a physical model as a consistent set of equations for predicting the motion of an electron in a given field. This we have achieved.

In this same year (1938) that Dirac put forward his point electron, Pryce* showed how one can modify the expression for the energy–momentum tensor so as to obtain a finite expression for the energy of the Coulomb field. He takes for his energy–momentum tensor

$$T'_{\alpha\beta} = T_{\alpha\beta} - \frac{\partial}{\partial x_\gamma} K_{\alpha\beta\gamma}, \tag{25.90}$$

where $T_{\alpha\beta}$ is the usual tensor of Eq. (17.35) and $K_{\alpha\beta\gamma}$ is a third-order tensor that is antisymmetric in the last two indices. It follows at once from the latter property and the vanishing of the divergence of $T_{\alpha\beta}$ that

$$\frac{\partial T'_{\alpha\beta}}{\partial x_\beta} = \frac{\partial T_{\alpha\beta}}{\partial x_\beta} - \frac{\partial^2 K_{\alpha\beta\gamma}}{\partial x_\beta\, \partial x_\gamma} = 0$$

and thus that $T'_{\alpha\beta}$ has also a vanishing divergence. As seen in Section 17 this property leads to the continuity equation (17.46) for the flow of electromagnetic energy.

By taking a suitable expression for $K_{\alpha\beta\gamma}$ in terms of the four-velocity and four-acceleration of the particle, Pryce was able to keep finite the energy and momentum associated with the electromagnetic field of the particle. In this way he was able to do away with the infinite electromagnetic energy of a point electron.

Dirac's theory has a very disturbing feature, shared by Lorentz's theory, that most of the solutions of the differential equations of Eq. (25.89) are run-away solutions. The electron continues to accelerate as $t \to \infty$, even when no field is applied. This has led to the imposition of an asymptotic condition for $t \to \infty$ that selects out the reasonable solution. This asymptotic condition serves also to compensate for the fact that the differential equations in Eq. (25.89) are of third order and one can apparently give arbitrary initial values not only to position and velocity but also to the acceleration. With the application of the asymptotic condition, only the initial position and velocity are assignable.

On applying the asymptotic conditions, however, we are led to a short time (of order 6×10^{-24} sec) violation of causality. The electron responds to an electromagnetic field a short time before it is applied. This feature will be discussed more fully in the next section.

* M. H. L. Pryce, *Proc. Roy. Soc.* (*London*), **A168**, 389 (1938).

26. Nonrelativistic Harmonic Oscillator, Natural Breadth of Spectral Lines, and Scattering of Light

To illustrate the use of the equation of motion of an electron when the self-force is included we shall first treat an electron of charge e and mass m elastically bound to some molecule. Thus we shall suppose that the only external forces acting on the electron are an elastic restoring force $-k\mathbf{r}$, where k is a positive real constant, and the Lorentz force due to an incident plane electromagnetic wave.

The equation of motion, Eq. (25.66), we developed in the last section from the Lorentz extended-electron model is the same as Eq. (25.89), which was obtained from the Dirac point-electron theory. To make the problem somewhat easier to solve and to interpret we shall use the nonrelativistic form of these equations as expressed by Eq. (25.71). In addition we shall suppose that v/c is very small, so that $\mathbf{v} \times \mathscr{H}/c$ can be neglected. The equation of motion then becomes

$$m\left(\dot{\mathbf{v}} - \frac{a'}{c}\ddot{\mathbf{v}}\right) + k\mathbf{r} = e\mathscr{E}, \tag{26.1}$$

where

$$a' = \frac{2e^2}{3mc^2} \tag{26.2}$$

is $\frac{4}{3}$ the classical electron radius $a = e^2/2mc^2$.

This is a linear nonhomogeneous differential equation for $\mathbf{r}$ as a function of t. The solution of this equation will consist of a particular solution plus the general solution of the homogeneous equation obtained by putting $\mathscr{E} = 0$, i.e., the general solution of the equation

$$m\left(\dot{\mathbf{v}} - \frac{a}{c}\ddot{\mathbf{v}}\right) + k\mathbf{r} = 0. \tag{26.3}$$

To obtain a first approximation to the solution of Eq. (26.3) we drop the self-force term and write

$$m\ddot{\mathbf{r}} + k\mathbf{r} = 0. \tag{26.4}$$

The general solution of this equation is

$$\mathbf{r} = \mathbf{A}\cos\omega_0 t + \mathbf{B}\sin\omega_0 t, \tag{26.5}$$

where

$$\omega_0 = 2\pi\nu_0 = \sqrt{\frac{k}{m}} \tag{26.6}$$

is the so-called *natural angular frequency* of the oscillator. We shall retain the name, using Eq. (26.6) as the definition of ω_0, even when dealing with Eqs. (26.1) and (26.3).

The time

$$\tau_0 = \frac{a'}{c}$$

that it takes light to travel the distance a' is about 6.2×10^{-24} sec. Introducing τ_0 and ω_0, we can rewrite Eq. (26.3) in the form

$$\ddot{\mathbf{r}} - \tau_0 \dddot{\mathbf{r}} + \omega_0^2 \mathbf{r} = 0, \tag{26.7}$$

which, being a linear homogeneous differential equation with constant coefficients, has as its general solution

$$\mathbf{r} = \mathbf{a}_1 e^{\alpha_1 t} + \mathbf{a}_2 e^{\alpha_2 t} + \mathbf{a}_3 e^{\alpha_3 t}. \tag{26.8}$$

Here $\mathbf{a}_1$, $\mathbf{a}_2$, $\mathbf{a}_3$ are arbitrary vectors (with possibly complex components)* and α_1, α_2, α_3 are roots of the equation

$$\alpha^2 - \frac{a}{c}\alpha^3 + \omega_0^2 = 0, \tag{26.9}$$

obtained on substituting from Eq. (26.8) into Eq. (26.7). These roots are†

$$\begin{aligned}
\alpha_1 &= \frac{c}{3a}\left(1 + 2\cosh\frac{\varphi}{3}\right), \\
\alpha_2 &= \frac{c}{3a}\left(1 - \cosh\frac{\varphi}{3} + i\sqrt{3}\sinh\frac{\varphi}{3}\right), \\
\alpha_3 &= \frac{c}{3a}\left(1 - \cosh\frac{\varphi}{3} - i\sqrt{3}\sinh\frac{\varphi}{3}\right),
\end{aligned} \tag{26.10}$$

where φ is determined by the equation

$$\cosh\varphi = 1 + \tfrac{27}{2}(\omega_0\tau_0)^2. \tag{26.11}$$

* The only requirement is that r have real components.

† See E. P. Adams and R. L. Hippisley, *Smithsonian Mathematical Formulae and Tables of Elliptic Functions*, Smithsonian Institution, Washington, D.C., 1947, pp. 9, 10.

For a resonance frequency in the optical region, $\omega_0 \simeq 3 \times 10^{15}$ rad/sec and thus $\omega_0\tau_0 \simeq 2 \times 10^{-9}$. We conclude therefore that cosh φ is almost equal to 1 and that consequently φ is very small. This means that to a very close approximation

$$\cosh \varphi = 1 + \tfrac{1}{2}\varphi^2. \tag{26.12}$$

Comparison of Eqs. (26.11) and (26.12) shows that

$$\varphi = 3\sqrt{3}\omega_0\tau_0 . \tag{26.13}$$

Moreover, by again using the fact that φ is very small we may write

$$\cosh \frac{\varphi}{3} \simeq 1 + \frac{1}{2}\left(\frac{\varphi}{3}\right)^2 = 1 + \frac{3}{2}\,\omega_0^2\tau_0^2,$$

$$\sinh \frac{\varphi}{3} \simeq \frac{\varphi}{3} = \sqrt{3}\,\omega_0\tau_0,$$

and thus from Eq. (26.10)

$$\begin{aligned} \alpha_1 &= \frac{1}{\tau_0}(1 + \omega_0^2\tau_0^2) \simeq \frac{1}{\tau_0}, \\ \alpha_2 &= i\omega_0 - \tfrac{1}{2}\gamma, \\ \alpha_3 &= -i\omega_0 - \tfrac{1}{2}\gamma, \end{aligned} \tag{26.14}$$

where

$$\gamma = \tau_0\omega_0^2 \ll \omega_0 . \tag{26.15}$$

With these values of the α's Eq. (26.8) becomes

$$\mathbf{r} = \mathbf{a}_1 e^{t/\tau_0} + e^{-\gamma t/2}(\mathbf{A} \cos \omega_0 t + \mathbf{B} \sin \omega_0 t), \tag{26.16}$$

where, since $\mathbf{r}$ is a real vector, we must require that the otherwise arbitrary vectors $\mathbf{a}_1$, $\mathbf{A}$, and $\mathbf{B}$ be real.

While the periodic terms in Eq. (26.16) oscillate with slowly decreasing amplitude, the first term increases rapidly as the time increases. Thus, unless $\mathbf{a}_1 = 0$, at large values of t the position vector $\mathbf{r}$ becomes

$$\mathbf{r} \simeq \mathbf{a}_1 e^{t/\tau_0},$$

and thus

$$\mathbf{v} = \frac{\mathbf{r}}{\tau_0},$$

$$\dot{\mathbf{v}} = \frac{\mathbf{r}}{\tau_0^2}.$$

As $t \to \infty$ we then have a "runaway" solution with the magnitudes of $\mathbf{r}$, $\mathbf{v}$, and $\dot{\mathbf{v}}$ becoming infinite.

To eliminate such "runaway" solutions and to obtain a satisfactory theory we augment our theory by imposing an *asymptotic condition** relating to the behavior of a solution as $t \to \infty$. For the nonrelativistic equation of motion given by Eq. (26.3) we can take as our asymptotic condition the requirement that the acceleration remains finite as $t \to \infty$.

As seen above, this requires in our case that $\mathbf{a}_1 = 0$ in Eq. (26.16) and hence that

$$\mathbf{r} = e^{-\gamma t/2}(\mathbf{A} \cos \omega_0 t + \mathbf{B} \sin \omega_0 t). \tag{26.17}$$

This solution differs from that obtained in Eq. (26.5), when the self-reaction term is omitted, by just the exponential decay factor $e^{-\gamma t/2}$.

The self-reaction term $-am\ddot{\mathbf{v}}/c$ was introduced into the equation of motion (26.1) (see Section 25) as a small correction term approximating for small acceleration the reaction of the field of a particle on the particle itself. The addition of this term, however, changes the differential equation from one of second order to one of third order, and therefore introduces an additional arbitrary constant vector into the general solution. One can therefore, apparently, specify as initial conditions not only the position and velocity but also the initial acceleration; however, the asymptotic condition that the acceleration remains finite, for finite applied forces, as $t \to \infty$ effectively specifies the initial acceleration in terms of the other two. We have, therefore, the same arbitrariness in the initial conditions as in the electrodynamics of test particles. The application of the asymptotic condition will generally assure, as in this case, that the inequalities (25.30), assumed in deriving Eq. (25.66), are met.

Radiation from an oscillating electron. Suppose the above charged particle, say an electron, with restoring force $-k\mathbf{r}$ is set in motion by having a constant electric field $\mathscr{E}_0$, that existed prior to $t = 0$, turned off at $t = 0$. The electron's position vector will thus satisfy Eq. (26.1) with

$$\mathscr{E} = \begin{cases} \mathscr{E}_0, & t < 0 \\ 0, & t \geq 0. \end{cases} \tag{26.18}$$

* Haag has pointed out the importance of such asymptotic conditions in the axioms of any field theory (see R. Haag, *Zeit. Naturf.*, **10a**, 752 (1955)). An instructive discussion of this and the other problems we are discussing is given in F. Rohrlich, *Classical Charged Particles*, Addison-Wesley, Reading, Mass., 1965, Chaps. 2 and 6.

Although we shall find it necessary to modify this assumption somewhat later on, we shall take the electron position to be fixed at $\mathbf{r} = \mathbf{r}_0$ prior to $t = 0$. Equation (26.1) reduces to $k\mathbf{r}_0 = e\boldsymbol{\mathscr{E}}_0$ during this interval and thus $\mathbf{r}_0 = (e/k)\boldsymbol{\mathscr{E}}_0$.

After $t = 0$ the differential equation reduces to Eq. (26.3) and will have a solution of the form of Eq. (26.15). In order to avoid the approximations made in evaluating the constants of that equation, however, we shall write that

$$\mathbf{r} = \mathbf{a}_1 e^{\beta t} + e^{-\gamma t/2}(\mathbf{A}\cos\omega t + \mathbf{B}\sin\omega t), \tag{26.19}$$

where β, γ, and ω are given approximately by Eqs. (26.15) and (26.16), while $\mathbf{a}_1$, $\mathbf{A}$, and $\mathbf{B}$ are determined by the boundary conditions. We shall for the moment ignore the asymptotic condition that would require that $\mathbf{a}_1 = 0$.

Since the inhomogeneous term in Eq. (26.1) is now discontinuous at $t = 0$, the highest derivative $\ddot{\mathbf{v}}$ in that equation must be discontinuous; however, $\dot{\mathbf{v}}$, $\mathbf{v}$, and $\mathbf{r}$, which may be obtained from $\ddot{\mathbf{v}}$ by successive integration, must be continuous. As initial conditions for the interval $t \geq 0$, therefore, we have (at $t = 0$)

$$\mathbf{r} = \mathbf{r}_0, \qquad \mathbf{v} = 0, \qquad \dot{\mathbf{v}} = 0. \tag{26.20}$$

Applying these three conditions to the solution given by Eq. (26.19), we obtain the following simultaneous equations:

$$\begin{aligned} \mathbf{a}_1 + \mathbf{A} &= \mathbf{r}_0, \\ \beta\mathbf{a}_1 - \frac{\gamma}{2}\mathbf{A} + \omega\mathbf{B} &= 0, \\ \beta^2\mathbf{a}_1 + \left(\frac{\gamma^2}{4} - \omega^2\right)\mathbf{A} - \gamma\omega\mathbf{B} &= 0. \end{aligned} \tag{26.21}$$

Solving these, we find that

$$\begin{aligned} \mathbf{a}_1 &= \frac{\tfrac{1}{4}\gamma^2 + \omega^2}{(\tfrac{1}{4}\gamma^2 + \omega^2) + \beta(\gamma + \beta)}\mathbf{r}_0 \simeq (\omega_0\tau_0)^2\mathbf{r}_0, \\ \mathbf{A} &= \frac{\beta(\gamma + \beta)}{(\tfrac{1}{4}\gamma^2 + \omega^2) + \beta(\gamma + \beta)}\mathbf{r}_0 \simeq \mathbf{r}_0, \\ \mathbf{B} &= \frac{\tfrac{1}{4}\gamma^2 - \omega^2 + \tfrac{1}{2}\beta\gamma}{(\tfrac{1}{4}\gamma^2 + \omega^2) + \beta(\gamma + \beta)}\frac{\beta\mathbf{r}_0}{\omega} \simeq -\frac{1}{2}\omega_0\tau_0\mathbf{r}_0, \end{aligned} \tag{26.22}$$

where the approximations given are based on Eqs. (26.15) and (26.16) and the fact that $\omega_0\tau_0 \ll 1$. If one associates a wavelength $\lambda_0 = 2\pi c/\omega_0$ with the natural resonance frequency ω_0 of the oscillator, one may replace the condition $\omega_0\tau_0 \ll 1$ by the condition $\lambda_0/a' = 2\pi/\omega_0\tau_0 \gg 1$. In other words, the natural resonance wavelength λ_0 must be very large compared to the classical electron radius a.

To meet the asymptotic condition that the acceleration be finite at infinity without spoiling the continuity conditions imposed at $t = 0$ we simply subtract $\mathbf{a}_1 e^{\beta t}$ $(\beta > 0)$ from the above solution. We are thus left with the solution

$$\mathbf{r} = \begin{cases} \mathbf{r}_0 - \mathbf{a}_1 e^{\beta t}, & t < 0 \\ e^{-\gamma t/2}[\mathbf{A}\cos\omega t + \mathbf{B}\sin\omega t], & t \geq 0. \end{cases} \tag{26.23}$$

This is the required solution, the $\mathbf{a}_1$, $\mathbf{A}$, and $\mathbf{B}$ being given by Eqs. (26.22).

We observe, however, a very strange aspect of this solution due to the presence of the term $-\mathbf{a}_1 e^{\beta t}$ for $t < 0$. The electron's behavior anticipates the change in field that is still to take place at $t = 0$. Our solution thus appears to contradict the principle of causality. This feature has been discussed at length in the literature* and while strange appears to be acceptable. As pointed out in the last section, the period $\tau_0 = a'/c$ of appreciable anticipation is extremely short, of the order of 10^{-23} sec. For the extended-electron model this period of significant anticipation corresponds roughly to the time required for a change in the field to travel from the edge to the center of the extended electron.

Since $\omega_0\tau_0 \ll 1$, we see from Eqs. (26.22) that $\mathbf{a}_1$ and $\mathbf{B}$ are negligibly large compared to $\mathbf{A}$; moreover, from Eq. (26.14), since $\gamma \ll \omega_0$, $\omega \simeq \omega_0$. To a very good approximation, therefore,

$$\mathbf{r} = \begin{cases} \mathbf{r}_0, & t < 0 \\ \mathbf{r}_0 e^{-\gamma t/2}\cos\omega_0 t, & t \geq 0, \end{cases} \tag{26.24}$$

and the velocity is

$$\mathbf{v} \simeq \begin{cases} 0, & t < 0 \\ -\omega_0 \mathbf{r}_0 e^{-\gamma t/2}\sin\omega_0 t, & t \geq 0. \end{cases} \tag{26.25}$$

We may express $\mathbf{v}$ as a Fourier integral by writing

$$\mathbf{v}(t) = \frac{1}{2\pi}\int_{-\infty}^{\infty} \mathbf{b}(\omega)e^{i\omega t}\,d\omega, \tag{26.26}$$

* See F. Rohrlich, *op. cit.*, pp. 24, 238, 263. Incidentally he points out that, strictly speaking, the asymptotic conditions include specification of the motion for $t \to -\infty$, as well as for $t \to \infty$. See also R. Haag, *Zeit. Naturf.*, **10a**, 752 (1955).

where

$$\mathbf{b}(\omega) = \int_{-\infty}^{\infty} \mathbf{v}(t) e^{-i\omega t}\, dt.$$

Using Eq. (26.25), we have

$$\mathbf{b}(\omega) \simeq -\omega_0 \mathbf{r}_0 \int_0^{\infty} e^{-(\frac{1}{2}\gamma + i\omega)t} \sin \omega_0 t\, dt$$

$$\simeq \frac{-\omega_0^2 \mathbf{r}_0}{[\frac{1}{2}\gamma + i(\omega - \omega_0)][\frac{1}{2}\gamma + i(\omega + \omega_0)]}; \tag{26.27}$$

therefore, the second time derivative of the electric dipole moment is

$$\ddot{\not{p}} = e\ddot{\mathbf{r}} = \frac{e}{2\pi} \frac{d}{dt} \int_{-\infty}^{\infty} \mathbf{b}(\omega) e^{i\omega t}\, d\omega = \frac{ei}{2\pi} \int_{-\infty}^{\infty} \omega \mathbf{b}(\omega) e^{i\omega t}\, d\omega. \tag{26.28}$$

We may now make use of the result obtained in Example 3, Section 20. Equation (26.28) can be made to agree with Eq. (20.52) by requiring that

$$i\omega \mathbf{b}(\omega) = -(2\pi\nu)^2 \mathbf{a}(\nu)$$

or, since $\omega = 2\pi\nu$,

$$\mathbf{a}(\nu) = \frac{-i}{\omega} \mathbf{b}(\omega).$$

Thus, by Eqs. (20.56) and (26.27) the total radiation per unit frequency range is

$$\frac{4(2\pi\nu)^4 e^2}{3c^3} \mathbf{a}(\nu) \cdot \mathbf{a}(-\nu) = \frac{4e^2}{3c^3} [-i\omega \mathbf{b}(\omega)] \cdot [i\omega \mathbf{b}(-\omega)]$$

$$= \frac{4e^2\omega^2\omega_0^4 r_0^2}{3c^3[\frac{1}{4}\gamma^2 + (\omega - \omega_0)^2][\frac{1}{4}\gamma^2 + (\omega + \omega_0)^2]}, \tag{26.29}$$

where, as in Example 3, the frequency ω refers only to the positive frequency range $0 \leq \omega < \infty$. Since $\frac{1}{4}\gamma^2 \ll \omega_0^2 \leq (\omega + \omega_0)^2$, the above distribution is very closely given by

$$f(\omega) \equiv \frac{4e^2\omega^2\omega_0^4 r_0^2}{3c^3(\omega + \omega_0)^2[\frac{1}{4}\gamma^2 + (\omega - \omega_0)^2]}. \tag{26.30}$$

The function $f(\omega)$ has a sharp maximum near $\omega = \omega_0$. In the neighborhood of $\omega = \omega_0$ we can put $\omega = \omega_0$, except in the term $(\omega - \omega_0)^2$, so that in the region of the maximum

$$f(\omega) \simeq \frac{e^2\omega_0^4 r_0^2}{3c^3[\frac{1}{4}\gamma^2 + (\omega - \omega_0)^2]}. \tag{26.31}$$

The maximum value of $f(\omega)$ is very nearly its value at $\omega = \omega_0$, namely,

$$f(\omega_0) = \frac{e^2\omega_0^4 r_0^2}{3c^3(\frac{1}{4}\gamma^2)} \ .$$

It falls off rapidly from this peak value and by Eq. (26.31) has only one-half this value at $\omega = \omega_0 \pm \gamma/2$. Thus γ represents the range of frequencies

$$\Delta\omega = \gamma \tag{26.32}$$

for which $f(\omega)$ exceeds half its maximum value, and therefore γ represents the *spectral width* in the frequency domain.

The change $\Delta\lambda$ in the wavelength corresponding to $\Delta\omega$ is

$$\Delta\lambda = \Delta\left(\frac{2\pi c}{\omega}\right) \simeq -\frac{2\pi c}{\omega^2}\Delta\omega,$$

or, by the use of Eqs. (26.16) and (26.32),

$$|\Delta\lambda| = \frac{2\pi c}{\omega_0^2}\gamma = 2\pi a', \tag{26.33}$$

where $a' = \frac{4}{3}a$ and a is the classical electron radius. The quantity $|\Delta\lambda|$ represents the width of the spectral line from our oscillating electron due to the reaction of the radiated field back on the electron.

This width $|\Delta\lambda|$ is seen to be required by the necessary damping of the electron's oscillatory motion as it radiates energy. It is referred to as the *natural breadth* of a spectral line. The line width given by Eq. (26.33) is very small compared to the line widths usually observed in the laboratory; thus for an electron

$$e = 4.803 \times 10^{-10} \quad \text{esu}, \tag{26.34}$$

$$m = 9.109 \times 10^{-28} \quad \text{gm}, \tag{26.35}$$

$$c = 2.9979 \times 10^{10} \quad \text{cm/sec}, \tag{26.36}$$

$$a' = \frac{2}{3}\frac{e^2}{mc^2} = 1.878 \times 10^{-13} \quad \text{cm}; \tag{26.37}$$

so that

$$2\pi a' = 1.180 \times 10^{-12}\,\text{cm} \simeq 0.0001\ \text{Å}.$$

The spectral widths of lines usually observed is very much greater than this on account of two main causes. One of these is the frequent interruption of the oscillation of an atom due to its collisions with other atoms. This gives rise to what is commonly referred to as *pressure broadening*.

The second is the Doppler shifts in the frequencies of the light coming from the individual atoms, due to their thermal motion. This effect is referred to as *Doppler broadening*.

Scattering. Returning now to Eq. (26.1), we may assume that the electron has incident upon it an electromagnetic wave with an electric vector given by

$$\mathscr{E} = \mathscr{E}_0 \cos(\boldsymbol{k} \cdot \mathbf{r} - ckt). \tag{26.38}$$

To obtain the motion of the electron we need to solve Eq. (26.1), which can be written

$$\ddot{\mathbf{r}} - \tau_0 \dddot{\mathbf{r}} + \omega_0^2 \mathbf{r} = \frac{e\mathscr{E}_0}{m} \cos(\boldsymbol{k} \cdot \mathbf{r} - ckt). \tag{26.39}$$

We shall suppose, however, that the displacement $\mathbf{r}$ of the electron from its equilibrium position at the origin satisfies at all times the inequality

$$\mathbf{k} \cdot \mathbf{r} \ll 1. \tag{26.40}$$

In this case, since $ck = 2\pi c/\lambda = \omega$, Eq. (26.39) becomes

$$\ddot{\mathbf{r}} - \tau_0 \dddot{\mathbf{r}} + \omega_0^2 \mathbf{r} = \frac{e\mathscr{E}_0}{m} \cos \omega t. \tag{26.41}$$

To find a particular solution of this equation, we let

$$\mathbf{r} = \mathbf{r}_0 \cos(\omega t - \delta) \tag{26.42}$$

in Eq. (26.41) and obtain the equation

$$(\omega_0^2 - \omega^2)\mathbf{r}_0 \cos(\omega t - \delta) - \tau_0 \omega^3 \mathbf{r}_0 \sin(\omega t - \delta) = \frac{e\mathscr{E}_0}{m} \cos \omega t. \tag{26.43}$$

Since this equality is assumed to hold for all t, we can require that the two sides of Eq. (26.43) and their time derivatives be equal at $t = 0$. We thereby obtain the equations

$$\begin{gathered} [(\omega_0^2 - \omega^2)\cos \delta + \tau_0 \omega^3 \sin \delta]\mathbf{r}_0 = \frac{e\mathscr{E}_0}{m}, \\ \tau_0 \omega^3 \cos \delta - (\omega_0^2 - \omega^2)\sin \delta = 0. \end{gathered} \tag{26.44}$$

The solution of these equations is

$$\mathbf{r}_0 = \frac{e\mathscr{E}_0/m}{[(\omega_0^2 - \omega^2)^2 + \tau_0^2 \omega^6]^{1/2}} \tag{26.45}$$

and

$$\tan \delta = \frac{\tau_0 \omega^3}{\omega_0^2 - \omega^2}. \tag{26.46}$$

The maximum value of $\boldsymbol{k} \cdot \mathbf{r}$ is equal to $\omega (r_0)_{max}/c$ and the maximum value of r_0 occurs for $\omega = \omega_0$. By Eq. (26.45) $(r_0)_{max} = e\mathscr{E}_0/m\tau_0\omega^3$ and thus

$$(\boldsymbol{k} \cdot \mathbf{r})_{max} = \frac{e\mathscr{E}_0}{cm\tau_0\omega_0^2} = \frac{e\mathscr{E}_0}{ma'\omega_0^2}. \tag{26.47}$$

The condition in Eq. (26.40) now becomes

$$\frac{e\mathscr{E}_0}{ma'\omega_0^2} \ll 1. \tag{26.48}$$

Remembering that $a' = 2e^2/3mc^2$ and putting $2\pi c/\omega_0 = \lambda_0$, we may rewrite this inequality in the form

$$\mathscr{E}_0 \ll \frac{8\pi^2 e}{3\lambda_0^2}. \tag{26.49}$$

This shows that the maximum field strength must be small in comparison with the field due to the particle at a distance equal to $\lambda_0/2$, where λ_0 is the wavelength at resonance. This condition, with e taken as the charge on the electron, will be fulfilled for most visible light sources.

The general solution of Eq. (26.41) is

$$\mathbf{r} = e^{-\gamma t/2}(\mathbf{A} \cos \omega_0 t + \mathbf{B} \sin \omega_0 t) + \mathbf{r}_0 \cos(\omega t - \delta), \tag{26.50}$$

where the first part is the general solution of the homogeneous equation. The first part represents the so-called *natural* vibration, and the last part, the *forced* vibration of the particle. The factor $e^{-\gamma t/2}$ in the first part makes this part approach zero for large t. This part of the solution is therefore also called the *transient* solution. After this part becomes negligible, the motion is represented by Eq. (26.42), with $\mathbf{r}_0$ and δ given by Eqs. (26.45) and (26.46).

The radiation per unit of time, after the establishment of the permanent regime, can be obtained by making use of the results found in Eq. (20.46); it is, by Eqs. (26.39) and (25.68),

$$I = \frac{e^2\omega^4 r_0^2}{3c^3} = \frac{e^2\omega^4\mathscr{E}_0^2}{3m^2c^3[(\omega_0^2 - \omega^2)^2 + \tau_0^2\omega^6]}$$

$$= \frac{3c^3\tau_0^2\omega^4\mathscr{E}_0^2}{4[(\omega_0^2 - \omega^2)^2 + \tau_0^2\omega^6]}. \tag{26.51}$$

The intensity of the incident electromagnetic wave, i.e., the energy falling per unit time on a unit area perpendicular to the direction of propagation, is equal to

$$\frac{c}{4\pi}|\mathscr{E} \times \mathscr{H}| = \frac{c\mathscr{E}_0^2}{4\pi}\cos^2(\boldsymbol{k}\cdot\mathbf{r} - ckt),$$

because $\mathscr{E}$ and $\mathscr{H}$ are equal in magnitude and phase and are mutually perpendicular. On the average, since the average of cosine squared is $\frac{1}{2}$, the energy per unit area per unit time is

$$I_0 = \frac{c\mathscr{E}_0^2}{8\pi}. \tag{26.52}$$

Thus from Eq. (26.51)

$$I = \frac{6\pi c^2\tau_0^2\omega^4 I_0}{(\omega_0^2 - \omega^2)^2 + \tau_0^2\omega^6} = \sigma I_0, \tag{26.53}$$

where

$$\sigma = \frac{6\pi c^2\tau_0^2\omega^4}{(\omega_0^2 - \omega^2)^2 + \tau_0^2\omega^6} \tag{26.54}$$

is the factor by which the incident intensity, the radiation energy per unit area per unit time, must be multiplied to obtain the *total* energy per unit time radiated by the particle. The factor σ must thus be dimensionally an area. The energy radiated by the particle, since its motion is periodic, must be derived from the incident radiation. It is customary, therefore, to speak of the radiation by the particle as the electromagnetic radiation *scattered* by it.

One can introduce a naïve mechanical picture of the particle's being merely an obstruction in the path of the electromagnetic wave, scattering whatever energy falls on it. On this picture σ is the cross-sectional area of this obstruction. Accordingly, on the basis of this picture one often speaks of σ as the *effective cross section for scattering*, or simply the *scattering cross section*.

In the neighborhood of $\omega = \omega_0$, σ becomes large, and the phenomenon relating to this large scattering is called *resonance fluorescence*. In this region we can then set

$$\omega_0^2 - \omega^2 = (\omega_0 + \omega)(\omega_0 - \omega) \simeq 2\omega_0(\omega_0 - \omega);$$

thus from Eq. (26.54)

$$\sigma = \frac{6\pi c^2\tau_0^2\omega_0^2}{4(\omega_0 - \omega)^2 + \tau_0^2\omega_0^4}. \tag{26.55}$$

For $\omega = \omega_0$, which is very near the resonance peak, this equation gives

$$\sigma = \frac{6\pi c^2}{\omega_0^2} = \frac{3}{2\pi}\lambda_0^2, \tag{26.56}$$

where, as before, $\lambda_0 = 2\pi c/\omega_0$.

27. Approximate Hamiltonian for a System of Interacting Particles

In electrostatics we were able to eliminate the fields of the particles from the Lagrangian of the system, obtaining according to Eq. (8.14)

$$L = \sum_i \frac{1}{2} m_i v_i^2 - \frac{1}{2}\sum_{j,k}{}' \frac{e_j e_k}{R_{jk}}, \tag{27.1}$$

where $R_{jk} = |\mathbf{r}_j - \mathbf{r}_k|$ is the distance between the jth and the kth particles. From this it follows at once, using Eq. (2.15), that

$$H = \sum_i \frac{p_i^2}{2m_i} + \frac{1}{2}\sum_{j,k}{}' \frac{e_j e_k}{R_{jk}}. \tag{27.2}$$

The prime on the summation sign indicates that terms for which $j = k$ are excluded. If an external field is present, the Hamiltonian becomes

$$H = \sum_i \left[\frac{p_i^2}{2m_i} + e_i\varphi(\mathbf{r}_i)\right] + \frac{1}{2}\sum_{j,k}{}' \frac{e_j e_k}{R_{jk}}, \tag{27.3}$$

where $\varphi(\mathbf{r})$ is the potential corresponding to the external field.

In advanced applications of electrodynamics, especially in quantum mechanics, it is often desirable to have expressions analogous to Eqs. (27.2) and (27.3), in which, however, relativistic effects are not completely neglected. We know that the fields expressing the interaction between the particles cannot be eliminated in an exact relativistic theory. It is possible, however, to obtain approximate expressions in which these fields are absent. To obtain these is the object of this section.*

* This was first done by C. G. Darwin, *Phil. Mag.*, **39**, 537 (1920).

We start with Eqs. (19.48) and (19.49), which for a system of particles become ($\mathbf{u}$ being replaced by $\mathbf{R}/R$)

$$\mathscr{E}(\mathbf{r}, t) = \frac{1}{2\pi}\sum_j e_j \iint \left(\frac{\mathbf{R}_j}{R_j^3} + \frac{\mathbf{R}_j}{R_j^2} - \frac{\mathbf{v}_j}{cR_j}\frac{1}{c}\frac{\partial}{\partial t}\right) e^{i\mu(t-\tau-R_j/c)}\, d\mu\, d\tau \quad (27.4)$$

and

$$\mathscr{H}(\mathbf{r}, t) = \frac{1}{2\pi c}\sum_j e_j \iint \frac{\mathbf{v}_j \times \mathbf{R}_j}{R_j^2}\left(\frac{1}{R_j} + \frac{1}{c}\frac{\partial}{\partial t}\right) e^{i\mu(t-\tau-R_j/c)}\, d\mu\, d\tau. \quad (27.5)$$

Here $\mathbf{R}_j = \mathbf{r} - \mathbf{r}_j$ and the summation is over all particles of the system. We may next proceed as we did in deriving Eq. (25.21) from Eq. (25.11). Instead of Eq. (25.14), we now obtain the equations

$$\mathscr{E}(\mathbf{r}, t) = \sum_j e_j \sum_{n=0}^{\infty} \frac{1}{n!}\left(-\frac{1}{c}\right)^n \left[\frac{d^n}{dt^n}(R_j^{n-3}\mathbf{R}_j) + \frac{1}{c}\frac{d^{n+1}}{dt^{n+1}}\left(R_j^{n-2}\mathbf{R}_j - \frac{R_j^{n-1}\mathbf{v}_j}{c}\right)\right] \quad (27.6)$$

and

$$\mathscr{H}(\mathbf{r}, t) = \sum_j \frac{e_j}{c} \sum_{n=0}^{\infty} \frac{1}{n!}\left(-\frac{1}{c}\right)^n \left[\frac{d^n}{dt^n}(R_j^{n-3}\mathbf{v}_j \times \mathbf{R}_j) + \frac{1}{c}\frac{d^{n+1}}{dt^{n+1}}(R_j^{n-2}\mathbf{v}_j \times \mathbf{R}_j)\right]. \quad (27.7)$$

For the purpose of obtaining the desired approximation, we shall make use of the following assumptions:

1. The velocities of the particles v_j are small compared to c; thus $v_j \ll c$. We can therefore write

$$1 \gg \frac{v_j}{c} \gg \left(\frac{v_j}{c}\right)^2 \gg \left(\frac{v_j}{c}\right)^3 \gg \cdots \quad (27.8)$$

and speak of these quantities as being of successively smaller *order of magnitude*. In the approximation to be applied, we shall drop all terms containing v_j/c to the third or higher power.

2. The vectors $\mathbf{R}_j = \mathbf{r} - \mathbf{r}_j$ do not change much in the time R_j/c that it takes a disturbance at $\mathbf{r}_j$ to be felt at $\mathbf{r}$. If we expand $\mathbf{R}_j(t - R_j/c)$ in a Taylor's series

$$\begin{aligned}\mathbf{R}_j\left(t - \frac{R_j}{c}\right) &= \mathbf{r} - \mathbf{r}_j\left(t - \frac{R_j}{c}\right) \\ &= \mathbf{r} - \left[\mathbf{r}_j(t) - \left(\frac{R_j}{c}\right)\mathbf{v}_j + \frac{1}{2}\left(\frac{R_j}{c}\right)^2\dot{\mathbf{v}}_j - \frac{1}{3!}\left(\frac{R_j}{c}\right)^3\ddot{\mathbf{v}}_j + \cdots\right] \\ &= \mathbf{R}_j(t) - \sum_{n=1}^{\infty}\frac{(-1)^n}{n!}\left(\frac{R_j}{c}\right)^n \mathbf{v}_j^{(n-1)},\end{aligned} \quad (27.9)$$

where $\mathbf{v}_j^{(k)}$ is the kth time derivative of $\mathbf{v}_j$, we assume we can drop all terms beyond some specified value of n.

To make it easier to explain the approximations that are made we shall refer to $(R_j/c)^{n-1}v_j^{(n-1)}/c$ as being "of the same *formal* order of magnitude as $(v_j/c)^n$." By this we mean that by reason of its form it can be expected to be, on the average, of the same order of magnitude as $(v_j/c)^n$. The retaining or dropping of terms to obtain our desired approximation will be based on this formal order-of-magnitude concept and thus will not always be justified in a strict sense.

3. The self-force terms are small compared to the inertial terms. Thus from Eq. (25.71) one concludes that

$$m\dot{v} \simeq e\mathscr{E} \simeq \frac{e^2}{R^2},$$

where R is sort of an average distance between the particles and $\simeq$ signifies an equality of the formal orders of magnitude. Multiplying both sides of this equation by R/mc^2, we have

$$\frac{R}{c^2}\dot{v} \simeq \frac{e^2}{mc^2R} \simeq \frac{a'}{R}.$$

The left-hand quantity by Assumption 2 has a formal order of v^2/c^2 and thus

$$\frac{a'}{R} \simeq \frac{v^2}{c^2} \ll 1. \qquad (27.10)$$

Multiplying the left-hand side of Eq. (25.71), namely, $m[\dot{v} - (a'/c)\ddot{v}]$, by R/mc^2, we have

$$\frac{R}{c^2}\dot{v} - \frac{a'}{R}\frac{R^2}{c^3}\ddot{v}.$$

The first term, corresponding to the original inertial term, has the same formal order of magnitude as v^2/c^2. By virtue of Eq. (27.10) the second term, corresponding to the self-force, has the formal order of magnitude of v^5/c^5. The self-force terms are thus down by three formal orders of magnitude from the inertial terms.

We wish to neglect the self-force of the electrons on themselves in order to have the equations of motion derivable from a Lagrangian containing only $\mathbf{r}_j$ and $\mathbf{v}_j$; i.e., the position coordinates and the velocities. Since the self-force terms involve $\ddot{\mathbf{v}}_j$, the corresponding Lagrangian would have to involve $\dot{\mathbf{v}}_j$ and this would require a fundamental change in the way the Lagrangian is customarily used (see Section 2).

In order to drop the self-force terms in the equations of motion we must be willing to approximate by retaining only terms that are not more than two formal orders of magnitude smaller than the dominant term. This means that we retain only terms that contain a multiplying factor of 1, v/c, or v^2/c^2 relative to the dominant term. We shall for simplicity say that our calculations are then correct to terms of the order of v^2/c^2. We can thus drop the self-force if we are content to let the Lagrangian, and hence also the force, the momentum, and the Hamiltonian be correct only to the order of $(v/c)^2$. Assumption 2 then says that $\mathbf{R}_j(t - R_j/c)$ is sufficiently well given by the truncated Taylor expansion

$$\mathbf{R}_j\left(t - \frac{R_j}{c}\right) \cong \mathbf{R}_j(t) + \frac{R_j}{c}\mathbf{v}_j - \frac{1}{2}\left(\frac{R_j}{c}\right)^2\dot{\mathbf{v}}_j.$$

This is certainly a reasonable assumption provided the R_j are required to be sufficiently small, and it supports our use of the idea of a formal order of magnitude in making our approximations.

The Lorentz force on the kth particle is

$$\mathbf{F}_k = e_k\left[\mathscr{E}(\mathbf{r}_k, t) + \frac{\mathbf{v}_k}{c} \times \mathscr{H}(\mathbf{r}_k, t)\right], \tag{27.11}$$

and we wish to calculate it to the order of v_j^2/c^2 in the velocities of the particles; it will be necessary, therefore, to calculate $\mathscr{E}(\mathbf{r}, t)$ to this same order, and $\mathscr{H}(\mathbf{r}, t)$ to the order of v_j/c.

It was shown in Section 25 that

$$\frac{d\mathbf{R}}{dt} = -\mathbf{v} \qquad \text{and} \qquad \frac{dR}{dt} = \frac{-\mathbf{R}\cdot\mathbf{v}}{R}, \tag{27.12a}$$

and the same, of course, applies to $\mathbf{R}_j$. The successive derivatives required in Eq. (27.6) are then

$$\frac{d^0}{dt^0}(R_j^{-3}\mathbf{R}_j) = R_j^{-3}\mathbf{R}_j,$$

$$\frac{1}{c}\frac{d}{dt}(R_j^{-2}\mathbf{R}_j) = \frac{1}{c}[2R_j^{-4}(\mathbf{R}_j\cdot\mathbf{v}_j)\mathbf{R}_j - R_j^{-2}\mathbf{v}_j], \tag{27.12b}$$

$$\frac{1}{c^2}\frac{d^2}{dt^2}(R_j^{-1}\mathbf{R}_j) = \frac{1}{c^2}[3R_j^{-5}(\mathbf{R}_j\cdot\mathbf{v}_j)^2\mathbf{R}_j - R_j^{-3}v_j^2\mathbf{R}_j + R_j^{-3}(\mathbf{R}_j\cdot\dot{\mathbf{v}}_j)\mathbf{R}_j - 2R_j^{-3}(\mathbf{R}_j\cdot\mathbf{v}_j)\mathbf{v}_j - R_j^{-1}\dot{\mathbf{v}}_j], \tag{27.12c}$$

$$\frac{1}{c^3}\frac{d^3}{dt^3}(\mathbf{R}_j) = \frac{-1}{c^3}\ddot{\mathbf{v}}_j,$$

and

$$\frac{1}{c}\frac{d}{dt}\left(\frac{R_j^{-1}\mathbf{v}_j}{c}\right) = \frac{1}{c^2}\,[R_j^{-3}(\mathbf{R}_j\cdot\mathbf{v}_j)\mathbf{v}_j + R_j^{-1}\dot{\mathbf{v}}_j], \tag{27.12d}$$

$$\frac{1}{c^2}\frac{d^2}{dt^2}\left(\frac{v_j}{c}\right) = \frac{-1}{c^3}\,\ddot{\mathbf{v}}_j.$$

We conclude from these first few special cases that

$$\frac{1}{c^n}\frac{d^n}{dt^n}(R_j^{n-3}\mathbf{R}_j) \qquad \text{and} \qquad \frac{1}{c^{n+1}}\frac{d^{n-1}}{dt^{n-1}}\left(\frac{R_j^{n-3}\mathbf{v}_j}{c}\right)$$

have the formal order of magnitude of $R_j^{-2}(v_j/c)^n$. Since we are seeking to be correct only to terms of the order of v^2/c^2, we shall limit the summations over n in Eqs. (27.6) and (27.7) as shown below:

$$\mathscr{E}(\mathbf{r}, t) = \sum_j e_j\Bigg[\sum_{n=0}^{2}\frac{1}{n!}\left(-\frac{1}{c}\right)^n\frac{d^n}{dt^n}(R_j^{n-3}\mathbf{R}_j) - \sum_{n=0}^{1}\frac{1}{n!}\left(-\frac{1}{c}\right)^{n+1}\frac{d^{n+1}}{dt^{n+1}}(R_j^{n-2}\mathbf{R}_j) - \frac{1}{c^2}\frac{d}{dt}(R_j^{-1}\mathbf{v}_j)\Bigg] \tag{27.13}$$

and

$$\mathscr{H}(\mathbf{r}, t) = \sum_j \frac{e_j}{c}\frac{\mathbf{v}_j\times\mathbf{R}_j}{R_j^3}. \tag{27.14}$$

With the application of Eqs. (27.12c) and (27.12d), Eq. (27.13) simplifies to

$$\mathscr{E}(\mathbf{r}, t) = \sum_j e_j\left\{\frac{\mathbf{R}_j}{R_j^3} + \frac{\mathbf{R}_j}{2c^2}\left[\frac{v_j^2}{R_j^3} - \frac{3(\mathbf{v}_j\cdot\mathbf{R}_j)^2}{R_j^5} - \frac{\dot{\mathbf{v}}_j\cdot\mathbf{R}_j}{R_j^3}\right] - \frac{\dot{\mathbf{v}}_j}{2c^2R_j}\right\}. \tag{27.15}$$

It is clear that in substituting from Eqs. (27.14) and (27.15) into Eq. (27.11) $\mathbf{r}$ is replaced by $\mathbf{r}_k$ and thus $\mathbf{R}_j = \mathbf{r} - \mathbf{r}_j$ is replaced by

$$\mathbf{r}_{kj} = \mathbf{r}_k - \mathbf{r}_j. \tag{27.16}$$

Furthermore, since we have excluded the self-force, the terms with $j = k$ are to be omitted; thus

$$\mathbf{F}_k = \sum_{j\neq k}\frac{e_ke_j}{r_{kj}^3}\mathbf{r}_{kj} + \sum_{j\neq k}\frac{e_ke_j}{2c^2}\left[\frac{v_j^2}{r_{kj}^3} - \frac{3(\mathbf{v}_j\cdot\mathbf{r}_{kj})^2}{r_{kj}^5}\right]\mathbf{r}_{kj} + \sum_{j\neq k}\frac{e_ke_j}{c^2}\mathbf{v}_k\times\frac{(\mathbf{v}_j\times\mathbf{r}_{kj})}{r_{kj}^3} - \sum_{j\neq k}\frac{e_ke_j}{2c^2}\left(\frac{\dot{\mathbf{v}}_j\cdot\mathbf{r}_{kj}}{r_{kj}^3}\mathbf{r}_{kj} + \frac{\dot{\mathbf{v}}_j}{r_{kj}}\right). \tag{27.17}$$

On the other hand, to terms of the order of v_j^2/c^2

$$\frac{d}{dt}\frac{m_k\mathbf{v}_k}{\sqrt{1-v_k^2/c^2}} = m_k\left[\frac{\dot{\mathbf{v}}_k}{(1-v_k^2/c^2)^{1/2}} + \frac{\mathbf{v}_k\mathbf{v}_k\cdot\dot{\mathbf{v}}_k}{c^2(1-v_k^2/c^2)^{3/2}}\right]$$
$$\simeq m_k\left[\dot{\mathbf{v}}_k + \frac{1}{c^2}\left(\mathbf{v}_k\mathbf{v}_k\cdot\dot{\mathbf{v}}_k + \frac{1}{2}v_k^2\dot{\mathbf{v}}_k\right)\right], \tag{27.18}$$

and as we have established earlier $\mathbf{F}_{\text{self}}$ is negligible.

Our equations of motion are now

$$m_k\left[\dot{\mathbf{v}}_k + \frac{1}{c^2}\left(\mathbf{v}_k\mathbf{v}_k\cdot\dot{\mathbf{v}}_k + \frac{1}{2}v_k^2\dot{\mathbf{v}}_k\right)\right] = \mathbf{F}_k, \tag{27.19}$$

where $\mathbf{F}_k$ is given by Eq. (27.17). The fact that $\ddot{v}_j$ and higher derivatives do not occur, which fact is assured by our assumptions, suggests the possibility of deriving these equations from a Lagrangian containing only coordinates and velocities.

For this purpose we start with the general Lagrangian for a system of particles with their fields (see Eqs. (24.1), (24.2), (24.3), and (24.4)),

$$L = -\sum_k m_kc^2\sqrt{1-\frac{v_k^2}{c^2}} + \sum_k\left[\frac{e_k}{c}\mathbf{A}(\mathbf{r}_k, t)\cdot\mathbf{v}_k - e_k\,\varphi(\mathbf{r}_k, t)\right]$$
$$+\frac{1}{8\pi}\int(\mathscr{E}^2 - \mathscr{H}^2)\,dV \tag{27.20}$$

and eliminate $\mathbf{A}$, φ, $\mathscr{E}$, and $\mathscr{H}$. The result is to be accurate to v_k^2/c^2.

We have with the help of Eq. (16.1)

$$\frac{1}{8\pi}\int\mathscr{E}^2\,dV = -\frac{1}{8\pi}\int\mathscr{E}\cdot\left(\nabla\varphi + \frac{1}{c}\frac{\partial\mathbf{A}}{\partial t}\right)dV$$
$$= -\frac{1}{8\pi}\int\nabla\cdot(\varphi\mathscr{E})\,dV + \frac{1}{8\pi}\int\left(\varphi\,\nabla\cdot\mathscr{E} - \frac{1}{c}\mathscr{E}\cdot\frac{\partial\mathbf{A}}{\partial t}\right)dV. \tag{27.21}$$

Assume that $\varphi\mathscr{E}$ goes to zero faster than R^{-2} as R approaches infinity, so that the surface integral into which $\int\nabla\cdot(\varphi\mathscr{E})\,dV$ transforms vanishes as $R\to\infty$; then, since $\nabla\cdot\mathscr{E} = 4\pi\rho$, we can reduce Eq. (27.21) to

$$\frac{1}{8\pi}\int\mathscr{E}^2\,dV = \frac{1}{2}\int\varphi\rho\,dV - \frac{1}{8\pi c}\int\mathscr{E}\cdot\frac{\partial\mathbf{A}}{\partial t}\,dV$$
$$= \frac{1}{2}\sum_k e_k\varphi(\mathbf{r}_k, t) - \frac{1}{8\pi c}\int\mathscr{E}\cdot\frac{\partial\mathbf{A}}{\partial t}\,dV. \tag{27.22}$$

Furthermore, by Eq. (16.1)

$$-\frac{1}{8\pi}\int \mathscr{H}^2\,dV = -\frac{1}{8\pi}\int \mathscr{H}\cdot\nabla\times\mathbf{A}\,dV$$
$$= -\frac{1}{8\pi}\int\{\nabla\cdot(\mathbf{A}\times\mathscr{H}) + \mathbf{A}\cdot\nabla\times\mathscr{H}\}\,dV. \quad (27.23)$$

Again, if $\mathbf{A}\times\mathscr{H}$ goes to zero faster than R^{-2} at infinity, the integral of the divergence vanishes as $R\rightarrow\infty$. Transforming the second integral by the use of the equation

$$\nabla\times\mathscr{H} = \frac{1}{c}\frac{\partial\mathscr{E}}{\partial t} + 4\pi\rho\frac{\mathbf{v}}{c}, \quad (27.24)$$

we have

$$-\frac{1}{8\pi}\int\mathscr{H}^2\,dV = -\frac{1}{8\pi}\int\mathbf{A}\cdot\nabla\times\mathscr{H}\,dV$$
$$= -\frac{1}{2}\int\frac{\rho\mathbf{v}}{c}\cdot\mathbf{A}\,dV - \frac{1}{8\pi c}\int\mathbf{A}\cdot\frac{\partial\mathscr{E}}{\partial t}\,dV$$
$$= -\frac{1}{2}\sum_k\frac{e_k}{c}\mathbf{A}(\mathbf{r}_k, t)\cdot\mathbf{v}_k - \frac{1}{8\pi c}\int\mathbf{A}\cdot\frac{\partial\mathscr{E}}{\partial t}\,dV. \quad (27.25)$$

We thus deduce that

$$\frac{1}{8\pi}\int(\mathscr{E}^2 - \mathscr{H}^2)\,dV = -\frac{1}{2}\sum_k\left[\frac{e_k}{c}\mathbf{A}(\mathbf{r}_k, t)\cdot\mathbf{v}_k - e_k\varphi(\mathbf{r}_k, t)\right]$$
$$-\frac{1}{8\pi c}\int\left(\mathscr{E}\cdot\frac{\partial\mathbf{A}}{\partial t} + \mathbf{A}\cdot\frac{\partial\mathscr{E}}{\partial t}\right)dt. \quad (27.26)$$

The last integral is equal to

$$-\frac{1}{8\pi c}\frac{d}{dt}\int\mathscr{E}\cdot\mathbf{A}\,dV.$$

It is therefore a total derivative and can be dropped without affecting the equations of motion (see Section 2). Substituting from Eq. (27.26) into the Lagrangian of Eq. (27.20) and retaining terms only to the order of v_k^2/c^2, wehave

$$L = -c^2\sum_k m_k + \sum_k\frac{1}{2}m_k v_k^2\left(1 + \frac{1}{4}\frac{v_k^2}{c^2}\right)$$
$$+\frac{1}{2}\sum_k\left[\frac{e_k}{c}\mathbf{A}(\mathbf{r}_k, t)\cdot\mathbf{v}_k - e_k\varphi(\mathbf{r}_k, t)\right]. \quad (27.27)$$

We need only the approximate expressions for $\mathbf{A}(\mathbf{r}_k, t)$ and $\varphi(\mathbf{r}_k, t)$ that are accurate to v/c and v^2/c^2, respectively. Starting with Eqs. (19.46) and (19.47), which in our case become

$$\varphi(\mathbf{r}, t) = \frac{1}{2\pi} \sum_j e_j \iint_{-\infty}^{\infty} \frac{1}{R_j} e^{i\mu(t-\tau-R_j/c)} \, d\mu \, d\tau \tag{27.28}$$

and

$$\mathbf{A}(\mathbf{r}, t) = \frac{1}{2\pi c} \sum_j e_j \iint_{-\infty}^{\infty} \frac{\mathbf{v}_j}{R_j} e^{i\mu(t-\tau-R_j/c)} \, d\mu \, d\tau, \tag{27.29}$$

we proceed just as in deriving Eqs. (27.14) and (27.15). In place of Eqs. (27.6) and (27.7) we obtain

$$\varphi(\mathbf{r}, t) = \sum_j e_j \sum_{n=0}^{\infty} \frac{1}{n!} \left(-\frac{1}{c}\right)^n \frac{d^n}{dt^n} (R_j^{n-1}) \tag{27.30}$$

and

$$\mathbf{A}(\mathbf{r}, t) = \sum_j \frac{e_j}{c} \sum_{n=0}^{\infty} \frac{1}{n!} \left(-\frac{1}{c}\right)^n \frac{d^n}{dt^n} (\mathbf{v}_j R_j^{n-1}). \tag{27.31}$$

Retaining only terms of the order of v_k^2/c^2, we have by Eq. (27.12a)

$$\begin{aligned} \varphi(\mathbf{r}, t) &= \sum_j e_j \left(\frac{1}{R_j} + \frac{1}{2c^2} \frac{d^2 R_j}{dt^2} \right) \\ &= \sum_j e_j \left\{ \frac{1}{R_j} + \frac{1}{2c^2} \left[\frac{v_j^2}{R_j} - \frac{\mathbf{R}_j \cdot \dot{\mathbf{v}}_j}{R_j} - \frac{(\mathbf{R}_j \cdot \mathbf{v}_j)^2}{R_j^3} \right] \right\} \end{aligned} \tag{27.32}$$

and

$$\mathbf{A}(\mathbf{r}, t) = \sum_j \frac{e_j}{c} \frac{\mathbf{v}_j}{R_j}. \tag{27.33}$$

Substituting these in Eq. (27.27) and again omitting self-interaction terms, we finally obtain

$$\begin{aligned} L = &-c^2 \sum_k m_k + \frac{1}{2} \sum_k m_k v_k^2 \left(1 + \frac{1}{4} \frac{v_k^2}{c^2}\right) \\ &- \frac{1}{2} \sum_{j,k}' \frac{e_k e_j}{r_{kj}} + \frac{1}{2} \sum_{j,k}' \frac{e_k e_j}{c^2} \frac{\mathbf{v}_k \cdot \mathbf{v}_j}{r_{kj}} \\ &- \sum_{j,k}' \frac{e_k e_j}{4c^2} \left[\frac{v_j^2}{r_{kj}} - \frac{\mathbf{r}_{kj} \cdot \dot{\mathbf{v}}_j}{r_{kj}} - \frac{(\mathbf{r}_{kj} \cdot \mathbf{v}_j)^2}{r_{kj}^3} \right], \end{aligned} \tag{27.34}$$

where again $\mathbf{r}_{kj}$ is the value of $\mathbf{R}_j$ at $\mathbf{r} = \mathbf{r}_k$ and the prime on the summation indicates that terms for which $j = k$ are to be omitted.

This expression for L is unsatisfactory in two ways: First it contains $\dot{\mathbf{v}}$, and second the last summation is not symmetrical with respect to j and k.

These blemishes can be removed as follows: We have

$$\frac{d\mathbf{r}_{kj}}{dt} = \dot{\mathbf{r}}_k - \dot{\mathbf{r}}_j = \mathbf{v}_k - \mathbf{v}_j \tag{27.35}$$

and

$$\frac{d}{dt}(\mathbf{r}_{kj} \cdot \mathbf{r}_{kj}) = 2\mathbf{r}_{kj} \cdot (\mathbf{v}_k - \mathbf{v}_j) = \frac{d}{dt} r_{kj}^2 = 2r_{kj} \frac{d}{dt} r_{kj},$$

so that

$$\frac{dr_{kj}}{dt} = \frac{\mathbf{r}_{kj} \cdot (\mathbf{v}_k - \mathbf{v}_j)}{r_{kj}}. \tag{27.36}$$

Using these equations, we find that

$$\begin{aligned}\frac{d}{dt}\frac{\mathbf{r}_{kj} \cdot \mathbf{v}_j}{r_{kj}} &= \frac{(\mathbf{v}_k - \mathbf{v}_j) \cdot \mathbf{v}_j}{r_{kj}} + \frac{\mathbf{r}_{kj} \cdot \dot{\mathbf{v}}_j}{r_{kj}} - \frac{(\mathbf{r}_{kj} \cdot \mathbf{v}_j)\mathbf{r}_{kj} \cdot (\mathbf{v}_k - \mathbf{v}_j)}{r_{kj}^3} \\ &= \left[\frac{\mathbf{v}_k \cdot \mathbf{v}_j}{r_{kj}} - \frac{(\mathbf{r}_{kj} \cdot \mathbf{v}_j)(\mathbf{r}_{kj} \cdot \mathbf{v}_k)}{r_{kj}^3}\right] - \left[\frac{v_j^2}{r_{kj}} - \frac{\mathbf{r}_{kj} \cdot \dot{\mathbf{v}}_j}{r_{kj}} - \frac{(\mathbf{r}_{kj} \cdot \mathbf{v}_j)^2}{r_{kj}^3}\right].\end{aligned} \tag{27.37}$$

The expression in the last parentheses is just the expression that occurs in Eq. (27.34). Substituting from Eq. (27.37) into Eq. (27.34) and dropping the total derivative of $\mathbf{r}_{kj} \cdot \mathbf{v}_j / r_{kj}$, we obtain finally

$$\begin{aligned}L = -c^2 \sum_k m_k + \frac{1}{2} \sum_k m_k v_k^2 \left(1 + \frac{1}{4}\frac{v_k^2}{c^2}\right) - \frac{1}{2} {\sum_{j,k}}' \frac{e_k e_j}{r_{kj}} \\ + {\sum_{j,k}}' \frac{e_k e_j}{4c^2}\left[\frac{\mathbf{v}_k \cdot \mathbf{v}_j}{r_{kj}} + \frac{(\mathbf{r}_{kj} \cdot \mathbf{v}_k)(\mathbf{r}_{kj} \cdot \mathbf{v}_j)}{r_{kj}^3}\right].\end{aligned} \tag{27.38}$$

The need of dropping a total derivative to obtain the desired expression for the Lagrangian given by Eq. (27.38) suggests that it would be more convenient to use potentials $\varphi'(\mathbf{r}, t)$ and $\mathbf{A}'(\mathbf{r}, t)$ obtained from the $\varphi(\mathbf{r}, t)$ and $\mathbf{A}(\mathbf{r}, t)$ by a gauge transformation. Thus, in accordance with Eq. (16.4), we take

$$\varphi'(\mathbf{r}, t) = \varphi(\mathbf{r}, t) - \frac{1}{c}\frac{\partial \chi}{\partial t} \tag{27.39}$$

and

$$\mathbf{A}'(\mathbf{r}, t) = \mathbf{A}(\mathbf{r}, t) + \nabla\chi, \tag{27.40}$$

where $\chi = \chi(\mathbf{r}, t)$. As we know from Section 16, the change to the primed potentials will not affect the field strengths $\mathscr{E}$ and $\mathscr{H}$ or the equations of motion. Let us take

$$\chi(\mathbf{r}, t) = -\frac{1}{2c}\sum_j e_j \frac{\mathbf{R}_j \cdot \mathbf{v}_j}{R_j}, \tag{27.41}$$

where it is understood that the motion of the particles is given explicitly so that

$$\mathbf{R}_j = \mathbf{R}_j(\mathbf{r}, t) = \mathbf{r} - \mathbf{r}_j(t) \tag{27.42}$$

and

$$\dot{\mathbf{R}}_j = \frac{\partial}{\partial t}\mathbf{R}_j(\mathbf{r}, t) = -\frac{\partial}{\partial t}\mathbf{r}_j(t) = -\mathbf{v}_j. \tag{27.43}$$

Equations (27.12a) will now hold provided d/dt is replaced by $\partial/\partial t$, and thus from Eq. (27.41)

$$-\frac{1}{c}\frac{\partial\chi}{\partial t} = -\frac{1}{2c^2}\left[\frac{v_j^2}{R_j} - \frac{\mathbf{R}_j \cdot \dot{\mathbf{v}}_j}{R_j} - \frac{(\mathbf{R}_j \cdot \mathbf{v}_j)^2}{R_j^3}\right] \tag{27.44}$$

and, since $\nabla R_j = \mathbf{R}_j/R_j$,

$$\begin{aligned}\nabla\chi &= -\frac{1}{2c}\sum_j e_j\left[\mathbf{R}_j \cdot \mathbf{v}_j \nabla\left(\frac{1}{R_j}\right) + \frac{1}{R_j}\nabla(\mathbf{R}_j \cdot \mathbf{v}_j)\right] \\ &= -\frac{1}{2c}\sum_j e_j\left[-\frac{(\mathbf{R}_j \cdot \mathbf{v}_j)\mathbf{R}_j}{R_j^3} + \frac{\mathbf{v}_j}{R_j}\right].\end{aligned} \tag{27.45}$$

With this choice of χ one has at once by Eqs. (27.32), (27.39), and (27.44)

$$\varphi'(\mathbf{r}, t) = \sum_j \frac{e_j}{R_j} \tag{27.46}$$

and, by Eqs. (27.33), (27.40), and (27.45),

$$\mathbf{A}'(\mathbf{r}, t) = \sum_j \frac{e_j}{2c}\left[\frac{\mathbf{v}_j}{R_j} + \frac{(\mathbf{R}_j \cdot \mathbf{v}_j)\mathbf{R}_j}{R_j^3}\right]. \tag{27.47}$$

With these expressions for the potentials Eq. (27.27) gives directly Eq. (27.38).*

* These potentials arise naturally if we replace the Lorentz condition $\nabla \cdot \mathbf{A} + (1/c)\, \partial\varphi/\partial t = 0$ by the condition $\nabla \cdot \mathbf{A} = 0$. (The assumption $\nabla \cdot \mathbf{A} = 0$ is used by G. Breit, *Phys. Rev.*, **39**, 616 (1932) for the same purpose.)

The Lagrangian of Eq. (27.38) leads to a momentum

$$\mathbf{p}_k = \frac{\partial L}{\partial \mathbf{v}_k} = m_k \mathbf{v}_k\left(1 + \frac{1}{2}\frac{v_k^2}{c^2}\right) + \sum_{j \neq k} \frac{e_k e_j}{2c^2}\left[\frac{\mathbf{v}_j}{r_{kj}} + \frac{\mathbf{r}_{kj}(\mathbf{r}_{kj} \cdot \mathbf{v}_j)}{r_{kj}^3}\right]. \tag{27.48}$$

It may be verified by straightforward calculations that the equations of motion

$$\frac{d\mathbf{p}_k}{dt} = \frac{\partial L}{\partial \mathbf{r}_k} \tag{27.49}$$

agree with Eq. (27.19).

The Hamiltonian is now obtained in the usual way, as follows:

$$\begin{aligned} H &= -L + \sum_k \mathbf{p}_k \cdot \mathbf{v}_k \\ &= c^2 \sum_k m_k + \sum_k \frac{1}{2} m_k v_k^2\left(1 + \frac{3}{4}\frac{v_k^2}{c^2}\right) + \frac{1}{2}\sum_{j,k}{}' \frac{e_k e_j}{r_{kj}} \\ &\quad + \sum_{j,k}{}' \frac{e_k e_j}{4c^2}\left[\frac{\mathbf{v}_k \cdot \mathbf{v}_j}{r_{kj}} + \frac{(\mathbf{r}_{kj} \cdot \mathbf{v}_k)(\mathbf{r}_{kj} \cdot \mathbf{v}_j)}{r_{kj}^3}\right], \end{aligned} \tag{27.50}$$

where the **v**'s have yet to be expressed in terms of the **p**'s.

From Eq. (27.48) we see that

$$\mathbf{v}_k = \frac{\mathbf{p}_k}{m_k} - \frac{1}{2}\frac{v_k^2}{c^2}\mathbf{v}_k - \sum_{j,k}{}' \frac{e_k e_j}{2m_k c^2}\left[\frac{\mathbf{v}_j}{r_{kj}} + \frac{\mathbf{r}_{kj}(\mathbf{r}_{kj} \cdot \mathbf{v}_j)}{r_{kj}^3}\right]. \tag{27.51}$$

For a system of elementary charged particles, $e_k e_j/2m_k c^2$ is of the order of magnitude of the classical electron radius $a = e^2/2mc^2$. Thus by Eq. (27.10) the terms in the summation are two formal orders of magnitude smaller than v_k. Since we want $\mathbf{v}_k$ correct to terms in v_j^2/c^2, the v's in the right-hand member need to be evaluated only to the zeroth order in v_j/c. To zeroth order

$$\mathbf{v}_k = \frac{\mathbf{p}_k}{m_k}, \tag{27.52}$$

so that to second order in v_j/c

$$\mathbf{v}_k = \frac{\mathbf{p}_k}{m_k} - \frac{1}{2}\frac{P_k^2}{m_k^3 c^2}\mathbf{p}_k - \sum_{j,k}{}' \frac{e_k e_j}{2m_k m_j c^2}\left[\frac{\mathbf{p}_j}{r_{kj}} + \frac{\mathbf{r}_{kj}(\mathbf{r}_{kj} \cdot \mathbf{p}_j)}{r_{kj}^3}\right]. \tag{27.53}$$

In substituting the value of $\mathbf{v}_k$ into Eq. (27.50), we need to use Eq. (27.53) only in the sum $\sum_k \frac{1}{2} m_k v_k^2$, because the other terms containing **v** are of the order of v_j^2/c^2. In these other terms Eq. (27.52) is a sufficient approximation.

We have from Eq. (27.53), correct to terms in v_j^2/c^2,

$$v_k^2 = \frac{p_k^2}{m_k^2} - \frac{p_k^4}{m_k^4 c^2} - \sum_{j,k}{}' \frac{e_k e_j}{m_k^2 m_j c^2}\left[\frac{\mathbf{p}_k \cdot \mathbf{p}_j}{r_{kj}} + \frac{(\mathbf{r}_{kj} \cdot \mathbf{p}_k)(\mathbf{r}_{kj} \cdot \mathbf{p}_j)}{r_{kj}^3}\right], \tag{27.54}$$

so that finally,

$$\begin{aligned}
H = {} & c^2 \sum_k m_k + \sum_k \frac{p_k^2}{2m_k} - \sum_k \frac{p_k^4}{2m_k^3 c^2} \\
& - \frac{1}{2}\sum_{k,j}{}' \frac{e_k e_j}{m_k m_j c^2}\left[\frac{\mathbf{p}_k \cdot \mathbf{p}_j}{r_{kj}} + \frac{(\mathbf{r}_{kj} \cdot \mathbf{p}_k)(\mathbf{r}_{kj} \cdot \mathbf{p}_j)}{r_{kj}^3}\right] + \frac{3}{8}\sum_k \frac{p_k^4}{m_k^3 c^2} \\
& + \frac{1}{2}\sum_{j,k}{}' \frac{e_k e_j}{r_{kj}} + \frac{1}{4}\sum_{j,k}{}' \frac{e_k e_j}{m_k m_j c^2}\left[\frac{\mathbf{p}_k \cdot \mathbf{p}_j}{r_{kj}} + \frac{(\mathbf{r}_{kj} \cdot \mathbf{p}_k)(\mathbf{r}_{kj} \cdot \mathbf{p}_j)}{r_{kj}^3}\right] \\
= {} & c^2 \sum_k m_k + \sum_k \frac{p_k^2}{2m_k}\left(1 - \frac{1}{4}\frac{p_k^2}{m_k^2 c^2}\right) + \frac{1}{2}\sum_{j,k}{}' \frac{e_k e_j}{r_{kj}} \\
& - \frac{1}{4}\sum_{j,k}{}' \frac{e_k e_j}{m_k m_j c^2}\left[\frac{\mathbf{p}_k \cdot \mathbf{p}_j}{r_{kj}} + \frac{(\mathbf{r}_{kj} \cdot \mathbf{p}_k)(\mathbf{r}_{kj} \cdot \mathbf{p}_j)}{r_{kj}^3}\right].
\end{aligned} \tag{27.55}$$

Problems

1. What assumptions are necessary to treat the dynamics of actual particles (such as electrons) whose charge is not negligibly small?

2. What is meant by the electromagnetic mass of the electron?

3. Why is the Lorentz model of the electron essentially inconsistent with the other postulates of electrodynamics?

4. Why is electrodynamics a successful theory despite the difficulties inherent in it?

5. Verify Eq. (25.11).

6. In treating the motion of an electron in a given external field what assumptions are we to make regarding the electron?

7. Give the main steps in the treatment by G. A. Schott of an electron, taking into consideration the relative motion of its parts. (See footnote, p. 285.)

8. Show that the lowest power of R occurring in d^sR^m/dt^s, when $\mathbf{v}$ is set equal to zero, is greater than or equal to $m - (\frac{1}{2})s$.

9. Show that Eq. (25.66) is the relativistically covariant form of Eq. (25.65).

10. What is the *self-force* of an electron? What is the first-order expression for it? Is it dependent on the charge distribution assumed for the finite electron?

11. How well does the *natural breadth* of a *spectral line* obtained from classical electrodynamics, as shown in this section, agree with experiments? What other sources of line broadening are there?

12. Does the presence of the exponentially growing terms in Eq. (26.16) represent a serious weakness in the classical theory of line broadening? Can one expect to remove this term by some relatively minor change in the assumptions relating to the electron structure?

13. Sketch roughly the behavior of $f(\omega)$ of Eq. (26.30) near $\omega = \omega_0$. How does it fall off for $|\omega - \omega_0| \to \infty$?

14. How does the scattering cross section behave for resonance fluorescence? How does it behave far from resonance?

15. Justify Eq. (27.15) in detail by carrying out the operations described.

16. Derive Eq. (27.27) from Eq. (27.20).

17. What are the limitations on the use of the Hamiltonian obtained in Eq. (27.55)?

18. Give the physical significance of each of the terms in Eq. (27.55).

19. Show that the Hamiltonian in Eq. (27.55) reduces to that for electrostatics when the appropriate assumptions are made. What are these assumptions?

Chapter VIII

Material Media

28. Classical Macroscopic Electrodynamics

Heretofore, we have considered chiefly single charges or groups of charges and the individual interactions of these charges with applied fields or with each other. Such considerations fall within the province of *microscopic electrodynamics*. We now wish to consider the interactions of fields and large aggregates of charges, where such aggregates are of macroscopic size. This subject may be called *macroscopic electrodynamics*, or the *electrodynamics of material media*.

Maxwell developed, phenomenologically, a set of field equations relating the macroscopic field vectors* $\mathscr{E}$, $\mathscr{B}$, $\mathscr{H}$, and $\mathscr{D}$, and their sources, the macroscopic charges and current densities ρ and $\mathbf{J}$. These equations, called *Maxwell's equations*, are given in Eqs. (28.49) to (28.52). Furthermore, the action of the field on a charged particle is given by the microscopic Lorentz force, Eq. (14.14), and on a small macroscopic charge by Eq. (28.67). Certain auxiliary equations called *constitutive relations*, Eqs. (28.86), which contain parameters characterizing the medium, are needed in order to solve Maxwell's equations.

If macroscopic electrodynamic quantities and laws are regarded as averages of the corresponding microscopic quantities and laws, and if Maxwell's equations and the constitutive relations are accepted as correct for macroscopic electrodynamics, then we would have to impose upon our

* We shall employ the same symbols for the electric field $\mathscr{E}$, magnetic field $\mathscr{H}$, and the charge density ρ as were used for the microscopic quantities; however, we shall introduce new symbols for the microscopic quantities.

microscopic theory the condition that the averaging process produces these accepted macroscopic results. The imposition of this condition seems to have been considered first by Lorentz,* and has also been discussed by a number of others.† In what follows, however, we shall proceed in a different manner to deduce from the classical microscopic theory both Maxwell's field equations and generalized constitutive relations. This derivation is nonrelativistic.

The action function $W = \int L\,dt$ for a system of point charges and their microscopic fields may be written

$$W = W_p + W_f + W_{pf}, \tag{28.1}$$

where W_p corresponds to the particles alone, W_f to the microscopic field alone, and W_{pf} to the interaction between the particles and the field. From Eq. (16.13)‡ we have

$$W_f = \frac{1}{8\pi}\iint(e^2 - h^2)\,dV\,dt, \tag{28.2}$$

where $\mathbf{e}$ and $\mathbf{h}$ are the microscopic electric and magnetic fields and the volume integration is over the entire three-space. Likewise, from (16.14) we have

$$W_{pf} = \int \sum_s e_s\left[\left(\frac{\mathbf{v}_s}{c}\right)\cdot\mathbf{a}(\mathbf{r}_s, t) - \phi(\mathbf{r}_s, t)\right] dt, \tag{28.3}$$

where $\mathbf{a}(\mathbf{r}_s, t)$ and $\phi(\mathbf{r}_s, t)$ are the microscopic field potentials evaluated at the position of the sth charge and $\mathbf{v}_s$ is the velocity of that charge. The summation with respect to s extends over all the charges. The microscopic field vectors $\mathbf{e}$ and $\mathbf{h}$ are related to the microscopic vector and scalar potentials ϕ and $\mathbf{a}$ by (see Eqs. (14.15) and (14.16))

$$\mathbf{e} = -\nabla\phi - \frac{1}{c}\frac{\partial \mathbf{a}}{\partial t} \tag{28.4}$$

and

$$\mathbf{h} = \nabla \times \mathbf{a}. \tag{28.5}$$

* H. A. Lorentz, *Collected Papers*, Vol. 3, M. Nijhoff, the Hague, Holland, 1936, p. 117.

† See, for example, J. H. Van Vleck, *The Theory of Electric and Magnetic Susceptibilities*, Clarendon Press, Oxford, 1932, Chap. 1.

‡ In what follows we shall usually use lower-case letters to denote microscopic variables, and the corresponding upper-case letter (script or regular) for the corresponding macroscopic quantities.

A macroscopic medium is characterized by the presence of groups of charged particles that remain stable in time; i.e., the charges of the group remain in a small region about their center of gravity. Let such groupings, which might be atoms, ions, or molecules (they shall hereafter be called molecules) be introduced into the microscopic description of the medium. Then the charges may be separated into two types—those that are in the molecules and those that are not.* The latter are called conduction (or free) charges. This separation of the charges into two types does not change the value of W and its parts, but W_{pf} may now be written

$$W_{pf} = W_{mf} + W_{cf}, \tag{28.6}$$

where W_{mf} is the interaction between the field and the molecular charges and W_{cf} is the interaction between the field and the conduction charges. Both W_{mf} and W_{cf} will have the form given in Eq. (28.3) except that the summation in the expression for W_{mf} extends over the molecular charges only, and in the expression for W_{cf} over the conduction charges only.

We shall neglect the self-force of individual charges and also the intramolecular forces between the charges of the same molecule. The field describing the remaining forces acting on the molecules and on the conduction charges is thus free of singularities in the immediate neighborhood of these molecules and charges. It is this field that largely determines the macroscopic properties of the medium.

We shall assume at this juncture that the field postulated above is expandable in a Taylor's series about some point $\mathbf{r}_n$ inside the molecule, which for definiteness we take to be its center of gravity. Thus it follows that the contribution of the nth molecule to W_{mf} can be expressed by

$$\begin{aligned} W_{mf,n} = \int L_{mf,n}\,dt = \sum_s e_s \int \Bigg\{ \frac{v_{ns}^i}{c} \bigg[a_i(\mathbf{r}_n, t) &+ \frac{\partial a_i}{\partial x_j}(\mathbf{r}_{ns} - \mathbf{r}_n)_j \\ &+ \frac{1}{2}\frac{\partial^2 a_i}{\partial x_j\,\partial x_k}(\mathbf{r}_{ns} - \mathbf{r}_n)_j(\mathbf{r}_{ns} - \mathbf{r}_n)_k + \cdots \bigg] \\ &- \bigg[\phi(\mathbf{r}_n, t) + \frac{\partial \phi}{\partial x_j}(\mathbf{r}_{ns} - \mathbf{r}_n)_j \\ &+ \frac{1}{2}\frac{\partial^2 \phi}{\partial x_j\,\partial x_k}(\mathbf{r}_{ns} - \mathbf{r}_n)_j(\mathbf{r}_{ns} - \mathbf{r}_n)_k + \cdots \bigg] \Bigg\}\,dt. \end{aligned} \tag{28.7}$$

* This is an oversimplified description, but it suffices for our general description of a material medium. A more accurate description would necessarily involve results borrowed from a quantum-mechanical treatment and would be inappropriate in a purely classical description.

In this expression $L_{mf,n}$ is the Lagrangian function for the interaction of the field and the charges of the nth molecule, $\mathbf{r}_{ns}$ is the position vector of the sth charge in the nth molecule, $\mathbf{v}_{ns}$ is its velocity measured in the inertial coordinate system of the observer (assumed at rest relative to the medium), and all partial derivatives are to be evaluated at $\mathbf{r}_n$. In accordance with our summation convention (see Section 2) a summation over the values 1, 2, 3 for the repeated latin indices i, j, and k is understood. W_{mf} will be obtained later by summing $W_{mf,n}$ over n. The summation index s now runs only over the charges of the nth molecule.

Letting $\mathbf{X}_s = \mathbf{r}_{ns} - \mathbf{r}_n = (X_s^1, X_s^2, X_s^3)$, we have

$$\mathbf{v}_{ns} = \mathbf{v}_n + \dot{X}_s, \tag{28.8}$$

where the dot represents differentiation with respect to time. It is convenient to introduce the modified molecular electric moments

$$q_n = \sum_s e_s, \tag{28.9a}$$

$$p_n^i = \sum_s e_s X_s^i, \tag{28.9b}$$

$$Q_n^{ij} = \frac{1}{2!} \sum_s e_s X_s^i X_s^j, \tag{28.9c}$$

$$Q_n^{ijk} = \frac{1}{3!} \sum_s e_s X_s^i X_s^j X_s^k, \tag{28.9d}$$

where q_n is the net charge of the nth molecule, P_n^i is its electric dipole moment vector, Q_n^{ij} is a modification of its electric quadrupole moment tensor, and Q_n^{ijk} is a modification of its electric octupole moment tensor. $L_{mf,n}$ can be written

$$\begin{aligned} L_{mf,n} = {} & \frac{1}{c} a_i(\mathbf{r}_n, t)[v_n^i q_n + \dot{P}_n^i] + \frac{1}{c}\frac{\partial a_i}{\partial x_j}\left[v_n^i P_n^j + \sum_s e_s \dot{X}_s^i X_s^j\right] \\ & + \frac{1}{c}\frac{\partial^2 a_i}{\partial x_j\,\partial x_k}\left[v_n^i Q_n^{jk} + \frac{1}{2!}\sum_s e_s \dot{X}_s^i X_s^j X_s^k\right] + \cdots \\ & - \left[\phi(\mathbf{r}_n, t) q_n + \frac{\partial \phi}{\partial x_j} P_n^j + \frac{\partial^2 \phi}{\partial x_j\,\partial x_k} Q_n^{jk} + \cdots\right]. \end{aligned} \tag{28.10}$$

In order to express $L_{mf,n}$ in a more suitable form it is necessary to establish several identities. Thus, starting with the identity

$$\frac{\partial a_i}{\partial x_j} v_n^i P_n^j = \frac{1}{2}\left(\frac{\partial a_i}{\partial x_j} - \frac{\partial a_j}{\partial x_i}\right)(v_n^i P_n^j - v_n^j P_n^i) + \frac{\partial a_j}{\partial x_i} v_n^i P_n^j, \tag{28.11}$$

we obtain the identity

$$\frac{\partial a_i}{\partial x_j} v_n^i P_n^j = (\mathbf{P}_n \times \mathbf{v}_n) \cdot (\nabla \times \mathbf{a}) + \frac{\partial a_j}{\partial x_i} v_n^i P_n^j. \tag{28.12}$$

If the magnetic moments of the nth molecule are defined by

$$M_n^i = \frac{1}{2c} \sum_s e_s (\mathbf{X}_s \times \dot{\mathbf{X}}_s)^i, \tag{28.13}$$

$$M_n^{ij} = \frac{1}{3c} \sum_s e_s X_s^i (\mathbf{X}_s \times \dot{\mathbf{X}}_s)^j, \tag{28.14}$$

where M_n is the magnetic dipole moment of the nth molecule (see Eq. (20.21)) and M_n^{ij} is related to its magnetic quadrupole moment. Using a similar technique to that employed in obtaining Eq. (28.12), one finds that

$$\frac{\partial^2 a_i}{\partial x_j \partial x_k} v_n^i Q_n^{jk} = \frac{1}{2} \sum_s e_s X_s^k (\mathbf{X}_s \times \mathbf{v}_n)^i \frac{\partial}{\partial x_k} (\nabla \times \mathbf{a})_i + \frac{\partial^2 a_j}{\partial x_i \partial x_k} v_n^i Q_n^{jk}, \tag{28.15}$$

$$\frac{1}{c} \frac{\partial a_i}{\partial x_j} \sum_s e_s \dot{X}_s^i X_s^j = \frac{1}{c} \frac{\partial a_i}{\partial x_j} \dot{Q}_n^{ij} + \mathbf{M}_n \cdot \nabla \times \mathbf{a}, \tag{28.16}$$

and

$$\frac{1}{2c} \frac{\partial^2 a_i}{\partial x_j \partial x_k} \sum_s e_s \dot{X}_s^i X_s^j X_s^k = \frac{1}{c} \frac{\partial^2 a_k}{\partial x_i \partial x_j} \dot{Q}_n^{ijk} + M_n^{ij} \frac{\partial}{\partial x_i} (\nabla \times \mathbf{a})_j. \tag{28.17}$$

It follows from Eqs. (28.10), (28.12), (28.15), (28.16), and (28.17) that

$$\begin{aligned} L_{mf} = \sum_n L_{mf,n} = \sum_n \Bigg\{ & \frac{1}{c} a_i(\mathbf{r}_n, t) v_n^i q_n + \frac{1}{c} \left[a_i(\mathbf{r}_n, t) \dot{P}_n^i + \frac{\partial a_j}{\partial x_i} v_n^i P_n^j \right] \\ & + \frac{1}{c} (\mathbf{P}_n \times \mathbf{v}_n) \cdot (\nabla \times \mathbf{a}) + \frac{1}{c} \frac{\partial a_i}{\partial x_j} \dot{Q}_n^{ij} + \frac{1}{c} \frac{\partial^2 a_j}{\partial x_i \partial x_k} Q_n^{jk} v_n^i \\ & + \mathbf{M}_n \cdot \nabla \times \mathbf{a} + \frac{1}{2!\, c} \sum_s e_s X_s^k (\mathbf{X}_s \times \mathbf{v}_n)^i \frac{\partial}{\partial x_k} (\nabla \times \mathbf{a})_i \\ & + M_n^{ij} \frac{\partial}{\partial x_i} (\nabla \times \mathbf{a})_j + \cdots \\ & - \left[\phi(\mathbf{r}_n, t) q_n + \frac{\partial \phi}{\partial x_i} P_n^i + \frac{\partial^2 \phi}{\partial x_i \partial x_j} Q_n^{ij} + \cdots \right] \Bigg\}. \end{aligned} \tag{28.18}$$

We now define the following microscopic quantities:

$$\rho_{cv} = \sum_n q_n \delta(\mathbf{r} - \mathbf{r}_n), \tag{28.19}$$

$$\mathbf{j}_{cv} = \sum_n q_n \left(\frac{\mathbf{v}_n}{c}\right) \delta(\mathbf{r} - \mathbf{r}_n), \tag{28.20}$$

$$\mathbf{p} = \sum_n P_n \delta(\mathbf{r} - \mathbf{r}_n), \tag{28.21}$$

$$q_{ij} = \sum_n Q_n^{ij} \delta(\mathbf{r} - \mathbf{r}_n), \tag{28.22}$$

$$\mathbf{f} = \frac{1}{c} \sum_n (\mathbf{P}_n \times \mathbf{v}_n) \delta(\mathbf{r} - \mathbf{r}_n), \tag{28.23}$$

$$\mathbf{m} = \sum_n \mathbf{M}_n \delta(\mathbf{r} - \mathbf{r}_n), \tag{28.24}$$

$$g_{ij} = \frac{1}{2c} \sum_{n,s} e_s X_s^i (\mathbf{X}_s \times \mathbf{v}_n)^j \delta(\mathbf{r} - \mathbf{r}_n), \tag{28.25}$$

$$m_{ij} = \sum_n M_n^{ij} \delta(\mathbf{r} - \mathbf{r}_n), \tag{28.26}$$

where ρ_{cv} is the microscopic convection charge density due to the net charges of the molecules, $\mathbf{j}_{cv}$ is the microscopic convection current density due to the motion of the molecules, $\mathbf{p}$ is the microscopic dipole moment density, and so on.

Integrating over the entire three-space and using the properties of $\delta(\mathbf{r} - \mathbf{r}_n)$, one obtains quite readily the identities

$$\sum_n q_n \phi(\mathbf{r}_n, t) = \int \rho_{cv} \phi(\mathbf{r}, t)\, dV, \tag{28.27}$$

$$(1/c) \sum_n q_n v_n^i a_i(\mathbf{r}_n, t) = \int \mathbf{j}_{cv} \cdot \mathbf{a}(\mathbf{r}, t)\, dV, \tag{28.28}$$

$$\sum_n \frac{\partial \phi}{\partial x_i} p_n^i = -\int \phi(\mathbf{r}, t)\, \nabla \cdot \mathbf{p}\, dV. \tag{28.29}$$

Since

$$\frac{\partial \mathbf{p}}{\partial t} = \sum_n \dot{\mathbf{P}}_n \delta(\mathbf{r} - \mathbf{r}_n) - \sum_n \mathbf{P}_n v_n^i \frac{\partial}{\partial x_i} \delta(\mathbf{r} - \mathbf{r}_n), \tag{28.30}$$

it also follows that

$$\sum_n \left[\dot{P}_n^i a_i(\mathbf{r}_n, t) + \frac{\partial a_j}{\partial x_i} v_n^i P_n^j \right] = \int \mathbf{a} \cdot \frac{\partial \mathbf{p}}{\partial t}\, dV. \tag{28.31}$$

Similar procedures lead to the following transformations:

$$\sum_n \mathbf{M}_n \cdot \nabla \times \mathbf{a} = \int \mathbf{a}(\mathbf{r}, t) \cdot \nabla \times \mathbf{m}\, dV, \tag{28.32}$$

$$\frac{1}{c} \sum_n (\mathbf{P}_n \times \mathbf{v}_n) \cdot \nabla \times \mathbf{a} = \int \mathbf{a}(\mathbf{r}, t) \cdot \nabla \times \mathbf{f}\, dV, \tag{28.33}$$

$$\sum_n M_n^{ij} \frac{\partial}{\partial x_i} (\nabla \times \mathbf{a})_j = -\int \mathbf{a}(\mathbf{r}, t) \cdot \nabla \times \frac{\partial}{\partial \mathbf{x}_i} (m_{ij} \mathbf{u}_j)\, dV, \tag{28.34}$$

$$\frac{1}{2c} \sum_{n,s} e_s X_s^i (\mathbf{X}_s \times \mathbf{v}_n)^j \frac{\partial}{\partial x_i} (\nabla \times \mathbf{a})_j = -\int \mathbf{a}(\mathbf{r}, t) \cdot \nabla \times \left(\mathbf{u}_j \frac{\partial}{\partial x_i} g_{ij}\right) dV, \tag{28.35}$$

$$\frac{1}{c} \sum_n \frac{\partial a_i}{\partial x_j} \dot{Q}_n^{ij} + \frac{1}{c} \sum_n \frac{\partial^2 a_j}{\partial x_i\, \partial x_k} Q_n^{jk} v_n^i = -\frac{1}{c} \int \mathbf{a} \cdot \frac{\partial}{\partial t} \left(\mathbf{u}_j \frac{\partial}{\partial x_i} g_{ij}\right) dV, \tag{28.36}$$

$$\sum_n Q_n^{ij} \frac{\partial^2 \phi}{\partial x_i\, \partial x_j} = \int \phi(\mathbf{r}, t) \frac{\partial^2 p_{ij}}{\partial x_i\, \partial x_j}\, dV, \tag{28.37}$$

where $\mathbf{u}_i$ is a unit vector along the ith axis.

By means of such transformations, L_{mf} may be written in the form

$$\begin{aligned} L_{mf} = \int \mathbf{a}(\mathbf{r}, t) \cdot \Big[\mathbf{j}_{cv} + \frac{1}{c} \frac{\partial \mathbf{p}}{\partial t} + \nabla \times \mathbf{m} + \nabla \times \mathbf{f} \\ - \frac{1}{c} \frac{\partial}{\partial t} \left(\frac{\partial q_{ij}}{\partial x_i} \mathbf{u}_j\right) - \nabla \times \left(\frac{\partial m_{ij}}{\partial x_i} \mathbf{u}_j\right) - \nabla \times \left(\frac{\partial g_{ij}}{\partial x_i} \mathbf{u}_j\right) + \cdots \Big] dV \\ - \int \phi(\mathbf{r}, t) \left[\rho_{cv} - \nabla \cdot \mathbf{p} + \frac{\partial^2 q_{ij}}{\partial x_i\, \partial x_j} - \cdots \right] dV. \end{aligned} \tag{28.38}$$

To obtain the field equations, Hamilton's principle is applied, and W is made stationary with respect to variation of the field, holding the trajectories of the charged particles fixed. Thus, one requires that $\delta W_p = 0$, and in the usual way one obtains the equations

$$\begin{aligned} \frac{1}{4\pi} \left[\nabla \times \mathbf{h} - \frac{1}{c} \frac{\partial \mathbf{e}}{\partial t} \right] = \mathbf{j}_c + \mathbf{j}_{cv} + \frac{1}{c} \frac{\partial \mathbf{p}}{\partial t} + \nabla \times \mathbf{m} + \nabla \times \mathbf{f} \\ - \frac{1}{c} \frac{\partial}{\partial t} \left(\frac{\partial q_{ij}}{\partial x_i} \mathbf{u}_j\right) - \nabla \times \left[\left(\frac{\partial m_{ij}}{\partial x_i} + \frac{\partial g_{ij}}{\partial x_i}\right) \mathbf{u}_j \right] + \cdots \end{aligned} \tag{28.39}$$

and

$$\frac{1}{4\pi} \nabla \cdot \mathbf{e} = \rho_c + \rho_{cv} - \nabla \cdot \mathbf{p} + \frac{\partial^2 q_{ij}}{\partial x_i\, \partial x_j} + \cdots. \tag{28.40}$$

We shall now combine the conduction and convection parts by introducing the total microscopic current density

$$\mathbf{j}' = \mathbf{j}_c + \mathbf{j}_{cv} \tag{28.41}$$

and the total microscopic charge density

$$\rho' = \rho_c + \rho_{cv}. \tag{28.42}$$

If the microscopic field vectors $\mathbf{b}$ and $\mathbf{d}$ are defined by

$$\mathbf{b} = \mathbf{h} - 4\pi(\mathbf{m} + \mathbf{f}) + 4\pi\left(\frac{\partial g_{ij}}{\partial x_i} + \frac{\partial m_{ij}}{\partial x_i}\right)\mathbf{u}_j - \cdots \tag{28.43}$$

and

$$\mathbf{d} = \mathbf{e} + 4\pi\mathbf{p} - 4\pi\frac{\partial q_{ij}}{\partial x_i}\mathbf{u}_j + \cdots, \tag{28.44}$$

then the field equations

$$\nabla \times \mathbf{b} - \frac{1}{c}\frac{\partial \mathbf{d}}{\partial t} = 4\pi\mathbf{j}' \tag{28.45}$$

and

$$\nabla \cdot \mathbf{d} = 4\pi\rho' \tag{28.46}$$

are obtained from Eqs. (28.39) and (28.40). From the definitions of $\mathbf{e}$ and $\mathbf{h}$ in Eqs. (28.4) and (28.5), it follows that

$$\nabla \times \mathbf{e} + \frac{1}{c}\frac{\partial \mathbf{h}}{\partial t} = 0 \tag{28.47}$$

and

$$\nabla \cdot \mathbf{h} = 0. \tag{28.48}$$

Equations (28.45) to (28.48) are the *microscopic field equations* for material media.

If any kind of linear averaging process which commutes with the operators $\partial/\partial x_i$ and $\partial/\partial t$ is applied to the above equations, *Maxwell's macroscopic field equations*

$$\nabla \times \mathscr{H} - \frac{1}{c}\frac{\partial \mathscr{D}}{\partial t} = 4\pi\mathbf{J}, \tag{28.49}$$

$$\nabla \cdot \mathscr{D} = 4\pi\rho, \tag{28.50}$$

$$\nabla \times \mathscr{E} + \frac{1}{c}\frac{\partial \mathscr{B}}{\partial t} = 0, \tag{28.51}$$

$$\nabla \cdot \mathscr{B} = 0, \tag{28.52}$$

are obtained, where the following identifications are made:

$$\bar{\mathbf{e}} = \mathscr{E}, \quad (28.53) \qquad \bar{\mathbf{b}} = \mathscr{H}, \quad (28.54)$$

$$\bar{\mathbf{h}} = \mathscr{B}, \quad (28.55) \qquad \bar{\mathbf{d}} = \mathscr{D}, \quad (28.56)$$

$$\bar{\mathbf{J}}' = \mathbf{J}, \quad (28.57) \qquad \bar{\rho}' = \rho. \quad (28.58)$$

(The bar above a microscopic quantity indicates the average of that quantity.) When the microscopic equations (28.43) and (28.44) are averaged, the generalized macroscopic equations*

$$\mathscr{H} = \mathscr{B} - 4\pi\mathbf{M} + 4\pi \frac{\partial M_{ij}}{\partial x_i} \mathbf{u}_j - \cdots \qquad (28.59)$$

$$\mathscr{D} = \mathscr{E} + 4\pi\mathbf{P} - 4\pi \frac{\partial Q_{ij}}{\partial x_i} \mathbf{u}_j + \cdots \qquad (28.60)$$

result, where

$$\bar{\mathbf{p}} = \mathbf{P}, \quad (28.61) \qquad \bar{\mathbf{m}} + \mathbf{f} = \mathbf{M}, \quad (28.62)$$

$$\bar{q}_{ij} = Q_{ij}, \quad (28.63) \qquad \bar{g}_{ij} + \bar{m}_{ij} = M_{ij}. \quad (28.64)$$

If the averaging process is applied to the microscopic equations (28.4) and (28.5) and if it is assumed that the averages of the microscopic potentials $\mathbf{a}$ and ϕ are the macroscopic potentials $\mathbf{A}$ and φ, respectively, the usual equations

$$\mathscr{E} = -\nabla\varphi - \frac{1}{c}\frac{\partial \mathbf{A}}{\partial t}, \qquad (28.65)$$

$$\mathscr{B} = \nabla \times \mathbf{A} \qquad (28.66)$$

are obtained.

Last, if both sides of the microscopic Lorentz force equation (14.14) is averaged, we obtain, for the force $\mathbf{F}$ on a small macroscopic charge q moving with velocity $\mathbf{v}$,

$$\mathbf{F} = q\left(\bar{\mathbf{e}} + \frac{\mathbf{v}}{c} \times \bar{\mathbf{h}}\right).$$

When Eqs. (28.53) and (28.55) are used, this yields the macroscopic Lorentz force equation

$$\mathbf{F} = q\left(\mathscr{E} + \frac{\mathbf{v}}{c} \times \mathscr{B}\right). \qquad (28.67)$$

* Similar results were obtained by J. Frenkel in *Lehrbuch der Electrodynamik*, Springer, Berlin, 1928, Vol. II, p. 25.

The macroscopic quantities, such as those occurring in Eq. (28.67), differ from the corresponding microscopic quantities is several important ways. First they usually represent some sort of average, and second, because of our initial assumptions, they ignore the intramolecular field, the field produced within a molecule by charges composing that molecule.*

A practical operational definition of the macroscopic fields $\mathscr{E}$ and $\mathscr{B}$ is provided by Eq. (28.67) when one lets the averaging process be that involved in the use of very small charged macroscopic bodies as test charges. The total force on such a body is the sum of the forces on the large number of charges composing the body. The force on the body thus depends in a statistical way on the field throughout its volume and is therefore essentially a measure of the average field over this volume. Since such a macroscopic test charge must necessarily be placed outside the molecules it is not affected by the intramolecular fields of the molecules. This property of the macroscopic test charge plus its automatic averaging of the microscopic field makes it a natural vehicle for defining the macroscopic fields.

Special cases

Case 1. *Microscopic electrostatics*

In this case, by assumption, there exists a coordinate system in which all of the charges are at rest, and the observer is assumed to be at rest with respect to this system. If the vector potential **A** and the field are assumed to be constant in time, the governing field equations are

$$\begin{aligned} &\nabla \times \mathscr{E} = 0, \\ &\nabla \cdot \mathscr{D} = \nabla \cdot \mathscr{E} + 4\pi \nabla \cdot \mathbf{P} - 4\pi \frac{\partial^2 Q_{ij}}{\partial x_i \, \partial x_j} + \cdots = 4\pi\rho. \end{aligned} \tag{28.68}$$

From these results and Eq. (28.65) one obtains Poisson's equation for the macroscopic scalar potential φ in a material medium:

$$\nabla^2 \varphi = -4\pi\left(\rho - \nabla \cdot \mathbf{P} + \frac{\partial^2 Q_{ij}}{\partial x_i \, \partial x_j} - \cdots\right). \tag{28.69}$$

* Calculation of such fields usually require, in any case, a resort to quantum mechanics.

Case 2. Macroscopic electrostatics

In this case the centers of gravity of all the molecules are at rest in some inertial coordinate system (the individual charges need not be at rest). We shall take the observer to be also at rest in this coordinate system. This means that the $\mathbf{v}_n$ in Eq. (28.20) vanish and therefore that the microscopic convection-current density $\mathbf{j}_{cv}$ is zero. Thus the macroscopic current density $\mathbf{J}$ is due to the conduction charges only. Also, $\mathbf{f} = 0$ and $g_{ij} = 0$ (see Eqs. (28.23) and (28.25)), with similar results for the higher-order terms of these types. The macroscopic field equations (28.49) to (28.52) and constitutive equations (28.59) and (28.60), however, retain the same forms, while Eqs. (28.62) and (28.64) reduce to $\bar{\mathbf{m}} = \mathbf{M}$ and $\bar{m}_{ij} = M_{ij}$, respectively.

Case 3. Moving media

In this case the centers of gravity of all the molecules are at rest in a primed coordinate system moving with velocity $\mathbf{v}$ with respect to the unprimed coordinate system of the observer; then $\mathbf{v}$ is the macroscopic velocity of the medium. Thus, $\mathbf{v}_n = \mathbf{v}$ for all n, and Eqs. (28.20) and (28.23) reduce to

$$\mathbf{j}_{cv} = \frac{\mathbf{v}}{c}\rho_{cv}, \qquad \mathbf{f} = \frac{1}{c}(\mathbf{p} \times \mathbf{v}). \tag{28.70}$$

Equation (28.45) can then be put in the form

$$\nabla \times \mathbf{h} - \frac{1}{c}\frac{\partial \mathbf{d}}{\partial t} = 4\pi\left[\mathbf{j}' + \nabla \times \mathbf{m} + \frac{1}{c}\nabla \times (\mathbf{p} \times \mathbf{v}) - \cdots\right]; \tag{28.71}$$

thus, for this case Eq. (28.49) can be written

$$\nabla \times \mathscr{B} - \frac{1}{c}\frac{\partial \mathscr{D}}{\partial t} = 4\pi\left[\mathbf{J} + \nabla \times \bar{\mathbf{m}} + \frac{1}{c}\nabla \times (\mathbf{P} \times \mathbf{v}) - \cdots\right]. \tag{28.72}$$

The contribution to the right-hand side of Eq. (28.72) of the term $(1/c)\,\nabla \times (\mathbf{P} \times \mathbf{v})$ was discovered by Röntgen and later Lorentz* labeled it the Röntgen current.

It is interesting to note that the macroscopic charge and current densities ρ and $\mathbf{J}$ are not simply averages of the microscopic quantities ρ_m and $\mathbf{j}_m$. This may be shown by applying the averaging process to the microscopic

* See a previous footnote for publications by Lorentz.

field equations (the Maxwell–Lorentz equations)

$$\nabla \times \mathbf{h} - \frac{1}{c}\frac{\partial \mathbf{e}}{\partial t} = 4\pi \mathbf{j}_m, \tag{28.73}$$

$$\nabla \cdot \mathbf{e} = 4\pi \rho_m, \tag{28.74}$$

$$\nabla \times \mathbf{e} + \frac{1}{c}\frac{\partial \mathbf{h}}{\partial t} = 0, \tag{28.75}$$

$$\nabla \cdot \mathbf{h} = 0, \tag{28.76}$$

which yields, when use is made of Eqs. (28.53) and (28.55), the macroscopic equations

$$\nabla \times \mathscr{B} - \frac{1}{c}\frac{\partial \mathscr{E}}{\partial t} = 4\pi \bar{\mathbf{j}}_m, \tag{28.77}$$

$$\nabla \cdot \mathscr{E} = 4\pi \bar{\rho}_m, \tag{28.78}$$

$$\nabla \times \mathscr{E} + \frac{1}{c}\frac{\partial \mathscr{B}}{\partial t} = 0, \tag{28.79}$$

$$\nabla \cdot \mathscr{B} = 0. \tag{28.80}$$

Equations (28.79) and (28.80) are identical with Eqs. (28.51) and (28.52) of Maxwell's equations, but a comparison of Eqs. (28.77) and (28.78) with Eqs. (28.49) and (28.50), making use of Eqs. (28.59) and (28.60), gives

$$\bar{\mathbf{j}}_m - \mathbf{J} = \nabla \times \left(\mathbf{M} - \frac{\partial M_{ij}}{\partial x_i}\mathbf{u}_j + \cdots\right) + \frac{1}{c}\frac{\partial}{\partial t}\left(\mathbf{P} - \frac{\partial Q_{ij}}{\partial x_i}\mathbf{u}_j + \cdots\right), \tag{28.81}$$

$$\bar{\rho}_m - \rho = -\nabla \cdot \mathbf{P} + \frac{\partial^2 Q_{ij}}{\partial x_i\, \partial x_j} - \cdots. \tag{28.82}$$

These equations express the differences between the averages of the microscopic charge and current densities and the macroscopic charge and current densities appearing in Maxwell's equations.

Lorentz explained the first-order terms of these differences by noting that when the microscopic charge density, for example, is averaged over a macroscopic volume element, all of the charges in the volume element are counted, including charges in molecules that have been cut by the edges of the volume element. He showed that such charges give rise to the term $-\nabla \cdot \mathbf{P}$ in Eq. (28.82). One might attempt to explain the higher-order effects in a similar manner, but one of the advantages of the preceding development is that such considerations are unnecessary. Here, only those molecules whose centers of gravity fall within the volume element contribute to the average, and they contribute their whole effect, despite the fact that some of the charges of these molecules lie in adjacent volume elements.

The field equations (28.77) to (28.80) have certain important differences from Maxwell's equations. Maxwell's equations do not completely determine the four field quantities $\mathscr{E}, \mathscr{B}, \mathscr{D}$, and $\mathscr{H}$ when ρ and $\mathbf{J}$ and the boundary conditions are known. Equations (28.77) to (28.80), however, contain only $\mathscr{B}$ and $\mathscr{E}$, and these quantities are determined by these equations when $\bar{\rho}_m$ and $\bar{\mathbf{j}}_m$ and the boundary conditions are known. The definitions here adopted for ρ and $\mathbf{J}$ (as averages of microscopic conduction and convection charge and current densities) are not in terms of simple macroscopic operations—such a definition would require that the macroscopic measuring device distinguish and count conduction and convection charges and ignore the other charges that are present in a macroscopic volume element. On the other hand, $\bar{\rho}_m$ and $\bar{\mathbf{j}}_m$ do have simple operational definitions in terms of the total charge in a volume element and charges passing through a certain unit surface per unit time. Unfortunately, while it is easy to conceive of such operational definitions, it is not a simple matter to make the required measurements, especially at interior points of a medium, in moving media, and so on. But in many cases some knowledge of the conduction and convection charge and current densities exists or can be assumed for the particular physical situation, and these values used in Maxwell's equations. Still, if $\bar{\rho}_m$ and $\bar{\mathbf{j}}_m$ can be determined by measurement or calculation, then the field equations (28.77) to (28.80) can be used, and presumably the same results obtained as when using Maxwell's field equations and the generalized relations (28.59) and (28.60).

The many successes obtained by use of Maxwell's equations and the relations

$$\mathscr{D} = \mathscr{E} + 4\pi\mathbf{P}', \qquad \mathscr{B} = \mathscr{H} + 4\pi\mathbf{M}' \tag{28.83}$$

that are usually used in place of Eqs. (28.59) and (28.60) indicate that these relations are sufficiently accurate for many physical situations. If these equations are regarded as defining $\mathbf{P}'$ and $\mathbf{M}'$, however, then measurement of $\mathbf{P}'$ and $\mathbf{M}'$ in accordance with these equations, in situations where the higher-order effects are significant, actually gives by Eqs. (28.59) and (28.60)

$$\mathbf{P}' = \frac{1}{4\pi}(\mathscr{D} - \mathscr{E}) = \mathbf{P} - \frac{\partial Q_{ij}}{\partial x_i}\mathbf{u}_j + \cdots, \tag{28.84}$$

$$\mathbf{M}' = \frac{1}{4\pi}(\mathscr{B} - \mathscr{H}) = \mathbf{M} - \frac{\partial M_{ij}}{\partial x_i}\mathbf{u}_j + \cdots, \tag{28.85}$$

and, if there is no independent check on **P** and **M**, there is no way of determining to what extent these values of **P** and **M** being used include the higher-order effects. Thus for a phenomenological treatment of electrodynamics in material media the simpler relations of Eq. (28.83) are usually completely satisfactory.

Phenomenological treatment. As pointed out above, Maxwell's equations, Eqs. (28.49) to (28.52), do not completely determine the field quantities $\mathscr{E}$, $\mathscr{B}$, $\mathscr{D}$, and $\mathscr{H}$ when ρ and **J** and the boundary conditions are known. To complete the description one needs to relate $\mathscr{D}$ to $\mathscr{E}$ and $\mathscr{H}$ to $\mathscr{B}$. This is done for *simple media* by assuming the *constitutive relations*

$$\mathscr{D} = \epsilon\mathscr{E}, \qquad \mathscr{B} = \mu\mathscr{H} \tag{28.86}$$

hold, where ϵ and μ are constants characteristic of the medium. One refers to ϵ as the *dielectric constant* and μ as the *permeability* of the medium. For a conductive medium, currents other than the applied currents flow and one needs to add the constitutive relation

$$\mathbf{J} = \sigma\mathscr{E}, \tag{28.87}$$

where σ is the *conductivity* of the medium. Equation (28.87) is just *Ohm's law* for an extended medium.

If ϵ, μ, and σ are dependent on the frequency ω, then the constitutive relations apply to the Fourier components of the field quantities and the medium is termed *dispersive*.

For anisotropic medium Eqs. (28.86) and (28.87) must be generalized as follows:

$$\begin{aligned} \mathscr{D}_i &= \epsilon_{ij}\mathscr{E}_j, \\ \mathscr{B}_i &= \mu_{ij}\mathscr{H}_j, \qquad (i = 1, 2, 3) \\ J_i &= \sigma_{ij}\mathscr{E}_j, \end{aligned} \tag{28.88}$$

where in accordance with the summation convention j is summed from 1 to 3. In this case, therefore, the dielectric constant, permeability, and the conductivity are second-order tensors.

The justification for assuming a particular set of constitutive relations is generally based not only on a knowledge of the structure of the material but also on the reasonableness of the results obtained and general agreement with experiment.

29. Moving Media

Maxwell's macroscopic equations for a material medium are given by Eqs. (28.49) to (28.52) and, together with the constitutive relations of Eqs. (28.86) and the appropriate boundary conditions serve to specify a unique macroscopic electromagnetic field for media at rest. Thus, for example, by taking the divergence of both sides of Eq. (28.49) and using Eq. (28.50), one obtains the *continuity equation*

$$\nabla \cdot \mathbf{J} + \frac{1}{c}\frac{\partial \rho}{\partial t} = 0. \tag{29.1}$$

We wish now to generalize the formulation to apply also to moving media. Before doing this, however, we shall express the macroscopic equations of Section 28 in covariant form.

The two homogeneous Maxwell equations expressed by Eqs. (28.51) and (28.52) differ from the homogeneous equations for empty space, Eqs. (16.5) and (16.6), merely by having $\mathscr{B}$ in place of $\mathscr{H}$. Thus the same four-dimensional equation

$$\frac{\partial F_{\alpha\beta}}{\partial x_\gamma} + \frac{\partial F_{\beta\gamma}}{\partial x_\alpha} + \frac{\partial F_{\gamma\alpha}}{\partial x_\beta} = 0 \tag{29.2}$$

as in Eq. (16.9) can be used for these equations, provided Eqs. (14.25) are replaced by the equations

$$\begin{aligned} F_{ij} &= -F_{ji} = \mathscr{B}_k, \qquad i, j, k = (1, 2, 3) \text{ in cyclic order} \\ F_{4j} &= -F_{j4} = i\mathscr{E}_j, \qquad j = 1, 2, 3. \end{aligned} \tag{29.3}$$

The inhomogeneous Maxwell equations (28.49) and (28.50) differ from those of Eqs. (16.23) and (16.24) for free space by having $\mathscr{D}$ written in place of $\mathscr{E}$. Thus the tensor form in Eq. (16.22) is to be replaced by the equation

$$\frac{\partial G_{\beta\gamma}}{\partial x_\gamma} = 4\pi s_\beta, \tag{29.4}$$

where

$$\begin{aligned} G_{ij} &= -G_{ji} = \mathscr{H}_k, \qquad i, j, k = (1, 2, 3) \text{ in cyclic order} \\ G_{4j} &= -G_{j4} = i\mathscr{D}_j, \qquad j = 1, 2, 3. \end{aligned} \tag{29.5}$$

This covariant formulation of Maxwell's macroscopic equations makes it clear that these equations are form invariant in all inertial coordinate systems. They thus apply equally well to a coordinate system in which the medium is in motion. The constitutive relations of Eqs. (28.86), however, need to be generalized.

In place of these constitutive relations we write*

$$\sum_\beta G_{\alpha\beta}u_\beta = \epsilon \sum_\beta F_{\alpha\beta}u_\beta , \tag{29.6}$$

$$F_{\alpha\beta}u_\gamma + F_{\beta\gamma}u_\alpha + F_{\gamma\alpha}u_\beta = \mu(G_{\alpha\beta}u_\gamma + G_{\beta\gamma}u_\alpha + G_{\gamma\alpha}u_\beta), \tag{29.7}$$

where, as in Eq. (11.24),

$$u_\alpha = \gamma\left(\frac{\mathbf{v}}{c}, i\right) \tag{29.8}$$

with $\gamma = [1 - (v/c)^2]^{-1/2}$ and $\mathbf{v}$ taken as the velocity of the medium. For the coordinate system in which the medium is at rest, assumed to be an inertial system, $\mathbf{v} = 0$ and hence from Eq. (29.8)

$$\begin{aligned} u_j &= 0, \qquad j = 1, 2, 3 \\ u_4 &= i. \end{aligned} \tag{29.9}$$

Thus the constitutive relations given in Eqs. (29.6) and (29.7) become

$$\begin{aligned} G_{i4} &= \epsilon F_{i4} , \\ F_{ij} &= \mu G_{ij} , \end{aligned}$$

which by Eqs. (29.3) and (29.5) are equivalent to the equations

$$\begin{aligned} \mathscr{D} &= \epsilon \mathscr{E}, \\ \mathscr{B} &= \mu \mathscr{H} \end{aligned} \tag{29.10}$$

given in Eq. (28.86).

Since Eqs. (29.6) and (29.7) hold in the inertial system in which the medium is at rest and are written in covariant form, they may be expected to hold in the inertial systems in which the medium is moving with some uniform velocity $\mathbf{v}$. There is implicit in this conclusion, however, the very natural assumption that the field is described in terms of the field tensors $F_{\alpha\beta}$ and $G_{\alpha\beta}$ and transforms accordingly.

* This formulation of the consitutive relations is that used by R. Becker in *Electromagnetic Fields and Interactions* (F. Sauter, ed.; transl. by A. W. Knudsen), Blackie, London and Glasgow, 1964, Vol. I, *Electromagnetic Theory and Relativity*, p. 374.

Making use of the relationship between the field tensors and the field vectors $\mathscr{E}$, $\mathscr{D}$, $\mathscr{H}$, and $\mathscr{B}$ set forth in Eqs. (29.3) and (29.5), we obtain from Eqs. (29.6) to (29.8) the three-dimensional constitutive relations

$$\mathscr{D} + \frac{\mathbf{v}}{c} \times \mathscr{H} = \epsilon\left(\mathscr{E} + \frac{\mathbf{v}}{c} \times \mathscr{B}\right), \tag{29.11}$$

$$\mathscr{B} - \frac{\mathbf{v}}{c} \times \mathscr{E} = \mu\left(\mathscr{H} - \frac{\mathbf{v}}{c} \times \mathscr{D}\right). \tag{29.12}$$

The homogeneous equation (29.2) can be solved, as in the microscopic theory, by introducing the potential 4-vector

$$\varphi_\alpha = (\mathbf{A}, i\varphi) \tag{29.13}$$

and requiring that

$$F_{\alpha\beta} = \frac{\partial \varphi_\beta}{\partial x_\alpha} - \frac{\partial \varphi_\alpha}{\partial x_\beta}. \tag{29.14}$$

The three-dimensional form of this last equation is expressed by the two equations

$$\mathscr{E} = -\nabla\varphi - \frac{1}{c}\frac{\partial \mathbf{A}}{\partial t}, \tag{29.15}$$

$$\mathscr{B} = \nabla \times \mathbf{A} \tag{29.16}$$

in agreement with Eqs. (14.15) and (14.16), but with $\mathscr{B}$ replacing $\mathscr{H}$.

Electromagnetic waves incident on a moving medium. Consider a plane electromagnetic wave propagating in free space in the positive Z direction and incident normally on the plane interface of a medium moving with a velocity v in the positive Z direction (v positive or negative). The set-up is illustrated in Figure 29.1. One is interested in determining the characteristics of the reflected and transmitted waves knowing the frequency, wavelength, and polarization of the incident wave.

Let the X-axis be taken parallel to the electric vector $\mathscr{E}$ and the Y-axis parallel to the magnetic vector $\mathscr{H}$ of the incident wave, the polarization being assumed fixed, and let the incident wave have an angular frequency ω and a propagation vector of magnitude k. The nonzero components of the $\mathscr{E}$ and $\mathscr{H}$ vectors are therefore given by (see Section 18)

$$\mathscr{E}_x = Ce^{i(kz-\omega t)}, \tag{29.17}$$

$$\mathscr{H}_y = Ce^{i(kz-\omega t)}, \tag{29.18}$$

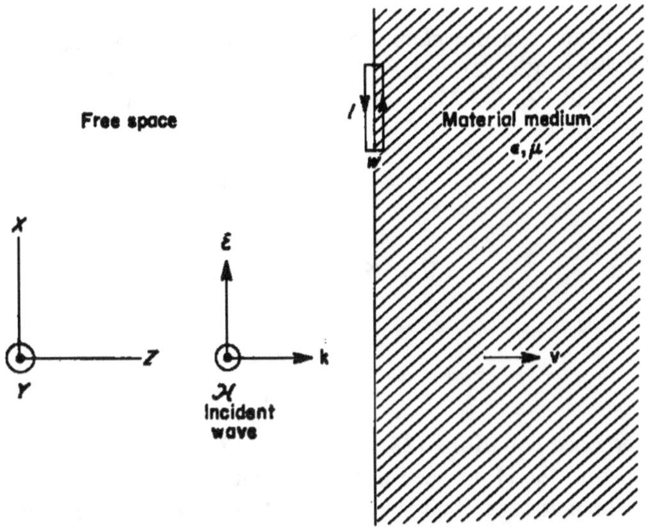

FIG. 29.1.. A plane electromagnetic wave incident normally on the plane interface of a material medium moving away with a velocity v.

where C is a complex constant. The instantaneous values of the fields are given by the real parts of the above quantities and only the real part has a direct physical significance.

To obtain a solution to the problem posed above it is convenient to obtain first a solution in the coordinate system in which the medium is at rest. In this system the incident wave is described by the nonzero components

$$\mathscr{E}'_x = C'_1 e^{i(k'z'-\omega' t')} = C'_1 e^{ik'_\mu x'_\mu}, \tag{29.19}$$

$$\mathscr{H}'_y = C'_1 e^{i(k'z'-\omega' t')} \tag{29.20}$$

with

$$k'_\mu = \left(k'_1, k'_2, k'_3, i\frac{\omega}{c}\right) = \left(0, 0, k', i\frac{\omega'}{c}\right). \tag{29.21}$$

The reflected wave is assumed to be described by the nonzero components

$$\mathscr{E}'_x = C'_2 e^{-i(k'z'+\omega' t')}, \tag{29.22}$$

$$\mathscr{H}'_y = -C'_2 e^{-i(k'z'+\omega' t')} \tag{29.23}$$

and the transmitted wave by

$$\mathscr{E}'_x = C'_3 e^{i(k'_t z'-\omega' t')}, \tag{29.24}$$

$$\mathscr{H}'_y = C'_4 e^{i(k'_t z'-\omega' t')}. \tag{29.25}$$

Note that in the primed system the time factor is always $e^{-i\omega' t'}$.

Since in the primed system the medium is at rest, we may use directly Eqs. (28.49) to (28.52) and Eq. (28.86). In order to include conductive media we add Eq. (28.87). For $\rho = 0$ and a time factor of $e^{-i\omega' t'}$ these equations reduce to

$$\nabla \times \mathscr{E}' = i \frac{\omega'}{c} \mu \mathscr{H}', \tag{29.26}$$

$$\nabla \times \mathscr{H}' = -i \frac{\omega'}{c} \epsilon^i \mathscr{E}', \tag{29.27}$$

where

$$\epsilon^i = \epsilon + i \frac{4\pi c}{\omega} \sigma. \tag{29.28}$$

The divergence equations (28.50) and (28.52) follow directly from these equations and therefore need not be separately imposed.

Using the fields specified in Eqs. (29.24) and (29.25), we have from Eq. (29.26) that

$$C_4' = \frac{c k_t'}{\mu \omega'} C_3' \tag{29.29}$$

and from Eq. (29.27)

$$C_4' = \frac{\omega' \epsilon^i}{c k_t'} C_3'. \tag{29.30}$$

For Eqs. (29.29) and (29.30) to be consistent we must have

$$k_t' = \frac{\omega'}{c} \sqrt{\epsilon^i \mu}, \tag{29.31}$$

which for a conductive medium is a complex number. For the incident and reflected waves $\epsilon^i = \epsilon = 1$ and $\mu = 1$, so

$$k' = \frac{\omega'}{c}. \tag{29.32}$$

To determine the phase velocity v_p' of the incident wave in the primed system one finds how the observer's position z' must change with time t' so that the phase $k'z' - \omega' t'$ in Eqs. (29.19) and (29.20) remains constant. Differentiating the phase with respect to t', we have

$$k' \frac{dz'}{dt'} - \omega' = 0,$$

and thus the phase velocity, by virtue of Eq. (29.32), is

$$v'_p \equiv \frac{dz'}{dt'} = \frac{\omega'}{k'} = c. \tag{29.33}$$

Similarly, the phase velocity for the reflected wave, as expressed by Eqs. (29.22) and (29.23), is $-c$.

For the transmitted wave of Eqs. (29.24) and (29.25) the phase is given by

$$\mathrm{Re}(k'_t)z' - \omega' t',$$

where $\mathrm{Re}(k'_t)$ is the real part of k'_t and thus the phase velocity is

$$v'_p = \frac{\omega'}{\mathrm{Re}(k'_t)} \tag{29.34}$$

within the medium. If the medium is nonconductive ($\sigma = 0$), $\epsilon^i = \epsilon$ and k'_t is real; therefore by Eq. (29.31)

$$v'_p = \frac{\omega'}{\mathrm{Re}(k'_t)} = \frac{\omega'}{k'_t} = \frac{c}{\sqrt{\epsilon\mu}} = \frac{c}{n}, \tag{29.35}$$

where $n = \sqrt{\epsilon\mu}$ is the *index of refraction* of the medium. From Eqs. (29.24) and (29.25) we see that when k'_t is complex ($\sigma \neq 0$) there is an attenuation factor $\exp(-\mathrm{Im}(k'_t)\, z')$, where $\mathrm{Im}(k'_t)$ is the imaginary part of k'_t, in these expressions for the electric and magnetic fields. The power per unit area carried by the transmitted wave is given by the Poynting vector (see Eq. (17.12)) and is proportional to the product $\mathscr{E}_x\mathscr{H}_y$. It thus has an attenuation factor $\exp(-2\,\mathrm{Im}(k'_t)\, z')$. The lost energy goes into heating the medium.

It is convenient to treat the ratio of $\mathscr{E}_x$ to $\mathscr{H}_y$ as a dimensionless *impedance*. Then for the incident wave the impedance is 1 and for the transmitted wave the impedance is

$$Z' = \frac{C'_3}{C'_4} = \sqrt{\frac{\mu}{\epsilon^i}}. \tag{29.36}$$

In the MKS system of units (see Appendix A10) this ratio has the dimensions of ohms and is a true impedance.

By applying Stokes' theorem to Eqs. (29.26) and (29.27) one can show that the tangential components of $\mathscr{E}'$ and $\mathscr{H}'$ must be continuous across the interface in the primed coordinate system in which the medium is at rest. One applies the theorem to the rectangular path shown in Figure 29.1 and then takes the limit as the width w of the rectangle approaches zero.

Taking the interface to be at $z = 0$, one obtains continuity of the tangential components of $\mathscr{E}'$ and $\mathscr{H}'$ by requiring that

$$C_1 + C_2' = C_3'$$

and

$$C_1' - C_2' = C_4' = C_3'/Z'$$

Letting

$$C_2' = R'C_1' \tag{29.37}$$

$$C_3' = T'C_1', \tag{29.38}$$

we have

$$T' = 1 + R'$$

$$T'/Z' = 1 - R';$$

thus

$$T' = \frac{2Z'}{Z' + 1} \tag{29.39}$$

$$R' = \frac{Z' - 1}{Z' + 1} \tag{29.40}$$

These are called, respectively, the *transmission coefficient* and the *reflection coefficient*. One observes that for an impedance match ($Z' = 1$ in this problem) the wave is totally transmitted ($T' = 1$ and $R' = 0$). The coefficient C_1' is arbitrary and merely specifies the strength of the incident wave.

To complete our solution we must transform all quantities to the unprimed system in which the medium has a velocity v. This is done by means of a Lorentz transformation involving the third and fourth components of the four-vectors. In this transformation the inner product $k_\mu' x_\mu'$ is an invariant and so $k_\mu' x_\mu' = k_\mu x_\mu$ or

$$k_3' z' - \omega' t' = k_3 z - \omega t. \tag{29.41}$$

In Eqs. (29.19) and (29.20) $k_3' = k'$ while in Eqs. (29.22) and (29.23) $k_3' = -k'$.

The transformation of the four-vector k_μ is given by

$$\begin{aligned} k_3 &= \gamma\left(k_3' - i\frac{v}{c}k_4'\right) = \gamma\left(k_3' + \frac{v}{c^2}\omega'\right), \\ k_4 &= \gamma\left(k_4' + i\frac{v}{c}k_3'\right), \end{aligned} \tag{29.42}$$

where $\gamma = (1 - v^2/c^2)^{-1/2}$. These transformation equations may be obtained from Eqs. (11.9) by observing that since the relative velocity between the coordinate systems is now along the Z-axes, rather than the X-axes, one needs to make a cyclic permutation of the space components; also, of course, one must treat ict as the fourth component of the four-vector x_α. From the last equation of Eq. (29.42) and from Eq. (29.21) we have

$$\omega = \gamma(\omega' + vk_3'). \tag{29.43}$$

One observes that the frequency shift is quite different for the incident wave ($k_3' = k'$) and for the reflected wave ($k_3' = -k'$). For the incident wave we have

$$k_3^{(1)} = \gamma\left(k' + \frac{v}{c^2}\omega'\right) = \gamma\left(1 + \frac{v}{c}\right)k', \tag{29.44}$$

$$\omega^{(1)} = \gamma\left(1 + \frac{v}{c}\right)\omega', \tag{29.45}$$

and thus the phase velocity $v_p^{(1)}$, by virtue of Eq. (29.34), is

$$v_p^{(1)} = \frac{\omega^{(1)}}{k_3^{(1)}} = \frac{\omega'}{k'} = c, \tag{29.46}$$

the same as in the unprimed system. However, the frequency ω and the propagation constant k are each multiplied by the factor

$$\gamma\left(1 + \frac{v}{c}\right) = \left(\frac{1 + v/c}{1 - v/c}\right)^{1/2} \tag{29.47}$$

relative to those in the primed system. For the reflected wave we likewise have

$$k_3^{(2)} = -\gamma\left(k' - \frac{v}{c^2}\omega'\right) = -\gamma\left(1 - \frac{v}{c}\right)k', \tag{29.48}$$

$$\omega^{(2)} = \gamma\left(1 - \frac{v}{c}\right)\omega', \tag{29.49}$$

and the phase velocity is

$$v_p^{(2)} = \frac{\omega^{(2)}}{k_3^{(2)}} = -\frac{\omega'}{k'} = -c, \tag{29.50}$$

which is again the same as in the primed system. The frequency relative to that in the primed system is multiplied by the factor

$$\gamma\left(1 - \frac{v}{c}\right) = \left(\frac{1 - v/c}{1 + v/c}\right)^{1/2}. \tag{29.51}$$

The above changes in frequency and wavelength are known as *Doppler shifts.* We observe, in particular, that the frequency $\omega^{(2)}$ of the reflected wave from a moving object is obtained from the frequency of the incident wave by multiplying by the factor

$$\frac{(1 - v/c)}{(1 + v/c)}. \tag{29.52}$$

It is clear this shift in frequency can be used in radar applications to distinguish moving objects from stationary ones and to measure the velocity of the former.

For the transmitted wave that is propagating inside the medium one has from Eqs. (29.42) and (29.43)

$$\begin{aligned} k_t &= \gamma\left(k_t' + \frac{v}{c^2}\omega'\right), \\ \omega_t &= \gamma(\omega' + vk_t'). \end{aligned} \tag{29.53}$$

For a nonconductive medium ($\sigma = 0$) one has by Eq. (29.35)

$$\begin{aligned} k_t &= \gamma\left(1 + \frac{v}{nc}\right)k_t', \\ \omega_t &= \gamma\left(1 + \frac{nv}{c}\right)\omega', \end{aligned} \tag{29.54}$$

and thus the phase velocity is

$$v_p = \frac{\omega_t}{k_t} = \frac{1 + nv/c}{1 + v/nc}\frac{\omega'}{k_t'} = \frac{1 + nv/c}{1 + v/nc}\frac{c}{n}. \tag{29.55}$$

To the first order in v/c one has that

$$v_p = \left[1 + \left(n - \frac{1}{n}\right)\frac{v}{c}\right]\frac{c}{n} = \frac{c}{n} + \left(1 - \frac{1}{n^2}\right)v. \tag{29.56}$$

Thus the wave is partially carried along by the medium and given an added velocity of $(1 - 1/n^2)v$ due to the velocity v of the medium. Fizeau obtained Eq. (29.56) as an empirical formula for the effect of a moving medium on the velocity of propagation.

The transformation of the components of the electric and magnetic fields outside the medium are governed by the inverse of Eqs. (15.5) with, however, a permutation of the components: $x \to z \to y \to x$ to account for the velocity now being in the Z direction. One has then

$$\mathscr{E}'_x = \gamma\left(\mathscr{E}_x - \frac{v}{c}\mathscr{H}_y\right), \qquad \mathscr{H}'_y = \gamma\left(\mathscr{H}_y - \frac{v}{c}\mathscr{E}_x\right),$$

and, therefore, the inverse transforms (obtained by replacing v by $-v$) are

$$\mathscr{E}_x = \gamma\left(\mathscr{E}'_x + \frac{v}{c}\mathscr{H}'_y\right), \qquad \mathscr{H}_y = \gamma\left(\mathscr{H}'_y + \frac{v}{c}\mathscr{E}'_x\right). \tag{29.57}$$

Applying Eqs. (29.57) to the incident wave of Eqs. (29.19) and (29.20) and making use of the invariance of the phase, namely, that

$$k'z' - \omega't' = k^{(1)}z - \omega^{(1)}t, \tag{29.58}$$

one has

$$\begin{aligned}\mathscr{E}_x^{(1)} &= \gamma\left(1 + \frac{v}{c}\right)\mathscr{E}'_x = \gamma\left(1 + \frac{v}{c}\right)C'_1 e^{i(k^{(1)}z - \omega^{(1)}t)} \\ &= C_1 e^{i(k^{(1)}z - \omega^{(1)}t)} = \mathscr{H}_y^{(1)}.\end{aligned} \tag{29.59}$$

Likewise for the reflected wave

$$\begin{aligned}\mathscr{E}_x^{(2)} &= \gamma\left(1 - \frac{v}{c}\right)\mathscr{E}'_x = \gamma\left(1 - \frac{v}{c}\right)C'_2 e^{-i(k^{(2)}z + \omega^{(2)}t)} = -\mathscr{H}_y^{(2)} \\ &= RC_1 e^{-i(k^{(2)}z + \omega^{(2)}t)},\end{aligned} \tag{29.60}$$

which defines the reflection coefficient R.

From Eqs. (29.37), (29.40), (29.59), and (29.60) we have

$$R = \frac{1 - v/c}{1 + v/c}R' = \frac{1 - v/c}{1 + v/c}\frac{Z' - 1}{Z' + 1} = \frac{Z_e - 1}{Z_e + 1}, \tag{29.61}$$

where

$$Z_e = \frac{Z' + v/c}{1 + (v/c)Z'}. \tag{29.62}$$

To obtain the transformation of the wave within the medium we use the transformation equations for the tensor components $F_{\alpha\beta}$ and $G_{\alpha\beta}$. These are obtained from Eqs. (15.4) by a permutation $1 \to 3 \to 2 \to 1$ of the indices and a change from v to $-v$ to obtain the inverse of the transformation; thus

$$F_{41} = \gamma\left(F'_{41} + i\frac{v}{c}F'_{31}\right),$$

$$G_{31} = \gamma\left(G'_{31} - i\frac{v}{c}G'_{41}\right)$$

or

$$\mathscr{E}_x = \gamma\left(\mathscr{E}'_x + \frac{v}{c}\mathscr{B}'_y\right) = \gamma\left(\mathscr{E}'_x + \frac{v}{c}\mu\mathscr{H}'_y\right), \tag{29.63}$$

$$\mathscr{H}_y = \gamma\left(\mathscr{H}'_y + \frac{v}{c}\mathscr{D}'_x\right) = \gamma\left(\mathscr{H}'_y + \frac{v}{c}\epsilon\mathscr{E}'_x\right). \tag{29.64}$$

Using Eqs. (29.24), (29.25), and (29.36) and the invariance of the phase, $k'_t z' - \omega' t' = k_t z - \omega_t t$, we have

$$\mathscr{E}_x = \gamma\left(1 + \frac{v}{c}\frac{\mu}{Z'}\right)\mathscr{E}'_x = \gamma\left(1 + \frac{v}{c}\frac{\mu}{Z'}\right)C'_3 e^{i(k_t z - \omega_t t)}$$
$$= TC_1 e^{i(k_t z - \omega_t t)} \tag{29.65}$$

and

$$\mathscr{H}_y = \gamma\left(1 + \frac{v}{c}\epsilon Z'\right)\mathscr{H}'_y = \gamma\left(1 + \frac{v}{c}\epsilon Z'\right)C'_4 e^{i(k_t z - \omega_t t)}$$
$$= Z^{-1}TC_1 e^{i(k_t z - \omega_t t)}, \tag{29.66}$$

which define T and Z.

From Eqs. (29.38), (29.39), (29.59), and (29.65)

$$T = \frac{1 + \frac{v}{c}\frac{\mu}{Z'}}{1 + \frac{v}{c}}T' = \frac{1 + \frac{v}{c}\frac{\mu}{Z'}}{1 + \frac{v}{c}}\frac{2Z'}{Z' + 1}, \tag{29.67}$$

and from Eqs. (29.36), (29.65), and (29.66)

$$Z = \frac{1 + \frac{v}{c}\frac{\mu}{Z'}}{1 + \frac{v}{c}\epsilon Z'}Z' = \frac{Z' + \frac{v}{c}\mu}{1 + \frac{v}{c}\epsilon Z'}. \tag{29.68}$$

For a lossless medium ($\sigma = 0$), by Eqs. (29.28) and (29.36), $Z' = \sqrt{\mu/\epsilon}$ and thus

$$\epsilon Z' = \sqrt{\epsilon\mu} = n = \frac{\mu}{Z'}. \tag{29.69}$$

Equations (29.67) and (29.68) then become

$$T = \frac{1 + n(v/c)}{1 + v/c} T' = \frac{1 + n(v/c)}{1 + v/c} \frac{2Z'}{Z' + 1} \tag{29.70}$$

and

$$Z = Z'. \tag{29.71}$$

Continuity of the tangential component of the electric field at the boundary requires, by Eqs. (29.59), (29.60), and (29.65), that $T = 1 + R$. From Eqs. (29.61) and (29.62)

$$1 + R = \frac{2Z_e}{Z_e + 1} = \frac{1 + (v/c)(1/Z')}{1 + v/c} \frac{2Z'}{Z' + 1}, \tag{29.72}$$

and, as seen from Eq. (29.67), this is equal to T only if $\mu = 1$ (or, of course, $v = 0$). The continuity of the tangential electric field, as seen earlier, is based on the application of Stokes' theorem to the thin rectangular loop shown in Figure 29.1, using Maxwell's equation (28.51). For the medium at rest the rate of change of the magnetic flux, the integral of $\mathscr{B}$ over the surface of the loop, approaches zero as the width of the rectangle approaches zero. On the other hand, when the boundary is moving, the loop being necessarily at rest in the unprimed coordinate system, the flux changes at a finite rate no matter how thin the rectangle is. This change in the flux is due to the change in the fractional part of the loop area for which $\mu \neq 1$.

The tangential component of the magnetic vector $\mathscr{H}$ is continuous at the boundary (see Eqs. (29.59), (29.60), and (29.66)) if $Z^{-1}T = 1 - R$. Again from Eqs. (29.61) and (29.62)

$$1 - R = \frac{2}{Z_e + 1} = \frac{1 + (v/c)Z'}{1 + v/c} \frac{2}{Z' + 1}, \tag{29.73}$$

and from Eqs. (29.67) and (29.68)

$$Z^{-1}T = \frac{1 + (v/c)\epsilon Z'}{1 + v/c} \frac{2}{Z' + 1}. \tag{29.74}$$

These two expressions are equal only if $\epsilon = 1$ (or again $v = 0$). Since usually $\epsilon \neq 1$ for a material medium, we conclude that the tangential magnetic field is generally discontinuous at the boundary of a moving medium.

Problems

1. Show in detail how Eq. (28.18) is derived from Eq. (28.10).

2. Give details on the derivation of Eq. (28.38) from Eq. (28.18).

3. Show that Eqs. (28.39) and (28.40) may be obtained by setting the variation of $W = W_p + W_{mf} + W_{ef} + W_f$ with respect to the potentials equal to zero.

4. Show that Maxwell equations for a material media without charges and currents are invariant under the following changes:

$$\mathscr{E} \to \mathscr{H}, \qquad \mathscr{H} \to -\mathscr{E}, \qquad \mathscr{D} \to \mathscr{B}, \qquad \mathscr{B} \to -\mathscr{D}.$$

5. Derive the wave equation for propagation in a conductive medium.

6. Show that for waves containing only frequencies satisfying the condition $\omega \ll \epsilon/\sigma$ the wave equation in problem 5 reduces to the equation

$$\nabla^2\mathscr{E} - \mu\sigma\frac{\partial\mathscr{E}}{\partial t} = 0.$$

Compare this equation with the heat flow and diffusion equations.

7. For pure metals the relaxation time $\tau = \epsilon/\sigma$ is of the order of 10^{-14} sec. For what frequencies will the *EM* waves satisfy the diffusion-type equation of problem 6?

8. (a) Show that the force F per unit area exerted by the electric field inside a capacitor on the bound surface charge of the dielectric is

$$F = \frac{1}{8\pi}(\mathscr{E}_0^2 - \mathscr{E}_1^2) = \frac{\varepsilon^2 - 1}{8\pi}\mathscr{E}_1^2$$

where $\mathscr{E}_0$ is the electric field outside the dielectric, $\mathscr{E}_1$ the electric field inside the dielectric, and ϵ is the dielectric constant of the medium.

(b) Show that the tension on the dielectric per unit area is not F, however, but may be obtained from F by adding the force on the other half of all the dipoles not entirely contained within the surface layer. Show that this additional force is just $P_1\mathscr{E}_1$, where P_1 is the polarization within the body of the dielectric.

(c) Show that the tension may be obtained directly from the force $P\,dE/dx$ per unit volume acting on the dipoles created in the material by the electric field and that this tension is

$$T = \frac{1}{8\pi}(\mathscr{E}_0 - \mathscr{E}_1)^2 = \frac{(\varepsilon - 1)^2}{8\pi}\mathscr{E}_1^2.$$

9. (a) Obtain the tension in the dielectric of problem 8 by considering the change in the stored electrical energy U with a mechanical stretching of the dielectric. If the thickness of the dielectric is designated by x, then

$$T = -\frac{\partial U}{\partial x} = \frac{\mathscr{E}_1^2}{8\pi}\left[\varepsilon(\varepsilon - 1) - \rho\frac{\partial \varepsilon}{\partial \rho}\right],$$

where ρ is the density.

(b) Show that if one takes account only of the change in the number of dipoles per unit volume due to an expansion, that $\varepsilon = 1 + Cx$ and that T in the equation above reduces to that obtained in problem 8.

10. Let an averaging operator A be defined by

$$Af(x, t) = \bar{f}(x, t) \equiv \int_{-\infty}^{\infty} K(u) f(x + u, t)\, du,$$

where the weighting function $K(x)$ is such that $K(x) \to 0$ as $|x| \to \infty$. Show that this operator commutes with the derivative operators $D_x \equiv \partial/\partial x$ and $D_t \equiv \partial/\partial t$, i.e., that

$$(DA - AD)f(x, t) = 0,$$

where D is D_x or D_t.

11. Show that the tangential components of $\mathscr{E}$ and $\mathscr{H}$ are generally continuous across a plane interface between two media, but that this is not necessarily true at the surface of a perfect conductor. Why is it also not generally true when there is a moving interface?

12. How does the Poynting theorem generalize for a moving dielectric medium? What are the expressions for the energy density and the flow of energy across a unit area of surface?

13. Determine the field of a stationary oscillating electric dipole in a moving dielectric medium.

14. Derive the Fresnel equations for reflection from a stationary dielectric medium having a small conductivity.

15. Derive the Fresnel equations for reflection from the surface of a nonconductive dielectric medium that is in motion. How do these simplify for a nonrelativistic velocity of the medium ($v \ll c$)?

Appendix A1

Concept of a Vector

Some physical quantities, such as displacement, velocity, acceleration, and force, require not only a magnitude but also a direction to describe them and are known as *vector quantities*. From such quantities we abstract the mathematical concept of a three-dimensional *vector*. This is a mathematical entity having a *magnitude* (given by a nonnegative real number) and a *direction* (described by two other real numbers). It can be represented by a directed line segment, such as OP in Figure A1.1, whose length is made proportional to the magnitude of the vector. The

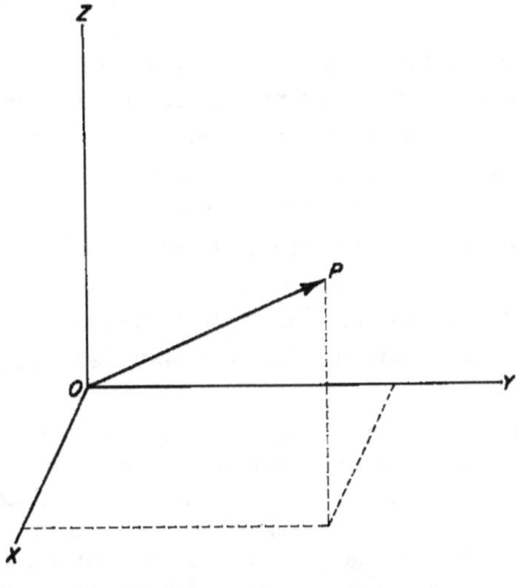

FIG. A1.1. The directed line segment (OP) as a simple example of a vector quantity.

mathematical nature of such a vector is contained in a set of postulates relating to two basic operations that may be performed with vectors: the *addition* of two vectors and the *multiplication of a vector by a number.** These operations will be explained more fully later.

A vector quantity may have other properties or requirements than those expressed by the vector associated with it. Thus a force involves also a line of application. Some vector quantities, however, are so completely given by the associated vector that we shall ignore the distinction between them.

An easily visualized vector quantity is the location of a point P in three-space relative to some point O called the *origin*. The corresponding vector is called a *position vector*. It can be represented by the directed line segment from O to P and labeled OP. Point O is then called the *origin* of the vector and point P its *end*, or *terminus*.

This position vector can also be represented by any directed line segment $O'P'$ that is parallel to OP and has the same length, because such a line segment equally well specifies the location of P relative to O. This is illustrative of a property of vectors: vectors may be moved about without change as long as their magnitude and direction remain the same. Even more generally, a position vector is specified by giving the spherical coordinates (r, θ, φ) of the point P with respect to the origin O, or of the point P' with respect to O'. In fact, any set of three quantities that determine both a direction and a length may be used to represent a position vector.†

It is usually convenient to use, as the three numbers defining a position vector, the lengths of the three projections of the directed line OP, or $O'P'$, on the three Cartesian coordinate axes. These are, of course, merely the Cartesian coordinates of the point P; in this connection, however, they are called the three *components* of the vector.

We shall denote a vector by a single letter in boldface type, or, when the coordinate system meant is understood, by its components. We shall use the symbol $\sim$ to denote the identity of vectors that are specified in different ways. Thus a position vector $\mathbf{r}$ may also be specified in a given

* Only these operations are used to define a *vector space;* however, it is customary to assume that two other operations are defined in three-dimensional vector analysis, namely, a *scalar product* and a *cross product*, which will be defined later.

† A position vector is not a true vector in that one needs to specify also the origin used. Thus it is not completely independent of the coordinate system. The concept of a true vector should be independent of any coordinate system, even though its concrete specification in terms of three quantities is basically a coordinate-system type description.

unprimed coordinate system by (x, y, z) and one may write

$$\mathbf{r} \sim (x, y, z). \tag{A1.1}$$

Another notation, which we shall often use, is to number the components consecutively, thus:

$$x = x_1, \qquad y = x_2, \qquad z = x_3. \tag{A1.2}$$

Here the same letter, designating what vector we are dealing with, is used for all the three components, with the numerical subscript indicating which of the components is being considered.

In this notation Eq. (A1.1) would be written $\mathbf{r} \sim (x_1, x_2, x_3)$. It is often convenient, when no confusion is likely to arise, to speak of the vector x_j, where one thinks of x_j as standing for the whole set of components (x_1, x_2, x_3). Thus Eq. (A1.1) can also be written

$$\mathbf{r} \sim x_j. \tag{A1.3}$$

The magnitude of a position vector $\mathbf{r}$ is usually referred to as its *length* and is designated by $|\mathbf{r}|$, or simply by r. In a Cartesian coordinate system the length of a position vector $\mathbf{r} \sim (x, y, z)$ is defined to be

$$r = (x^2 + y^2 + z^2)^{1/2}.$$

A rotation of axes involves a change to another coordinate system, which we shall designate as the primed coordinate system, without a change in the length of any position vectors; thus

$$r = (x^2 + y^2 + z^2)^{1/2} = (x'^2 + y'^2 + z'^2)^{1/2} = r'. \tag{A1.4}$$

In a rotation of coordinate axes the new coordinates are linear homogeneous functions of the old:

$$\begin{aligned} x' &= a_{11}x + a_{12}y + a_{13}z, \\ y' &= a_{21}x + a_{22}y + a_{23}z, \\ z' &= a_{31}x + a_{32}y + a_{33}z. \end{aligned}$$

The a's are constants which, in order that the transformation of coordinates may be a rotation, must satisfy certain well-known conditions. For a discussion of these conditions see Section A3. These equations may be written in the form

$$\begin{aligned} x_1' &= a_{11}x_1 + a_{12}x_2 + a_{13}x_3, \\ x_2' &= a_{21}x_1 + a_{22}x_2 + a_{23}x_3, \\ x_3' &= a_{31}x_1 + a_{32}x_2 + a_{33}x_3, \end{aligned}$$

which may be written as

$$x'_j = \sum_{k=1}^{3} a_{jk} x_k, \qquad j = 1, 2, 3.$$

In the future, in accordance with the usual summation convention, we shall not write down the summation sign, but will assume that *a summation is implied whenever a literal subscript occurs twice in the same term.* Our three equations of transformation are thus condensed to

$$x'_j = a_{jk} x_k, \tag{A1.5}$$

the constants a_{jk} being certain known quantities for each rotation. We shall also assume that all unrepeated indices, such as j in this equation, take on all the values they are ever permitted to assume (here the values 1, 2, and 3) and that there is an equation corresponding to each set of values these indices assume.

We are now in a position to define a vector in general. We shall call a *vector* any mathematical construct having three components (in a three-dimensional space, otherwise the number equal to the number of dimensions of the space under consideration), each having a special reference to one of the Cartesian coordinate axes, and transforming as the components of a position vector, when the coordinate axes are rotated. Thus, if $\mathbf{A}$ is a vector,

$$\mathbf{A} \sim (A_x, A_y, A_z) = (A_1, A_2, A_3) \sim A_j \tag{A1.6}$$

are the various ways of expressing it in terms of its components. When the coordinate axes are changed and the components of a position vector are transformed according to Eq. (A1.5), the new components of the vector $\mathbf{A}$ will be

$$A'_j = a_{jk} A_k, \tag{A1.7}$$

where the coefficients a_{jk} are the same as in the equation of transformation of the position vector.

The *negative* of a given vector $\mathbf{A}$ is a vector, designated as $-\mathbf{A}$, all components of which are the negatives of the components of the given vector. Another way of saying this is that the negative of a vector has the same magnitude but a direction opposite to the direction of the given vector. In general, we may write that if $\mathbf{B} = -\mathbf{A}$, then

$$B_j = -A_j \qquad \text{or} \qquad \mathbf{B} \sim (-A_x, -A_y, -A_z). \tag{A1.8}$$

If $\mathbf{r}_{PQ}$ is the vector representing the displacement of Q from P, then $\mathbf{r}_{QP}$

is the displacement of P from Q and

$$\mathbf{r}_{QP} = -\mathbf{r}_{PQ}. \tag{A1.9}$$

A *zero vector*, or a *null vector*, is a vector all components of which are zero; it will be designated by **0** or, more often, simply by 0.

If we have a number of points, say the points A, B, C, D, and E of Figure A1.2, we may trace them in some order by taking a succession of

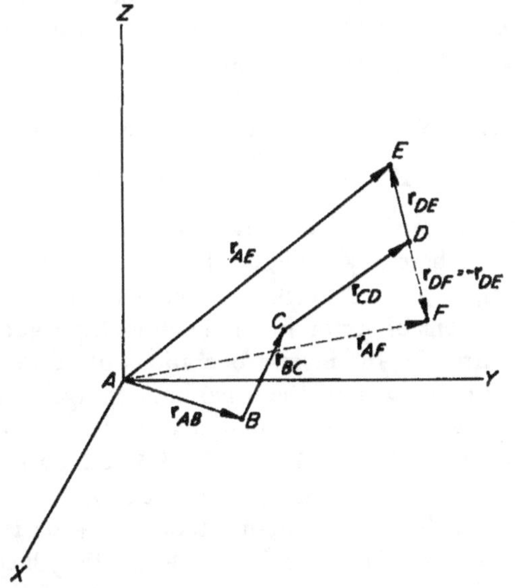

FIG. A1.2. The sum of vectors as represented by directed line segments and the meaning of the negative of a vector.

displacements, say $\mathbf{r}_{AB}$, $\mathbf{r}_{BC}$, $\mathbf{r}_{CD}$, $\mathbf{r}_{DE}$, as in the figure, arriving at point E; or, we may proceed directly from A to E, along the vector $\mathbf{r}_{AE}$, arriving at the same point. This simple fact illustrates the idea of vector addition and can be stated symbolically thus:

$$\mathbf{r}_{AE} = \mathbf{r}_{AB} + \mathbf{r}_{BC} + \mathbf{r}_{CD} + \mathbf{r}_{DE}, \tag{A1.10}$$

which is easy to remember by the sequence of the subscripts. Vector $\mathbf{r}_{AE}$ is called the *sum* of the other vectors in Eq. (A1.10).

A difference is defined with the help of Eq. (A1.9). To subtract a vector is to add its negative. Thus, if in the sequence of vectors in Eq. (A1.10) one vector had occurred with the opposite sign, say if the last term had been $-\mathbf{r}_{DE}$, we could replace it by its equal $\mathbf{r}_{ED} = \mathbf{r}_{DF}$ and thus obtain for the sum the displacements $\mathbf{r}_{AF}$ (see Figure A1.2).

In the general case, however, we define the sum of vectors through their components. Thus $\mathbf{C} = \mathbf{A} + \mathbf{B}$ is that vector whose components C_j are given by

$$C_j = A_j + B_j. \tag{A1.11}$$

This definition of the sum of vectors agrees, of course, with Eq. (A1.10), which is a specialization of it to the case of displacement vectors.

When we add several, say n, equal vectors we obtain a vector that we may denote by prefixing the number n. Thus, $n\mathbf{A}$ will be a vector with components nA_j. We generalize this procedure by writing

$$\mathbf{B} = k\mathbf{A} \qquad \text{and} \qquad B_j = kA_j \tag{A1.12}$$

as equivalent equations. This defines $k\mathbf{A}$ as a vector with components kA_j. The quantity k is here any quantity of ordinary algebra, i.e., a number. Such quantities, in contradistinction to vectors, are called *scalars*.

For the operations of vector addition and multiplication by a scalar (as well as, of course, for the inverse operations of vector subtraction and division of vectors by scalars) the usual laws of algebra hold. These can all be easily proved from the definitions of the operations. Thus, for example, to prove that $k(\mathbf{A} + \mathbf{B}) = k\mathbf{A} + k\mathbf{B}$ we show that corresponding components of each side of this equation are equal. Thus, in accordance with Eqs. (A1.12), the jth component of $k(\mathbf{A} + \mathbf{B})$ is $k(\mathbf{A} + \mathbf{B})_j$ and $(\mathbf{A} + \mathbf{B})_j$, by Eq. (A1.11), is equal to $A_j + B_j$; the jth component of the left-hand side of our equation is therefore equal to $k(A_j + B_j)$. By using first Eq. (A1.11) and then Eqs. (A1.12) the jth component of the right-hand side of our equation is seen to be equal to $kA_j + kB_j$. The corresponding components are therefore equal by the distributive law of ordinary algebra, and this, by definition, insures the equality of the vectors.

We can now define the differentiation of a vector with respect to a scalar. Thus, if $\mathbf{A} = \mathbf{A}(t)$, where t is some scalar variable, we may define $d\mathbf{A}/dt$ as follows:

$$\frac{d\mathbf{A}}{dt} \equiv \lim_{h \to 0} \frac{\mathbf{A}(t + h) - \mathbf{A}(t)}{h}. \tag{A1.13}$$

By taking components, we see that the equations

$$\mathbf{B} = \frac{d\mathbf{A}}{dt} \qquad \text{and} \qquad B_j = \frac{dA_j}{dt} \tag{A1.14}$$

are equivalent.

From this definition follows at once the distributive law for the differentiation of a sum of two vectors with respect to a given scalar variable:

$$\frac{d}{dt}(\mathbf{A} + \mathbf{B}) = \frac{d\mathbf{A}}{dt} + \frac{d\mathbf{B}}{dt}. \tag{A1.15}$$

The process of differentiation with respect to a scalar enables us to construct new vectors. If $\mathbf{r}(t)$ is the position vector of a particle at the time t, then $\mathbf{r}(t + h) - \mathbf{r}(t)$ is the subsequent displacement of the particle in a time h. It is thus clear that

$$\dot{\mathbf{r}}(t) \equiv \frac{d\mathbf{r}}{dt} = \operatorname*{Lim}_{h\to 0} \frac{\mathbf{r}(t + h) - \mathbf{r}(t)}{h} \tag{A1.16}$$

is the instantaneous rate of displacement of the particle at the time t. This is what we refer to as the *velocity* of the particle and is usually designated by the vector $\mathbf{v}$. Similarly, by differentiating once more we obtain the *acceleration*

$$\dot{\mathbf{v}} = \ddot{\mathbf{r}} = \frac{d\mathbf{v}}{dt} = \frac{d^2\mathbf{r}}{dt^2}. \tag{A1.17}$$

It is usually designated by the vector $\mathbf{a}$.

Appendix A2

Vector Algebra

We have seen that corresponding to a position vector $\mathbf{r}$ there is a scalar, the length of the vector, which we designated by r and which satisfies the relation

$$r^2 = x^2 + y^2 + z^2. \tag{A2.1}$$

This relation is not changed when the coordinate system is rotated, and the value of r in the new coordinate system is, of course, the same as in the old.

Analogously we can define the *magnitude* A (also designated by $|\mathbf{A}|$), or the *length*, of any vector $\mathbf{A}$ by the equation

$$|\mathbf{A}|^2 = A^2 = A_x^2 + A_y^2 + A_z^2 = A_j^2, \tag{A2.2}$$

where in the last expression, since the index j can be said to have occurred twice ($A_j^2 = A_jA_j$), a summation with respect to j is implied. Since the components of $\mathbf{A}$, when the coordinate system is rotated, are required to transform like x, y, and z, the length of a vector will be transformed like r; that is, the magnitude of a vector is unchanged by a rotation of coordinate axes. The quantities that have this property are called *invariants* under the transformation considered. We shall regard invariants as scalars and shall use the two terms interchangeably.

We shall say that a given vector is a *unit vector* if it has unit magnitude. The components of such a vector are pure numbers and have no dimensions, because otherwise the magnitude of the vector would depend on the system of units used. All unit vectors differ only in direction. We say therefore that a unit vector determines a direction, and is determined by a direction. It is for this reason sometimes called the *direction vector*. Thus, a direction is determined by

$$\mathbf{u} \quad \text{when} \quad u^2 = 1. \tag{A2.3}$$

As typical direction vectors we have unit vectors along the three coordinate axes in a particular coordinate system. We shall call these vectors *fundamental unit vectors* and shall designate them by $\mathbf{i}$, $\mathbf{j}$, and $\mathbf{k}$, respectively. Their components are

$$\mathbf{i} = (1, 0, 0), \qquad \mathbf{j} = (0, 1, 0), \qquad \mathbf{k} = (0, 0, 1). \tag{A2.4}$$

It is often useful to represent a given vector as a linear combination of the fundamental unit vectors. This is easily done if we know the components of the vector. In fact, if the components of a vector are A_x, A_y, and A_z, it can be written in the form

$$\mathbf{A} = A_x\mathbf{i} + A_y\mathbf{j} + A_z\mathbf{k}, \tag{A2.5}$$

which can be verified by taking the components of both sides of the equation.*

We shall consider two types of products of vectors. The first type, called the *scalar product*, is indicated by a dot, and is defined as follows:

$$\mathbf{A} \cdot \mathbf{B} = A_xB_x + A_yB_y + A_zB_z = A_jB_j. \tag{A2.6}$$

Since A_j and B_j transform alike, $\mathbf{A} \cdot \mathbf{B}$ will transform as A_j^2 of Eq. (A2.2), that is, it will be invariant. Equation (A2.6) may be regarded as a generalization of Eq. (A2.2), which could be written as

$$A^2 = \mathbf{A} \cdot \mathbf{A}.$$

We can now define the angle θ between two vectors $\mathbf{A}$ and $\mathbf{B}$ by the equation

$$\mathbf{A} \cdot \mathbf{B} = AB \cos\theta, \qquad 0 \leq \theta \leq \pi. \tag{A2.7}$$

In the special case when one of the two vectors is a unit vector, we have

$$\mathbf{A} \cdot \mathbf{u} = A \cos\theta = A_u, \tag{A2.8}$$

which is the projection of A on the direction of $\mathbf{u}$. In particular, we have

$$\mathbf{A} \cdot \mathbf{i} = A_x, \qquad \mathbf{A} \cdot \mathbf{j} = A_y, \qquad \mathbf{A} \cdot \mathbf{k} = A_z. \tag{A2.9}$$

When $\mathbf{A}$ is perpendicular to $\mathbf{B}$, we write $\mathbf{A} \perp \mathbf{B}$ and, since $\cos\theta = 0$, we

* If the fundamental unit vectors were not dimensionless the components of a vector would have different dimensions than the vector itself. This point is often overlooked in the treatment of physical vectors.

conclude from Eq. (A2.7) that

$$\mathbf{A} \perp \mathbf{B} \Rightarrow \mathbf{A} \cdot \mathbf{B} = 0; \tag{A2.10}$$

when **A** is parallel to **B**, we write $\mathbf{A} \parallel \mathbf{B}$ and, since $\cos\theta = 1$, conclude that

$$\mathbf{A} \parallel \mathbf{B} \Rightarrow \mathbf{A} \cdot \mathbf{B} = AB. \tag{A2.11}$$

The symbol $\Rightarrow$ is to be read "implies that." From these results we conclude that

$$\begin{aligned} \mathbf{i} \cdot \mathbf{i} = \mathbf{j} \cdot \mathbf{j} = \mathbf{k} \cdot \mathbf{k} = 1, \\ \mathbf{i} \cdot \mathbf{j} = \mathbf{j} \cdot \mathbf{i} = \mathbf{j} \cdot \mathbf{k} = \mathbf{k} \cdot \mathbf{j} = \mathbf{k} \cdot \mathbf{i} = \mathbf{i} \cdot \mathbf{k} = 0. \end{aligned} \tag{A2.12}$$

The scalar product of two vectors obeys the commutative and distributive laws of algebra. Thus, from the definition in Eq. (A2.6), it follows directly that

$$\mathbf{A} \cdot \mathbf{B} = \mathbf{B} \cdot \mathbf{A}. \tag{A2.13}$$

The distributive law

$$\mathbf{A} \cdot (\mathbf{B} + \mathbf{C}) = \mathbf{A} \cdot \mathbf{B} + \mathbf{A} \cdot \mathbf{C}, \tag{A2.14}$$

can be proved as follows. The scalar product, from its definition, is independent of the choice of coordinate axes. Take the X-axis along vector **A**, so that $\mathbf{A} = A\mathbf{i}$. The two expressions that have to be proved equal to each other become then $A\mathbf{i} \cdot (\mathbf{B} + \mathbf{C})$ and $A\mathbf{i} \cdot \mathbf{B} + A\mathbf{i} \cdot \mathbf{C}$; these, in accordance with Eq. (A2.9), become $A(\mathbf{B} + \mathbf{C})_x$ and $AB_x + AC_x$. The first of these reduces, with the help of Eq. (A1.12), to $A(B_x + C_x)$ and by ordinary algebra is seen to be equal to the second expression.

From Eqs. (A2.13) and (A2.14) it also follows that

$$(\mathbf{B} + \mathbf{C}) \cdot \mathbf{A} = \mathbf{B} \cdot \mathbf{A} + \mathbf{C} \cdot \mathbf{A},$$

because

$$(\mathbf{B} + \mathbf{C}) \cdot \mathbf{A} = \mathbf{A} \cdot (\mathbf{B} + \mathbf{C}) = \mathbf{A} \cdot \mathbf{B} + \mathbf{A} \cdot \mathbf{C} = \mathbf{B} \cdot \mathbf{A} + \mathbf{C} \cdot \mathbf{A}.$$

As a simple example of the use of commutative and associative laws we derive the expression in Eq. (A2.6) for the scalar product starting from Eqs. (A2.5) and (A2.12):

$$\begin{aligned} \mathbf{A} \cdot \mathbf{B} &= (A_x\mathbf{i} + A_y\mathbf{j} + A_z\mathbf{k}) \cdot (B_x\mathbf{i} + B_y\mathbf{j} + B_z\mathbf{k}) \\ &= A_xB_x\mathbf{i} \cdot \mathbf{i} + A_yB_y\mathbf{j} \cdot \mathbf{j} + A_zB_z\mathbf{k} \cdot \mathbf{k} \\ &\quad + (A_xB_y + A_yB_x)\mathbf{i} \cdot \mathbf{j} + (A_yB_z + A_zB_y)\mathbf{j} \cdot \mathbf{k} + (A_zB_x + A_xB_z)\mathbf{k} \cdot \mathbf{i} \\ &= A_xB_x + A_yB_y + A_zB_z = A_iB_i. \end{aligned}$$

The other type of a product of two vectors **A** and **B** is called the *vector product*, or *cross product*, and is designated by **A** × **B**. It is a vector perpendicular to the plane of the two vectors and having the magnitude $AB \sin \theta$, where $0 \leq \theta \leq \pi$ is the smaller of the angles between the vectors. Thus, if **n** of Figure A2.1 is a unit vector, normal to the plane of

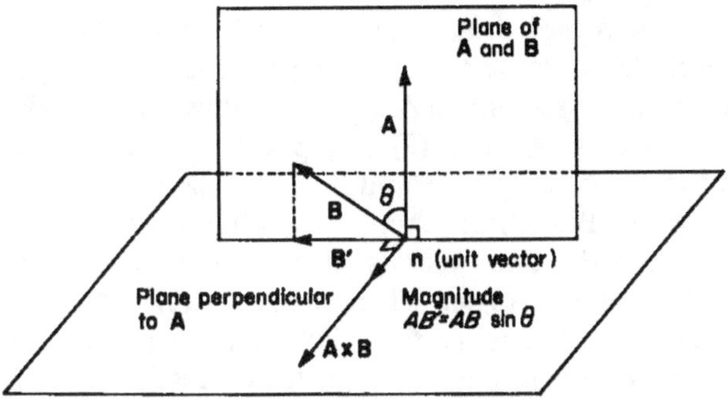

FIG. A2.1. The relationship of the cross product **A** × **B** to the vectors **A** and **B**.

A and **B**, and directed in such a way that looking in the direction of **n** the rotation from **A** to **B** through the angle θ appears positive (clockwise), then

$$\mathbf{A} \times \mathbf{B} = [AB \sin \theta]\mathbf{n}. \tag{A2.15}$$

The vector product is therefore the directed area of the parallelogram formed by **A** and **B**. If the two vectors are parallel, their vector product is zero; in particular, we have

$$\begin{aligned} &\mathbf{i} \times \mathbf{i} = \mathbf{j} \times \mathbf{j} = \mathbf{k} \times \mathbf{k} = 0, \\ &\mathbf{i} \times \mathbf{j} = \mathbf{k}, \qquad \mathbf{j} \times \mathbf{i} = -\mathbf{k}, \qquad \mathbf{j} \times \mathbf{k} = \mathbf{i}, \\ &\mathbf{k} \times \mathbf{j} = -\mathbf{i}, \qquad \mathbf{k} \times \mathbf{i} = \mathbf{j}, \qquad \mathbf{i} \times \mathbf{k} = -\mathbf{j}. \end{aligned} \tag{A2.16}$$

The commutative law of multiplication *does not hold* for vector products, as is easily seen from the definition (A2.15), and in fact

$$\mathbf{A} \times \mathbf{B} = -\mathbf{B} \times \mathbf{A}. \tag{A2.17}$$

The associative law of multiplication also does not hold. Thus, **A** × (**B** × **C**) is perpendicular to **B** × **C**, which is perpendicular to the plane of **B** and **C**; therefore **A** × (**B** × **C**) lies in the plane of **B** and **C**. On the other hand (**A** × **B**) × **C** lies in the plane of **A** and **B**. In general, therefore, the two expressions are not equal.

The distributive law of multiplication *does* hold, so that

$$\begin{aligned} \mathbf{A} \times (\mathbf{B} + \mathbf{C}) &= \mathbf{A} \times \mathbf{B} + \mathbf{A} \times \mathbf{C}, \\ (\mathbf{A} + \mathbf{B}) \times \mathbf{C} &= \mathbf{A} \times \mathbf{C} + \mathbf{B} \times \mathbf{C}. \end{aligned} \tag{A2.18}$$

We shall prove the first form of Eq. (A2.18); the second then follows with the help of Eq. (A2.17). From the definition of $\mathbf{A} \times \mathbf{B}$ it follows that this product can be regarded as being formed by performing the following three operations upon vector **B**: (1) projection of **B** upon the plane perpendicular to **A** (see Figure A2.1), (2) positive rotation about vector **A** through an angle of 90°, and (3) multiplication by A. If we apply these operations to the vectors **B**, **C**, and $\mathbf{B} + \mathbf{C} = \mathbf{D}$, we shall obtain $\mathbf{A} \times \mathbf{B}$, $\mathbf{A} \times \mathbf{C}$, $\mathbf{A} \times \mathbf{D}$. But these operations will not affect the geometrical relation between the three vectors, namely, the fact that **A**, **B**, and **D** are three sides of a triangle. Therefore $\mathbf{A} \times \mathbf{B}$, $\mathbf{A} \times \mathbf{C}$, and $\mathbf{A} \times \mathbf{D}$ will still be sides of a triangle, and thus Eq. (A2.18) is proved.

We can apply these results to obtain the expression for $\mathbf{A} \times \mathbf{B}$ in terms of the components. We have

$$\begin{aligned} \mathbf{A} \times \mathbf{B} &= (A_x\mathbf{i} + A_y\mathbf{j} + A_z\mathbf{k}) \times (B_x\mathbf{i} + B_y\mathbf{j} + B_z\mathbf{k}) \\ &= A_xB_x\,\mathbf{i} \times \mathbf{i} + A_yB_y\,\mathbf{j} \times \mathbf{j} + A_zB_z\,\mathbf{k} \times \mathbf{k} \\ &\quad + A_xB_y\,\mathbf{i} \times \mathbf{j} + A_yB_z\,\mathbf{j} \times \mathbf{k} + A_zB_x\,\mathbf{k} \times \mathbf{i} \\ &\quad + A_yB_x\,\mathbf{j} \times \mathbf{i} + A_zB_y\,\mathbf{k} \times \mathbf{j} + A_xB_z\,\mathbf{i} \times \mathbf{k} \\ &= (A_yB_z - A_zB_y)\mathbf{i} + (A_zB_x - A_xB_z)\mathbf{j} + (A_xB_y - A_yB_x)\mathbf{k} \qquad \text{(A2.19)} \\ &= \begin{vmatrix} \mathbf{i} & \mathbf{j} & \mathbf{k} \\ A_x & A_y & A_z \\ B_x & B_y & B_z \end{vmatrix}, \qquad \text{(A2.20)} \end{aligned}$$

which could be taken as the general definition of $\mathbf{A} \times \mathbf{B}$, and from which all the other properties easily follow.

With the help of scalar and vector products we can build numerous combinations of several vectors. The most important are the *scalar triple product*, $\mathbf{A} \cdot \mathbf{B} \times \mathbf{C}$, and the *vector triple product*, $\mathbf{A} \times (\mathbf{B} \times \mathbf{C})$. The first is easily calculated:

$$\begin{aligned} \mathbf{A} \cdot \mathbf{B} \times \mathbf{C} &= \mathbf{A} \cdot \begin{vmatrix} \mathbf{i} & \mathbf{j} & \mathbf{k} \\ B_x & B_y & B_z \\ C_x & C_y & C_z \end{vmatrix} \\ &= \begin{vmatrix} \mathbf{A} \cdot \mathbf{i} & \mathbf{A} \cdot \mathbf{j} & \mathbf{A} \cdot \mathbf{k} \\ B_x & B_y & B_z \\ C_x & C_y & C_z \end{vmatrix} = \begin{vmatrix} A_x & A_y & A_z \\ B_x & B_y & B_z \\ C_x & C_y & C_z \end{vmatrix}. \qquad \text{(A2.21)} \end{aligned}$$

Geometrically the scalar triple product is the volume of a parallelpiped the three adjacent edges of which are the vectors **A**, **B**, and **C**. From Eq. (A2.21) it follows that

$$\mathbf{A} \cdot \mathbf{B} \times \mathbf{C} = \mathbf{B} \cdot \mathbf{C} \times \mathbf{A} = \mathbf{C} \cdot \mathbf{A} \times \mathbf{B}. \qquad \text{(A2.22)}$$

The dot and the cross in the scalar triple product may be interchanged; thus,

$$\begin{aligned} \mathbf{A} \cdot \mathbf{B} \times \mathbf{C} &= \mathbf{C} \cdot \mathbf{A} \times \mathbf{B} = \mathbf{C} \cdot (\mathbf{A} \times \mathbf{B}), && \text{by Eq. (A2.22)}, \\ &= (\mathbf{A} \times \mathbf{B}) \cdot \mathbf{C} = \mathbf{A} \times \mathbf{B} \cdot \mathbf{C}, && \text{by Eq. (A2.13)}. \qquad \text{(A2.23)} \end{aligned}$$

Interchange of the order of any two letters requires the change of sign; this follows directly from Eqs. (A2.22) and (A2.17), or from (A2.21). In the special case, therefore, when two of the vectors are the same the scalar triple product is zero, because an interchange of the two identical vectors makes the product equal to its own negative:

$$\mathbf{A} \cdot \mathbf{B} \times \mathbf{B} = \mathbf{A} \cdot \mathbf{B} \times \mathbf{A} = \mathbf{A} \cdot \mathbf{A} \times \mathbf{B} = 0. \qquad \text{(A2.24)}$$

The vector triple product $\mathbf{A} \times (\mathbf{B} \times \mathbf{C})$, as has been shown in the discussion following Eq. (A2.17), lies in the plane of **B** and **C** and can be therefore represented in the form $\alpha\mathbf{B} + \beta\mathbf{C}$, where α and β are scalars. Further, the triple vector product contains each of the vectors linearly; therefore α must contain **A** and **C** linearly, while β must contain **A** and **B** linearly. Thus, we must have

$$\mathbf{A} \times (\mathbf{B} \times \mathbf{C}) = a\, \mathbf{A} \cdot \mathbf{C}\, \mathbf{B} + b\, \mathbf{A} \cdot \mathbf{B}\, \mathbf{C},$$

where a and b are scalars independent of the choice of **A**, **B**, and **C**. It is easiest to determine these by taking special cases; thus, for example,

$$\mathbf{i} \times (\mathbf{i} \times \mathbf{j}) = \mathbf{i} \times \mathbf{k} = -\mathbf{j} = a\, \mathbf{i} \cdot \mathbf{j}\, \mathbf{i} + b\, \mathbf{i} \cdot \mathbf{i}\, \mathbf{j} = b\mathbf{j},$$

so that $b = -1$. Analogously we determine that $a = 1$; therefore, we have, finally,

$$\mathbf{A} \times (\mathbf{B} \times \mathbf{C}) = \mathbf{A} \cdot \mathbf{C}\, \mathbf{B} - \mathbf{A} \cdot \mathbf{B}\, \mathbf{C}, \qquad \text{(A2.25)}$$

which can also be proved by direct substitution in terms of components.

We leave it to the reader to prove the following expansion formulas for the quadrupole products:

$$(\mathbf{A} \times \mathbf{B}) \cdot (\mathbf{C} \times \mathbf{D}) = \mathbf{A} \cdot \mathbf{C}\, \mathbf{B} \cdot \mathbf{D} - \mathbf{A} \cdot \mathbf{D}\, \mathbf{B} \cdot \mathbf{C}, \qquad \text{(A2.26)}$$

which has an important special case,

$$(\mathbf{A} \times \mathbf{B})^2 = A^2B^2 - (\mathbf{A} \cdot \mathbf{B})^2 \qquad \text{(A2.27)}$$

and

$$(\mathbf{A} \times \mathbf{B}) \times (\mathbf{C} \times \mathbf{D}) = (\mathbf{A} \cdot \mathbf{B} \times \mathbf{D})\mathbf{C} - (\mathbf{A} \cdot \mathbf{B} \times \mathbf{C})\mathbf{D}$$
$$= (\mathbf{A} \cdot \mathbf{C} \times \mathbf{D})\mathbf{B} - (\mathbf{B} \cdot \mathbf{C} \times \mathbf{D})\mathbf{A}. \quad \text{(A2.28)}$$

EXAMPLE 1. Let us consider the problem of expressing a given vector **D**, in terms of three other given vectors **A**, **B**, and **C**. We can write

$$\mathbf{A} = A_x\mathbf{i} + A_y\mathbf{j} + A_z\mathbf{k},$$
$$\mathbf{B} = B_x\mathbf{i} + B_y\mathbf{j} + B_z\mathbf{k},$$
$$\mathbf{C} = C_x\mathbf{i} + C_y\mathbf{j} + C_z\mathbf{k},$$
$$\mathbf{D} = D_x\mathbf{i} + D_y\mathbf{j} + D_z\mathbf{k}.$$

Now, the first three equations can, in general, be solved for **i**, **j**, and **k** in terms of **A**, **B**, and **C**, and these substituted into the last equation to give the desired result. The process fails when the first three equations are not independent, that is, when the determinant of their coefficients vanishes. In accordance with Eq. (A2.21) this becomes

$$\mathbf{A} \cdot \mathbf{B} \times \mathbf{C} = 0.$$

Thus, any vector **D** may be expressed in terms of three arbitrary vectors **A**, **B**, and **C**, except when the volume of the parallelpiped having these vectors as adjacent edges is zero; in other words, except when the three vectors are in the same plane.

EXAMPLE 2. Consider transformation of vectors to a rotating coordinate system. Let us choose the Z-axis along the axis of rotation in such a way that, when looking in the positive Z direction, the direction of rotation should appear to be clockwise (see Figure A2.2).

Consider any point P, coordinates of which in the original coordinate system are (x, y, z). If the coordinate system is rotated through any angle θ, the new coordinates of the point, (x', y', z'), will be

$$x' = x \cos\theta + y \sin\theta,$$
$$y' = -x \sin\theta + y \cos\theta, \quad \text{(A2.29)}$$
$$z' = z.$$

We have seen that a vector is characterized by the fact that its components transform as coordinates; therefore, the new components of any vector

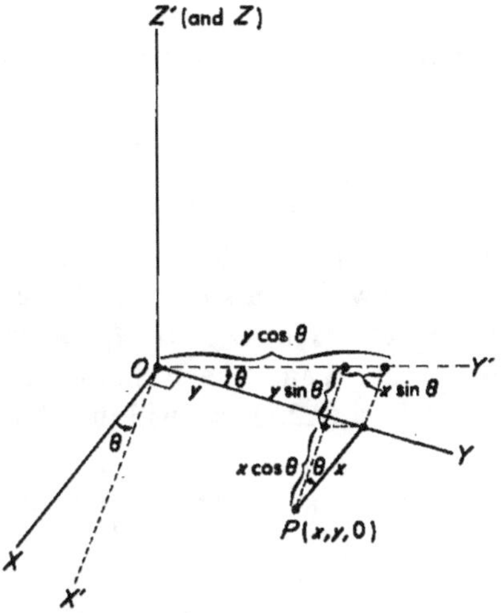

FIG. A2.2. Coordinates (x', y', z') of a point P (fixed in the unprimed coordinate system) relative to axes OX', OY', OZ' that have been rotated about the Z-axis through an angle θ relative to the unprimed system.

A, expressed in terms of the old, will be

$$\begin{aligned} A'_x &= A_x \cos\theta + A_y \sin\theta, \\ A'_y &= -A_x \sin\theta + A_y \cos\theta, \\ A'_z &= A_z. \end{aligned} \tag{A2.30}$$

When coordinate axes are rotating with the angular speed of $\omega = d\theta/dt$, the components of a vector **A** will change continually. By taking the differential of Eq. (A2.30) we obtain

$$\begin{aligned} dA'_x &= -A_x \sin\theta\, d\theta + A_y \cos\theta\, d\theta = A'_y\, d\theta, \\ dA'_y &= -A_x \cos\theta\, d\theta - A_y \sin\theta\, d\theta = -A'_x\, d\theta, \\ dA'_z &= 0, \end{aligned}$$

and therefore

$$\frac{dA'_x}{dt} = \omega A'_y, \qquad \frac{dA'_y}{dt} = -\omega A'_x, \qquad \frac{dA'_z}{dt} = 0.$$

If we take $d\boldsymbol{\theta}$ and $\boldsymbol{\omega}$ to be vectors in the positive Z direction (along the axis of rotation), then these equations can be summarized in the vector equations

$$d\mathbf{A}' = \mathbf{A}' \times d\boldsymbol{\theta} \qquad \text{or} \qquad \frac{d\mathbf{A}'}{dt} = -\boldsymbol{\omega} \times \mathbf{A}', \tag{A2.31}$$

which are easily verified by direct substitution.

So far we have considered the case when the coordinate system is rotated, but the vectors remain unchanged. Consider now the opposite case. Let the coordinate system remain unchanged, while all vectors are rotated with the angular velocity $\boldsymbol{\omega}$. This case is realized when we have a rotating body and the vectors considered pertain to the points of the body, being therefore fixed with respect to the body. In such a case the relation (A2.30) still holds, but now $\mathbf{A}'$ must be regarded as constant and $\mathbf{A}$ as variable. Solving the equations for (A_x, A_y, A_z), we must evidently obtain similar relations, except that $\mathbf{A}$ and $\mathbf{A}'$ will change places and the sign of θ will change. Instead of Eqs. (A2.31) we will then have

$$d\mathbf{A} = -\mathbf{A} \times d\boldsymbol{\theta} \quad \text{and} \quad \frac{d\mathbf{A}}{dt} = \boldsymbol{\omega} \times \mathbf{A}, \tag{A2.32}$$

of which

$$d\mathbf{r} = d\boldsymbol{\theta} \times \mathbf{r}, \quad \mathbf{v} = \boldsymbol{\omega} \times \mathbf{r}, \quad \frac{d\mathbf{v}}{dt} = \boldsymbol{\omega} \times \mathbf{v} \tag{A2.33}$$

are special cases.

Appendix A3

Vector Differential Calculus

Frequently we have the case when a certain scalar quantity is associated with each point of space. As examples we may mention gas pressure, temperature, or electrostatic potential. Such quantities are called *scalar function of position*. In general they are also functions of time. We shall designate them by φ, $\varphi(\mathbf{r})$, or $\varphi(\mathbf{r}, t)$, the latter when we wish to emphasize their dependence upon time t.

Consider the difference in φ for two points very near each other. By a limiting process we can then obtain the differential

$$d\varphi = \frac{\partial \varphi}{\partial x_i} dx_i = \frac{\partial \varphi}{\partial \mathbf{r}} \cdot d\mathbf{r}, \tag{A3.1}$$

where $\partial\varphi/\partial\mathbf{r}$ has the components $\partial\varphi/\partial x_i$. We shall now prove that these components transform as coordinates, and that, therefore, $\partial\varphi/\partial\mathbf{r}$ is a vector. To do this, we must first digress to consider in more detail the transformation of coordinates corresponding to a rotation.

We have seen in Section A1 that a rotation is given by Eq. (A1.5),

$$x'_j = a_{jk} x_k, \tag{A3.2}$$

provided that the coefficients a_{jk} are suitably chosen for each rotation. For all rotations, however, the choice must be such that the expression for the length of a vector remains invariant, that is, in accordance with Eq. (A1.4),

$$x'^2_j = x^2_j. \tag{A3.3}$$

Solving Eq. (A3.2) for x_k, we obtain the reciprocal form of the *same* transformation

$$x_k = b_{kj}x'_j, \tag{A3.4}$$

where b_{kj} are again constants. We shall now investigate the relation between the a_{ij} and the b_{ij}.

Substitution from Eq. (A3.4) into Eq. (A3.2) gives

$$x'_i = a_{ik}x_k = a_{ik}b_{kj}x'_j,$$

which must hold for an arbitrary set of numbers x'_j. By making all but one of these numbers equal to zero, we easily obtain the result

$$a_{ik}b_{kj} = \delta_{ij}. \tag{A3.5}$$

The symbol δ_{ij} is used to designate unity or zero, according as i is equal, or is not equal, to j; thus,*

$$\delta_{ij} = \begin{cases} 1, & \text{whenever} \quad i = j, \\ 0, & \text{whenever} \quad i \neq j. \end{cases} \tag{A3.6}$$

Similarly, by substitution from Eq. (A3.2) into Eq. (A3.4), we obtain

$$b_{ki}a_{ij} = \delta_{kj}. \tag{A3.7}$$

These interesting relations are true for all linear homogeneous transformations, as we have made no use of Eq. (A3.3), which characterizes a rotation. Substituting now from Eq. (A3.2) into Eq. (A3.3), we get

$$x_i'^2 = a_{ik}x_ka_{ij}x_j = (a_{ik}a_{ij})x_kx_j = x_i^2, \tag{A3.8}$$

from which it follows that

$$a_{ik}a_{ij} = \delta_{kj}. \tag{A3.9}$$

Finally, multiplying the last equation by b_{jm} (and, of course, summing over all the values of j) we obtain

$$\delta_{kj}b_{jm} = a_{ik}a_{ij}b_{jm},$$

or, with the help of Eq. (A3.5),

$$b_{km} = a_{ik}(a_{ij}b_{jm}) = a_{ik}\delta_{im} = a_{mk}. \tag{A3.10}$$

This relation is characteristic of a rotation. With its help Eq. (A3.4) becomes

$$x_k = a_{ik}x'_i. \tag{A3.11}$$

Returning now to the $\partial\varphi/\partial x_i$, we may with the help of Eq. (A3.2) write it as follows:

$$\frac{\partial\varphi}{\partial x_k} = \frac{\partial\varphi}{\partial x'_i}\frac{\partial x'_i}{\partial x_k} = a_{ik}\frac{\partial\varphi}{\partial x'_i}. \tag{A3.12}$$

* δ_{ij} is called the Kronecker delta.

But the derivatives $\partial\varphi/\partial x_i'$ are the components of $\partial\varphi/\partial\mathbf{r}$ in the new coordinate system. We see that they are connected with the components in the old coordinate system by the same type of relation as in Eq. (A3.11), and this shows that $\partial\varphi/\partial\mathbf{r}$ is a vector.

There are various ways of writing this vector, all of which are convenient on various occasions; we write here the most important ones:

$$\begin{aligned}\frac{\partial\varphi}{\partial\mathbf{r}} \equiv \operatorname{grad}\varphi &= \mathbf{i}\frac{\partial\varphi}{\partial x} + \mathbf{j}\frac{\partial\varphi}{\partial y} + \mathbf{k}\frac{\partial\varphi}{\partial z} \\ &= \left(\mathbf{i}\frac{\partial}{\partial x} + \mathbf{j}\frac{\partial}{\partial y} + \mathbf{k}\frac{\partial}{\partial z}\right)\varphi \\ &= \nabla\varphi \sim \left(\frac{\partial}{\partial x}, \frac{\partial}{\partial y}, \frac{\partial}{\partial z}\right)\varphi. \end{aligned} \tag{A3.13}$$

The operator

$$\nabla = \left(\mathbf{i}\frac{\partial}{\partial x} + \mathbf{j}\frac{\partial}{\partial y} + \mathbf{k}\frac{\partial}{\partial z}\right) \tag{A3.14}$$

is called *del*, or *nabla*; while the vector $\nabla\varphi$ is usually called the *gradient of* φ.

Analogously to $\partial\varphi/\partial\mathbf{r}$, but with the help of a vector $\mathbf{A}$ other than $\mathbf{r}$, we can construct the vector

$$\frac{\partial\varphi}{\partial\mathbf{A}} = \mathbf{i}\frac{\partial\varphi}{\partial A_x} + \mathbf{j}\frac{\partial\varphi}{\partial A_y} + \mathbf{k}\frac{\partial\varphi}{\partial A_z}, \tag{A3.15}$$

for any vector $\mathbf{A}$, provided that $\varphi = \varphi(\mathbf{A})$. This vector, however, has no special name.

The operator ∇ acting upon a scalar gives a vector; it has, thus, some of the abstract properties of a vector, which will become evident in what follows. But it is also a linear differential operator and has the properties of such an operator. In particular, we have

$$\nabla(a\varphi) = a\,\nabla\varphi, \qquad \text{where } a \text{ is a constant,} \tag{A3.16}$$

$$\nabla(\varphi + \psi) = \nabla\varphi + \nabla\psi, \tag{A3.17}$$

$$\nabla(\varphi\psi) = \varphi\,\nabla\psi + \psi\,\nabla\varphi. \tag{A3.18}$$

It is then supposed that we have a *vector function of position*, say $\mathbf{A} = \mathbf{A}(\mathbf{r})$, which associates a vector with each point of a limited or unlimited portion of space. Such a vector function of position is also called a *vector field*.

Regarded as a vector the operator ∇ may be multiplied by another vector, either in scalar or vector multiplication. It is clear, therefore, that we can form $\nabla \cdot \mathbf{A}$ and $\nabla \times \mathbf{A}$. The first of these expressions is called the *divergence* of $\mathbf{A}$; the second is called the *curl*, or the *rotation*, of $\mathbf{A}$. We have, therefore, the following relations:

$$\nabla \cdot \mathbf{A} = \operatorname{div} \mathbf{A} = \left(\frac{\partial}{\partial x}\mathbf{i} + \frac{\partial}{\partial y}\mathbf{j} + \frac{\partial}{\partial z}\mathbf{k}\right) \cdot \mathbf{A}$$

$$= \frac{\partial A_x}{\partial x} + \frac{\partial A_y}{\partial y} + \frac{\partial A_z}{\partial z} = \frac{\partial A_i}{\partial x_i} \tag{A3.19}$$

and

$$\nabla \times \mathbf{A} = \operatorname{curl} \mathbf{A} = \operatorname{rot} \mathbf{A}$$

$$= \left(\frac{\partial}{\partial x}\mathbf{i} + \frac{\partial}{\partial y}\mathbf{j} + \frac{\partial}{\partial z}\mathbf{k}\right) \times (A_x\mathbf{i} + A_y\mathbf{j} + A_z\mathbf{k})$$

$$= \left(\frac{\partial A_z}{\partial y} - \frac{\partial A_y}{\partial z}\right)\mathbf{i} + \left(\frac{\partial A_x}{\partial z} - \frac{\partial A_z}{\partial x}\right)\mathbf{j} + \left(\frac{\partial A_y}{\partial x} - \frac{\partial A_x}{\partial y}\right)\mathbf{k}$$

$$= \begin{vmatrix} \mathbf{i} & \mathbf{j} & \mathbf{k} \\ \dfrac{\partial}{\partial x} & \dfrac{\partial}{\partial y} & \dfrac{\partial}{\partial z} \\ A_x & A_y & A_z \end{vmatrix}. \tag{A3.20}$$

Let us consider the expression for the difference of $\mathbf{A}(\mathbf{r})$ between two points very near each other. By a limiting process we then obtain the differential

$$d\mathbf{A} = \frac{\partial \mathbf{A}}{\partial x_i}\,dx_i = \left(dx_i \frac{\partial}{\partial x_i}\right)\mathbf{A} = (d\mathbf{r} \cdot \nabla)\mathbf{A}. \tag{A3.21}$$

The expression $d\mathbf{r} \cdot \nabla$ is evidently a scalar operator, operating on $\mathbf{A}$. We can also form an operator $\mathbf{B} \cdot \nabla$ with any vector $\mathbf{B}$. When applied to a scalar $\mathbf{B} \cdot \nabla$ requires no parentheses because

$$(\mathbf{B} \cdot \nabla)\varphi = \left(B_i \frac{\partial}{\partial x_i}\right)\varphi = B_i \frac{\partial \varphi}{\partial x_i} = \mathbf{B} \cdot (\nabla\varphi) = \mathbf{B} \cdot \nabla\varphi. \tag{A3.22}$$

In simplifying the result of an application of the operator ∇ to a product we make use of the fact that we can act with it on each factor in turn, keeping other factors constant, according to a generalization of Eq. (A3.18). Thus, for example, we can write

$$\nabla \cdot \varphi \mathbf{A} = \nabla_\varphi \cdot \varphi \mathbf{A} + \nabla_\mathbf{A} \cdot \varphi \mathbf{A},$$

where the subscript attached to ∇ shows on which function it is to act, the other factors being considered constant. The dot after ∇ can refer only to another vector; therefore,

$$\nabla \cdot \varphi \mathbf{A} = \mathbf{A} \cdot \nabla \varphi + \varphi \nabla \cdot \mathbf{A}. \tag{A3.23}$$

This method of derivation of formulas is simple and is easily remembered; we shall use it almost exclusively. However, each formula so obtained may be proved more rigorously with the help of components. For example, by Eq. (A3.19), we have

$$\nabla \cdot \varphi \mathbf{A} = \partial(\varphi \mathbf{A})_i / \partial x_i,$$

which by Eq. (A1.12) becomes

$$\begin{aligned} \nabla \cdot \varphi \mathbf{A} &= \partial(\varphi A_i)/\partial x_i \\ &= \frac{\partial \varphi}{\partial x_i} A_i + \varphi \frac{\partial A_i}{\partial x_i}. \end{aligned}$$

Thus, finally, by Eqs. (A3.19) and (A3.22)

$$\nabla \cdot \varphi \mathbf{A} = \mathbf{A} \cdot \nabla \varphi + \varphi \nabla \cdot \mathbf{A}.$$

In a way similar to the derivation of Eq. (A3.23) the above symbolic method gives us at once

$$\nabla \times \varphi \mathbf{A} = \varphi \nabla \times \mathbf{A} + (\nabla \varphi) \times \mathbf{A} = \varphi \nabla \times \mathbf{A} - \mathbf{A} \times \nabla \varphi \tag{A3.24}$$

and

$$\mathbf{B} \cdot \nabla(\varphi \mathbf{A}) = \mathbf{B} \cdot (\nabla \varphi)\mathbf{A} + \varphi \mathbf{B} \cdot \nabla \mathbf{A}. \tag{A3.25}$$

Divergence of a vector product can also be easily found; thus,

$$\begin{aligned} \nabla \cdot \mathbf{A} \times \mathbf{B} &= \nabla_\mathbf{A} \cdot \mathbf{A} \times \mathbf{B} + \nabla_\mathbf{B} \cdot \mathbf{A} \times \mathbf{B} \\ &= \mathbf{B} \cdot \nabla_\mathbf{A} \times \mathbf{A} - \mathbf{A} \cdot \nabla_\mathbf{B} \times \mathbf{B} \\ &= \mathbf{B} \cdot \nabla \times \mathbf{A} - \mathbf{A} \cdot \nabla \times \mathbf{B}. \end{aligned} \tag{A3.26}$$

Here the convenience of attaching a subscript to ∇ is very clear; after the subscript is once attached, ∇ can be treated exactly like any other vector. Thus we were able to make use of Eqs. (A2.17) and (A2.22).

Slightly more difficult is the calculation of the curl of a vector product. Using Eq. (A2.25) and the above scheme, we have

$$\begin{aligned}\nabla \times (\mathbf{A} \times \mathbf{B}) &= \nabla_A \times (\mathbf{A} \times \mathbf{B}) + \nabla_B \times (\mathbf{A} \times \mathbf{B})\\ &= \nabla_A \cdot \mathbf{B}\,\mathbf{A} - \nabla_A \cdot \mathbf{A}\,\mathbf{B} + \nabla_B \cdot \mathbf{B}\,\mathbf{A} - \nabla_B \cdot \mathbf{A}\,\mathbf{B}\\ &= \mathbf{B} \cdot \nabla\mathbf{A} - \mathbf{B}\,\nabla \cdot \mathbf{A} + \mathbf{A}\,\nabla \cdot \mathbf{B} - \mathbf{A} \cdot \nabla\mathbf{B}. \qquad \text{(A3.27)}\end{aligned}$$

Still more difficult is the following calculation of the gradient of a scalar product:

$$\nabla\,\mathbf{A} \cdot \mathbf{B} = \nabla_A\,\mathbf{A} \cdot \mathbf{B} + \nabla_B\,\mathbf{A} \cdot \mathbf{B}.$$

By Eq. (A2.25) the first term becomes

$$\nabla_A\,\mathbf{A} \cdot \mathbf{B} = \nabla_A \cdot \mathbf{B}\,\mathbf{A} + \mathbf{B} \times (\nabla_A \times \mathbf{A}) = \mathbf{B} \cdot \nabla\mathbf{A} + \mathbf{B} \times (\nabla \times \mathbf{A}),$$

and by an interchange of **A** and **B** the second term becomes

$$\nabla_B\,\mathbf{A} \cdot \mathbf{B} = \mathbf{A} \cdot \nabla\mathbf{B} + \mathbf{A} \times (\nabla \times \mathbf{B}).$$

We therefore have

$$\nabla\,\mathbf{A} \cdot \mathbf{B} = \mathbf{B} \cdot \nabla\mathbf{A} + \mathbf{A} \cdot \nabla\mathbf{B} + \mathbf{B} \times (\nabla \times \mathbf{A}) + \mathbf{A} \times (\nabla \times \mathbf{B}). \qquad \text{(A3.28)}$$

Finally, for completeness, we may add that

$$\mathbf{C} \cdot \nabla(\mathbf{A} \times \mathbf{B}) = (\mathbf{C} \cdot \nabla\mathbf{A}) \times \mathbf{B} + \mathbf{A} \times (\mathbf{C} \cdot \nabla\mathbf{B}) \qquad \text{(A3.29)}$$

and

$$\mathbf{B} \cdot \mathbf{C} \times (\nabla \times \mathbf{A}) = \mathbf{C} \cdot (\mathbf{B} \cdot \nabla\mathbf{A}) - \mathbf{B} \cdot (\mathbf{C} \cdot \nabla\mathbf{A}). \qquad \text{(A3.30)}$$

For convenience of reference we collect our results:

$$\begin{aligned}
\nabla(\varphi\psi) &= \varphi\,\nabla\psi + \psi\,\nabla\varphi, && \text{(A3.18)}\\
\nabla \cdot \varphi\mathbf{A} &= \varphi\,\nabla \cdot \mathbf{A} + \mathbf{A} \cdot \nabla\varphi, && \text{(A3.23)}\\
\nabla \times \varphi\mathbf{A} &= \varphi\,\nabla \times \mathbf{A} - \mathbf{A} \times \nabla\varphi, && \text{(A3.24)}\\
\mathbf{B} \cdot \nabla(\varphi\mathbf{A}) &= \mathbf{B} \cdot \nabla\varphi\,\mathbf{A} + \varphi\mathbf{B} \cdot \nabla\mathbf{A}, && \text{(A3.25)}\\
\nabla \cdot \mathbf{A} \times \mathbf{B} &= \mathbf{B} \cdot \nabla \times \mathbf{A} - \mathbf{A} \cdot \nabla \times \mathbf{B}, && \text{(A3.26)}\\
\nabla \times (\mathbf{A} \times \mathbf{B}) &= \mathbf{B} \cdot \nabla\mathbf{A} - \mathbf{B}\nabla \cdot \mathbf{A} + \mathbf{A}\nabla \cdot \mathbf{B} - \mathbf{A} \cdot \nabla\mathbf{B}, && \text{(A3.27)}\\
\nabla\,\mathbf{A} \cdot \mathbf{B} &= \mathbf{A} \cdot \nabla\mathbf{B} + \mathbf{B} \cdot \nabla\mathbf{A} + \mathbf{A} \times (\nabla \times \mathbf{B})\\
&\qquad + \mathbf{B} \times (\nabla \times \mathbf{A}), && \text{(A3.28)}\\
\mathbf{C} \cdot \nabla(\mathbf{A} \times \mathbf{B}) &= (\mathbf{C} \cdot \nabla\mathbf{A}) \times \mathbf{B} + \mathbf{A} \times (\mathbf{C} \cdot \nabla\mathbf{B}), && \text{(A3.29)}\\
\mathbf{B} \cdot \mathbf{C} \times (\nabla \times \mathbf{A}) &= \mathbf{C} \cdot (\mathbf{B} \cdot \nabla\mathbf{A}) - \mathbf{B} \cdot (\mathbf{C} \cdot \nabla\mathbf{A}). && \text{(A3.30)}
\end{aligned}$$

If $f(\varphi)$ is a scalar function of φ, a function of position, then

$$\nabla f(\varphi) = \left(\mathbf{i}\frac{\partial f}{\partial x} + \mathbf{j}\frac{\partial f}{\partial y} + \mathbf{k}\frac{\partial f}{\partial z}\right)$$
$$= \frac{\partial f}{\partial \varphi}\left(\mathbf{i}\frac{\partial \varphi}{\partial x} + \mathbf{j}\frac{\partial \varphi}{\partial y} + \mathbf{k}\frac{\partial \varphi}{\partial z}\right)$$
$$= f'(\varphi)\,\nabla\varphi. \quad \text{(A3.31)}$$

Similarly, it can be shown that

$$\nabla \cdot \mathbf{A}(\varphi) = \mathbf{A}'(\varphi) \cdot \nabla\varphi, \quad \text{(A3.32)}$$
$$\nabla \times \mathbf{A}(\varphi) = \nabla\varphi \times \mathbf{A}'(\varphi), \quad \text{(A3.33)}$$

and

$$\mathbf{B} \cdot \nabla\mathbf{A}(\varphi) = \mathbf{A}'(\varphi)\,\mathbf{B} \cdot \nabla\varphi. \quad \text{(A3.34)}$$

Second derivatives are formed by a double application of the operator ∇. Thus, for example, we have

$$\nabla \cdot (\nabla\varphi) = \frac{\partial}{\partial x_i}(\nabla\varphi)_i = \frac{\partial^2 \varphi}{\partial x_i^2} = (\nabla \cdot \nabla)\varphi = \nabla^2\varphi = \Delta\varphi, \quad \text{(A3.35)}$$

the last two being frequent notations for this important expression. The operator

$$\nabla^2 = \Delta = \left(\frac{\partial^2}{\partial x^2} + \frac{\partial^2}{\partial y^2} + \frac{\partial^2}{\partial z^2}\right) = \frac{\partial^2}{\partial x_i^2}$$

is called the *Laplace* operator.* It can also be applied to a vector, thus

$$\nabla^2\mathbf{A} = \frac{\partial^2 \mathbf{A}}{\partial x^2} + \frac{\partial^2 \mathbf{A}}{\partial y^2} + \frac{\partial^2 \mathbf{A}}{\partial z^2}. \quad \text{(A3.36)}$$

The following formulas are obtained quite readily:

$$\nabla \times (\nabla \times \mathbf{A}) = \nabla\nabla \cdot \mathbf{A} - \nabla^2\mathbf{A}, \quad \text{(A3.37)}$$
$$\nabla \times \nabla\varphi = 0, \quad \text{(A3.38)}$$
$$\nabla \cdot \nabla \times \mathbf{A} = (\nabla \times \nabla) \cdot \mathbf{A} = 0, \quad \text{(A3.39)}$$
$$\nabla^2(\varphi\psi) = \varphi\,\nabla^2\psi + \psi\,\nabla^2\varphi + 2\nabla\varphi \cdot \nabla\psi, \quad \text{(A3.40)}$$
$$\mathbf{A} \cdot \nabla(\mathbf{B} \cdot \nabla\varphi) - \mathbf{B} \cdot \nabla(\mathbf{A} \cdot \nabla\varphi) = (\mathbf{A} \cdot \nabla\mathbf{B} - \mathbf{B} \cdot \nabla\mathbf{A}) \cdot \nabla\varphi. \quad \text{(A3.41)}$$

Some special cases. We have that

$$\nabla(r^2) = \nabla(x^2 + y^2 + z^2) = 2(x\mathbf{i} + y\mathbf{j} + z\mathbf{k}) = 2\mathbf{r}. \quad \text{(A3.42)}$$

* It is also called the *Laplacian*.

Accordingly, with the help of Eq. (A3.31), we have

$$\nabla f(r) = \frac{df(r)}{d(r^2)}\nabla(r^2) = f'(r)\frac{dr}{d(r^2)}\nabla(r^2) = \frac{f'(r)}{r}\mathbf{r}. \qquad \text{(A3.43)}$$

Analogously, we can show that

$$\nabla \cdot \mathbf{A}(r) = \frac{\mathbf{A}'(r)}{r} \cdot \mathbf{r} \qquad \text{(A3.44)}$$

and

$$\nabla \times \mathbf{A}(r) = \mathbf{r} \times \frac{\mathbf{A}'(r)}{r}. \qquad \text{(A3.45)}$$

To obtain $\nabla \times \mathbf{r}$ we note that according to Eq. (A3.42)

$$\nabla \times \nabla r^2 = \nabla \times (2\mathbf{r}) = 2\nabla \times \mathbf{r},$$

while, in accordance with Eq. (A3.38), $\nabla \times \nabla(r^2) = 0$; therefore,

$$\nabla \times \mathbf{r} = 0. \qquad \text{(A3.46)}$$

From this it follows that

$$\nabla \times f(r)\,\mathbf{r} = f\nabla \times \mathbf{r} + \nabla f \times \mathbf{r} = \nabla f \times \mathbf{r} = \frac{f'(r)}{r}\mathbf{r} \times \mathbf{r} = 0. \qquad \text{(A3.47)}$$

The divergence of $\mathbf{r}$ is calculated directly; it is

$$\nabla \cdot \mathbf{r} = \frac{\partial x_i}{\partial x_i} = 3. \qquad \text{(A3.48)}$$

Accordingly, one has that

$$\nabla \cdot f(r)\,\mathbf{r} = \mathbf{r} \cdot \nabla f + f\nabla \cdot \mathbf{r} = 3f + \mathbf{r} \cdot \nabla f = 3f + \mathbf{r} \cdot \left(\frac{f'}{r}\mathbf{r}\right)$$

$$= 3f + rf' = \frac{1}{r^2}(r^3 f)', \qquad \text{(A3.49)}$$

where the prime indicates differentiation with respect to r.

When the Laplace operator acts upon a function of r, we obtain by Eqs. (A3.43) and (A3.49),

$$\nabla^2 f(r) = \nabla \cdot \nabla f(r) = \nabla \cdot \left(\frac{\mathbf{r} f'(r)}{r}\right) = \frac{1}{r^2}(r^2 f')'. \qquad \text{(A3.50)}$$

Finally, an interesting special case is obtained by putting $\mathbf{r}$ instead of $\mathbf{A}$ into Eq. (A3.21); thus, we have

$$d\mathbf{r} = d\mathbf{r} \cdot \nabla\mathbf{r}. \tag{A3.51}$$

In this equation $d\mathbf{r}$ is entirely arbitrary; therefore, we conclude that for any vector $\mathbf{B}$

$$\mathbf{B} \cdot \nabla\mathbf{r} = \mathbf{B}. \tag{A3.52}$$

Other special cases with $\mathbf{A} = \mathbf{A}(r)$ and $f = f(r)$ are

$$\mathbf{r} \cdot \nabla\mathbf{A} = r\mathbf{A}', \tag{A3.53}$$

$$\nabla(\mathbf{r} \cdot \mathbf{A}) = \mathbf{A} + \frac{\mathbf{r} \cdot \mathbf{A}'\,\mathbf{r}}{r}, \tag{A3.54}$$

$$\nabla \times (\mathbf{A} \times \mathbf{r}) = 2\mathbf{A} + r\mathbf{A}' - \frac{\mathbf{r} \cdot \mathbf{A}'\,\mathbf{r}}{r}, \tag{A3.55}$$

$$\nabla \cdot (\mathbf{A} \times \mathbf{r}) = 0, \tag{A3.56}$$

$$\nabla(\mathbf{A} \cdot \mathbf{r}) = \mathbf{A}, \qquad \text{for constant } \mathbf{A}, \tag{A3.57}$$

$$\mathbf{B} \cdot \nabla(f\mathbf{r}) = f\mathbf{B} + \frac{f'}{r}\mathbf{B} \cdot \mathbf{r}\,\mathbf{r}, \qquad \text{for arbitrary } \mathbf{B}. \tag{A3.58}$$

If a particle moves in a region in which a scalar function of position $\varphi(\mathbf{r}, t)$ is defined, then the differential change in φ due to a change in both t and $\mathbf{r}$ is given by

$$d\varphi = \frac{\partial\varphi}{\partial t}\,dt + (d\varphi)_{t=\text{const}} \,.$$

The last term is obtained from Eq. (A3.1); therefore, we have

$$d\varphi = \frac{\partial\varphi}{\partial t}\,dt + d\mathbf{r} \cdot \nabla\varphi \tag{A3.59a}$$

or, by dividing by dt,

$$\frac{d\varphi}{dt} = \frac{\partial\varphi}{\partial t} + \mathbf{v} \cdot \nabla\varphi. \tag{A3.59b}$$

Similarly, with the help of Eq. (A3.21), we obtain the equations

$$d\mathbf{A} = \frac{\partial\mathbf{A}}{\partial t}\,dt + d\mathbf{r} \cdot \nabla\mathbf{A} \tag{A3.60a}$$

and

$$\frac{d\mathbf{A}}{dt} = \frac{\partial\mathbf{A}}{\partial t} + \mathbf{v} \cdot \nabla\mathbf{A}. \tag{A3.60b}$$

We may therefore write the operator equation

$$\frac{d}{dt} = \frac{\partial}{\partial t} + \mathbf{v} \cdot \boldsymbol{\nabla}, \tag{A3.61}$$

which is seen to be applicable to moving particles.

It is necessary at times (see Section 9) to work with functions of $\mathbf{R} = \mathbf{r} - \mathbf{r}'$. These can alternatively be considered as functions of the position vectors $\mathbf{r} \sim (x, y, z)$ and $\mathbf{r}' \sim (x', y', z')$. The nabla operator $\boldsymbol{\nabla}$ is still given by Eq. (A3.14) with the added understanding that (in taking the partial derivatives) $\mathbf{r}'$ is to remain fixed. Similarly we can define a nabla operator

$$\boldsymbol{\nabla}' = \left(\mathbf{i}\frac{\partial}{\partial x'} + \mathbf{j}\frac{\partial}{\partial y'} + \mathbf{k}\frac{\partial}{\partial z'}\right) \tag{A3.62}$$

with the understanding that now $\mathbf{r}$ is to be treated as a constant.

Since

$$\frac{\partial R}{\partial x} = \frac{\partial}{\partial x}[(x - x')^2 + (y - y')^2 + (z - z')^2]^{1/2} = \frac{x - x'}{R}$$

and

$$\frac{\partial R}{\partial x'} = \frac{-(x - x')}{R},$$

we see that

$$\boldsymbol{\nabla} R = \frac{\mathbf{R}}{R} = -\boldsymbol{\nabla}' R. \tag{A3.63}$$

We note also that

$$\boldsymbol{\nabla} \cdot \mathbf{R} = \frac{\partial}{\partial x}(x - x') + \frac{\partial}{\partial y}(y - y') + \frac{\partial}{\partial z}(z - z') = 3, \tag{A3.64a}$$

while

$$\boldsymbol{\nabla}' \cdot \mathbf{R} = \frac{\partial}{\partial x'}(x - x') + \frac{\partial}{\partial y'}(y - y') + \frac{\partial}{\partial z'}(z - z') = -3. \tag{A3.64b}$$

For a general scalar function f of $R = |\mathbf{r} - \mathbf{r}'|$ the gradient is

$$\boldsymbol{\nabla} f(R) = f'(R)\,\boldsymbol{\nabla} R = f'(R)\frac{\mathbf{R}}{R}, \tag{A3.65}$$

which is just Eq. (A3.43) with $\mathbf{r}$ replaced by $\mathbf{R}$. Likewise we have, for

$\mathbf{A}(R)$, a vector function of R,

$$\nabla \cdot \mathbf{A}(R) = \mathbf{A}'(R) \cdot \nabla R = \mathbf{A}'(R) \cdot \frac{\mathbf{R}}{R} \tag{A3.66}$$

and

$$\nabla \times \mathbf{A}(R) = \nabla R \times \mathbf{A}'(R) = \frac{\mathbf{R}}{R} \times \mathbf{A}'(R). \tag{A3.67}$$

Comparing Eqs. (A3.64a), (A3.65), (A3.66), and (A3.67) with the analogous equations (A3.48), (A3.43), (A3.44), and (A3.45), we conclude that the formulas for ∇ operating on a function of R can be obtained from those for functions of r by merely replacing r by R. By replacing ∇ by ∇' in Eqs. (A3.65), (A3.66), and (A3.67) and using Eq. (A3.63), we see that ∇' gives in each case the negative of ∇. We are led to conclude, therefore, that for operations on functions of $R = |\mathbf{r} - \mathbf{r}'|$

$$\nabla' = -\nabla. \tag{A3.68}$$

Miscellaneous Theorems

Theorem 1. *The necessary and sufficient condition for the existence of a relation*

$$\varphi(u, v) = 0 \tag{A3.69}$$

between two given functions

$$u = f(x, y, z) \tag{A3.70}$$

and

$$v = g(x, y, z) \tag{A3.71}$$

*(with $\varphi(u, v)$ not a mere constant) is that**

$$\nabla u \times \nabla v \equiv 0. \tag{A3.72}$$

On the other hand, if $\nabla u \times \nabla v \not\equiv 0$, Eqs. (A3.70) *and* (A3.71) *can be solved for two of the variables x, y, and z in terms of the third and of u and v.*

* Here, as elsewhere, the symbol $\equiv$ means identically equal; i.e., equal for all values of the variables, x, y, and z. It will be used to replace the ordinary equality symbol $=$ only where it seems important to emphasize that we mean identically equal.

Proof. If a function $\varphi(u, v)$ satisfying Eq. (A3.69) exists, then

$$\frac{\partial \varphi}{\partial x} = \frac{\partial \varphi}{\partial u}\frac{\partial u}{\partial x} + \frac{\partial \varphi}{\partial v}\frac{\partial v}{\partial x} = 0, \qquad \text{etc.,}$$

or*

$$\nabla \varphi = \frac{\partial \varphi}{\partial u}\nabla u + \frac{\partial \varphi}{\partial v}\nabla v = 0;$$

so that

$$\begin{aligned} \nabla \varphi \times \nabla v &= \frac{\partial \varphi}{\partial u}\nabla u \times \nabla v = 0, \\ \nabla u \times \nabla \varphi &= \frac{\partial \varphi}{\partial v}\nabla u \times \nabla v = 0. \end{aligned} \tag{A3.73}$$

Equations (A3.73) show that $\nabla u \times \nabla v = 0$; for otherwise $\partial \varphi / \partial u \equiv 0$ and $\partial \varphi / \partial v \equiv 0$, and hence φ would be merely a constant.

To prove the converse we suppose that Eq. (A3.72) is satisfied and show that a relationship of the form of Eq. (A3.69) must exist between u and v. If $\nabla u \equiv 0$ or $\nabla v \equiv 0$, the case is trivial. Thus, if $\nabla u \equiv 0$, $u = c$, and

$$\varphi(u, v) = u - c = 0$$

is of the form of Eq. (A3.69). We may assume, therefore, that $\nabla u \not\equiv 0$ and $\nabla v \not\equiv 0$.

Since $\nabla u \not\equiv 0$, f of Eq. (A3.70) contains at least one of the independent variables x, y, and z. This variable we may designate by x_i, in which case

$$\frac{\partial f}{\partial x_i} \not\equiv 0. \tag{A3.74}$$

The latter is a necessary and sufficient condition for the possibility of solving Eq. (A3.70) for x_i (see the discussion of implicit functions in any standard text in advanced calculus); thus

$$x_i = F(u, x, y, z), \qquad \text{with} \quad \frac{\partial F}{\partial x_i} \equiv 0. \tag{A3.75}$$

The latter condition is used to indicate that F is not a function of the variable x_i for which we solved. Substituting x_i into Eq. (A3.71), we obtain

$$v = G(u, x, y, z), \qquad \text{with} \quad \frac{\partial G}{\partial x_i} \equiv 0. \tag{A3.76}$$

* This is a generalization of Eq. (A3.31). More generally, if $\varphi = \varphi(u, v, w)$, we have

$$\nabla \varphi = \frac{\partial \varphi}{\partial u}\nabla u + \frac{\partial \varphi}{\partial v}\nabla v + \frac{\partial \varphi}{\partial w}\nabla w.$$

Of course, if $g(x, y, z)$ does not contain x_i, $G \equiv g$, and G does not contain u; but in that case $\partial g/\partial x_i \equiv 0$, which is contrary to our assumptions. This can be seen as follows. From Eq. (A3.76) we have

$$\nabla v = \frac{\partial G}{\partial u}\nabla u + \nabla G, \tag{A3.77}$$

where $\nabla G \equiv (\partial G/\partial x, \partial G/\partial y, \partial G/\partial z)$. Since now $\nabla u \times \nabla v \equiv 0$, ∇u is parallel to ∇v, so that

$$\nabla v = \lambda \nabla u, \tag{A3.78}$$

where λ is some function of position; one component of this equation, by use of Eq. (A3.71), is

$$\frac{\partial v}{\partial x_i} = \lambda \frac{\partial f}{\partial x_i}.$$

Since we have assumed that $\nabla v \not\equiv 0$, λ does not equal zero, as can be seen from Eq. (A3.78); and hence, by use of inequality (A3.74), $\partial g/\partial x_i \not\equiv 0$, so that $g(x, y, z)$ contains the variable x_i. Also, since F always contains u, when we substitute for x_i in $g(x, y, z)$ we will necessarily introduce u. Thus G is a function of u.

We next show that $\nabla G \equiv 0$; i.e., that G is not a function of x, y, or z. We first multiply both sides of Eq. (A3.70) by $\nabla u \times$:

$$\nabla u \times \nabla G = \nabla u \times \left(\nabla v - \frac{\partial G}{\partial u}\nabla u\right) = 0;$$

so that ∇G is parallel to ∇u, or

$$\nabla G = \mu \nabla u, \tag{A3.79}$$

where μ is some function of position. This implies that

$$\mu \frac{\partial f}{\partial x_i} \equiv \frac{\partial G}{\partial x_i} \equiv 0,$$

by Eqs. (A3.76). Thus, using inequality (A3.74), $\mu \equiv 0$, and hence by Eq. (A3.79) $\nabla G \equiv 0$. The function G is therefore a function of u only, giving

$$v = G(u),$$

or

$$\varphi(u, v) \equiv v - G(u) = 0, \tag{A3.80}$$

which proves the theorem.

On the other hand, if we start with

$$\nabla u \times \nabla v \not\equiv 0, \tag{A3.81}$$

then $\nabla u \not\equiv 0$ and $\nabla v \not\equiv 0$, and we may proceed as before, obtaining Eqs. (A3.76) and (A3.77). From the latter, and the assumption (A3.81),

$$\nabla u \times \nabla G = \nabla u \times \nabla v \not\equiv 0;$$

so that

$$\nabla G \not\equiv 0. \tag{A3.82}$$

Since, however, by Eqs. (A3.76), $\partial G/\partial x_i \equiv 0$, there must be at least one other of the three variables (x, y, z), say x_j, for which

$$\frac{\partial G}{\partial x_j} \not\equiv 0. \tag{A3.83}$$

Equation (A3.76) can then be solved for x_j; thus

$$x_j = H(u, v, x, y, z), \qquad \text{with} \quad \frac{\partial H}{\partial x_i} \equiv \frac{\partial H}{\partial x_j} \equiv 0, \tag{A3.84}$$

which upon substitution into Eq. (A3.75) must give

$$x_i = K(u, v, x, y, z), \qquad \text{with} \quad \frac{\partial K}{\partial x_i} \equiv \frac{\partial K}{\partial x_j} \equiv 0. \tag{A3.85}$$

The equations (A3.84) and (A3.85), obviously, are the expression of the second part of our theorem.

Moreover, let us consider the x_i and x_j components of Eq. (A3.77). We have

$$\frac{\partial g}{\partial x_i} = \frac{\partial G}{\partial u}\frac{\partial f}{\partial x_i} + \frac{\partial G}{\partial x_i} = \frac{\partial G}{\partial u}\frac{\partial f}{\partial x_i}$$

by Eq. (A3.76) and

$$\frac{\partial g}{\partial x_j} = \frac{\partial G}{\partial u}\frac{\partial f}{\partial x_j} + \frac{\partial G}{\partial x_j};$$

therefore

$$\begin{aligned}\frac{\partial f}{\partial x_i}\frac{\partial g}{\partial x_j} - \frac{\partial f}{\partial x_j}\frac{\partial g}{\partial x_i} &= \frac{\partial f}{\partial x_i}\left(\frac{\partial G}{\partial u}\frac{\partial f}{\partial x_j} + \frac{\partial G}{\partial x_j}\right) - \frac{\partial f}{\partial x_j}\left(\frac{\partial G}{\partial u}\frac{\partial f}{\partial x_i}\right)\\ &= \frac{\partial f}{\partial x_i}\frac{\partial G}{\partial x_j}.\end{aligned}$$

However, each of the factors on the right was assumed to be not zero;

hence the variables x_i and x_j must be that pair of variables for which the condition

$$\frac{\partial f}{\partial x_i}\frac{\partial g}{\partial x_j} - \frac{\partial f}{\partial x_j}\frac{\partial g}{\partial x_i} \not\equiv 0 \tag{A3.86}$$

is satisfied. The left-hand member of Eq. (A3.86) is one of the components of $\nabla u \times \nabla v$; hence we see that it is possible to solve for only those pairs of the variables x, y, and z which correspond to nonvanishing components of $\nabla u \times \nabla v$.

Theorem 2. *The necessary and sufficient condition for the existence of a relation*

$$\varphi(u, v, w) = 0 \tag{A3.87}$$

between three given functions

$$\begin{aligned} u &= f(x, y, z), \\ v &= g(x, y, z), \\ w &= h(x, y, z), \end{aligned} \tag{A3.88}$$

(with $\varphi(u, v, w)$ not a mere constant) is that the triple scalar product

$$\nabla u \cdot \nabla v \times \nabla w \equiv 0 \tag{A3.89}$$

for all values of x, y, and z.

If, on the other hand, Eq. (A3.89) *does not hold, then Eqs.* (A3.88) *can be solved for x, y, and z in terms of u, v, and w.*

Proof. If $\varphi(u, v, w)$ satisfying Eq. (A3.87) exists, then

$$\nabla\varphi = \frac{\partial\varphi}{\partial u}\nabla u + \frac{\partial\varphi}{\partial v}\nabla v + \frac{\partial\varphi}{\partial w}\nabla w = 0$$

from which we can deduce (multiply by $\cdot\,(\nabla v \times \nabla w)$, etc.)

$$\frac{\partial\varphi}{\partial u}\nabla u \cdot \nabla v \times \nabla w = \frac{\partial\varphi}{\partial v}\nabla u \cdot \nabla v \times \nabla w = \frac{\partial\varphi}{\partial w}\nabla u \cdot \nabla v \times \nabla w = 0.$$

Therefore Eq. (A3.89) holds, unless

$$\frac{\partial\varphi}{\partial u} \equiv \frac{\partial\varphi}{\partial v} \equiv \frac{\partial\varphi}{\partial w} \equiv 0,$$

which would imply that φ is a mere constant.

In proving the sufficiency of Eq. (A3.89) we may, as in the proof of Theorem 1, dismiss the trivial cases of any of the three gradients, ∇u, ∇v, or ∇w being identically zero. Further, if $\nabla u \times \nabla v$, $\nabla v \times \nabla w$, or $\nabla w \times \nabla u$ is identically zero, the case is again simple. Thus, for example, if $\nabla u \times \nabla u \equiv 0$, then by Theorem 1, $v = G(u)$, and

$$\varphi(u, v, w) \equiv v - G(u) = 0$$

is of the form of Eq. (A3.87).

We thus suppose that

$$\nabla u \times \nabla v \not\equiv 0, \qquad \nabla v \times \nabla w \not\equiv 0, \qquad \nabla w \times \nabla u \not\equiv 0; \tag{A3.90}$$

which implies at once that

$$\nabla u \not\equiv 0, \qquad \nabla v \not\equiv 0, \qquad \nabla w \not\equiv 0. \tag{A3.91}$$

By Theorem 1 the first of the inequalities (A3.90) implies that the first two of the Eqs. (A3.88) can be solved for two of the variables (x, y, z), say x_i and x_j, giving

$$x_j = H(u, v, x, y, z), \qquad \text{with } \frac{\partial H}{\partial x_i} \equiv \frac{\partial H}{\partial x_j} \equiv 0 \tag{A3.92}$$

and

$$x_i = K(u, v, x, y, z), \qquad \text{with } \frac{\partial K}{\partial x_i} \equiv \frac{\partial K}{\partial x_j} \equiv 0, \tag{A3.93}$$

with (see discussion of Eq. (A3.86))

$$\frac{\partial f}{\partial x_i}\frac{\partial g}{\partial x_j} - \frac{\partial f}{\partial x_j}\frac{\partial g}{\partial x_i} \not\equiv 0. \tag{A3.94}$$

Upon substitution from Eqs. (A3.92) and (A3.93) into the last of Eqs. (A3.88) we obtain

$$w = F(u, v, x, y, z), \qquad \text{with } \frac{\partial F}{\partial x_i} \equiv \frac{\partial F}{\partial x_j} \equiv 0. \tag{A3.95}$$

We therefore have

$$\nabla w = \frac{\partial F}{\partial u}\nabla u + \frac{\partial F}{\partial v}\nabla v + \nabla F, \tag{A3.96}$$

where

$$\nabla F = \left(\frac{\partial F}{\partial x}, \frac{\partial F}{\partial y}, \frac{\partial F}{\partial z}\right).$$

If now Eq. (A3.89) is assumed to hold, we obtain

$$\nabla u \times \nabla v \cdot \nabla F = \nabla u \times \nabla v \cdot \left(\nabla w - \frac{\partial F}{\partial u}\nabla u - \frac{\partial F}{\partial v}\nabla v\right)$$

$$= \nabla u \times \nabla v \cdot \nabla w = 0;$$

so that ∇F is in the plane of ∇u and ∇v, and hence

$$\nabla F = \alpha \nabla u + \beta \nabla v, \tag{A3.97}$$

where α and β are any functions of position. From this we obtain

$$\begin{aligned} \frac{\partial F}{\partial x_i} &= \alpha \frac{\partial f}{\partial x_i} + \beta \frac{\partial g}{\partial x_i} = 0, \\ \frac{\partial F}{\partial x_j} &= \alpha \frac{\partial f}{\partial x_j} + \beta \frac{\partial g}{\partial x_j} = 0, \end{aligned} \tag{A3.98}$$

by Eqs. (A3.94). Thus $\alpha \equiv \beta \equiv 0$, unless

$$\frac{\partial f}{\partial x_i}\frac{\partial g}{\partial x_j} - \frac{\partial f}{\partial x_j}\frac{\partial g}{\partial x_i} = 0,$$

but this latter is forbidden by the inequality (A3.94). It follows that $\alpha \equiv \beta \equiv 0$ and therefore by Eq. (A3.97) that $\nabla F = 0$; so that $F = F(u, v)$. This proves the theorem.

On the other hand, if $\nabla u \times \nabla v \cdot \nabla w \not\equiv 0$, Eq. (A3.96) gives

$$\nabla u \times \nabla v \cdot \nabla w = \nabla u \times \nabla v \cdot \nabla F \not\equiv 0$$

so that $\nabla F \not\equiv 0$. Since, however, $\partial F/\partial x_i \equiv \partial F/\partial x_j \equiv 0$, F must contain the third variable, which we will label x_k, from the set $\{x, y, z\}$; that is,

$$w = F(u, v, x_k) \qquad \text{with} \quad \frac{\partial F}{\partial x_k} \not\equiv 0. \tag{A3.99}$$

We can then solve this equation for x_k, obtaining

$$x_k = x_k(u, v, w). \tag{A3.100}$$

Equations (A3.92) and (A3.93) now can be rewritten in the forms

$$x_j = H(u, v, x_k) \tag{A3.101}$$

and

$$x_i = K(u, v, x_k). \tag{A3.102}$$

By using Eq. (A3.100) to substitute for x_k in Eqs. (A3.101) and (A3.102) we obtain expression for x_i, x_j, and x_k in terms of u, v, and w; therefore we can write

$$x = x(u, v, w), \qquad y = y(u, v, w), \qquad z = z(u, v, w), \quad \text{(A3.103)}$$

where the right-hand sides represent functions of u, v, and w.

Theorem 3. *It is possible to find any desired number of functions* $\alpha(\mathbf{r})$, $\beta(\mathbf{r})$, $\gamma(\mathbf{r})$, *etc., such that*

$$\mathbf{A} \cdot \nabla\alpha = \mathbf{A} \cdot \nabla\beta = \mathbf{A} \cdot \nabla\gamma = \cdots = 0, \quad \text{(A3.104)}$$

with $\mathbf{A} = \mathbf{A}(\mathbf{r})$ *any given vector function of position.*

Proof. Consider any one-parameter family of curves on an arbitrary surface not parallel at any point to the vector **A**. Through each point of each curve draw a line that is everywhere in the direction of **A**, thereby constructing a surface for each curve α of the family. Let the one-parameter family of surfaces thus obtained be represented by

$$f(\mathbf{r}, \alpha) = 0. \quad \text{(A3.105)}$$

For any point $\mathbf{r}$ in space Eq. (A3.105) is an implicit equation for α, giving the value of the parameter α for that surface of the family of surfaces that passes through the point $\mathbf{r}$. In fact we may solve Eq. (A3.105) to obtain α as an explicit function of $\mathbf{r}$ and may thus write

$$\alpha = F(\mathbf{r}). \quad \text{(A3.106)}$$

If we move from some point $\mathbf{r}$ in the direction of **A** to some adjacent point $\mathbf{r} + d\mathbf{r}$, then we have

$$d\mathbf{r} = \mathbf{A}\, ds, \quad \text{(A3.107)}$$

where s is the distance along **A** measured from any suitable point. This movement is clearly along one of the surfaces α of the family of surfaces, and thus $d\alpha$ is zero for a $d\mathbf{r}$ given by Eq. (A3.107). Since α as given by Eq. (A3.106) is a scalar function of position and is not a function of time, we have (see Eq. (A3.59a))

$$d\alpha = d\mathbf{r} \cdot \nabla\alpha = ds\, \mathbf{A} \cdot \nabla\alpha = 0$$

for any ds. Thus, we have

$$\mathbf{A} \cdot \nabla\alpha = 0, \quad \text{(A3.108)}$$

with α defined by Eq. (A3.105) or Eq. (A3.106).

Evidently there will be a solution for each family of curves one selects.

Theorem 4. *Every vector function of position* $\mathbf{A}(\mathbf{r})$ *can be written in the form*

$$\mathbf{A}(\mathbf{r}) = \mu\, \nabla\alpha \times \nabla\beta, \tag{A3.109}$$

with $\mathbf{A} \cdot \nabla\alpha = \mathbf{A} \cdot \nabla\beta = 0$, *where* α, β, *and* μ *are functions of* $\mathbf{r}$.

Proof. By Theorem 3 we can construct the two families of surfaces

$$\alpha = \alpha(\mathbf{r}) \qquad \text{and} \qquad \beta = \beta(\mathbf{r}) \tag{A3.110}$$

such that on the surfaces of one family α is a constant, while on the surfaces of the other family β is a constant, and such that

$$\mathbf{A} \cdot \nabla\alpha = \mathbf{A} \cdot \nabla\beta = 0. \tag{A3.111}$$

If the two families (A3.110) are distinct, we have

$$\nabla\alpha \times \nabla\beta \not\equiv 0. \tag{A3.112}$$

This can be seen as follows: Suppose $\nabla\alpha \times \nabla\beta \equiv 0$; then, by Theorem 1, $\alpha = f(\beta)$, so that the set of points for which β has a given value β_0 coincides exactly with the set of points for which α has the value $\alpha_0 = f(\beta_0)$. Thus all points on a surface of one family are also on a surface of the other family. The two families thus consist of the same surfaces, and are therefore the same family of surfaces.

With Eqs. (A3.111) and inequality (A3.112) holding, we have

$$\mathbf{A} \times (\nabla\alpha \times \nabla\beta) = \mathbf{A} \cdot \nabla\beta\, \nabla\alpha - \mathbf{A} \cdot \nabla\alpha\, \nabla\beta = 0$$

by Eq. (A3.101). Thus $\mathbf{A}$ is parallel to $\nabla\alpha \times \nabla\beta$ everywhere, and hence Eq. (A3.109) holds for some scalar function of position μ.

Theorem 5. *If*

$$\nabla \times \mathbf{A} = 0, \tag{A3.113}$$

then

$$\mathbf{A} = \nabla\varphi \tag{A3.114}$$

for some φ, *and conversely.*

Proof. The proof of this theorem, except for minor changes, has already been given in Section 6.

Theorem 6. *If $u(\mathbf{r})$ and $v(\mathbf{r})$ are two independent functions of position (i.e., $\nabla u \times \nabla v \not\equiv 0$), then for every differentiable pair of functions $f = f(u, v)$ and $g = g(u, v)$ there exist a $\mu = \mu(u, v)$ and a $\varphi = \varphi(u, v)$ such that*

$$\mu(f \nabla u + g \nabla v) = \nabla\varphi. \tag{A3.115}$$

Proof. Equation (A3.115) by Theorem 5 is equivalent to

$$\nabla \times \mu(f \nabla u + g \nabla v) = 0 \tag{A3.116}$$

or by expansion to

$$-\nabla u \times \nabla(\mu f) - \nabla v \times \nabla(\mu g) = 0. \tag{A3.117}$$

Now

$$\begin{aligned} \nabla(\mu f) &= \frac{\partial(\mu f)}{\partial u}\nabla u + \frac{\partial(\mu f)}{\partial v}\nabla v, \\ \nabla(\mu g) &= \frac{\partial(\mu g)}{\partial u}\nabla u + \frac{\partial(\mu g)}{\partial v}\nabla v; \end{aligned} \tag{A3.118}$$

therefore Eq. (A3.117) becomes

$$\nabla u \times \nabla v \frac{\partial(\mu f)}{\partial v} + \nabla v \times \nabla u \frac{\partial(\mu g)}{\partial u} = 0.$$

Since $\nabla u \times \nabla v \not\equiv 0$, we have

$$\frac{\partial(\mu f)}{\partial v} - \frac{\partial(\mu g)}{\partial u} = 0, \tag{A3.119}$$

which is therefore the necessary and sufficient condition that μ must satisfy in order that Eq. (A3.115) holds. It remains to prove that there exists a $\mu = \mu(u, v)$ satisfying Eq. (A3.119).

Let $F = F(u, v, \mu)$ be a function satisfying the equation

$$g\frac{\partial F}{\partial u} - f\frac{\partial F}{\partial v} + \mu\left(\frac{\partial f}{\partial v} - \frac{\partial g}{\partial u}\right)\frac{\partial F}{\partial \mu} = 0, \tag{A3.120}$$

in which u, v, and μ are treated as independent variables. Then, setting any solution of this equation for which $\partial F/\partial\mu \not\equiv 0$, equal to zero, establishes a relation between u, v, and μ of the form

$$F(u, v, \mu) = 0, \tag{A3.121}$$

which requires that

$$dF = \frac{\partial F}{\partial u}\,du + \frac{\partial F}{\partial v}\,dv + \frac{\partial F}{\partial \mu}\,d\mu = 0. \qquad \text{(A3.122)}$$

Equation (A3.121) clearly defines μ as a function of u and v; moreover, if we hold v fixed and vary u, we have from Eq. (A3.122)

$$\frac{\partial F}{\partial u}(du)_v + \frac{\partial F}{\partial \mu}(d\mu)_v = 0,$$

where $(du)_v$ and $(d\mu)_v$ are the differentials for a variation of u and not v. Thus the partial derivative of $\mu(u, v)$ with respect to u is given by

$$\frac{\partial \mu(u, v)}{\partial u} = \frac{(d\mu)_v}{(du)_v} = -\frac{\partial F/\partial u}{\partial F/\partial \mu}. \qquad \text{(A3.123)}$$

Similarly, we determine that

$$\frac{\partial \mu(u, v)}{\partial v} = -\frac{\partial F/\partial v}{\partial F/\partial \mu}. \qquad \text{(A3.124)}$$

Dividing Eq. (A3.120) by $\partial F/\partial \mu$ and using Eqs. (A3.123) and (A3.124), we obtain

$$-g\frac{\partial \mu}{\partial u} + f\frac{\partial \mu}{\partial v} + \mu\left(\frac{\partial f}{\partial v} - \frac{\partial g}{\partial u}\right) = 0,$$

which is equivalent to Eq. (A3.119). Thus a solution for μ in Eq. (A3.119) is given by the implicit function for μ given by Eq. (A3.121), where $F(u, v, \mu)$ is some solution of Eq. (A3.120).

Since u, v, and μ are independent variables in Eq. (A3.120), they can be treated exactly as we treated x, y, and z in Theorem 3. In fact, if we let

$$\nabla' F = \left(\frac{\partial F}{\partial u}, \frac{\partial F}{\partial v}, \frac{\partial F}{\partial \mu}\right),$$

then Eq. (A3.120) is of the form

$$\mathbf{A} \cdot \nabla F = 0, \qquad \text{(A3.125)}$$

where

$$A_x = g, \qquad A_y = -f, \qquad A_z = \mu\left(\frac{\partial f}{\partial y} - \frac{\partial g}{\partial x}\right).$$

As was shown in Theorem 3, Eq. (A3.125) always has any number of solutions. Thus $F(u, v, \mu)$ satisfying Eq. (A3.120) exists, which proves the theorem.

Theorem 7. *The necessary and sufficient condition for* A(r) *to be representable as*

$$\mathbf{A}(\mathbf{r}) \equiv \lambda(\mathbf{r}) \, \nabla\varphi(\mathbf{r}) \tag{A3.126}$$

is that

$$\mathbf{A} \cdot \nabla \times \mathbf{A} \equiv 0. \tag{A3.127}$$

Proof. It is easy to show that this condition is necessary. Thus, from Eq. (A3.126), we deduce that

$$\nabla \times \mathbf{A} \equiv \nabla \times (\lambda \, \nabla\varphi) \equiv \nabla\lambda \times \nabla\varphi + \lambda \, \nabla \times \nabla\varphi \equiv \nabla\lambda \times \nabla\varphi;$$

so that

$$\mathbf{A} \cdot \nabla \times \mathbf{A} \equiv \lambda \, \nabla\varphi \cdot (\nabla\lambda \times \nabla\varphi) = 0.$$

To prove that condition (A3.127) is sufficient, we proceed as follows: By Theorem 4, we can write

$$\nabla \times \mathbf{A} = \mu \nabla\alpha \times \nabla\beta; \tag{A3.128}$$

so that the assumed condition (A3.127) becomes (since $\mu \not\equiv 0$)

$$\mathbf{A} \cdot (\nabla\alpha \times \nabla\beta) = 0,$$

which shows that **A** is in the plane of $\nabla\alpha$ and $\nabla\beta$. This can be written

$$\mathbf{A} = f \, \nabla\alpha + g \, \nabla\beta, \tag{A3.129}$$

in which case, by Eqs. (A3.24) and (A3.38),

$$\nabla \times \mathbf{A} = \nabla \times (f \, \nabla\alpha) + \nabla \times (g \, \nabla\beta) = \nabla f \times \nabla\alpha + \nabla g \times \nabla\beta. \tag{A3.130}$$

On the other hand, Eq. (A3.128) requires that

$$\nabla\alpha \cdot \nabla \times \mathbf{A} = \nabla\beta \cdot \nabla \times \mathbf{A} = 0;$$

so that, using Eq. (A3.130),

$$\nabla\alpha \cdot (\nabla f \times \nabla\alpha + \nabla g \times \nabla\beta) = \nabla\alpha \cdot \nabla g \times \nabla\beta = 0$$

and

$$\nabla\beta \cdot (\nabla f \times \nabla\alpha + \nabla g \times \nabla\beta) = \nabla\beta \cdot \nabla f \times \nabla\alpha = 0.$$

By Theorem 2, the last two equations show that

$$f = f(\alpha, \beta) \qquad \text{and} \qquad g = g(\alpha, \beta),$$

and hence Eq. (A3.129) becomes

$$\mathbf{A} = f(\alpha, \beta) \, \nabla\alpha + g(\alpha, \beta) \, \nabla\beta. \tag{A3.131}$$

By Theorem 6, therefore, there exists a $\mu = \mu(\alpha, \beta)$ such that

$$\mu \mathbf{A} = \mu(f \nabla \alpha + g \nabla \beta) = \nabla \varphi(\alpha, \beta).$$

Putting $\mu = 1/\lambda$, we obtain finally Eq. (A3.126), which proves the theorem.

Theorem 8. *If*

$$\nabla \cdot \mathbf{A} = 0, \tag{A3.132}$$

then

$$\mathbf{A} = \nabla \times \mathbf{B} = \nabla u \times \nabla \beta \qquad (\mathbf{B} = u \nabla \beta) \tag{A3.133}$$

for some u and β, and conversely.

Proof. By Theorem 4 we can write

$$\mathbf{A} = \mu \nabla \alpha \times \nabla \beta, \tag{A3.134}$$

and, if Eq. (A3.132) holds, then

$$\nabla \cdot \mathbf{A} = \nabla \cdot (\mu \nabla \alpha \times \nabla \beta) = \nabla \mu \cdot (\nabla \alpha \times \nabla \beta) + \mu \nabla \cdot (\nabla \alpha \times \nabla \beta) = 0.$$

The second term vanishes because

$$\nabla \cdot (\nabla \alpha \times \nabla \beta) = \nabla \cdot \nabla \times (\alpha \nabla \beta) = 0;$$

therefore Eq. (A3.132) reduces to

$$\nabla \mu \cdot \nabla \alpha \times \nabla \beta = 0.$$

This, by Theorem 2, means that

$$\mu = \mu(\alpha, \beta). \tag{A3.135}$$

We may therefore put

$$\mu = \frac{\partial u}{\partial \alpha} \qquad \text{or} \qquad u = \int \mu(\alpha, \beta)\, d\alpha, \tag{A3.136}$$

where in the integration β is treated as a parameter. This shows that $u = u(\alpha, \beta)$ and therefore that

$$\mu \nabla \alpha = \frac{\partial u}{\partial \alpha} \nabla \alpha = \nabla u - \frac{\partial u}{\partial \beta} \nabla \beta.$$

It then follows that

$$\mathbf{A} = \mu \nabla \alpha \times \nabla \beta = \left(\nabla u - \frac{\partial u}{\partial \beta} \nabla \beta\right) \times \nabla \beta = \nabla u \times \nabla \beta = \nabla \times (u \nabla \beta),$$

which is Eq. (A3.133).

If on the other hand $\mathbf{A} = \nabla \times \mathbf{B}$, then it is readily seen directly that

$$\nabla \cdot \mathbf{A} = \nabla \cdot \nabla \times \mathbf{B} = 0.$$

Theorem 9. *Every vector* **A** *can be put in the form*

$$\mathbf{A} = u\,\nabla\beta + \nabla\varphi. \tag{A3.137}$$

Proof. Since, for every **A**, $\nabla \cdot (\nabla \times \mathbf{A}) = 0$, by Theorem 8,

$$\nabla \times \mathbf{A} = \nabla u \times \nabla\beta = \nabla \times (u\,\nabla\beta);$$

so that for every **A** there exist two functions u and β such that

$$\nabla \times (\mathbf{A} - u\,\nabla\beta) = 0. \tag{A3.138}$$

Applying Theorem 7, we then have

$$\mathbf{A} - u\,\nabla\beta = \nabla\varphi$$

for some φ, which is, of course, the same as Eq. (A3.137).

Finally, we shall give without proof the *Helmholtz theorem.*

Theorem 10. Any single-valued vector field **F** that together with its derivatives is finite and continuous and vanishes at infinity can be written

$$\mathbf{F} = \nabla \times \mathbf{A} + \nabla\varphi, \tag{A3.139}$$

This theorem shows that any such vector field **F** can be written as the sum of a *solenoidal* field $\nabla \times \mathbf{A}$ (one whose divergence vanishes everywhere) and an *irrotational* field $\nabla\varphi$ (one whose curl vanishes everywhere).

Appendix A4

Transformation of Integrals

We have seen in the discussion of vector products that $\mathbf{A} \times \mathbf{B}$ is a vector, the magnitude of which is equal to the area of the parallelogram of which $\mathbf{A}$ and $\mathbf{B}$ are two adjacent sides (when their origins are made to coincide), and the direction of which is normal to the plane of $\mathbf{A}$ and $\mathbf{B}$. Thus the vector $\mathbf{A} \times \mathbf{B}$ may be said to represent a portion of the plane of $\mathbf{A}$ and $\mathbf{B}$. Since, however, its negative $\mathbf{B} \times \mathbf{A}$ represents equally well this same portion, we say that $\mathbf{A} \times \mathbf{B}$ represents a directed portion of the plane; that is, a portion with the positive sense specified for the normal. More generally, any portion of a plane, or a small portion of any surface, may be represented by a vector having the magnitude equal to the area of the surface and a direction along the normal to the surface. For any surface $d\mathbf{S} = \mathbf{n}\, dS$ is a vector representing an element of it, where dS is the area of the element, and $\mathbf{n}$ is a unit vector parallel to the normal to the surface. As to the sign of the normal, we shall adhere to the following convention. When the surface is a closed surface, each element $d\mathbf{S}$ of the closed surface will have the direction from the inside of the enclosed volume outward through the surface, which we shall call for short the *outward direction*. When $d\mathbf{S}$ is taken isolated, or is a portion of an open surface, then the direction of $d\mathbf{S}$ will be determined by the direction in which its boundary curve is traced, and will be such that this direction will appear clockwise when looking in the direction of $d\mathbf{S}$ (see Figure A4.1). To a finite portion of a curved surface we may also assign a vector; this vector is obtained by summing the elementary vectors $d\mathbf{S}$ for all its portions and may be symbolized by $\int d\mathbf{S}$. Integrals involving $d\mathbf{S}$ are called *surface integrals*. A more general type of a surface integral is $\int \varphi\, d\mathbf{S}$, where each element of the surface is multiplied, before summing, by the value of φ at the point.

Other surface integrals are $\int \mathbf{A} \cdot d\mathbf{S}$, $\int \mathbf{A} \times d\mathbf{S}$, etc., definitions of which present no new difficulty.

Every surface integral taken over a closed surface, in which the integrand is a differentiable function of position, *can be represented as a volume*

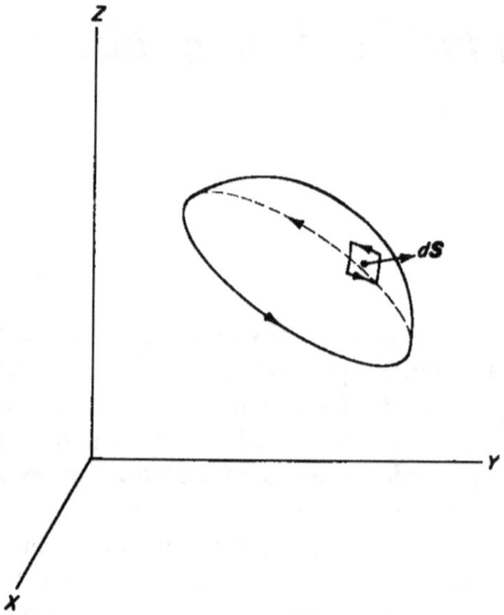

FIG. A4.1. Relationship of a surface element $d\mathbf{S}$ to the direction in which its boundary curve is traced.

integral, taken over the space enclosed by the surface. Consider, for example,

$$\mathbf{B} = \int \varphi \, d\mathbf{S}, \tag{A4.1}$$

taken over a closed surface, say the surface represented in Figure A4.2. Take a component of this vector, say the Z component:

$$B_z = \int \varphi \, dS_z. \tag{A4.2}$$

Here dS_z is the Z component of $d\mathbf{S}$; its magnitude is equal to dS multiplied by the cosine of the angle between $d\mathbf{S}$ and the Z-axis or, in other words, to the projection of the surface element upon the XY-plane. We can choose all the surface elements in such a way that their projections on the XY-plane are the rectangles $dx\ dy$, as shown in the figure for one surface element. Our equation then becomes

$$B_z = \iint \varphi \, dx \, dy, \qquad \text{taken over the closed surface.} \qquad \text{(A4.3)}$$

The range of integration will consist of two parts, the boundary between which is indicated in the figure by the dotted line. For the lower part,

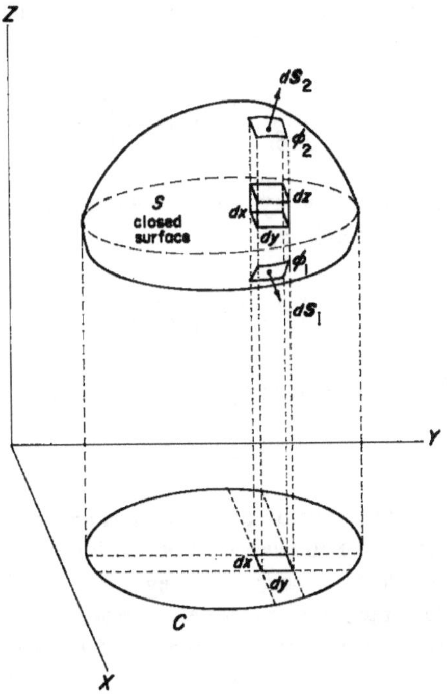

FIG. A4.2. Relationship of the elements involved in a surface integral over a closed surface S to those involved in a volume integral over the enclosed region.

which we shall call part 1, dS_z is negative; that is, for this part $d\mathbf{S}$ is directed toward the plane XY. For the upper part, part 2, dS_z is positive. The projections of both parts on the XY-plane have the same boundary C, bounding a portion of the plane F. Equation (A4.3) can therefore be written as follows:

$$B_z = \iint (\varphi_2 - \varphi_1) \, dx \, dy, \qquad \text{taken over } F, \qquad \text{(A4.4)}$$

where φ_1 and φ_2 are the values of φ on the lower and the upper parts of S, respectively, corresponding to the same x and y. The expression $(\varphi_2 - \varphi_1)$, however, for fixed x and y, is equal to

$$\int \frac{\partial \varphi}{\partial z}\, dz,$$

taken for the range of z contained within the closed surface. Thus, we can write Eq. (A4.4) in the form

$$B_z = \iiint \frac{\partial \varphi}{\partial z}\, dx\, dy\, dz = \int \frac{\partial \varphi}{\partial z}\, dV, \tag{A4.5}$$

taken over the space enclosed by the surface. Similar expressions hold for all the components; therefore, we have*

$$\int \varphi\, d\mathbf{S} = \int \nabla\varphi\, dV. \tag{A4.6}$$

If written in the form $\int d\mathbf{S}\, \varphi = \int dV\, \nabla\varphi$, this equation can be symbolically represented as

$$d\mathbf{S} \sim dV\, \nabla, \tag{A4.7}$$

which turns out to be applicable to any surface integral over a closed surface. While our derivation assumed particularly simple conditions, the result obtained is correct whenever the integrand is a differentiable function of position. If the integrand has one or more points of discontinuity inside the surface, these may be surrounded by small surfaces, which can then be considered as a part of the boundary of the remaining space within the original surface. To this boundary we can then apply the formula found above.

One of the applications of the relation (A4.7) is

$$\int \mathbf{A} \cdot d\mathbf{S} = \int \nabla \cdot \mathbf{A}\, dV, \tag{A4.8}$$

which is known as *Gauss' theorem.* We will, however, call by this name any formula obtained by the application of the symbolic relation (A4.7).

* Here, as elsewhere in the appendices, we omit specifying the volume, or surface, over which the integration is performed because only one volume, or surface, is involved.

EXAMPLES

$$\int \mathbf{A} \times d\mathbf{S} = -\int \nabla \times \mathbf{A}\, dV, \tag{A4.9}$$

$$\int \mathbf{A}\,\mathbf{B} \cdot d\mathbf{S} = \int d\mathbf{S} \cdot \mathbf{B}\,\mathbf{A} = \int dV\, \nabla \cdot \mathbf{B}\,\mathbf{A}, \qquad \text{where } \nabla \text{ acts on both } \mathbf{A} \text{ and } \mathbf{B},$$

$$= \int dV\, \nabla_{\mathbf{B}} \cdot \mathbf{B}\,\mathbf{A} + \int dV\, \nabla_{\mathbf{A}} \cdot \mathbf{B}\,\mathbf{A}$$

$$= \int \mathbf{A}\nabla \cdot \mathbf{B}\, dV + \int \mathbf{B} \cdot \nabla\mathbf{A}\, dV. \tag{A4.10}$$

Integrals involving the vector differential $d\mathbf{r} = (dx, dy, dz)$, such as

$$\int_{\mathbf{r}_1}^{\mathbf{r}_2} \varphi(\mathbf{r})\, d\mathbf{r}, \qquad \int_{\mathbf{r}_1}^{\mathbf{r}_2} \mathbf{A}(\mathbf{r}) \cdot d\mathbf{r}, \qquad \int_{\mathbf{r}_1}^{\mathbf{r}_2} \mathbf{A} \times d\mathbf{r},$$

are called *line integrals*. They generally depend not only on the *end points* $\mathbf{r}_1$ and $\mathbf{r}_2$ but also on the path the terminus of $\mathbf{r}$ takes between $\mathbf{r}_1$ and $\mathbf{r}_2$. With the path specified only one component is needed to specify $\mathbf{r}$ fully, and one can write

$$\mathbf{r} = \mathbf{f}_1(x) = \mathbf{f}_2(y) = \mathbf{f}_3(z),$$

where $\mathbf{f}_1$, $\mathbf{f}_2$, and $\mathbf{f}_3$ are vector functions determined by the path. If the end points are $\mathbf{r}_1 = (x_1, y_1, z_1)$ and $\mathbf{r}_2 = (x_2, y_2, z_2)$, the first integral above is seen to be the vector

$$\left(\int_{x_1}^{x_2} \varphi(\mathbf{r})\, dx, \quad \int_{y_1}^{y_2} \varphi(\mathbf{r})\, dy, \quad \int_{z_1}^{z_2} \varphi(\mathbf{r})\, dz\right)$$

$$= \left(\int_{x_1}^{x_2} \varphi(\mathbf{f}_1(x))\, dx, \quad \int_{y_1}^{y_2} \varphi(\mathbf{f}_2(y))\, dy, \quad \int_{z_1}^{z_2} \varphi(\mathbf{f}_3(z))\, dz\right).$$

Of particular importance is a line integral around a closed path ($\mathbf{r}_1 = \mathbf{r}_2$). It is fully specified by the integrand and the path taken. To emphasize that it is taken around a closed path we replace the integral sign by $\oint$. We shall now show that *each line integral taken around a closed curve C*, in which the integrand is a differentiable function of position (scalar or vector), *may be represented as a surface integral* taken over any open surface bounded by the curve. Consider, for example, the line integral

$$\mathbf{B} = \oint \varphi\, d\mathbf{r}, \tag{A4.11}$$

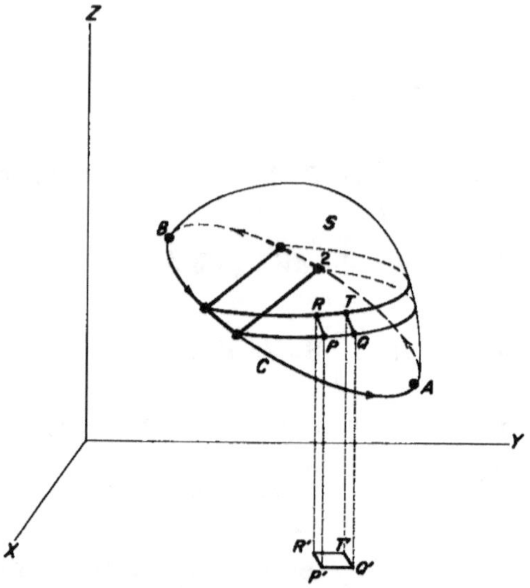

FIG. A4.3. A surface S bounded by a contour C and cut by planes parallel to the XY plane showing the projection of a typical surface element $PQRT$ on the XY plane.

where the closed path C and the surface S it bounds are related as shown in Figure A4.3. Consider a component of vector **B**, say B_z; then, as seen earlier,

$$B_z = \oint \varphi \, dz. \tag{A4.12}$$

Let A and B be points on the curve C corresponding to the minimum and maximum values, respectively, of z. For a simple closed curve like that represented in the figure there are two points (1 and 2) on the curve for each value of z, between the minimum and maximum values of z corresponding to points A and B. Since in the integration (A4.12) the differentials dz corresponding to point 1 are negative and those to point 2 are positive, we may write

$$\begin{aligned} B_z &= \int_A^B [\varphi_2(\mathbf{r}) - \varphi_1(\mathbf{r})] \, dz \\ &= \int_A^B [\varphi_2(\mathbf{f}_3(z)) - \varphi_1(\mathbf{f}_3(z))] \, dz, \end{aligned} \tag{A4.13}$$

where φ_1 and φ_2 are the values of φ on the two parts of the curve into which the original curve was divided by points A and B. The integration here is the ordinary integration of integral calculus, the lower and the upper limits being, respectively, the lowest and the highest values of z on the original curve.

Since the original curve C is the boundary of some surface S, as is indicated in the figure, then to each value of z there will correspond a curve on the surface (such as the curve 1–2), the curve along which the plane $z =$ constant cuts the surface. The difference $\varphi_2 - \varphi_1$ may then be represented as

$$\varphi_2 - \varphi_1 = \int_1^2 \delta\varphi = \int_1^2 \left(\frac{\partial\varphi}{\partial x}\,\delta x + \frac{\partial\varphi}{\partial y}\,\delta y\right), \tag{A4.14}$$

the integration being along the curve 1–2, along which z is constant, $\delta\varphi$, δx, and δy denoted changes along the curve of constant z. Thus we have

$$B_z = \iint \left(\frac{\partial\varphi}{\partial x}\,\delta x\,dz + \frac{\partial\varphi}{\partial y}\,\delta y\,dz\right), \tag{A4.15}$$

where integration is to be extended over the surface S, which is easily seen to be one of the components of the vector equation

$$\mathbf{B} = \int \varphi\,d\mathbf{r} = \int d\mathbf{S} \times \nabla\varphi. \tag{A4.16}$$

Although proved with many specializations this formula holds in general for differentiable integrands. Further generalization, which also turns out to be correct, is obtained by noting that Eq. (A4.16) follows from the following equivalence of symbolic forms:

$$d\mathbf{r} \sim d\mathbf{S} \times \nabla. \tag{A4.17}$$

As an application of this equivalence one has the very important equation

$$\int \mathbf{A} \cdot d\mathbf{r} = \int \nabla \times \mathbf{A} \cdot d\mathbf{S}, \tag{A4.18}$$

which is known as the *Stokes' theorem.* We shall apply this name, however, to any relation derived with the help of the symbolic relation (A4.17).

EXAMPLE 1. A direct application of the symbolic equivalence (A4.17) yields the equation

$$\int \mathbf{A} \times d\mathbf{r} = -\int d\mathbf{r} \times \mathbf{A} = -\int (d\mathbf{S} \times \nabla) \times \mathbf{A} = -\int (d\mathbf{S} \times \nabla_{\mathbf{A}}) \times \mathbf{A}.$$

Expanding the vector triple product using Eqs. (A2.25) and (A3.28), we have

$$\begin{aligned}\int \mathbf{A} \times d\mathbf{r} &= \int [\nabla_{\mathbf{A}} \cdot \mathbf{A}\, d\mathbf{S} - \nabla_{\mathbf{A}}(\mathbf{A} \cdot d\mathbf{S})] \\ &= \int [\nabla \cdot \mathbf{A}\, d\mathbf{S} - d\mathbf{S} \times (\nabla \times \mathbf{A}) - d\mathbf{S} \cdot \nabla \mathbf{A}] \\ &= \int \nabla \cdot \mathbf{A}\, d\mathbf{S} + \int (\nabla \times \mathbf{A}) \times d\mathbf{S} - \int d\mathbf{S} \cdot \nabla \mathbf{A}. \qquad \text{(A4.19)}\end{aligned}$$

An interesting special case of this formula,

$$\tfrac{1}{2}\int \mathbf{r} \times d\mathbf{r} = \int d\mathbf{S}, \qquad \text{(A4.20)}$$

is obtained by replacing **A** by **r**.

If this surface S is divided into strips by lines of constant z and if these strips are further subdivided into small quadrangles, a typical one of which is shown in Figure A4.3, then these quadrangles may be conveniently chosen as the surface elements that reduce in the limit to the differential surface element dS of a surface integral over S. Let us consider a typical one of these quadrangles, one having a vertex at the point (x, y, z) on the surface. Let us designate the coordinates of the adjacent vertex on the line of constant z passing through this point by $(x + \delta x, y + \delta y, z)$, and the adjacent vertex on the next line of constant z by $(x + dx, y + dy, z + dz)$; the last vertex is then conveniently chosen to have the coordinates $(x + dx + \delta x, y + dy + \delta y, z + dz)$. Projection of the surface element on the YZ-plane is then a parallelogram with slightly curved sides. In the limit of infinitesimal displacements* δy and dz the area of this parallelogram becomes $dS_x = \delta y\, dz$. Similarly, since for δx and dz positive dS will be negative, we have in the limit $dS = -\delta x\, dz$. Equation (A4.15) can thus be written as

$$B_z = \int \left(\frac{\partial \varphi}{\partial y}\, dS_x - \frac{\partial \varphi}{\partial x}\, dS_y\right).$$

EXAMPLE 2. Suppose one is given a moving fluid and any stationary closed surface S lying within the fluid. What is the rate of increase of the quantity of fluid in the volume V bounded by S?

* We here succumb to the temptation to shorten proofs by considering integrals as merely sums, in a generalized sense, of infinitesimals. In a rigorous treatment we would first obtain finite sums and then take the limit of smaller and smaller divisions to obtain Riemann integrals.

If the density of material is $\rho(\mathbf{r}, t)$, then the quantity of material within V is given by

$$w \equiv \int_V \rho(\mathbf{r}, t)\, dV.$$

Since the region V over which the integration is to be performed does not vary with time, the rate of change of w is given by

$$\frac{d}{dt}\int_V \rho(\mathbf{r}, t)\, dV = \int_V \frac{\partial \rho(\mathbf{r}, t)}{\partial t}\, dV. \qquad \text{(A4.21)}$$

If the material is moving with the velocity $\mathbf{v}(\mathbf{r}, t)$, the amount of the material going out through an element of surface per unit of time is $\rho\mathbf{v} \cdot d\mathbf{S}$, so that the rate at which the material is flowing out through the surface is

$$\int \rho\mathbf{v} \cdot d\mathbf{S}. \qquad \text{(A4.22)}$$

The sum of the two, namely

$$\int \frac{\partial \rho}{\partial t}\, dV + \int \rho\mathbf{v} \cdot d\mathbf{S} = \int \left(\frac{\partial \rho}{\partial t} + \nabla \cdot (\rho\mathbf{v})\right) dV, \qquad \text{(A4.23)}$$

represents the rate at which the material is created within the surface. If the material is neither created nor destroyed, the integral (A4.23) must be zero for every choice of the volume V. This can be true only if the integrand vanishes everywhere, and hence

$$\frac{\partial \rho}{\partial t} + \nabla \cdot (\rho\mathbf{v}) = 0 \qquad \text{(A4.24)}$$

is a necessary and sufficient condition for the conservation of matter. This is the very important *continuity equation* of hydrodynamics.

Appendix A5

Tensors

In this section we shall not be bound by the limitation of the number of dimensions of space to three, but shall consider the more general case of the space of n dimensions. Each vector A_α will now have n components ($\alpha = 1, 2, \ldots, n$), each component having reference to one of the coordinate axes. We shall again restrict ourselves to rotations of the coordinate axes, so that*

$$x'_\alpha = a_{\alpha\beta} x_\beta \qquad (\alpha = 1, 2, \ldots, n), \tag{A5.1}$$

where, in agreement with Eqs. (A3.5), (A3.7), and (A3.9), the $a_{\alpha\beta}$ satisfy the equations

$$a_{\alpha\gamma} a_{\beta\gamma} = a_{\gamma\alpha} a_{\gamma\beta} = \delta_{\alpha\beta}. \tag{A5.2}$$

A vector is defined, as before, to be a set of numbers (components) that transform under a rotation of the coordinate system like the coordinates of a point in the space. Thus a vector **A**, having components A_α in the unprimed coordinate system, will have components

$$A'_\alpha = a_{\alpha\beta} A_\beta \tag{A5.3}$$

in the primed coordinate system.

From the linear combinations of the components of a vector other vectors may be built up. Thus we could have

$$B_\alpha = b_{\alpha\beta} A_\beta \qquad (\alpha = 1, 2, \ldots, n), \tag{A5.4}$$

* By the summation convention we have adopted in the body of the text a repeated index implies summation with respect to that index; thus

$$a_{\alpha\beta} x_\beta = \sum_\beta a_{\alpha\beta} x_\beta .$$

where $b_{\alpha\beta}$ is any set of n^2 numbers. In the primed coordinate system given by Eq. (A5.1) we would then have

$$B'_\alpha = b'_{\alpha\beta} A'_\beta, \tag{A5.5}$$

where the $b'_{\alpha\beta}$ are a new set of numbers properly related to the $b_{\alpha\beta}$. This relation can be easily determined in the following way. From Eq. (A5.4) and the equivalent of Eq. (A5.3) we have

$$B'_\alpha = a_{\alpha\beta} B_\beta = a_{\alpha\beta} b_{\beta\gamma} A_\gamma, \tag{A5.6}$$

while from Eqs. (A5.3) and (A5.5)

$$B'_\alpha = b'_{\alpha\beta} a_{\beta\gamma} A_\gamma.$$

Thus we must have

$$a_{\alpha\beta} b_{\beta\gamma} A_\gamma = b'_{\alpha\beta} a_{\beta\gamma} A_\gamma. \tag{A5.7}$$

If the $b_{\alpha\beta}$ are arbitrarily assigned, the n^2 quantities $b'_{\alpha\beta}$ need only satisfy the n equations (A5.7), for each coordinate system and each vector A_α. We may thus impose further restrictions on the $b'_{\alpha\beta}$. The most interesting-case is the one in which we demand that these quantities should be independent of the choice of the vector A_α. In this case there will be a law of transformation between $b'_{\alpha\beta}$ and $b_{\alpha\beta}$ depending only upon the transformation of the coordinate system. Instead of Eq. (A5.7) we must then have

$$a_{\alpha\beta} b_{\beta\gamma} = b'_{\alpha\beta} a_{\beta\gamma}, \tag{A5.8}$$

Replacing β by μ, γ by ν, multiplying by $a_{\beta\nu}$, and using Eq. (A5.2), we obtain

$$b'_{\alpha\beta} = a_{\alpha\mu} a_{\beta\nu} b_{\mu\nu}. \tag{A5.9}$$

A set of n^2 quantities transforming in accordance with Eq. (A5.9) is called a *tensor of second rank*, or simply a *tensor*.

More generally, a set of n^m quantities $B_{\alpha\beta\ldots\kappa}$, where each of the m subscripts takes on independently the values from 1 to n, is called a *tensor of* mth *rank* provided that, when the coordinate system is rotated, the quantities transform in accordance with the law of transformation

$$B'_{\alpha\beta\cdots\kappa} = a_{\alpha\mu} a_{\beta\nu} \cdots a_{\kappa\sigma} B_{\mu\nu\cdots\sigma}. \tag{A5.10}$$

It is of course obvious that Eq. (A5.9) is a special case of Eq. (A5.10), the case of $m = 2$. Equation (A5.3) is another special case, the case of $m = 1$; thus vectors are tensors of the first rank. When $m = 0$, Eq. (A5.10) becomes $B' = B$; so that scalars may be called tensors of rank zero.

Our definition of a tensor, in view of the preceding discussion, may be stated as a theorem: *If for an arbitrary choice of a vector A_α the set of quantities $B_\alpha = b_{\alpha\beta}A_\beta$ constitutes a vector, then the set $b_{\alpha\beta}$ is a tensor.* This theorem can be generalized as follows: *If for an arbitrary choice of a vector A_α the set of quantities*

$$B_{\beta\gamma\ldots} = B_{\alpha\beta\gamma\ldots}A_\alpha \tag{A5.11}$$

is a tensor of some rank m, then the set of quantities $B_{\alpha\beta\gamma\ldots}$ is a tensor of rank $m + 1$. The individual quantities of the set are called *components* of the tensor.

As a special case of this theorem we have that, if

$$A_\alpha B_\alpha = C = \text{scalar} \tag{A5.12}$$

for arbitrary vector B_α, then A_α is a vector. A direct proof of this can be readily supplied by the reader.

Two tensors $A_{\alpha\beta\cdots\kappa}$ and $B_{\alpha\beta\cdots\kappa}$ are said to be *equal* if their corresponding components are equal. If the components are equal in one coordinate system, since they will transform in exactly the same way, they will be equal in all coordinate systems.

The sum, or difference, of two given tensors $A_{\alpha\beta\cdots\kappa}$ and $B_{\alpha\beta\cdots\kappa}$ is defined as a tensor $C_{\alpha\beta\cdots\kappa}$ with components equal to the sum, or difference, respectively, of the corresponding components of the given tensors; thus

$$C_{\alpha\beta\cdots\kappa} = A_{\alpha\beta\cdots\kappa} \pm B_{\alpha\beta\cdots\kappa} \qquad (\text{for all } \alpha, \beta, \ldots, \kappa).$$

If we transform to another coordinate system and designate by primes the transformed tensors, we have

$$\begin{aligned} C'_{\alpha\beta\cdots\kappa} &= a_{\alpha\mu}a_{\beta\nu}\cdots a_{\kappa\sigma}C_{\mu\nu\cdots\sigma} \\ &= a_{\alpha\mu}a_{\beta\nu}\cdots a_{\kappa\sigma}(A_{\mu\nu\cdots\sigma} \pm B_{\mu\nu\cdots\sigma}) = A'_{\alpha\beta\cdots\kappa} \pm B'_{\alpha\beta\cdots\kappa}, \end{aligned}$$

showing that the same relations exist between the new components of the tensors.

The product of a given tensor by a scalar is defined as a tensor each of whose components is obtained by multiplying the corresponding component of the given tensor by the scalar.

A tensor, all of whose components are zero in one coordinate system, has zero components in all coordinate systems, as can be seen from Eq. (A5.10). Such a tensor is referred to as a *null-tensor*, or *zero-tensor*, and is designated by the symbol 0.

By transposing all terms to the left, a tensor equation can be written in the form $T_{\alpha\beta\cdots\kappa} = 0$, which asserts that all components of a certain tensor are zero. From the above, it follows that the components will be zero in all coordinate systems. This means that a tensor equation, if valid in one coordinate system, is valid in all.

The right-hand member of Eq. (A5.4) is a special combination of two tensors (one of second rank, the other of the first rank) giving as a result another tensor. This is a special case of application of two general processes known as the *multiplication* and the *contraction* of tensors.

If we have two tensors $A_{\alpha\beta\dots}$ and $B_{\phi\pi\dots}$, of ranks k and m, respectively, and if we multiply in pairs the components of the first with the components of the second; we obtain n^{k+m} quantities

$$C_{\alpha\beta\cdots\phi\pi\cdots} = A_{\alpha\beta\dots}B_{\phi\pi\dots}, \tag{A5.13}$$

which constitute a tensor of rank $k + m$. The proof consists in observing that

$$\begin{aligned} C'_{\alpha'\beta'\cdots\phi'\pi'\cdots} &= A'_{\alpha'\beta'\dots}B'_{\phi'\pi'\dots} \\ &= a_{\alpha'\alpha}a_{\beta'\beta}\cdots A_{\alpha\beta\dots}a_{\phi'\phi}a_{\pi'\pi}\cdots B_{\phi\pi\dots} \\ &= a_{\alpha'\alpha}a_{\beta'\beta}\cdots a_{\phi'\phi}a_{\pi'\pi}\cdots C_{\alpha\beta\cdots\phi\pi\cdots}. \end{aligned} \tag{A5.14}$$

This process of forming new tensors from two (and, by an easy generalization, from any number of) given tensors is called the process of *multiplication of tensors*, or sometimes *outer multiplication.*

If we have a tensor $A_{\alpha\beta\phi\pi\dots}$ of rank k, and we replace one of its literal indices by one used elsewhere, so that two of its literal indices become equal, thus implying summation with respect to that index, we obtain a set of n^{k-2} quantities, corresponding to the remaining $k - 2$ free indices,

$$B_{\beta\pi\dots} = A_{\alpha\beta\alpha\pi\dots}, \tag{A5.15}$$

in which we assumed, for the sake of illustration, that ϕ was replaced by α. This set of quantities forms a tensor of rank $k - 2$. For the proof we only need to observe that

$$\begin{aligned} B'_{\beta'\pi'\dots} &= A'_{\alpha'\beta'\alpha'\pi'\dots} \\ &= a_{\alpha'\alpha}a_{\beta'\beta}a_{\alpha'\phi}a_{\pi'\pi}\cdots A_{\alpha\beta\phi\pi\dots} \\ &= \delta_{\alpha\phi}a_{\beta'\beta}a_{\pi'\pi}\cdots A_{\alpha\beta\phi\pi\dots} \qquad \text{on account of Eq. (A5.2),} \\ &= a_{\beta'\beta}a_{\pi'\pi}\cdots A_{\alpha\beta\alpha\pi\dots} \\ &= a_{\beta'\beta}a_{\pi'\pi}\cdots B_{\beta\pi\dots}. \end{aligned} \tag{A5.16}$$

This process is known as the *contraction of tensors.* It can be immediately generalized. Thus, by setting m of the indices of a tensor of rank k ($k \geq 2m$) equal to m other indices in pairs we obtain n^{k-2m} quantities, which then constitute a tensor of rank $k - 2m$.

The two processes may be easily combined. Thus, after constructing a tensor of fourth rank

$$C_{\alpha\beta\phi\pi} = A_{\alpha\beta}B_{\phi\pi},$$

we can replace ϕ by β, obtaining a tensor of second rank:

$$D_{\alpha\pi} = C_{\alpha\beta\beta\pi} = A_{\alpha\beta}B_{\beta\pi}. \tag{A5.17}$$

This second-rank tensor is sometimes called the *inner product* of the tensors $A_{\alpha\beta}$ and $B_{\phi\pi}$. Similarly the scalar C of Eq. (A5.12) is the inner product of two vectors. It can be regarded as a contraction of the tensor $C_{\alpha\beta} = A_\alpha B_\beta$. Another interesting case is the scalar $A_{\alpha\beta}A_{\alpha\beta}$ formed by twice contracting the square of the tensor $A_{\alpha\beta}$.

Two types of symmetry are of frequent occurrence among the tensors found in physics. The first type is defined by the relation

$$A_{\alpha\beta} = A_{\beta\alpha}. \tag{A5.18}$$

Tensors satisfying this condition are said to be *symmetric.* The second type satisfies the condition

$$A_{\alpha\beta} = -A_{\beta\alpha} \tag{A5.19}$$

and are said to be *antisymmetric*, or *skew-symmetric.* It is evident that this property may refer to any pair, or any number, of indices. Thus, if

$$B_{\beta\phi\pi} = B_{\beta\pi\phi},$$

the tensor is said to be symmetric with respect to the last two indices. If, however,

$$B_{\beta\phi\pi} = B_{\beta\pi\phi} = B_{\phi\beta\pi} = B_{\pi\phi\beta},$$

the tensor is said to be symmetric with respect to all three indices. Similar generalization for skew-symmetry is obvious. The importance of symmetry is due, at least in part, to the fact that it is preserved by coordinate transformations. Thus, for example, suppose that Eq. (A5.18) holds; then

$$A'_{\alpha'\beta'} = a_{\alpha'\alpha}a_{\beta'\beta}A_{\alpha\beta} = a_{\beta'\beta}a_{\alpha'\alpha}A_{\beta\alpha} = A'_{\beta'\alpha'}. \tag{A5.20}$$

It is interesting to note that each tensor can be expressed as a sum of a symmetric and an antisymmetric term thus, for any $A_{\alpha\beta}$

$$A_{\alpha\beta} = \tfrac{1}{2}(A_{\alpha\beta} + A_{\beta\alpha}) + \tfrac{1}{2}(A_{\alpha\beta} - A_{\beta\alpha}), \tag{A5.21}$$

of which the first member of the right-hand side is obviously symmetric, while the second is antisymmetric.

One of the most important transformations of the symmetric tensors is the so-called transformation to the principal axes. Consider the problem of finding a vector **y** such that

$$I = A_{\alpha\beta} y_\alpha y_\beta \tag{A5.22}$$

has its largest value, while

$$R^2 = y_\alpha y_\alpha = \text{constant.} \tag{A5.23}$$

Equation (A5.23) is the equation of a sphere in n-dimensional space, so that this condition means that the *magnitude*, or the *length*, of the vector y_β, which is here designated by R, is kept fixed.* Under these conditions the scalar I must possess a certain largest value for one or more directions of the vector y_β. To find these directions we must put $\delta I = 0$, subject to condition (A5.23). We have

$$\delta I = A_{\alpha\beta} y_\alpha \, \delta y_\beta + A_{\alpha\beta} y_\beta \, \delta y_\alpha = 2 A_{\alpha\beta} y_\beta \, \delta y_\alpha, \tag{A5.24}$$

since $A_{\alpha\beta} = A_{\beta\alpha}$. On the other hand

$$R^2 = 2 y_\alpha \, \delta y_\alpha = 0. \tag{A5.25}$$

Setting $\delta I = 0$, and using the method of Lagrangian multipliers to take care of the condition (A5.25), we obtain

$$\delta I - \lambda \, \delta R^2 = 2(A_{\alpha\beta} y_\beta - \lambda y_\alpha) \, \delta y_\alpha = 0. \tag{A5.26}$$

In this equation all δy_α can be treated as independent; therefore Eq. (A5.26) can be satisfied if, and only if,

$$A_{\alpha\beta} y_\beta = \lambda y_\alpha. \tag{A5.27}$$

The problem thus reduces to the question of finding a vector y_α which would be parallel to the vector $A_{\alpha\beta} y_\beta$.

Equation (A5.27), which is really a set of n equations (obtained by putting $\alpha = 1, 2, \ldots, n$), can be written in the form

$$(A_{\alpha\beta} - \lambda \delta_{\alpha\beta}) y_\beta = 0, \tag{A5.28}$$

which is then seen to be a set of n linear homogeneous equations in n unknowns $y_\beta = (y_1, y_2, \ldots, y_n)$. Such a set of equations has solutions,

* Condition (A5.23) is often referred to as the *normalization condition* and is usually specialized to the condition $y_\alpha^2 = 1$ by taking $R = 1$.

other than the trivial one $y_\beta = 0$ if, and only if,

$$f(\lambda) \equiv \det(A_{\alpha\beta} - \lambda\delta_{\alpha\beta}) = \begin{vmatrix} A_{11}-\lambda & A_{12} & A_{13} & \cdots & A_{1n} \\ A_{21} & A_{22}-\lambda & A_{23} & \cdots & A_{2n} \\ \cdot & & & & \\ \cdot & & & & \\ \cdot & & & & \\ A_{n1} & A_{n2} & A_{n3} & \cdots & A_{nn}-\lambda \end{vmatrix} = 0. \quad \text{(A5.29)}$$

It is evident that $f(\lambda) = 0$ is an equation of nth degree in λ. It will thus have exactly n roots, $\lambda_1, \lambda_2, \ldots, \lambda_n$, some of which may be equal to each other. These quantities, which for a real and symmetric $A_{\alpha\beta}$ may be proved to be all real, are called *characteristic values*, *principal values*, *proper values*, or *eigenvalues* of the tensor $A_{\alpha\beta}$.

A solution of Eq. (A5.28) corresponding to λ_k will be designated by y_β^k. It will be termed a *characteristic vector*, *principal vector*, etc; of the tensor $A_{\alpha\beta}$ corresponding to the characteristic value λ_k.

We need to distinguish two cases. In the first, all the characteristic values λ_k are different; in the second, some, or all, of them are the same. In the first case, there is just one characteristic vector y_β^k, satisfying Eq. (A5.28) and the normalization condition (A5.23), for each λ_k. Consider, moreover, the eigenvectors associated with two different eigenvalues λ_k and λ_m; then

$$A_{\alpha\beta} y_\beta^k = \lambda_k y_\alpha^k \qquad \text{(not summed over } k\text{)}$$

and

$$A_{\alpha\beta} y_\beta^m = \lambda_m y_\alpha^m \qquad \text{(not summed over } m\text{)}.$$

Multiplying the first of these equations by y_α^m and the second by y_α^k, and subtracting the second from the first, we find that

$$(\lambda_k - \lambda_m) y_\alpha^k y_\alpha^m = A_{\alpha\beta} y_\beta^k y_\alpha^m - A_{\alpha\beta} y_\beta^m y_\alpha^k = (A_{\alpha\beta} - A_{\beta\alpha}) y_\beta^k y_\alpha^m = 0,$$

because $A_{\alpha\beta}$ is symmetric. Since by assumption $\lambda_m \neq \lambda_k$, we must have

$$y_\alpha^k y_\alpha^m = 0. \qquad \text{(A5.30)}$$

The left-hand member of Eq. (A5.30) is just the scalar product of the vectors y_α^k and y_α^m; therefore, that equation requires the n distinct vectors obtained as solutions of Eq. (A5.28) to be mutually perpendicular.

The directions of these n vectors can be taken as the directions of a new set of coordinate axes and are customarily referred to as the *principal axes* of the tensor $A_{\alpha\beta}$. The components of the tensor in this new coordinate system will be designated as $A'_{\alpha\beta}$. Suppose y^k_α is the kth solution of Eq. (A5.27) in the original coordinate system; then in the new coordinate system it will lie along the kth coordinate axis. Its new components will all vanish, except the kth, which must clearly equal R; thus we have

$$y'^k_\alpha = R\delta_{\alpha k}. \tag{A5.31}$$

In the new primed coordinate system Eq. (A5.27) is written $A'_{\alpha\beta}y'^k_\beta = \lambda_k y'^k_\alpha$, and by Eq. (A5.31) this reduces to

$$A'_{\alpha k} = \lambda_k \delta_{\alpha k}.$$

We see, therefore, that

$$A'_{\alpha\beta} = 0, \qquad \text{if} \quad \alpha \neq \beta$$

and

$$A'_{\alpha\alpha} = \lambda_\alpha, \qquad \text{for} \quad \alpha = 1, 2, \ldots, n.$$

Thus, the matrix representing $A'_{\alpha\beta}$ takes the *diagonal* form:

$$A'_{\alpha\beta} = \begin{bmatrix} \lambda_1 & 0 & 0 & \cdots \\ 0 & \lambda_2 & 0 & \cdots \\ 0 & 0 & \lambda_3 & \cdots \\ \cdots & \cdots & \cdots & \cdots \end{bmatrix}. \tag{A5.32}$$

Suppose next that a pair of the characteristic values, say λ_k and λ_m, are equal, but all other pairs are unequal. In this case the direction of y^k_α is not completely defined. In fact, let y^k_α and y^m_α be any two characteristic vectors satisfying Eq. (A5.27) corresponding to the same value of λ, namely $\lambda = \lambda_k = \lambda_m$. Then, as becomes evident by a direct substitution,

$$Y_\alpha = Ky^k_\alpha + My^m_\alpha, \tag{A5.33}$$

for arbitrary K and M, is also a solution of Eq. (A5.27) for the same value of λ. Each of the vectors Y_α is, however, perpendicular to any solution of Eq. (A5.27) with a different λ; thus, only the plane of all vectors of the form (A5.33) is defined. We can choose, however, any two vectors in this plane that are perpendicular to each other as a pair of coordinate axes. These, together with the axes defined by the other characteristic values, become the principal axes of $A_{\alpha\beta}$.

More generally, for every set of equal eigenvalues there is defined a linear manifold having a dimension equal to the number of equal eigenvalues. In this manifold we can choose, quite arbitrarily, a set of orthogonal axes; these together with all other axes (either uniquely defined by

Eq. (A5.27), or chosen in a similar way for other sets of equal roots) constitute a set of principal axes for the tensor $A_{\alpha\beta}$. In the coordinate system formed by these principal axes the tensor will have the form (A5.32), with each eigenvalue occurring the proper number of times.

As another example of a transformation of coordinate system let us consider a rotation of axes through an angle θ in the plane of two unit

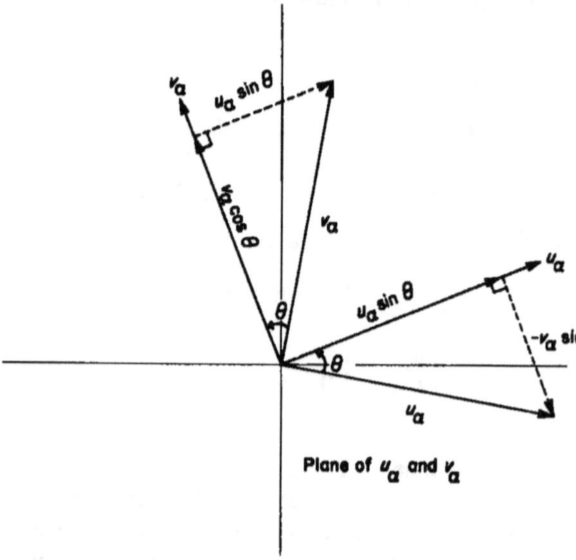

FIG. A5.1. Effect of a rotation in the plane of two unit vectors, u_α and v_α.

vectors u_α and v_α, which we shall suppose perpendicular to each other. We thus have

$$u_\alpha^2 = v_\alpha^2 = 1 \qquad \text{and} \qquad u_\alpha v_\alpha = 0. \tag{A5.34}$$

We shall take the rotation of axes to be away from u_α toward v_α; that is, counterclockwise in Figure A5.1.

Let U_α and V_α be unit vectors that have the same components in the old (unprimed) coordinate system as u_α and v_α have, respectively, in the new (primed) coordinate after the rotation of axes; i.e., $U_\alpha = u'_\alpha$ and $V_\alpha = v'_\alpha$. Clearly U_α and V_α are obtained by rotating u_α and v_α in the plane determined by these vectors through an angle of $-\theta$. Since U_α lies in the plane of u_α and v_α, it is a linear combination of these two vectors. In fact, from Figure A5.1 it is seen that

$$U_\alpha = u_\alpha \cos\theta - v_\alpha \sin\theta = u'_\alpha; \tag{A5.35a}$$

similarly we have

$$V_\alpha = u_\alpha \sin\theta - v_\alpha \cos\theta = v'_\alpha. \tag{A5.35b}$$

Moreover, any vector x_α perpendicular to the plane of rotation is unchanged. This means that

$$(x_\alpha u_\alpha = 0 \quad \text{and} \quad x_\alpha v_\alpha = 0) \Rightarrow x'_\alpha = x_\alpha. \tag{A5.36}$$

We are now interested in finding the transformation coefficients $a_{\alpha\beta}$ in Eq. (A5.1) that correspond to the rotation of axes specified above. In the first place these coefficients must give

$$u'_\alpha = a_{\alpha\beta} u_\beta = u_\alpha \cos\theta - v_\alpha \sin\theta, \tag{A5.37}$$

$$v'_\alpha = a_{\alpha\beta} v_\beta = u_\alpha \sin\theta + v_\alpha \cos\theta. \tag{A5.38}$$

Further, they must give

$$x'_\alpha = a_{\alpha\beta} x_\beta = x_\alpha, \qquad \text{whenever} \quad x_\alpha u_\alpha = x_\alpha v_\alpha = 0. \tag{A5.39}$$

The last equation can be written in the form

$$(a_{\alpha\beta} - \delta_{\alpha\beta}) x_\beta = 0, \qquad \text{whenever} \quad x_\beta u_\beta = x_\beta v_\beta = 0.$$

This can only be the case if

$$a_{\alpha\beta} - \delta_{\alpha\beta} = m_\alpha u_\beta + n_\alpha v_\beta, \tag{A5.40}$$

where m_α and n_α are constants.

Multiplying Eq. (A5.40) first by u_β and then by v_β and summing, we find by Eq. (A5.34) that

$$a_{\alpha\beta} u_\beta = (\delta_{\alpha\beta} + m_\alpha u_\beta + n_\alpha v_\beta) u_\beta = u_\alpha + m_\alpha \tag{A5.41}$$

and

$$a_{\alpha\beta} v_\beta = (\delta_{\alpha\beta} + m_\alpha u_\beta + n_\alpha v_\beta) v_\beta = v_\alpha + n_\alpha. \tag{A5.42}$$

Substitution from Eqs. (A5.37) and (A5.38) into Eqs. (A5.41) and (A5.42), respectively, now gives

$$m_\alpha = u_\alpha \cos\theta - v_\alpha \sin\theta - u_\alpha \tag{A5.43}$$

and

$$n_\alpha = u_\alpha \sin\theta + v_\alpha \cos\theta - v_\alpha. \tag{A5.44}$$

Finally, substitution from Eqs. (A5.43) and (A5.44) into Eq. (A5.40) gives us the desired result:

$$a_{\alpha\beta} = \delta_{\alpha\beta} - (u_\alpha u_\beta + v_\alpha v_\beta)(1 - \cos\theta) + (u_\alpha v_\beta - v_\alpha u_\beta)\sin\theta. \tag{A5.45}$$

The nabla operator ∇ can be defined in an n-dimensional space as a vector operator having in a given unprimed coordinate system the components $\partial/\partial x_\alpha$, where the x_α are the corresponding Cartesian coordinates. With this definition we have no difficulty in interpreting the symbols $\nabla\phi$, $\nabla \cdot \mathbf{A}$, $\nabla^2\phi$, and $\nabla^2\mathbf{A}$; thus

$$\nabla\phi \sim \frac{\partial\phi}{\partial x_\alpha},$$

$$\nabla \cdot \mathbf{A} = \frac{\partial A_\alpha}{\partial x_\alpha},$$

$$\nabla^2\phi = \frac{\partial^2}{\partial x_\alpha^2}\phi = \nabla \cdot \nabla\phi,$$

$$\nabla^2\mathbf{A} \sim \frac{\partial^2}{\partial x_\alpha^2} A_\beta,$$

where α, β can take on the values $1, 2, \ldots, n$ and, as usual, repeated indices are summed.

A difficulty arises, however, in attempting to interpret $\nabla \times \mathbf{A}$. The nearest we can come to this is the following analogue:

$$\nabla \times \mathbf{A} \sim \frac{\partial A_\beta}{\partial x_\alpha} - \frac{\partial A_\alpha}{\partial x_\beta} \equiv F_{\alpha\beta}$$

This, as can be easily shown, is an antisymmetric tensor of second rank, and not a vector. Since $F_{\alpha\beta} = -F_{\beta\alpha}$, the number of independent components of this tensor in the space of n dimensions is $\frac{1}{2}n(n - 1)$. This is equal to n only for $n = 0$ and $n = 3$. Thus, in the space of three dimensions, and only in such a space, has $F_{\alpha\beta}$ as many independent components as a vector. In this special case we can put

$$F_{12} = -F_{21} = F_3, \qquad F_{23} = -F_{32} = F_1, \qquad F_{31} = -F_{13} = F_2$$

and show that for a rotation of coordinate system $\mathbf{F} = (F_1, F_2, F_3)$ transforms like a vector. This is not surprising because $\mathbf{F}$ is just $\nabla \times \mathbf{A}$, the curl of $\mathbf{A}$.

Note, however, how $\mathbf{F} = \nabla \times \mathbf{A}$ behaves for an *inversion* of the coordinate system expressed by

$$x_1 \rightarrow -x_1, \qquad x_2 \rightarrow -x_2, \qquad x_3 \rightarrow -x_3. \tag{A5.46}$$

If $\mathbf{A}$ is a true vector, we have

$$A_k \rightarrow -A_k \tag{A5.47}$$

and thus

$$\begin{aligned} F_k = \frac{\partial A_j}{\partial x_i} - \frac{\partial A_i}{\partial x_j} &\rightarrow \left(-\frac{\partial}{\partial x_i}\right)(-A_j) - \left(-\frac{\partial}{\partial x_j}\right)(-A_i) \\ &= \frac{\partial A_j}{\partial x_i} - \frac{\partial A_i}{\partial x_j} = F_k, \end{aligned} \tag{A5.48}$$

where i, j, and k are 1, 2, 3 in some cyclic order. Vector-like quantities that transform like F_k in Eq. (A5.48) under the inversion of coordinates of (A5.46) are called *pseudovectors.*

We may also speak of *pseudoscalars* and *pseudotensors.* These transform as scalars and tensors, respectively, for a rotation of the coordinate system but involve an additional change of sign for an inversion of the coordinate system. Note that an inversion converts a right-handed coordinate system into a left-handed one, and vice versa.

Of the formulas of the last section we need only the generalization of Eq. (A4.8). Let us denote by $d\Omega$ the generalization of the volume element dV. Then $d\Omega = dx_1\, dx_2 \cdots dx_n$. The analog of $d\mathbf{S}$ will be an element of an $(n - 1)$-dimensional manifold, which, like the surface S, is defined by a single relation between the coordinates x_α.

Such an $(n - 1)$-dimensional manifold is called a *hypersurface.* The analog of $d\mathbf{S}$ will be a vector having components dS_α. Thus, dS_1 will be the projection of the element of the hypersurface upon the $(n - 1)$-dimensional coordinate hyperplane formed by the $n - 1$ coordinate axes $x_2, x_3, \ldots, x_n$, etc. A generalization of the Gauss theorem, holding for any n, is

$$\int_S A_\alpha\, dS_\alpha = \int_\Omega \frac{\partial A}{\partial x_\alpha}\, d\Omega, \tag{A5.49}$$

where the hypersurface S is the boundary of the n-dimensional volume Ω.

Appendix A6

Curvilinear Coordinates

Curvilinear coordinates, which we shall designate by (u, v, w), are usually introduced by the set of equations

$$\begin{aligned} u &= u(x, y, z), \\ v &= v(x, y, z), \\ w &= w(x, y, z), \end{aligned} \tag{A6.1}$$

where, either directly or by virtue of some supplementary conditions, the functions are single-valued. Of course, the coordinate system can be also defined by the set of equations

$$\begin{aligned} x &= f_1(u, v, w), \\ y &= f_2(u, v, w), \\ z &= f_3(u, v, w). \end{aligned} \tag{A6.2}$$

In order that these may be a consequence of Eqs. (A6.1) it is necessary and sufficient, by Theorem 2 of Section A3, that

$$\nabla u \cdot \nabla v \times \nabla w \not\equiv 0. \tag{A6.3}$$

The expression $\nabla u \cdot \nabla v \cdot \nabla w$ can also be written

$$\nabla u \cdot \nabla v \times \nabla w \equiv \begin{vmatrix} \frac{\partial u}{\partial x} & \frac{\partial u}{\partial y} & \frac{\partial u}{\partial z} \\ \frac{\partial v}{\partial x} & \frac{\partial v}{\partial y} & \frac{\partial v}{\partial z} \\ \frac{\partial w}{\partial x} & \frac{\partial w}{\partial y} & \frac{\partial w}{\partial z} \end{vmatrix}$$

$$\equiv \frac{\partial(u, v, w)}{\partial(x, y, z)} \equiv J'; \tag{A6.4}$$

$\partial(u, v, w)/\partial(x, y, z)$ is an abbreviated notation for the determinant, called the *Jacobian of* (u, v, w) *with respect to* (x, y, z).

It is obvious from the symmetry of the relation between the two sets of variables that Eqs. (A6.2) will be solvable for (u, v, w) when and only when

$$\begin{vmatrix} \dfrac{\partial f_1}{\partial u} & \dfrac{\partial f_1}{\partial v} & \dfrac{\partial f_1}{\partial w} \\ \dfrac{\partial f_2}{\partial u} & \dfrac{\partial f_2}{\partial v} & \dfrac{\partial f_2}{\partial w} \\ \dfrac{\partial f_3}{\partial u} & \dfrac{\partial f_3}{\partial v} & \dfrac{\partial f_3}{\partial w} \end{vmatrix} \equiv \frac{\partial(x, y, z)}{\partial(u, v, w)} \equiv J \not\equiv 0. \tag{A6.5}$$

It can be shown further that

$$JJ' = 1, \tag{A6.6}$$

which is an often useful relation. In what follows we shall assume that neither J nor J' is identically zero.

If we start with some point (u, v, w) and then, keeping u constant, permit v and w to vary arbitrarily, the locus of the points so obtained will be the surface

$$u(x, y, z) = u = \text{constant}. \tag{A6.7}$$

This surface may be called the *coordinate surface* u = *constant*. Evidently we have a single parameter family of these surfaces, the parameter being the value of the constant u. In the same way we obtain two other families of surfaces v = constant, and w = constant. Thus each given point (u, v, w) defines a set of three coordinate surfaces intersecting at the point.

The surfaces

$$\begin{aligned} v(x, y, z) &= v, \\ w(x, y, z) &= w \end{aligned} \tag{A6.8}$$

intersect along a line defined by the pair of equations (A6.8). Along this line only u varies, and when it has the value u the line passes through the point (u, v, w). This line may be called the u-axis (of course, in general, curved) passing through the point (u, v, w). Similarly, by permitting v and w to vary in turn, while the other two coordinates are kept fixed, we obtain the v-axis and the w-axis. Thus, for each given point there is a set of coordinate axes intersecting at the point.

When the coordinates of a point change from (u, v, w) to $(u + du, v, w)$, the point will move some distance along the u-axis, say dr_u, which will be a function of u, v, w, and du. If we neglect the higher powers of the small quantity du, we will have

$$dr_u = U\,du,$$

where $U = U(u, v, w)$. Similar expressions will be obtained for very small displacements along the other two axes. Thus, with an obvious extension of the notation,

$$dr_u = U\,du, \qquad dr_v = V\,dv, \qquad dr_w = W\,dw. \tag{A6.9}$$

We shall use $\mathbf{u}$, $\mathbf{v}$, and $\mathbf{w}$ to denote the unit vectors tangent to the coordinate axes (the u-, v-, and w-axis, respectively). Equations (A6.9) are then equivalent to the set

$$\frac{\partial \mathbf{r}}{\partial u} = U\mathbf{u}, \qquad \frac{\partial \mathbf{r}}{\partial v} = V\mathbf{v}, \qquad \frac{\partial \mathbf{r}}{\partial w} = W\mathbf{w}, \tag{A6.10}$$

or to the equation

$$d\mathbf{r} = \mathbf{u}U\,du + \mathbf{v}V\,dv + \mathbf{w}W\,dw. \tag{A6.11}$$

It is obvious that Eqs. (A6.10) can be used to define the functions U, V, W and the unit vectors $\mathbf{u}$, $\mathbf{v}$, $\mathbf{w}$.

Since, as can be readily seen from Eqs. (A2.21) and (A6.5),

$$J \equiv \frac{\partial \mathbf{r}}{\partial u} \cdot \frac{\partial \mathbf{r}}{\partial v} \times \frac{\partial \mathbf{r}}{\partial w} \equiv UVW\,\mathbf{u} \cdot \mathbf{v} \times \mathbf{w}, \tag{A6.12}$$

it follows by Eqs. (A6.10) and our assumption about J that

$$U \not\equiv 0, \qquad V \not\equiv 0, \qquad W \not\equiv 0, \qquad \mathbf{u} \cdot \mathbf{v} \times \mathbf{w} \not\equiv 0. \tag{A6.13}$$

Therefore, by Example 1 of Appendix A2, every vector $\mathbf{A}$ may be expressed in terms of our unit vectors in the form*

$$\mathbf{A} = A_u\mathbf{u} + A_v\mathbf{v} + A_w\mathbf{w} \tag{A6.14}$$

of which Eq. (A6.11) is a special case.

If we put

$$U\,\nabla u = \mathbf{a}, \qquad V\,\nabla v = \mathbf{b}, \qquad W\,\nabla w = \mathbf{c}, \tag{A6.15}$$

we have another set of three vectors for which, by virtue of inequalities (A6.3) and (A6.13),

$$\mathbf{a} \cdot \mathbf{b} \times \mathbf{c} \not\equiv 0. \tag{A6.16}$$

* This would, of course, not be true at singular points where $\mathbf{u} \cdot \mathbf{v} \times \mathbf{w} = 0$, and this limitation should be understood in the subsequent development.

Every vector may be also expressed, therefore, as a linear combination of these vectors. Thus, we shall write

$$\mathbf{A} = A_a\mathbf{a} + A_b\mathbf{b} + A_c\mathbf{c}. \tag{A6.17}$$

The two sets of vectors **u**, **v**, **w** and **a**, **b**, **c** are what are called reciprocal sets of vectors; i.e., they are connected by the relations

$$\begin{aligned} &\mathbf{a}\cdot\mathbf{u} = \mathbf{b}\cdot\mathbf{v} = \mathbf{c}\cdot\mathbf{w} = 1 \\ &\mathbf{a}\cdot\mathbf{v} = \mathbf{a}\cdot\mathbf{w} = \mathbf{b}\cdot\mathbf{w} = \mathbf{b}\cdot\mathbf{u} = \mathbf{c}\cdot\mathbf{u} = \mathbf{c}\cdot\mathbf{v} = 0. \end{aligned} \tag{A6.18}$$

These relations can be easily established. Thus, for example, using Eqs. (A6.10) and (A6.15), we have

$$\begin{aligned} \mathbf{a}\cdot\mathbf{u} &= U\,\nabla u\cdot\frac{1}{U}\frac{\partial\mathbf{r}}{\partial u} = \nabla u\cdot\frac{\partial\mathbf{r}}{\partial u} \\ &= \frac{\partial u}{\partial x}\frac{\partial x}{\partial u} + \frac{\partial u}{\partial y}\frac{\partial y}{\partial u} + \frac{\partial u}{\partial z}\frac{\partial z}{\partial u} = \frac{\partial u}{\partial u} = 1. \end{aligned}$$

Similarly, we find that

$$\begin{aligned} \mathbf{a}\cdot\mathbf{v} &= \frac{U}{V}\nabla u\cdot\frac{\partial\mathbf{r}}{\partial v} \\ &= \frac{U}{V}\left(\frac{\partial u}{\partial x}\frac{\partial x}{\partial v} + \frac{\partial u}{\partial y}\frac{\partial y}{\partial v} + \frac{\partial u}{\partial z}\frac{\partial z}{\partial v}\right) = \frac{U}{V}\frac{\partial u}{\partial v} = 0. \end{aligned}$$

Further, we have

$$\mathbf{a}\times(\mathbf{v}\times\mathbf{w}) = \mathbf{a}\cdot\mathbf{w}\,\mathbf{v} - \mathbf{a}\cdot\mathbf{v}\,\mathbf{w} = 0,$$

by Eqs. (A6.18), so that **a** is parallel to $\mathbf{v}\times\mathbf{w}$ and consequently

$$\mathbf{a} = \lambda\,\mathbf{v}\times\mathbf{w}.$$

To determine λ we note that $\mathbf{a}\cdot\mathbf{u} = 1$ and thus that

$$\mathbf{u}\cdot\lambda(\mathbf{v}\times\mathbf{w}) = \lambda(\mathbf{u}\cdot\mathbf{v}\times\mathbf{w}) = 1.$$

We then have

$$\mathbf{a} = \lambda\,\mathbf{v}\times\mathbf{w} = \frac{\mathbf{v}\times\mathbf{w}}{\mathbf{u}\cdot\mathbf{v}\times\mathbf{w}}.$$

In a similar way we can express each of the six vectors in terms of the three vectors of the other set, obtaining

$$\begin{aligned} &\mathbf{a} = \frac{\mathbf{v}\times\mathbf{w}}{\mathbf{u}\cdot\mathbf{v}\times\mathbf{w}}, \qquad \mathbf{b} = \frac{\mathbf{w}\times\mathbf{u}}{\mathbf{u}\cdot\mathbf{v}\times\mathbf{w}}, \qquad \mathbf{c} = \frac{\mathbf{u}\times\mathbf{v}}{\mathbf{u}\cdot\mathbf{v}\times\mathbf{w}}; \\ &\mathbf{u} = \frac{\mathbf{b}\times\mathbf{c}}{\mathbf{a}\cdot\mathbf{b}\times\mathbf{c}}, \qquad \mathbf{v} = \frac{\mathbf{c}\times\mathbf{a}}{\mathbf{a}\cdot\mathbf{b}\times\mathbf{c}}, \qquad \mathbf{w} = \frac{\mathbf{a}\times\mathbf{b}}{\mathbf{a}\cdot\mathbf{b}\times\mathbf{c}}. \end{aligned} \tag{A6.19}$$

The conditions (A6.18) are, obviously, automatically satisfied when the vectors are thus expressed.

We must remember, however, in spite of the symmetry that seems to exist, that while **u**, **v**, and **w** are unit vectors and thus

$$|\mathbf{u}|^2 = |\mathbf{v}|^2 = |\mathbf{w}|^2 = 1, \tag{A6.20}$$

the set **a**, **b**, **c** are not in general unit vectors.

With the help of Eqs. (A6.18), the coefficients in each of the equations (A6.14) and (A6.17) are easily expressed in terms of the scalar products of **A** and the vectors of the other set. Thus, we easily obtain

$$\begin{aligned} A_u &= \mathbf{a} \cdot \mathbf{A}, \qquad A_v = \mathbf{b} \cdot \mathbf{A}, \qquad A_w = \mathbf{c} \cdot \mathbf{A}; \\ A_a &= \mathbf{u} \cdot \mathbf{A}, \qquad A_b = \mathbf{v} \cdot \mathbf{A}, \qquad A_c = \mathbf{w} \cdot \mathbf{A}. \end{aligned} \tag{A6.21}$$

We shall now express some commonly occurring operations of vector analysis in terms of the curvilinear coordinates (u, v, w). Let $\phi = \phi(u, v, w)$; then

$$\nabla\phi = \frac{\partial\phi}{\partial u}\nabla u + \frac{\partial\phi}{\partial v}\nabla v + \frac{\partial\phi}{\partial w}\nabla w,$$

which becomes, with the help of Eqs. (A6.15),

$$\nabla\phi = \frac{\mathbf{a}}{U}\frac{\partial\phi}{\partial u} + \frac{\mathbf{b}}{V}\frac{\partial\phi}{\partial v} + \frac{\mathbf{c}}{W}\frac{\partial\phi}{\partial w}. \tag{A6.22}$$

Multiplying this by $\mathbf{A}\cdot$ and using Eqs. (A6.21), we obtain

$$\mathbf{A}\cdot\nabla\phi = \frac{A_u}{U}\frac{\partial\phi}{\partial u} + \frac{A_v}{V}\frac{\partial\phi}{\partial v} + \frac{A_w}{W}\frac{\partial\phi}{\partial w}. \tag{A6.23}$$

To compute $\nabla \times \mathbf{A}$ we need to have $\nabla \times \mathbf{a}$, etc. These can be obtained by making use of Eqs. (A6.15) and (A6.22) as follows:

$$\begin{aligned} \nabla \times \mathbf{a} &= \nabla \times (U\nabla u) = U\nabla \times \nabla u - \nabla u \times \nabla U \\ &= -\frac{\mathbf{a}}{U} \times \nabla U = -\frac{\mathbf{a}}{U} \times \left(\frac{\mathbf{a}}{U}\frac{\partial U}{\partial u} + \frac{\mathbf{b}}{V}\frac{\partial U}{\partial v} + \frac{\mathbf{c}}{W}\frac{\partial U}{\partial w}\right) \\ &= \frac{\mathbf{c}\times\mathbf{a}}{WU}\frac{\partial U}{\partial w} - \frac{\mathbf{a}\times\mathbf{b}}{UV}\frac{\partial U}{\partial v}. \end{aligned}$$

On the other hand, by Eqs. (A6.19), we have

$$\mathbf{a} \times \mathbf{b} = \mathbf{a} \cdot \mathbf{b} \times \mathbf{c}\, \mathbf{w},$$

and, by Eqs. (A6.15), (A6.4), and (A6.6),

$$\mathbf{a}\cdot\mathbf{b}\times\mathbf{c} = UVW\,\nabla u\cdot\nabla v\times\nabla w = \frac{UVW}{J} \tag{A6.24}$$

We thus have

$$\mathbf{a}\times\mathbf{b} = \frac{UVW}{J}\,\mathbf{w}, \tag{A6.25a}$$

and similarly

$$\mathbf{b}\times\mathbf{c} = \frac{UVW}{J}\,\mathbf{u}, \qquad \mathbf{c}\times\mathbf{a} = \frac{UVW}{J}\,\mathbf{v}. \tag{A6.25b}$$

Using these results in the above expression for $\nabla\times\mathbf{a}$ we finally obtain

$$\nabla\times\mathbf{a} = \frac{1}{J}\left(V\frac{\partial U}{\partial w}\mathbf{v} - W\frac{\partial U}{\partial v}\mathbf{w}\right); \tag{A6.26}$$

similarly, we find that

$$\nabla\times\mathbf{b} = \frac{1}{J}\left(W\frac{\partial V}{\partial u}\mathbf{w} - U\frac{\partial V}{\partial w}\mathbf{u}\right) \tag{A6.27}$$

and

$$\nabla\times\mathbf{c} = \frac{1}{J}\left(U\frac{\partial W}{\partial v}\mathbf{u} - V\frac{\partial W}{\partial u}\mathbf{v}\right). \tag{A6.28}$$

We now have

$$\begin{aligned}\nabla\times\mathbf{A} &= \nabla\times(A_a\mathbf{a} + A_b\mathbf{b} + A_c\mathbf{c})\\ &= (A_a\nabla\times\mathbf{a} - \mathbf{a}\times\nabla A_a)\\ &\quad + (A_b\nabla\times\mathbf{b} - \mathbf{b}\times\nabla A_b) + (A_c\nabla\times\mathbf{c} - \mathbf{c}\times\nabla A_c);\end{aligned}$$

so that

$$\begin{aligned}\mathbf{a}\cdot\nabla\times\mathbf{A} &= A_a\,\mathbf{a}\cdot\nabla\times\mathbf{a} + A_b\,\mathbf{a}\cdot\nabla\times\mathbf{b}\\ &\quad + A_c\,\mathbf{a}\cdot\nabla\times\mathbf{c} - \mathbf{a}\times\mathbf{b}\cdot\nabla A_b + \mathbf{c}\times\mathbf{a}\cdot\nabla A_c.\end{aligned}$$

Using the relations (A6.18), Eqs. (A6.25) to (A6.28), and Eq. (A6.23), this becomes

$$\begin{aligned}\mathbf{a}\cdot\nabla\times\mathbf{A} &= A_b\left(-\frac{U}{J}\frac{\partial V}{\partial w}\right) + A_c\left(\frac{U}{J}\frac{\partial W}{\partial v}\right) - \frac{UVW}{J}(\mathbf{w}\cdot\nabla A_b - \mathbf{v}\cdot\nabla A_c)\\ &= \frac{U}{J}\left[\left(A_c\frac{\partial W}{\partial v} + W\frac{\partial A_c}{\partial v}\right) - \left(A_b\frac{\partial V}{\partial w} + V\frac{\partial A_b}{\partial w}\right)\right]\\ &= \frac{U}{J}\left[\frac{\partial}{\partial v}(WA_c) - \frac{\partial}{\partial w}(VA_b)\right].\end{aligned}$$

Here we have used the relations*

$$\mathbf{w}\cdot\nabla A_b = \frac{\partial A_b}{\partial w} \qquad \text{and} \qquad \mathbf{v}\cdot\nabla A_c = \frac{\partial A_c}{\partial v}.$$

In the same way we obtain $\mathbf{b}\cdot\nabla\times\mathbf{A}$ and $\mathbf{c}\cdot\nabla\times\mathbf{A}$; therefore, as is seen by substituting $\nabla\times\mathbf{A}$ for $\mathbf{A}$ in Eqs. (A6.14) and (A6.21),

$$\nabla\times\mathbf{A} = \frac{U}{J}\left[\frac{\partial}{\partial v}(WA_c) - \frac{\partial}{\partial w}(VA_b)\right]\mathbf{u} + \frac{V}{J}\left[\frac{\partial}{\partial w}(UA_a) - \frac{\partial}{\partial u}(WA_c)\right]\mathbf{v} + \frac{W}{J}\left[\frac{\partial}{\partial u}(VA_b) - \frac{\partial}{\partial v}(UA_a)\right]\mathbf{w}. \qquad \text{(A6.29)}$$

To obtain $\nabla\cdot\mathbf{A}$ we need first to obtain $\nabla\cdot\mathbf{u}$, $\nabla\cdot\mathbf{v}$, and $\nabla\cdot\mathbf{w}$. This can be done using Eqs. (A6.25) as follows:

$$\nabla\cdot\mathbf{u} = \nabla\cdot\frac{J\mathbf{b}\times\mathbf{c}}{UVW} = \frac{J}{UVW}\nabla\cdot\mathbf{b}\times\mathbf{c} + \mathbf{b}\times\mathbf{c}\cdot\nabla\left(\frac{J}{UVW}\right)$$

$$= \frac{J}{UVW}(\mathbf{c}\cdot\nabla\times\mathbf{b} - \mathbf{b}\cdot\nabla\times\mathbf{c}) + \frac{UVW}{J}\mathbf{u}\cdot\nabla\left(\frac{J}{UVW}\right).$$

Now, with the help of Eqs. (A6.27), (A6.28), and (A6.23),

$$\nabla\cdot\mathbf{u} = \frac{J}{UVW}\left(\frac{W}{J}\frac{\partial V}{\partial u} + \frac{V}{J}\frac{\partial W}{\partial u}\right) + \frac{VW}{J}\frac{\partial}{\partial u}\left(\frac{J}{UVW}\right)$$

$$= \frac{1}{J}\left[\frac{J}{UVW}\frac{\partial}{\partial u}(VW) + VW\frac{\partial}{\partial u}\left(\frac{J}{UVW}\right)\right];$$

so that finally

$$\nabla\cdot\mathbf{u} = \frac{1}{J}\frac{\partial}{\partial u}\left(\frac{J}{U}\right). \qquad \text{(A6.30a)}$$

By cyclic permutation of $\mathbf{u}$, $\mathbf{v}$, $\mathbf{w}$ and U, V, W we then must have also

$$\nabla\cdot\mathbf{v} = \frac{1}{J}\frac{\partial}{\partial v}\left(\frac{J}{V}\right), \qquad \nabla\cdot\mathbf{w} = \frac{1}{J}\frac{\partial}{\partial w}\left(\frac{J}{W}\right). \qquad \text{(A6.30b)}$$

* The first of these follows from the fact that each side represents the rate of change of A_b as one moves along the w axis. The second follows similarly.

Applying these results and Eq. (A6.23), we have

$$\begin{aligned}\nabla \cdot \mathbf{A} &= \nabla \cdot (A_u \mathbf{u} + A_v \mathbf{v} + A_w \mathbf{w}) \\ &= (A_u \nabla \cdot \mathbf{u} + \mathbf{u} \cdot \nabla A_u) + (A_v \nabla \cdot \mathbf{v} + \mathbf{v} \cdot \nabla A_v) \\ &\quad + (A_w \nabla \cdot \mathbf{w} + \mathbf{w} \cdot \nabla A_w) \\ &= \left[\frac{A_u}{J}\frac{\partial}{\partial u}\left(\frac{J}{U}\right) + \frac{1}{U}\frac{\partial A_u}{\partial u}\right] + \left[\frac{A_v}{J}\frac{\partial}{\partial v}\left(\frac{J}{V}\right) + \frac{1}{V}\frac{\partial A_v}{\partial v}\right] \\ &\quad + \left[\frac{A_w}{J}\frac{\partial}{\partial w}\left(\frac{J}{W}\right) + \frac{1}{W}\frac{\partial A_w}{\partial w}\right] \\ &= \frac{1}{J}\left[\frac{\partial}{\partial u}\left(\frac{JA_u}{U}\right) + \frac{\partial}{\partial v}\left(\frac{JA_v}{V}\right) + \frac{\partial}{\partial w}\left(\frac{JA_w}{W}\right)\right]. \qquad \text{(A6.31)}\end{aligned}$$

To avoid excessive complexity we shall now restrict ourselves to *orthogonal* coordinate systems. These may be defined by the condition that

$$(dr)^2 = U^2\,(du)^2 + V^2\,(dv)^2 + W^2\,(dw)^2. \qquad \text{(A6.32)}$$

Comparing this with Eq. (A6.11), and remembering that

$$|\mathbf{u}|^2 = |\mathbf{v}|^2 = |\mathbf{w}|^2 = 1, \qquad \text{(A6.33)}$$

we see that we must now have in addition

$$\mathbf{u} \cdot \mathbf{v} = \mathbf{v} \cdot \mathbf{w} = \mathbf{w} \cdot \mathbf{u} = 0. \qquad \text{(A6.34)}$$

From Eq. (A6.34) it follows that

$$\mathbf{u} \times (\mathbf{v} \times \mathbf{w}) = (\mathbf{u} \cdot \mathbf{w})\mathbf{v} - (\mathbf{u} \cdot \mathbf{v})\mathbf{w} = 0$$

and hence that

$$\mathbf{u} = \lambda(\mathbf{v} \times \mathbf{w}).$$

Multiplying this equation on the left by **u**, we find that

$$\mathbf{u} \cdot \mathbf{u} = 1 = \lambda(\mathbf{u} \cdot \mathbf{v} \times \mathbf{w}),$$

which determines λ. We thus find that

$$\mathbf{u} = \lambda(\mathbf{v} \times \mathbf{w}) = \frac{\mathbf{v} \times \mathbf{w}}{\mathbf{u} \cdot \mathbf{v} \times \mathbf{w}},$$

with similar equations for **v** and **w**. Thus, using Eqs. (A6.19),

$$\mathbf{a} = \mathbf{u}, \qquad \mathbf{b} = \mathbf{v}, \qquad \mathbf{c} = \mathbf{w}. \qquad \text{(A6.35)}$$

From Eqs. (A6.12) and (A6.24) now follows

$$J^2 = (UVW)^2;$$

and hence

$$\mathbf{u} \cdot \mathbf{v} \times \mathbf{w} = \mathbf{a} \cdot \mathbf{b} \times \mathbf{c} = \frac{J}{UVW} = \pm 1.$$

If the naming of the vectors **u**, **v**, **w** is done in such a way that they form a right-handed set, then

$$\mathbf{u} \cdot \mathbf{v} \times \mathbf{w} = \mathbf{a} \cdot \mathbf{b} \times \mathbf{c} = 1,$$

$$\mathbf{a} = \mathbf{b} \times \mathbf{c}, \qquad \mathbf{b} = \mathbf{c} \times \mathbf{a}, \qquad \mathbf{c} = \mathbf{a} \times \mathbf{b}, \tag{A6.36}$$

and

$$J = UVW. \tag{A6.37}$$

Since **a**, **b**, and **c** are now an orthogonal right-handed set of unit vectors, Eqs. (A6.15) may be used to define U, V, and W. Of course, the functions u, v, and w must satisfy the orthogonality conditions, namely

$$\nabla u \cdot \nabla v = \nabla v \cdot \nabla w = \nabla w \cdot \nabla u = 0. \tag{A6.38}$$

In terms of the orthogonal vectors **a**, **b**, **c** the two sets of equations in Eqs. (A6.21) now become one:

$$A_u = \mathbf{a} \cdot \mathbf{A}, \qquad A_v = \mathbf{b} \cdot \mathbf{A}, \qquad A_w = \mathbf{c} \cdot \mathbf{A}. \tag{A6.39}$$

Equations (A6.22) and (A6.23) remain unchanged, while there is but slight simplification of Eq. (A6.29), which becomes

$$\nabla \times \mathbf{A} = \frac{\mathbf{a}}{VW}\left[\frac{\partial}{\partial v}(WA_w) - \frac{\partial}{\partial w}(VA_v)\right] + \frac{\mathbf{b}}{WU}\left[\frac{\partial}{\partial w}(UA_u) - \frac{\partial}{\partial u}(WA_w)\right] + \frac{\mathbf{c}}{UV}\left[\frac{\partial}{\partial u}(VA_v) - \frac{\partial}{\partial v}(UA_u)\right].$$

This can be written in the form of a determinant:

$$\nabla \times \mathbf{A} = \frac{1}{UVW}\begin{vmatrix} U\mathbf{a} & V\mathbf{b} & W\mathbf{c} \\ \dfrac{\partial}{\partial u} & \dfrac{\partial}{\partial v} & \dfrac{\partial}{\partial w} \\ UA_u & VA_v & WA_w \end{vmatrix}. \tag{A6.40}$$

Likewise Eq. (A6.31) becomes

$$\nabla \cdot \mathbf{A} = \frac{1}{UVW}\left[\frac{\partial}{\partial u}(VWA_u) + \frac{\partial}{\partial v}(WUA_v) + \frac{\partial}{\partial w}(UVA_w)\right]. \tag{A6.41}$$

Equation (A6.22) indicates that in curvilinear coordinates one may regard the operator ∇ as being

$$\nabla \equiv \frac{\mathbf{a}}{U}\frac{\partial}{\partial u} + \frac{\mathbf{b}}{V}\frac{\partial}{\partial v} + \frac{\mathbf{c}}{W}\frac{\partial}{\partial w}. \tag{A6.42}$$

In Eqs. (A6.40) and (A6.41), however, this identification in dubious; nevertheless, let us test the applicability of this operator in finding $\nabla \cdot \mathbf{A}$. We have

$$\nabla \cdot \mathbf{A} = \nabla \cdot (A_u\mathbf{a} + A_v\mathbf{b} + A_w\mathbf{c}).$$

Expansion of the first term gives

$$\begin{aligned}
\nabla \cdot (A_u\mathbf{a}) &= \mathbf{a} \cdot \nabla A_u + A_u \nabla \cdot \mathbf{a} \\
&= \mathbf{a} \cdot \left(\frac{\mathbf{a}}{U}\frac{\partial A_u}{\partial u} + \frac{\mathbf{b}}{V}\frac{\partial A_u}{\partial v} + \frac{\mathbf{c}}{W}\frac{\partial A_u}{\partial w}\right) \\
&\quad + A_u\left(\frac{\mathbf{a}}{U}\frac{\partial}{\partial u} + \frac{\mathbf{b}}{V}\frac{\partial}{\partial v} + \frac{\mathbf{c}}{W}\frac{\partial}{\partial w}\right) \cdot \mathbf{a} \\
&= \frac{1}{U}\frac{\partial A_u}{\partial u} + A_u\left(\frac{\mathbf{a}}{U} \cdot \frac{\partial \mathbf{a}}{\partial u} + \frac{\mathbf{b}}{V} \cdot \frac{\partial \mathbf{a}}{\partial v} + \frac{\mathbf{c}}{W} \cdot \frac{\partial \mathbf{a}}{\partial w}\right).
\end{aligned}$$

Thus, to be able to use Eq. (A6.42) easily, one must have available the values of the nine derivatives $\partial\mathbf{a}/\partial u$, $\partial\mathbf{a}/\partial v$, . . . , $\partial\mathbf{c}/\partial w$. These we shall now evaluate.

Since $\mathbf{a}$ is not a function of time, we have by Eqs. (A3.60a), (A6.11), and (A6.35) that

$$d\mathbf{a} = d\mathbf{r} \cdot \nabla\mathbf{a} = (\mathbf{a}U\,du + \mathbf{b}V\,dv + \mathbf{c}W\,dw) \cdot \nabla\mathbf{a},$$

which requires that

$$\frac{\partial \mathbf{a}}{\partial u} = U\,\mathbf{a} \cdot \nabla\mathbf{a}, \qquad \frac{\partial \mathbf{a}}{\partial v} = V\,\mathbf{b} \cdot \nabla\mathbf{a}, \qquad \frac{\partial \mathbf{a}}{\partial w} = W\,\mathbf{c} \cdot \nabla\mathbf{a}. \tag{A6.43}$$

Now, since $a^2 = 1$, we have by Eq. (A3.28)

$$\nabla(a^2) = 2\mathbf{a} \cdot \nabla\mathbf{a} + 2\mathbf{a} \times (\nabla \times \mathbf{a}) = 0; \tag{A6.44}$$

so that, using Eqs. (A6.26), (A6.35), and (A6.37),

$$\begin{aligned}
\mathbf{a} \cdot \nabla\mathbf{a} = -\mathbf{a} \times (\nabla \times \mathbf{a}) &= -\mathbf{a} \times \left(\frac{\mathbf{b}}{WU}\frac{\partial U}{\partial w} - \frac{\mathbf{c}}{UV}\frac{\partial U}{\partial v}\right) \\
&= -\left(\frac{\mathbf{b}}{UV}\frac{\partial U}{\partial v} + \frac{\mathbf{c}}{UW}\frac{\partial U}{\partial w}\right).
\end{aligned} \tag{A6.45}$$

By cyclic permutations we can write down $\mathbf{b} \cdot \nabla\mathbf{b}$ and $\mathbf{c} \cdot \nabla\mathbf{c}$.

Since by Eq. (A3.30) for any vector **g**

$$\mathbf{a} \cdot (\mathbf{g} \cdot \nabla \mathbf{a}) = \mathbf{g} \cdot \mathbf{a} \times (\nabla \times \mathbf{a}) + \mathbf{g} \cdot (\mathbf{a} \cdot \nabla \mathbf{a}),$$

we see, on comparing this with Eq. (A6.44), that $\mathbf{a} \cdot (\mathbf{g} \cdot \nabla \mathbf{a}) = 0$. By symmetry we generalize this to

$$\mathbf{a} \cdot (\mathbf{g} \cdot \nabla \mathbf{a}) = 0, \qquad \mathbf{b} \cdot (\mathbf{g} \cdot \nabla \mathbf{b}) = 0, \qquad \mathbf{c} \cdot (\mathbf{g} \cdot \nabla \mathbf{c}) = 0. \tag{A6.46}$$

These equations merely express the fact that the changes in the unit vectors **a**, **b**, and **c** are perpendicular respectively to **a**, **b**, and **c**. We should also note that, by virtue of Eqs. (A6.26) to (A6.28),

$$\mathbf{a} \cdot \nabla \times \mathbf{a} = 0, \qquad \mathbf{b} \cdot \nabla \times \mathbf{b} = 0, \qquad \mathbf{c} \cdot \nabla \times \mathbf{c} = 0. \tag{A6.47}$$

Consider now the application of Eq. (A3.28)

$$0 = \nabla \mathbf{a} \cdot \mathbf{b} = \mathbf{a} \cdot \nabla \mathbf{b} + \mathbf{b} \cdot \nabla \mathbf{a} + \mathbf{a} \times (\nabla \times \mathbf{b}) + \mathbf{b} \times (\nabla \times \mathbf{a}) \tag{A6.48}$$

and of Eq. (A3.27)

$$\nabla \times \mathbf{c} = \nabla \times (\mathbf{a} \times \mathbf{b}) = \mathbf{a}\, \nabla \cdot \mathbf{b} - \mathbf{b}\, \nabla \cdot \mathbf{a} - \mathbf{a} \cdot \nabla \mathbf{b} + \mathbf{b} \cdot \nabla \mathbf{a}. \tag{A6.49}$$

Multiplying Eq. (A6.48) by $\mathbf{a}\,\cdot$ and $\mathbf{b}\,\cdot$ we obtain, respectively,

$$\mathbf{a} \cdot (\mathbf{a} \cdot \nabla \mathbf{b}) + \mathbf{a} \cdot (\mathbf{b} \cdot \nabla \mathbf{a}) + \mathbf{a} \cdot \mathbf{b} \times (\nabla \times \mathbf{a}) = 0$$

and

$$\mathbf{b} \cdot (\mathbf{a} \cdot \nabla \mathbf{b}) + \mathbf{b} \cdot (\mathbf{b} \cdot \nabla \mathbf{a}) + \mathbf{b} \cdot \mathbf{a} \times (\nabla \times \mathbf{b}) = 0.$$

With the help of Eqs. (A6.46), (A6.26), and (A6.27), together with Eqs. (A6.35) and (A6.36), these equations lead to

$$\begin{aligned} \mathbf{a} \cdot (\mathbf{a} \cdot \nabla \mathbf{b}) &= -(\mathbf{a} \times \mathbf{b}) \cdot \nabla \times \mathbf{a} = -\mathbf{c} \cdot \nabla \times \mathbf{a} \\ &= -\mathbf{c} \cdot \left(\frac{\mathbf{b}}{WU} \frac{\partial U}{\partial w} - \frac{\mathbf{c}}{UV} \frac{\partial U}{\partial v} \right) = \frac{1}{UV} \frac{\partial U}{\partial v} \end{aligned} \tag{A6.50}$$

and

$$\begin{aligned} \mathbf{b} \cdot (\mathbf{b} \cdot \nabla \mathbf{a}) &= (\mathbf{a} \times \mathbf{b}) \cdot \nabla \times \mathbf{b} = \mathbf{c} \cdot \nabla \times \mathbf{b} \\ &= \mathbf{c} \cdot \left(\frac{\mathbf{c}}{UV} \frac{\partial V}{\partial u} - \frac{\mathbf{a}}{VW} \frac{\partial V}{\partial w} \right) = \frac{1}{UV} \frac{\partial V}{\partial u}. \end{aligned} \tag{A6.51}$$

Multiplying Eqs. (A6.48) and (A6.49) by $\mathbf{c}\,\cdot$ we obtain, respectively, using Eq. (A6.47),

$$\begin{aligned} \mathbf{c} \cdot (\mathbf{a} \cdot \nabla \mathbf{b}) + \mathbf{c} \cdot (\mathbf{b} \cdot \nabla \mathbf{a}) &= -(\mathbf{c} \times \mathbf{a}) \cdot \nabla \times \mathbf{b} + (\mathbf{b} \times \mathbf{c}) \cdot \nabla \times \mathbf{a} \\ &= -\mathbf{b} \cdot \nabla \times \mathbf{b} + \mathbf{a} \cdot \nabla \times \mathbf{a} = 0 \end{aligned}$$

and

$$\mathbf{c} \cdot (\mathbf{a} \cdot \nabla \mathbf{b}) - \mathbf{c} \cdot (\mathbf{b} \cdot \nabla \mathbf{a}) = -\mathbf{c} \cdot \nabla \times \mathbf{c} = 0;$$

so that

$$\mathbf{c} \cdot (\mathbf{a} \cdot \nabla \mathbf{b}) = \mathbf{c} \cdot (\mathbf{b} \cdot \nabla \mathbf{a}) = 0. \tag{A6.52}$$

Collecting our results, Eqs. (A6.46), (A6.50), (A6.51), and (A6.52), we have

$$\mathbf{a} \cdot (\mathbf{a} \cdot \nabla \mathbf{b}) = \frac{1}{UV}\frac{\partial U}{\partial v}, \qquad \mathbf{b} \cdot (\mathbf{a} \cdot \nabla \mathbf{b}) = 0, \qquad \mathbf{c} \cdot (\mathbf{a} \cdot \nabla \mathbf{b}) = 0; \tag{A6.53}$$

and

$$\mathbf{a} \cdot (\mathbf{b} \cdot \nabla \mathbf{a}) = 0, \qquad \mathbf{b} \cdot (\mathbf{b} \cdot \nabla \mathbf{a}) = \frac{1}{UV}\frac{\partial V}{\partial u}, \qquad \mathbf{c} \cdot (\mathbf{b} \cdot \nabla \mathbf{a}) = 0. \tag{A6.54}$$

These show, respectively, that

$$\mathbf{a} \cdot \nabla \mathbf{b} = \frac{\mathbf{a}}{UV}\frac{\partial U}{\partial v} \qquad \text{and} \qquad \mathbf{b} \cdot \nabla \mathbf{a} = \frac{\mathbf{b}}{VU}\frac{\partial V}{\partial u}. \tag{A6.55}$$

By analogy, remembering Eqs. (A6.43) and (A6.45), we can now write the whole set of nine equations:

$$\begin{aligned}
&\frac{\partial \mathbf{a}}{\partial u} = -\left(\frac{\mathbf{b}}{V}\frac{\partial U}{\partial v} + \frac{\mathbf{c}}{W}\frac{\partial U}{\partial w}\right), \qquad \frac{\partial \mathbf{a}}{\partial v} = \frac{\mathbf{b}}{U}\frac{\partial V}{\partial u}, \qquad \frac{\partial \mathbf{a}}{\partial w} = \frac{\mathbf{c}}{U}\frac{\partial W}{\partial u}, \\
&\frac{\partial \mathbf{b}}{\partial u} = \frac{\mathbf{a}}{V}\frac{\partial U}{\partial v}, \qquad \frac{\partial \mathbf{b}}{\partial v} = -\left(\frac{\mathbf{c}}{W}\frac{\partial V}{\partial w} + \frac{\mathbf{a}}{U}\frac{\partial V}{\partial u}\right), \qquad \frac{\partial \mathbf{b}}{\partial w} = \frac{\mathbf{c}}{V}\frac{\partial W}{\partial v}, \\
&\frac{\partial \mathbf{c}}{\partial u} = \frac{\mathbf{a}}{W}\frac{\partial U}{\partial w}, \qquad \frac{\partial \mathbf{c}}{\partial v} = \frac{\mathbf{b}}{W}\frac{\partial V}{\partial w}, \qquad \frac{\partial \mathbf{c}}{\partial w} = -\left(\frac{\mathbf{a}}{U}\frac{\partial W}{\partial u} + \frac{\mathbf{b}}{V}\frac{\partial W}{\partial v}\right).
\end{aligned} \tag{A6.56}$$

With the help of these equations we can complete the calculation of $\nabla \cdot \mathbf{A}$ started above:

$$\begin{aligned}
\nabla \cdot (A_u \mathbf{a}) &= \frac{1}{U}\frac{\partial A_u}{\partial u} + A_u\left(\frac{\mathbf{a}}{U} \cdot \frac{\partial \mathbf{a}}{\partial u} + \frac{\mathbf{b}}{V} \cdot \frac{\partial \mathbf{a}}{\partial v} + \frac{\mathbf{c}}{W} \cdot \frac{\partial \mathbf{a}}{\partial w}\right) \\
&= \frac{1}{U}\frac{\partial A_u}{\partial u} + A_u\left(\frac{1}{VU}\frac{\partial V}{\partial u} + \frac{1}{UW}\frac{\partial W}{\partial u}\right) \\
&= \frac{1}{UVW}\left(VW\frac{\partial A_u}{\partial u} + WA_u\frac{\partial V}{\partial u} + VA_u\frac{\partial W}{\partial u}\right) \\
&= \frac{1}{UVW}\frac{\partial}{\partial u}(VWA_u).
\end{aligned}$$

Similar equations hold for $\nabla \cdot (A_v\mathbf{b})$ and $\nabla \cdot (A_w\mathbf{c})$. Adding these equations we obtain Eq. (A6.41).

To obtain $\mathbf{A} \cdot \nabla\mathbf{B}$ we may proceed as follows, using Eq. (A6.42):

$$\begin{aligned}\mathbf{A} \cdot \nabla\mathbf{B} &= \mathbf{A} \cdot \nabla(B_u\mathbf{a} + B_v\mathbf{b} + B_w\mathbf{c}) \\ &= \mathbf{a}\,\mathbf{A} \cdot \nabla B_u + \mathbf{b}\,\mathbf{A} \cdot \nabla B_v + \mathbf{c}\,\mathbf{A} \cdot \nabla B_w \\ &\quad + B_u\,\mathbf{A} \cdot \nabla\mathbf{a} + B_v\,\mathbf{A} \cdot \nabla\mathbf{b} + B_w\,\mathbf{A} \cdot \nabla\mathbf{c} \\ &= \mathbf{a}\,\mathbf{A} \cdot \nabla B_u + \mathbf{b}\,\mathbf{A} \cdot \nabla B_v + \mathbf{c}\,\mathbf{A} \cdot \nabla B_w \\ &\quad + B_u\left(\frac{\mathbf{A} \cdot \mathbf{a}}{U}\frac{\partial\mathbf{a}}{\partial u} + \frac{\mathbf{A} \cdot \mathbf{b}}{V}\frac{\partial\mathbf{a}}{\partial v} + \frac{\mathbf{A} \cdot \mathbf{c}}{W}\frac{\partial\mathbf{a}}{\partial w}\right) + \cdots.\end{aligned}$$

We now use our Eqs. (A6.56) and, after some calculation, find that

$$\begin{aligned}\mathbf{A} \cdot \nabla\mathbf{B} &= \mathbf{a}\,\mathbf{A} \cdot \nabla B_u + \mathbf{b}\,\mathbf{A} \cdot \nabla B_v + \mathbf{c}\,\mathbf{A} \cdot \nabla B_w \\ &\quad + \frac{A_u}{U}\mathbf{B} \times (\mathbf{a} \times \nabla U) + \frac{A_v}{V}\mathbf{B} \times (\mathbf{b} \times \nabla V) + \frac{A_w}{W}\mathbf{B} \times (\mathbf{c} \times \nabla W).\end{aligned} \tag{A6.57}$$

The expressions occurring here may be evaluated at once by means of Eqs. (A6.22) and (A6.23).

Important expressions involving second derivatives are $\nabla^2\phi$, $\nabla\nabla \cdot \mathbf{A}$, $\nabla \times \nabla \times \mathbf{A}$, and $\nabla^2\mathbf{A}$. Since $(\nabla\phi)_u = \mathbf{a} \cdot \nabla\phi = (1/U)\partial\phi/\partial u$, etc., we have immediately, by Eq. (A6.41),

$$\nabla^2\phi = \nabla \cdot (\nabla\phi) = \frac{1}{UVW}\left[\frac{\partial}{\partial u}\left(\frac{VW}{U}\frac{\partial\phi}{\partial u}\right) + \frac{\partial}{\partial v}\left(\frac{WU}{V}\frac{\partial\phi}{\partial v}\right) + \frac{\partial}{\partial w}\left(\frac{UV}{W}\frac{\partial\phi}{\partial w}\right)\right] \tag{A6.58}$$

Using Eqs. (A6.39) and (A6.23), we have

$$\begin{aligned}\nabla\nabla \cdot \mathbf{A} &= (\nabla\nabla \cdot \mathbf{A})_u\mathbf{a} + (\nabla\nabla \cdot \mathbf{A})_v\mathbf{b} + (\nabla\nabla \cdot \mathbf{A})_w\mathbf{c} \\ &= (\mathbf{a} \cdot (\nabla\nabla \cdot \mathbf{A}))\mathbf{a} + (\mathbf{b} \cdot (\nabla\nabla \cdot \mathbf{A}))\mathbf{b} + (\mathbf{c} \cdot (\nabla\nabla \cdot \mathbf{A}))\mathbf{c}\end{aligned} \tag{A6.59}$$

and

$$\begin{aligned}\mathbf{a} \cdot (\nabla\nabla \cdot \mathbf{A}) &= \frac{1}{U}\frac{\partial}{\partial u}(\nabla \cdot \mathbf{A}), \\ \mathbf{b} \cdot (\nabla\nabla \cdot \mathbf{A}) &= \frac{1}{V}\frac{\partial}{\partial v}(\nabla \cdot \mathbf{A}), \\ \mathbf{c} \cdot (\nabla\nabla \cdot \mathbf{A}) &= \frac{1}{W}\frac{\partial}{\partial w}(\nabla \cdot \mathbf{A}),\end{aligned} \tag{A6.60}$$

which can be evaluated with the help of Eq. (A6.41). Similarly, one has

$$\nabla \times \nabla \times \mathbf{A}$$
$$= \mathbf{a}\,\mathbf{a}\cdot(\nabla \times \nabla \times \mathbf{A}) + \mathbf{b}\,\mathbf{b}\cdot(\nabla \times \nabla \times \mathbf{A}) + \mathbf{c}\,\mathbf{c}\cdot(\nabla \times \nabla \times \mathbf{A}). \tag{A6.61}$$

This can be evaluated, with the help of Eq. (A6.40), since

$$\begin{aligned}\mathbf{a}\cdot(\nabla \times \nabla \times \mathbf{A}) &= \frac{1}{VW}\left\{\frac{\partial}{\partial v}[W(\nabla \times \mathbf{A})_w] - \frac{\partial}{\partial w}[V(\nabla \times \mathbf{A})_v]\right\}\\ &= \frac{1}{VW}\left\{\frac{\partial}{\partial v}\left[\frac{W}{UV}\left(\frac{\partial}{\partial u}(VA_v) - \frac{\partial}{\partial v}(UA_u)\right)\right]\right.\\ &\quad\left. - \frac{\partial}{\partial w}\left[\frac{V}{WU}\left(\frac{\partial}{\partial w}(UA_u) - \frac{\partial}{\partial u}(WA_w)\right)\right]\right\},\end{aligned} \tag{A6.62}$$

with corresponding expressions for the other two components.

To calculate $\nabla^2\mathbf{A}$ we may either apply the operator

$$\nabla^2 = \frac{1}{UVW}\left[\frac{\partial}{\partial u}\left(\frac{VW}{U}\frac{\partial}{\partial u}\right) + \frac{\partial}{\partial v}\left(\frac{WU}{V}\frac{\partial}{\partial v}\right) + \frac{\partial}{\partial w}\left(\frac{UV}{W}\frac{\partial}{\partial w}\right)\right] \tag{A6.63}$$

as it appears in Eq. (A6.58) to each term of

$$\mathbf{A} = A_u\mathbf{a} + A_v\mathbf{b} + A_w\mathbf{c}, \tag{A6.64}$$

and make use of Eqs. (A6.56); or we may treat each term of Eq. (A6.64), with use of Eqs. (A3.37), (A3.23), and (A3.24), as follows:

$$\begin{aligned}\nabla^2 A_u\mathbf{a} &= \nabla\,\nabla\cdot(A_u\mathbf{a}) - \nabla \times \nabla \times (A_u\mathbf{a})\\ &= \nabla(A_u\,\nabla\cdot\mathbf{a} + \mathbf{a}\cdot\nabla A_u) - \nabla \times (A_u\,\nabla \times \mathbf{a} - \mathbf{a} \times \nabla A_u).\end{aligned}$$

These can be further expanded by applying Eqs. (A3.18), (A3.28), (A3.24), and (A3.27) to the four terms, respectively. We thus obtain

$$\nabla^2 A_u\mathbf{a} = A_u(\nabla\,\nabla\cdot\mathbf{a} - \nabla \times \nabla \times \mathbf{a}) + 2(\nabla A_u \cdot \nabla)\mathbf{a} + \mathbf{a}\,\nabla^2 A_u, \tag{A6.65a}$$

or, using Eq. (A3.37),

$$\nabla^2 A_u\mathbf{a} = A_u\,\nabla^2\mathbf{a} + 2\nabla A_u \cdot \nabla\mathbf{a} + \mathbf{a}\,\nabla^2 A_u. \tag{A6.65b}$$

The latter is a natural generalization of Eq. (A3.40) and could have been obtained, of course, quite easily by applying that equation to the components of $\nabla^2 A_u\mathbf{a}$.

Since we have shown how to obtain $\nabla \nabla \cdot \mathbf{A}$ and $\nabla \times \nabla \times \mathbf{A}$ in curvilinear coordinates, we can obtain from $\nabla^2\mathbf{a} = \nabla \nabla \cdot \mathbf{a} - \nabla \times \nabla \times \mathbf{a}$ and therefore

$$\begin{aligned}\nabla^2\mathbf{a} = &-\left[\frac{U^2}{V^2}\left(\frac{\partial \log U}{\partial v}\right)^2 + \frac{U^2}{W^2}\left(\frac{\partial \log U}{\partial w}\right)^2 + \left(\frac{\partial \log V}{\partial u}\right)^2 + \left(\frac{\partial \log W}{\partial u}\right)^2\right]\frac{\mathbf{a}}{U^2}\\ &+\left[\frac{\partial^2}{\partial u\,\partial v}\log\frac{U}{V} + \frac{\partial \log U}{\partial v}\frac{\partial \log W}{\partial u}\right.\\ &\left.-\frac{\partial \log V}{\partial u}\frac{\partial \log W}{\partial v} + \frac{\partial \log W}{\partial u}\frac{\partial \log W}{\partial v}\right]\frac{\mathbf{b}}{UV}\\ &+\left[\frac{\partial^2}{\partial u\,\partial w}\log\frac{U}{W} + \frac{\partial \log U}{\partial w}\frac{\partial \log V}{\partial u}\right.\\ &\left.-\frac{\partial \log W}{\partial u}\frac{\partial \log V}{\partial w} + \frac{\partial \log V}{\partial u}\frac{\partial \log V}{\partial w}\right]\frac{\mathbf{c}}{UW}. \qquad \text{(A6.66)}\end{aligned}$$

Examples. For the *spherical polar coordinate system* Eqs. (A6.2) become, when we put $u = r$, $v = \theta$, $w = \phi$,

$$\begin{aligned} x &= r \sin\theta \cos\phi,\\ y &= r \sin\theta \sin\phi, \qquad \text{(A6.67)}\\ z &= r\cos\theta. \end{aligned}$$

The three families of coordinate surfaces are

$$\begin{aligned} (x^2 + y^2 + z^2)^{1/2} &= r, \qquad \text{with } r \geq 0;\\ \tan^{-1}\left(\frac{x^2 + y^2}{z^2}\right)^{1/2} &= \theta, \qquad \text{with } 0 \leq \theta \leq \pi; \qquad \text{(A6.68)}\\ \tan^{-1}\left(\frac{y}{x}\right) &= \phi, \qquad \text{with } 0 \leq \phi < 2\pi. \end{aligned}$$

The first is a family of spheres of variable radius r; the second is a family of cones having their vertices at the origin, their axes along the Z-axis, and an angle θ between their generating lines and the Z-axis; the third is a family of half-planes having the Z-axis as their edges and making an angle ϕ with the XZ-plane. The coordinate axes at any point P (not on the Z-axis) are the following curves relative to the sphere passing through P:

a radius of the sphere, a meridian circle, a latitude circle. These are readily seen to be perpendicular to one another.

For these spherical coordinates the functions U, V, W of Eqs. (A6.9) and the Jacobian of Eq. (A6.12) are seen to be given by

$$U = 1, \qquad V = r, \qquad W = r\sin\theta, \qquad J = r^2\sin\theta; \tag{A6.69}$$

therefore by Eqs. (A6.26), (A6.30), and (A6.56),

$$\nabla\times\mathbf{a} = 0, \qquad \nabla\times\mathbf{b} = \frac{\mathbf{c}}{r}, \qquad \nabla\times\mathbf{c} = \frac{1}{r}(\mathbf{a}\cot\theta - \mathbf{b}), \tag{A6.70}$$

$$\nabla\cdot\mathbf{a} = \frac{2}{r}, \qquad \nabla\cdot\mathbf{b} = \frac{\cot\theta}{r}, \qquad \nabla\cdot\mathbf{c} = 0, \tag{A6.71}$$

$$\begin{aligned} &\frac{\partial\mathbf{a}}{\partial r} = 0, \qquad \frac{\partial\mathbf{a}}{\partial\theta} = \mathbf{b}, \qquad \frac{\partial\mathbf{a}}{\partial\phi} = \mathbf{c}\sin\theta, \\ &\frac{\partial\mathbf{b}}{\partial r} = 0, \qquad \frac{\partial\mathbf{b}}{\partial\theta} = -\mathbf{a}, \qquad \frac{\partial\mathbf{b}}{\partial\phi} = \mathbf{c}\cos\theta, \\ &\frac{\partial\mathbf{c}}{\partial r} = 0, \qquad \frac{\partial\mathbf{c}}{\partial\theta} = 0, \qquad \frac{\partial\mathbf{c}}{\partial\phi} = -(\mathbf{a}\sin\theta + \mathbf{b}\cos\theta). \end{aligned} \tag{A6.72}$$

Further, by Eqs. (A6.22), (A6.41), (A6.40), and (A6.58), we determine that

$$\nabla\psi = \mathbf{a}\frac{\partial\psi}{\partial r} + \frac{\mathbf{b}}{r}\frac{\partial\psi}{\partial\theta} + \frac{\mathbf{c}}{r\sin\theta}\frac{\partial\psi}{\partial\phi}, \tag{A6.73}$$

$$\nabla\cdot\mathbf{A} = \frac{1}{r^2}\frac{\partial(A_r r^2)}{\partial r} + \frac{1}{r\sin\theta}\frac{\partial(A_\theta\sin\theta)}{\partial\theta} + \frac{1}{r\sin\theta}\frac{\partial A_\phi}{\partial\phi}, \tag{A6.74}$$

$$\nabla\times\mathbf{A} = \frac{\mathbf{a}}{r\sin\theta}\left[\frac{\partial}{\partial\theta}(A_\phi\sin\theta) - \frac{\partial A_\theta}{\partial\phi}\right] + \frac{\mathbf{b}}{r}\left[\frac{1}{\sin\theta}\frac{\partial A_r}{\partial\phi} - \frac{\partial}{\partial r}(A_\phi r)\right] + \frac{\mathbf{c}}{r}\left(\frac{\partial}{\partial r}(A_\theta r) - \frac{\partial A_r}{\partial\theta}\right), \tag{A6.75}$$

$$\nabla^2\psi = \frac{1}{r^2}\frac{\partial}{\partial r}\left(r^2\frac{\partial\psi}{\partial r}\right) + \frac{1}{r^2\sin\theta}\frac{\partial}{\partial\theta}\left(\sin\theta\frac{\partial\psi}{\partial\theta}\right) + \frac{1}{r^2\sin^2\theta}\frac{\partial^2\psi}{\partial\phi^2}. \tag{A6.76}$$

For the *cylindrical coordinate system* we put $u = u$, $v = \phi$, $w = z$, and obtain

$$x = u \cos \phi,$$
$$y = u \sin \phi, \tag{A6.77}$$
$$z = z.$$

The three families of surfaces are given by

$$\begin{aligned} (x^2 + y^2)^{1/2} &= u, &\quad &\text{with} \quad u \geq 0, \\ \tan^{-1}(y/x) &= \phi, &\quad &\text{with} \quad 0 \leq \phi < 2\pi, \\ z &= z, &\quad &\text{with} \quad -\infty < z < \infty. \end{aligned} \tag{A6.78}$$

These are, respectively, cylinders of radius u with their axes along the Z-axis, half-planes with their edges along the Z-axis and making an angle ϕ with the XZ-plane, and planes parallel to the XY-plane intersecting the Z-axis at z. The coordinate axes at any point P (not on the Z-axis) are the following lines passing through P: a circle in a plane perpendicular to the Z-axis, a radial line of this circle, and a line parallel to the Z-axis, the coordinate system is an orthogonal system with

$$U = 1, \qquad V = u, \qquad W = 1, \tag{A6.79}$$

and

$$J = u. \tag{A6.80}$$

We find therefore (see again Eqs. (A6.26), (A6.27), (A6.28), (A6.30), and (A6.56)) that

$$\nabla \times \mathbf{a} = 0, \qquad \nabla \times \mathbf{b} = \frac{\mathbf{c}}{u}, \qquad \nabla \times \mathbf{c} = 0; \tag{A6.81}$$

$$\nabla \cdot \mathbf{a} = \frac{1}{u}, \qquad \nabla \cdot \mathbf{b} = 0, \qquad \nabla \cdot \mathbf{c} = 0; \tag{A6.82}$$

$$\begin{aligned} &\frac{\partial \mathbf{a}}{\partial u} = 0, \qquad \frac{\partial \mathbf{a}}{\partial \phi} = \mathbf{b}, \qquad \frac{\partial \mathbf{a}}{\partial z} = 0, \\ &\frac{\partial \mathbf{b}}{\partial u} = 0, \qquad \frac{\partial \mathbf{b}}{\partial \phi} = -\mathbf{a}, \qquad \frac{\partial \mathbf{b}}{\partial z} = 0, \\ &\frac{\partial \mathbf{c}}{\partial u} = 0, \qquad \frac{\partial \mathbf{c}}{\partial \phi} = 0, \qquad \frac{\partial \mathbf{c}}{\partial z} = 0; \end{aligned} \tag{A6.83}$$

and (see Eqs. (A6.22), (A6.41), (A6.40), and (A6.58)) we also have

$$\nabla\psi = \mathbf{a}\frac{\partial\psi}{\partial u} + \frac{\mathbf{b}}{u}\frac{\partial\psi}{\partial\phi} + \mathbf{c}\frac{\partial\psi}{\partial z}, \tag{A6.84}$$

$$\nabla\cdot\mathbf{A} = \frac{1}{u}\frac{\partial}{\partial u}(uA_u) + \frac{1}{u}\frac{\partial A_\phi}{\partial\phi} + \frac{\partial A_z}{\partial z}, \tag{A6.85}$$

$$\nabla\times\mathbf{A} = \mathbf{a}\left(\frac{1}{u}\frac{\partial A_z}{\partial\phi} - \frac{\partial A_\phi}{\partial z}\right) + \mathbf{b}\left(\frac{\partial A_u}{\partial z} - \frac{\partial A_z}{\partial u}\right)$$
$$+ \mathbf{c}\left(\frac{A_\phi}{u} + \frac{\partial A_\phi}{\partial u} - \frac{1}{u}\frac{\partial A_u}{\partial\phi}\right), \tag{A6.86}$$

$$\nabla^2\psi = \frac{1}{u}\frac{\partial}{\partial u}\left(u\frac{\partial\psi}{\partial u}\right) + \frac{1}{u^2}\frac{\partial^2\psi}{\partial\phi^2} + \frac{\partial^2\psi}{\partial z^2}. \tag{A6.87}$$

An element of volume in this coordinate system is an infinitesimal parallelepiped having as three of its edges

$$d\mathbf{r}_1 = \frac{\partial\mathbf{r}}{\partial u}\,du, \qquad d\mathbf{r}_2 = \frac{\partial\mathbf{r}}{\partial v}\,dv, \qquad d\mathbf{r}_3 = \frac{\partial\mathbf{r}}{\partial w}\,dw.$$

Using Eqs. (A6.10) and (A6.12), we obtain for its volume

$$d\mathbf{r}_1\cdot d\mathbf{r}_2\times d\mathbf{r}_3 = UVW\,\mathbf{u}\cdot\mathbf{v}\times\mathbf{w}\,du\,dv\,dw = J\,du\,dv\,dw, \tag{A6.88}$$

which, when coordinates are orthogonal, becomes

$$UVW\,du\,dv\,dw. \tag{A6.89}$$

If a surface S is given by an equation

$$f(\mathbf{r}) = \text{constant}, \tag{A6.90}$$

then, for any displacement on the surface,

$$df = d\mathbf{r}\cdot\nabla f = (U\mathbf{u}\,du + V\mathbf{v}\,dv + W\mathbf{w}\,dw)\cdot\nabla f = 0, \tag{A6.91}$$

which establishes a relation between du, dv, and dw.

Let two displacements on the surface be chosen in such a way that for the first, say $d\mathbf{r}_1$, $dv = 0$; while for the second, say $d\mathbf{r}_2$, $du = 0$. We have then from Eq. (A6.91) for the first displacement

$$W\,dw_1 = -\frac{\mathbf{u}\cdot\nabla f}{\mathbf{w}\cdot\nabla f}\,U\,du,$$

while for the second

$$W\,dw_2 = -\frac{\mathbf{v}\cdot\nabla f}{\mathbf{w}\cdot\nabla f}\,V\,dv.$$

Thus we have

$$d\mathbf{r}_1 = (\mathbf{w}\cdot\nabla f\,\mathbf{u} - \mathbf{u}\cdot\nabla f\,\mathbf{w})\frac{U\,du}{\mathbf{w}\cdot\nabla f}$$

and

$$d\mathbf{r}_2 = (\mathbf{w}\cdot\nabla f\,\mathbf{v} - \mathbf{v}\cdot\nabla f\,\mathbf{w})\frac{V\,dv}{\mathbf{w}\cdot\nabla f}.$$

The element of area, therefore, corresponding to the differentials du and dv on the surface S is

$$d\mathbf{S} = d\mathbf{r}_1 \times d\mathbf{r}_2 = \frac{J\,du\,dv}{W\,\mathbf{w}\cdot\nabla f}(\mathbf{a}\,\mathbf{u}\cdot\nabla f + \mathbf{b}\,\mathbf{v}\cdot\nabla f + \mathbf{c}\,\mathbf{w}\cdot\nabla f)$$

$$= \frac{J\,du\,dv}{W\,\mathbf{w}\cdot\nabla f}\,\nabla f, \tag{A6.92}$$

where we have used Eqs. (A6.18) and (A6.19). As an example, if the surface S is the coordinate surface

$$w = \text{constant},$$

then $f = w$ and, by Eq. (A6.42), $\nabla f = \mathbf{c}/W$. Thus by Eqs. (A6.18)

$$\mathbf{u}\cdot\nabla f = \mathbf{v}\cdot\nabla f = 0$$

and hence by Eq. (A6.92)

$$d\mathbf{S} = \frac{J\,du\,dv}{W}\,\mathbf{c}. \tag{A6.93}$$

Appendix A7

Dyadics

Equation (A5.4), which for three dimensions becomes

$$B_j = b_{jk}A_k, \qquad j = 1, 2, 3, \tag{A7.1}$$

assigns to each vector $\mathbf{A} \sim A_j$ another vector $\mathbf{B} \sim B_j$. We may therefore write

$$\mathbf{B} = \mathbf{f}(\mathbf{A}). \tag{A7.2}$$

We are dealing here with the case in which the components of **B** in every coordinate system are linear homogeneous functions of the components of **A**. In such a case **B** is said to be a linear homogeneous function of **A**.

If two such functions of **A** give the same **B**, that is, if

$$\mathbf{f}_1(\mathbf{A}) = \mathbf{f}_2(\mathbf{A}) \tag{A7.3}$$

for all **A**, then the corresponding tensors are the same. This can be seen as follows. Let the corresponding tensors be b_{jk} and b'_{jk}, then

$$b_{jk}A_k = b'_{jk}A_k$$

or

$$(b_{jk} - b'_{jk})A_k = 0 \tag{A7.4}$$

for all A_k. Since the A_k are arbitrary, this can only be true if

$$b_{jk} - b'_{jk} \equiv 0,$$

which proves the assertion.

In general, it is not possible to express b_{jk} as an outer product of two vectors; i.e., in general there do not exist two vectors $\mathbf{M} \sim M_j$ and $\mathbf{N} \sim N_j$ such that

$$b_{jk} = M_jN_k.$$

It is possible, however, to choose three pairs of vectors **M**, **N**; **M′**, **N′**; and **M″**, **N″** in many different ways so that

$$b_{jk} = M_j N_k + M'_j N'_k + M''_j N''_k. \qquad \text{(A7.5)}$$

We shall now show that if **u**, **v**, **w** are three unit vectors orthogonal to each other, we can find three vectors **P**, **Q**, **R** such that

$$b_{jk} = P_j u_k + Q_j v_k + R_j w_k. \qquad \text{(A7.6)}$$

The conditions on **u**, **v**, and **w** are

$$\begin{aligned} u_j^2 = v_j^2 = w_j^2 = 1, \\ u_j v_j = v_j w_j = w_j u_j = 0; \end{aligned} \qquad \text{(A7.7)}$$

therefore, if Eq. (A7.6) holds,

$$b_{jk} u_k = P_j u_k^2 + Q_j v_k u_k + R_j w_k u_k.$$

This, using Eq. (A7.7), becomes $P_j = b_{jk} u_k$. This and similar equations in v_k and w_k give us

$$P_j = b_{jk} u_k, \qquad Q_j = b_{jk} v_k, \qquad R_j = b_{jk} w_k, \qquad \text{(A7.8)}$$

provided there exist three such vectors satisfying Eq. (A7.6). To prove that the vectors thus found will satisfy Eq. (A7.6), we must recall the result of Example 1 of Appendix A2, p. 368. There it was found that any vector can be expressed in terms of three other vectors, provided the triple scalar product of these vectors is not zero. This latter condition is satisfied by our three orthogonal vectors **u**, **v**, and **w**, so that, for any vector **A**, we can write

$$\mathbf{A} = a\mathbf{u} + b\mathbf{v} + c\mathbf{w}. \qquad \text{(A7.9)}$$

The coefficients a, b, and c can be easily found by multiplying Eq. (A7.9) by $\mathbf{u}\cdot$, $\mathbf{v}\cdot$, and $\mathbf{w}\cdot$, successively, and making use of Eqs. (A7.7). We thus find that

$$a = \mathbf{u}\cdot\mathbf{A}, \qquad b = \mathbf{v}\cdot\mathbf{A}, \qquad c = \mathbf{w}\cdot\mathbf{A}, \qquad \text{(A7.10)}$$

and therefore that

$$\mathbf{A} = \mathbf{u}\cdot\mathbf{A}\,\mathbf{u} + \mathbf{v}\cdot\mathbf{A}\,\mathbf{v} + \mathbf{w}\cdot\mathbf{A}\,\mathbf{w}. \qquad \text{(A7.11)}$$

With b_{jk} given by Eq. (A7.6) and values of P_j, Q_j, R_j found in Eqs. (A7.8), we have

$$\begin{aligned} B_j = b_{jk} A_k &= (P_j u_k + Q_j v_k + R_j w_k) A_k \\ &= (b_{ji} u_i u_k + b_{ji} v_i v_k + b_{ji} w_i w_k) A_k \\ &= b_{ji}(u_i u_k A_k + v_i v_k A_k + w_i w_k A_k). \end{aligned} \qquad \text{(A7.12)}$$

This equation will be valid and therefore Eqs. (A7.8) will be sufficient conditions for Eq. (A7.6), provided

$$u_i u_k A_k + v_i v_k A_k + w_i w_k A_k = A_i \; ; \tag{A7.13}$$

but this is merely Eq. (A7.11) written in terms of components.

Equation (A7.12) can be written

$$\begin{aligned} \mathbf{B} &= \mathbf{P}\,\mathbf{u}\cdot\mathbf{A} + \mathbf{Q}\,\mathbf{v}\cdot\mathbf{A} + \mathbf{R}\,\mathbf{w}\cdot\mathbf{A} \\ &= (\mathbf{P}\mathbf{u} + \mathbf{Q}\mathbf{v} + \mathbf{R}\mathbf{w})\cdot\mathbf{A} = \boldsymbol{\Phi}\cdot\mathbf{A}, \end{aligned} \tag{A7.14}$$

where

$$\boldsymbol{\Phi} \equiv \mathbf{P}\mathbf{u} + \mathbf{Q}\mathbf{v} + \mathbf{R}\mathbf{w}. \tag{A7.15}$$

Thus the vector **B** may be looked upon as the result of the operator $\boldsymbol{\Phi}$ acting upon **A**. $\boldsymbol{\Phi}$ is called a *dyadic*. We may speak of b_{jk} as components of the dyadic $\boldsymbol{\Phi}$ and write

$$\Phi_{jk} = b_{jk} = P_j u_k + Q_j v_k + R_j w_k. \tag{A7.16}$$

More generally, any expression of the form

$$\boldsymbol{\Phi} = \mathbf{P}_1\mathbf{Q}_1 + \mathbf{P}_2\mathbf{Q}_2 + \cdots + \mathbf{P}_m\mathbf{Q}_m, \tag{A7.17}$$

where $\mathbf{P}_1, \mathbf{P}_2, \ldots, \mathbf{P}_m$ and $\mathbf{Q}_1, \mathbf{Q}_2, \ldots, \mathbf{Q}_m$ are vectors, is called a dyadic; while the special case ($m = 1$)

$$\boldsymbol{\Phi} = \mathbf{P}\mathbf{Q}, \tag{A7.18a}$$

or in component form

$$\Phi_{jk} = P_j Q_k, \tag{A7.18b}$$

is called a *dyad*.

Given any dyadic $\boldsymbol{\Phi}$ expressed as in Eq. (A7.17) we may *dot* it on the right by any vector **A** to obtain a new vector

$$\mathbf{A}' = \boldsymbol{\Phi}\cdot\mathbf{A} = \mathbf{P}_1(\mathbf{Q}_1\cdot\mathbf{A}) + \mathbf{P}_2(\mathbf{Q}_2\cdot\mathbf{A}) + \cdots + \mathbf{P}_m(\mathbf{Q}_m\cdot\mathbf{A}) \tag{A7.19}$$

expressed in terms of $\mathbf{P}_1, \mathbf{P}_2, \ldots, \mathbf{P}_m$. Thus $\boldsymbol{\Phi}$ can be looked upon as an operator that takes each vector **A** into the corresponding vector $\mathbf{A}'$. Analogously by dotting $\boldsymbol{\Phi}$ on the left by **A** we obtain another vector

$$\mathbf{A}'' = \mathbf{A}\cdot\boldsymbol{\Phi} = (\mathbf{A}\cdot\mathbf{P}_1)\mathbf{Q}_1 + (\mathbf{A}\cdot\mathbf{P}_2)\mathbf{Q}_2 + \cdots + (\mathbf{A}\cdot\mathbf{P}_m)\mathbf{Q}_m \tag{A7.20}$$

expressed in terms of $\mathbf{Q}_1, \mathbf{Q}_2, \ldots, \mathbf{Q}_m$.

Two dyadics $\boldsymbol{\Phi}$ and $\boldsymbol{\Psi}$ are said to be *equal* if

$$\boldsymbol{\Phi}\cdot\mathbf{A} = \boldsymbol{\Psi}\cdot\mathbf{A} \tag{A7.21}$$

for all **A**. From what has been said above this means that $\Phi_{jk} = \Psi_{jk}$, and therefore that

$$\mathbf{A} \cdot \mathbf{\Phi} = \mathbf{A} \cdot \mathbf{\Psi} \tag{A7.22}$$

for all **A**. A dyadic $\mathbf{\Phi}$ is zero if $\mathbf{\Phi} \cdot \mathbf{A} = 0$ for all **A**; this implies that $\Phi_{jk} = 0$. Addition of dyadics is defined by requiring that

$$\mathbf{\Psi} = \mathbf{\Phi}_1 + \mathbf{\Phi}_2, \qquad \Psi_{jk} = (\Phi_1)_{jk} + (\Phi_2)_{jk}. \tag{A7.23}$$

be equivalent equations.

Dyads may be regarded as a kind of outer product of two vectors. Ordinary rules of algebra, except the commutative rule of multiplication, apply to this product. Thus, for example, a general distributive law holds:

$$(\mathbf{P} + \mathbf{Q})(\mathbf{R} + \mathbf{S}) = \mathbf{PR} + \mathbf{QR} + \mathbf{PS} + \mathbf{QS}. \tag{A7.24}$$

This can be proved by noting that for **A** we have

$$\begin{aligned}(\mathbf{P} + \mathbf{Q})(\mathbf{R} + \mathbf{S}) \cdot \mathbf{A} &= (\mathbf{P} + \mathbf{Q})(\mathbf{R} \cdot \mathbf{A} + \mathbf{S} \cdot \mathbf{A}) \\ &= \mathbf{PR} \cdot \mathbf{A} + \mathbf{QR} \cdot \mathbf{A} + \mathbf{PS} \cdot \mathbf{A} + \mathbf{QS} \cdot \mathbf{A} \\ &= (\mathbf{PR} + \mathbf{QR} + \mathbf{PS} + \mathbf{QS}) \cdot \mathbf{A}.\end{aligned}$$

Any dyadic $\mathbf{\Phi}$ given by Eq. (A7.17) may be reduced to the form given by Eq. (A7.15). This can be proved by letting

$$\mathbf{Q}_1 = \alpha_1\mathbf{u} + \beta_1\mathbf{v} + \gamma_1\mathbf{w}, \qquad \mathbf{Q}_2 = \alpha_2\mathbf{u} + \beta_2\mathbf{v} + \gamma_2\mathbf{w}, \tag{A7.25}$$

etc.; then

$$\begin{aligned}\mathbf{P}_1\mathbf{Q}_1 + \mathbf{P}_2\mathbf{Q}_2 + \cdots &= \mathbf{P}_1(\alpha_1\mathbf{u} + \beta_1\mathbf{v} + \gamma_1\mathbf{w}) + \mathbf{P}_2(\alpha_2\mathbf{u} + \beta_2\mathbf{v} + \gamma_2\mathbf{w}) + \cdots \\ &= (\alpha_1\mathbf{P}_1 + \alpha_2\mathbf{P}_2 + \cdots)\mathbf{u} + (\beta_1\mathbf{P}_1 + \beta_2\mathbf{P}_2 + \cdots)\mathbf{v} \\ &\quad + (\gamma_1\mathbf{P}_1 + \cdots)\mathbf{w} \\ &= \mathbf{Pu} + \mathbf{Qv} + \mathbf{Rw},\end{aligned} \tag{A7.26}$$

where

$$\mathbf{P} = \alpha_1\mathbf{P}_1 + \alpha_2\mathbf{P}_2 + \cdots, \qquad \mathbf{Q} = \beta_1\mathbf{P}_1 + \beta_2\mathbf{P}_2 + \cdots,$$
$$\mathbf{R} = \gamma_1\mathbf{P}_1 + \gamma_2\mathbf{P}_2 + \cdots.$$

If the dyadic $\mathbf{\Phi}$ is given in any form, it is easy to express it in the form given by Eq. (A7.15) by observing that, since **u**, **v**, and **w** are mutually perpendicular, if $\mathbf{\Phi} = \mathbf{Pu} + \mathbf{Qv} + \mathbf{Rw}$,

$$\mathbf{P} = \mathbf{\Phi} \cdot \mathbf{u}, \qquad \mathbf{Q} = \mathbf{\Phi} \cdot \mathbf{v}, \qquad \mathbf{R} = \mathbf{\Phi} \cdot \mathbf{w}. \tag{A7.27}$$

These are, of course, merely Eqs. (A7.8) in another form.

Similarly we can write

$$\mathbf{\Phi} = \mathbf{u}\mathbf{P}' + \mathbf{v}\mathbf{Q}' + \mathbf{w}\mathbf{R}', \tag{A7.28}$$

where

$$\mathbf{P}' = \mathbf{u}\cdot\mathbf{\Phi}, \qquad \mathbf{Q}' = \mathbf{v}\cdot\mathbf{\Phi}, \qquad \mathbf{R}' = \mathbf{w}\cdot\mathbf{\Phi}. \tag{A7.29}$$

If $\mathbf{\Phi}$ is first dotted on the right by the vector $\mathbf{B}$ and the resulting vector dotted on the left by the vector $\mathbf{A}$, we obtain the scalar

$$\begin{aligned}\mathbf{A}\cdot(\mathbf{\Phi}\cdot\mathbf{B}) &= \mathbf{A}\cdot(\mathbf{P}\,\mathbf{u}\cdot\mathbf{B} + \mathbf{Q}\,\mathbf{v}\cdot\mathbf{B} + \mathbf{R}\,\mathbf{w}\cdot\mathbf{B})\\ &= (\mathbf{A}\cdot\mathbf{P})(\mathbf{u}\cdot\mathbf{B}) + (\mathbf{A}\cdot\mathbf{Q})(\mathbf{v}\cdot\mathbf{B}) + (\mathbf{A}\cdot\mathbf{R})(\mathbf{w}\cdot\mathbf{B})\\ &= (\mathbf{A}\cdot\mathbf{P}\,\mathbf{u} + \mathbf{A}\cdot\mathbf{Q}\,\mathbf{v} + \mathbf{A}\cdot\mathbf{R}\,\mathbf{w})\cdot\mathbf{B}\\ &= (\mathbf{A}\cdot\mathbf{\Phi})\cdot\mathbf{B}.\end{aligned}$$

Thus the parentheses are unnecessary, and we can write

$$\mathbf{A}\cdot\mathbf{\Phi}\cdot\mathbf{B} = (\mathbf{A}\cdot\mathbf{P})(\mathbf{u}\cdot\mathbf{B}) + (\mathbf{A}\cdot\mathbf{Q})(\mathbf{v}\cdot\mathbf{B}) + (\mathbf{A}\cdot\mathbf{R})(\mathbf{w}\cdot\mathbf{B}). \tag{A7.30}$$

Similarly by applying the formal operations that are obviously called for, we have

$$\mathbf{A}\cdot\mathbf{\Phi}\times\mathbf{B} = (\mathbf{A}\cdot\mathbf{P})(\mathbf{u}\times\mathbf{B}) + (\mathbf{A}\cdot\mathbf{Q})(\mathbf{v}\times\mathbf{B}) + (\mathbf{A}\cdot\mathbf{R})(\mathbf{w}\times\mathbf{B}), \tag{A7.31}$$

$$\begin{aligned}\mathbf{C}\times(\mathbf{A}\cdot\mathbf{\Phi}\times\mathbf{B}) = (\mathbf{A}\cdot\mathbf{P})\mathbf{C}\times(\mathbf{u}\times\mathbf{B}) &+ (\mathbf{A}\cdot\mathbf{Q})\mathbf{C}\times(\mathbf{v}\times\mathbf{B})\\ &+ (\mathbf{A}\cdot\mathbf{R})\mathbf{C}\times(\mathbf{w}\times\mathbf{B}),\end{aligned} \tag{A7.32}$$

$$\begin{aligned}\mathbf{C}\cdot(\mathbf{A}\cdot\mathbf{\Phi}\times\mathbf{B}) &= (\mathbf{A}\cdot\mathbf{P})(\mathbf{C}\cdot\mathbf{u}\times\mathbf{B}) + (\mathbf{A}\cdot\mathbf{Q})(\mathbf{C}\cdot\mathbf{v}\times\mathbf{B})\\ &\quad + (\mathbf{A}\cdot\mathbf{R})(\mathbf{C}\cdot\mathbf{w}\times\mathbf{B})\\ &= \mathbf{C}\cdot(\mathbf{A}\cdot\mathbf{\Phi})\times\mathbf{B},\end{aligned} \tag{A7.33}$$

$$\begin{aligned}(\mathbf{A}\cdot\mathbf{\Phi})^2 = (\mathbf{A}\cdot\mathbf{\Phi})\cdot(\mathbf{A}\cdot\mathbf{\Phi}) &= (\mathbf{A}\cdot\mathbf{P}\,\mathbf{u} + \mathbf{A}\cdot\mathbf{Q}\,\mathbf{v} + \mathbf{A}\cdot\mathbf{R}\,\mathbf{w})^2\\ &= (\mathbf{A}\cdot\mathbf{P})^2 + (\mathbf{A}\cdot\mathbf{Q})^2 + (\mathbf{A}\cdot\mathbf{R})^2,\end{aligned} \tag{A7.34}$$

$$(\mathbf{A}\cdot\mathbf{\Phi}\times\mathbf{B})^2 = [(\mathbf{A}\cdot\mathbf{\Phi})\times\mathbf{B}]^2 = (\mathbf{A}\cdot\mathbf{\Phi})^2B^2 - (\mathbf{A}\cdot\mathbf{\Phi}\cdot\mathbf{B})^2. \tag{A7.35}$$

A dyadic obtained from a given dyadic by interchanging the order of the factors of all its dyads is called the *conjugate* of the given dyadic. Thus the conjugate of $\mathbf{\Phi} = \mathbf{P}\mathbf{u} + \mathbf{Q}\mathbf{v} + \mathbf{R}\mathbf{w}$ is

$$\mathbf{\Phi}_c = \mathbf{u}\mathbf{P} + \mathbf{v}\mathbf{Q} + \mathbf{w}\mathbf{R}. \tag{A7.36}$$

For this dyadic

$$(\Phi_c)_{jk} = u_jP_k + v_jQ_k + w_jR_k = \Phi_{kj}\,\cdot \tag{A7.37}$$

This leads to naming the dyadic for which

$$\mathbf{\Phi}_c = \mathbf{\Phi} \qquad \text{or} \qquad \Phi_{kj} = \Phi_{jk} \tag{A7.38}$$

selfconjugate, or *symmetric;* that is, the corresponding tensor is symmetric. Similarly, if

$$\mathbf{\Phi}_c = -\mathbf{\Phi}, \qquad \text{i.e.,} \qquad \Phi_{kj} = -\Phi_{jk}, \tag{A7.39}$$

the dyadic is called *anticonjugate*, or *antisymmetric*.

Let the unit vectors along the three Cartesian axes be designated by $\hat{\mathbf{1}}$, $\hat{\mathbf{2}}$, and $\hat{\mathbf{3}}$ with $\hat{\mathbf{j}}$ designating the jth one of these. It is then clear that any vector $\mathbf{P}$ can be written

$$\mathbf{P} = \sum_{i=1}^{3} P_i \hat{\mathbf{i}},$$

and therefore that its jth component is

$$P_j = \mathbf{P} \cdot \hat{\mathbf{j}}.$$

Likewise any dyadic $\mathbf{\Phi}$ can be written

$$\mathbf{\Phi} = \sum_{i,j=1}^{3} \Phi_{ij} \hat{\mathbf{i}}\hat{\mathbf{j}}, \tag{A7.40}$$

and thus its ijth component is given by

$$\Phi_{ij} = \hat{\mathbf{i}} \cdot \mathbf{\Phi} \cdot \hat{\mathbf{j}}. \tag{A7.41}$$

The agreement of this with Eq. (A7.16) is seen as follows:

$$\begin{aligned} \hat{\mathbf{j}} \cdot \mathbf{\Phi} \cdot \hat{\mathbf{k}} &= \hat{\mathbf{j}} \cdot (\mathbf{P}\mathbf{u} + \mathbf{Q}\mathbf{v} + \mathbf{R}\mathbf{w}) \cdot \hat{\mathbf{k}} \\ &= (\hat{\mathbf{j}} \cdot \mathbf{P})(\mathbf{u} \cdot \hat{\mathbf{k}}) + (\hat{\mathbf{j}} \cdot \mathbf{Q})(\mathbf{v} \cdot \hat{\mathbf{k}}) + (\hat{\mathbf{j}} \cdot \mathbf{R})(\mathbf{w} \cdot \hat{\mathbf{k}}) \\ &= P_j u_k + Q_j v_k + R_j w_k = \Phi_{jk}. \end{aligned}$$

We have seen in Appendix A5 that a coordinate system can always be found such that a real symmetric tensor is reduced to diagonal form. Since a dyadic is essentially a second-order tensor expressed in a somewhat different form, it is not surprizing that a real symmetric dyadic $\mathbf{\Phi}$ can also always be reduced to diagonal form; i.e.,

$$\Phi_{ij} = 0, \qquad \text{if} \quad i \neq j.$$

For such a coordinate system Eq. (A7.40) reduces to

$$\mathbf{\Phi} = \Phi_{11}\hat{\mathbf{1}}\hat{\mathbf{1}} + \Phi_{22}\hat{\mathbf{2}}\hat{\mathbf{2}} + \Phi_{33}\hat{\mathbf{3}}\hat{\mathbf{3}},$$

or in the more conventional form

$$\mathbf{\Phi} = \Phi_{11}\mathbf{ii} + \Phi_{22}\mathbf{jj} + \Phi_{33}\mathbf{kk}, \tag{A7.42}$$

which can be called the diagonal form of a dyadic.

A special case of such a symmetrical dyadic, namely one that reduces to $\Phi_{\alpha\beta} = \delta_{\alpha\beta}$, is called the *unit dyadic*, the *identical dyadic*, or the *idemfactor*. It can be written as

$$\mathbf{I} = \mathbf{ii} + \mathbf{jj} + \mathbf{kk}. \tag{A7.43}$$

With this dyadic

$$\mathbf{I} \cdot \mathbf{A} = \mathbf{A} \cdot \mathbf{I} = \mathbf{i\,i} \cdot \mathbf{A} + \mathbf{j\,j} \cdot \mathbf{A} + \mathbf{k\,k} \cdot \mathbf{A} = \mathbf{A}, \tag{A7.44}$$

which explains the names. This property is, of course, preserved in a coordinate transformation. Thus, if $\mathbf{i} \to \mathbf{u}$, $\mathbf{j} \to \mathbf{v}$, $\mathbf{k} \to \mathbf{w}$, the dyadic $\mathbf{I}$ goes over into

$$\mathbf{I} \to \mathbf{uu} + \mathbf{vv} + \mathbf{ww}, \tag{A7.45}$$

and therefore

$$\mathbf{I} \cdot \mathbf{A} = \mathbf{u\,u} \cdot \mathbf{A} + \mathbf{v\,v} \cdot \mathbf{A} + \mathbf{w\,w} \cdot \mathbf{A} = \mathbf{A}, \tag{A7.46}$$

by Eq. (A7.11).

Suppose a dyadic

$$\mathbf{\Phi} = \mathbf{A}_1\mathbf{P}_1 + \mathbf{A}_2\mathbf{P}_2 + \mathbf{A}_3\mathbf{P}_3,$$

where $\mathbf{A}_i$ and $\mathbf{P}_i$ are vectors, is a function of some parameter λ; then we may write

$$\mathbf{\Phi}(\lambda) = \sum_{i=1}^{3} \mathbf{A}_i(\lambda)\,\mathbf{P}_i(\lambda). \tag{A7.47}$$

The derivative of $\mathbf{\Phi}$ with respect to λ is defined as

$$\frac{d\mathbf{\Phi}}{d\lambda} = \lim_{\Delta\lambda\to 0}\left[\frac{\mathbf{\Phi}(\lambda+\Delta\lambda) - \mathbf{\Phi}(\lambda)}{\Delta\lambda}\right]. \tag{A7.48}$$

We can therefore show that

$$\begin{aligned}
\frac{d\mathbf{\Phi}}{d\lambda} &= \sum_{i=1}^{3} \lim_{\Delta\lambda\to 0}\left\{\frac{1}{\Delta\lambda}\left[\mathbf{A}_i(\lambda+\Delta\lambda)\,\mathbf{P}_i(\lambda+\Delta\lambda) - \mathbf{A}_i(\lambda)\,\mathbf{P}_i(\lambda)\right]\right\} \\
&= \sum_{i=1}^{3} \lim_{\Delta\lambda\to 0}\left\{\frac{[\mathbf{A}_i(\lambda+\Delta\lambda) - \mathbf{A}_i(\lambda)]}{\Delta\lambda}\,\mathbf{P}_i(\lambda+\Delta\lambda)\right. \\
&\qquad \left. + \mathbf{A}_i(\lambda)\,\frac{[\mathbf{P}_i(\lambda+\Delta\lambda) - \mathbf{P}_i(\lambda)]}{\Delta\lambda}\right\} \\
&= \sum_{i=1}^{3}\left[\frac{d\mathbf{A}_i(\lambda)}{d\lambda}\,\mathbf{P}_i(\lambda) + \mathbf{A}_i(\lambda)\,\frac{d\mathbf{P}_i(\lambda)}{d\lambda}\right].
\end{aligned} \tag{A7.49}$$

In particular, if a dot over a letter is used to represent a time derivative, we can write

$$\dot{\Omega} = \sum_{i=1}^{3} (\dot{\mathbf{A}}_i\mathbf{P}_i + \mathbf{A}_i\dot{\mathbf{P}}_i). \tag{A7.50}$$

These formulas show that a dyad $\mathbf{A}_i\mathbf{P}_i$ may be differentiated as one would any product, except that the order of the factors must here be preserved.

Appendix A8

Orthogonal Functions: Fourier Series, Integrals, and Transforms

A set of functions

$$\phi_1(x), \quad \phi_2(x), \quad \phi_3(x), \quad \ldots, \quad \phi_n(x), \tag{A8.1}$$

where n is finite, is said to be an *orthogonal set of functions* over an interval a to b, $b > a$ (a, b real) if none of them are identically zero and if they are mutually *orthogonal*,

$$\int_a^b \phi_i^*(x)\phi_j(x)\,dx = 0, \qquad i \neq j, \quad i, j = 1, 2, \ldots, n, \tag{A8.2}$$

over this interval. We may also have an infinite set of orthogonal functions,

$$\phi_1(x), \quad \phi_2(x), \quad \ldots, \quad \phi_n(x), \quad \ldots,$$

satisfying Eq. (A8.2). The asterisk indicates the complex conjugate of the quantity to which it is attached. Obviously, when functions are all real, the asterisk can be left out. If we have

$$\int_a^b \phi_i^*(x)\phi_i(x)\,dx = 1, \qquad i = 1, 2, \ldots, n, \tag{A8.3}$$

the functions are said to be *normalized.* For an orthogonal normalized (or orthonormal) set of functions Eqs. (A8.2) and (A8.3) can be combined:

$$\int_a^b \phi_i^*(x)\phi_j(x)\,dx = \delta_{ij}, \qquad i, j = 1, 2, \ldots, n, \tag{A8.4}$$

where δ_{ij} is defined in Eq. (A3.6).

If the functions of the set (A8.1) are quadratically integrable, then, since they are not identically zero and $\phi_i^*(x)\phi_i(x) \geq 0$, one has

$$\int_a^b \phi_i^*(x)\phi_i(x)\,dx = N_i^2 > 0 \tag{A8.5}$$

where N_i is a positive real finite number. Thus the functions may be normalized without destroying their orthogonality by dividing each function by the corresponding N_i.

The functions of an orthogonal set are *linearly independent* of each other. This means that no linear combination of them is identically zero without all the coefficients being zero. For suppose that

$$\sum_{m=1}^{n} a_m \phi_m(x) = 0, \tag{A8.6}$$

with at least one of the a_m not equal to zero. Multiplying this equation by ϕ_k^* and integrating between the limits a and b, we have*

$$\sum_{m=1}^{n} a_m \int \phi_k^*(x)\phi_m(x)\,dx = 0. \tag{A8.7}$$

Due to the orthogonality of the $\phi_s(x)$ as expressed by Eq. (A8.2) all the terms for which $m \neq k$ vanish and one is left with

$$a_k \int \phi_k^*(x)\phi_k(x)\,dx = 0. \tag{A8.8}$$

Since this equation must hold for $k = 1, 2, \ldots, n$, either we must have

$$\int \phi_m^*(x)\phi_m(x)\,dx = \int |\phi_m(x)|^2\,dx = 0 \tag{A8.9}$$

for some m or all the a_k must be zero. The first would require that $\phi_m(x)$ be identically zero, which is excluded by our definition of an orthogonal set of functions. The second possibility is excluded by our assumption that at least one of the a_m is not zero. We thus conclude that the conditions for linear dependence cannot be met, and hence that the functions constitute a linearly independent set of functions.

Let $f(x)$ be an arbitrary function defined between the same limits a and b and let us try to express it by

$$f(x) = \sum_{m=1}^{n} a_m \phi_m(x), \tag{A8.10}$$

* We shall generally omit writing in the limits of integration, except when confusion would otherwise result.

where the ϕ's satisfy Eq. (A8.4). If this equation is to hold, we must have

$$\int \phi_k^*(x)f(x)\,dx = \sum_{m=1}^{n} a_m \int \phi_k^*(x)\phi_m(x)\,dx$$
$$= \sum_{m=1}^{n} a_m \delta_{km} = a_k. \tag{A8.11}$$

The coefficients a_m in Eq. (A8.10) are thus completely defined; however, even with coefficients given by Eq. (A8.11), Eq. (A8.10) need not hold.

Suppose that $f(x)$ is any given function, and that a_k are calculated by means of Eq. (A8.11); then, in general,

$$f(x) = \sum_{1}^{n} a_m \phi_m(x) + g(x), \tag{A8.12}$$

where $g(x)$ is some function defined by this equation. We then have

$$\int \phi_k^*(x)g(x)\,dx = \int \phi_k^*(x)f(x)\,dx - \sum_{m=1}^{n} a_m \int \phi_k^*(x)\phi_m(x)\,dx$$
$$= a_k - \sum_{m=1}^{n} a_m \delta_{km} = a_k - a_k = 0. \tag{A8.13}$$

This shows that $g(x)$ is orthogonal to every function of the set $\{\phi_m(x)\}$.

If the set (A8.1) is complete, that is, if no function (other than the zero function*) exists outside of the set which is orthogonal to all functions of the set; then $g(x)$ must be zero and Eq. (A8.10) holds. Conversely, if Eq. (A8.10), with coefficients determined by Eq. (A8.11), holds for all functions $f(x)$ and for all x between a and b, then $g(x) \equiv 0$ and the set is complete. Exceptions will arise when the integrals defining a_k, Eq. (A8.11), do not converge or when the set of functions is infinite and the series in Eq. (A8.10) does not converge.

Consider now the integral

$$\int f^*(x)f(x)\,dx = \int \left\{ \sum_{m=1}^{n} a_m^* \phi_m^*(x) + g^*(x) \right\} \left\{ \sum_{k=1}^{n} a_k \phi_k(x) + g(x) \right\} dx$$
$$= \sum_{m,k=1}^{n} a_m^* a_k \int \phi_m^*(x)\phi_k(x)\,dx + \sum_{m=1}^{n} a_m^* \int \phi_m^*(x)g(x)\,dx$$
$$+ \sum_{k=1}^{n} a_k \int \phi_k(x)g^*(x)\,dx + \int g^*(x)g(x)\,dx.$$

The second summation vanishes by Eq. (A8.13). The third also vanishes

* The *zero function* has a value of zero for all x.

because the integrals occurring in it are merely the complex conjugates of the integrals in the second member. Using Eq. (A8.4), we now obtain

$$\int f^*(x)f(x)\,dx = \sum_{m,k=1}^{n} a_m^* a_k \delta_{mk} + \int g^*(x)g(x)\,dx$$

or

$$\int f^*(x)f(x)\,dx = \sum_{k=1}^{n} a_k^* a_k + \int g^*(x)g(x)\,dx. \tag{A8.14}$$

Therefore, since each term of this equation is positive,

$$\sum_{k=1}^{n} a_k^* a_k \leq \int f^*(x)f(x)\,dx, \tag{A8.15}$$

where the equality holds only if $g(x) = 0$, i.e., when Eq. (A8.10) is valid. This can, therefore, be used as a test of the validity of Eq. (A8.10).

One can rather quickly convince oneself that a finite set of functions cannot form a complete set. Since we shall from now on be concerned mainly with questions of completeness, we shall usually deal with infinite sets and Eq. (A8.10) will be considered to be replaced by

$$f(x) = \sum_{m=1}^{\infty} a_m \phi_m(x).$$

In addition, we shall write $\sum_m$ to mean a sum over all m, which for the infinite set means $\sum_{m=1}^{\infty}$.

Substitution from Eq. (A8.11) into Eq. (A8.10) gives

$$\begin{aligned} f(x) &= \sum_m \phi_m(x) \int \phi_m^*(x')f(x')\,dx' \\ &= \int_a^b f(x')\left\{\sum_m \phi_m^*(x')\phi_m(x)\right\} dx', \end{aligned} \tag{A8.16}$$

which for a complete orthogonal set must hold for any function $f(x)$. As a special case, let us choose $f(x)$ as follows: Let α and β be two numbers such that $a < \alpha < \beta < b$, and let

$$f(x) = \begin{cases} 1, & \text{if } \alpha < x < \beta, \\ 0, & \text{otherwise.} \end{cases}$$

Then Eq. (A8.16) becomes

$$\int_\alpha^\beta \left[\sum_m \phi_m^*(x')\phi_m(x)\right] dx' = \begin{cases} 1, & \text{if } \alpha < x < \beta, \\ 0, & \text{otherwise.} \end{cases} \tag{A8.17}$$

This equation, when x is assumed given, must hold for all α and β satisfying the inequalities $a < \alpha < \beta < b$. The integrand must therefore be zero for all values of x' other than $x' = x$, because the integral is zero for any limits α and β both smaller or both larger than x. On the other hand, Eq. (A8.17) requires that for the limit $\alpha = x - \varepsilon$ and $\beta = x + \varepsilon$ (where ε is ever so small) the integral is 1.

We summarize these findings by writing that

$$\sum_m \phi_m^*(x')\phi_m(x) = \delta(x' - x) \tag{A8.18}$$

holds for all x and x' in the interval a to b. The function $\delta(x)$ is defined by the properties

$$\delta(x) = 0, \qquad x \neq 0 \tag{A8.19a}$$

and

$$\int_{-\infty}^{\infty} \delta(x)\, dx = \int_{-\varepsilon}^{\varepsilon} \delta(x)\, dx = 1, \tag{A8.19b}$$

and is called the *Dirac delta function*,* or simply the *delta function*. Although not a proper function of mathematics, it is often used in physics, and is very convenient. Two further properties of the delta function are often useful:

$$\int_{-\infty}^{\infty} f(x)\delta(x-\lambda)\, dx = \int_{\lambda-\varepsilon}^{\lambda+\varepsilon} f(x)\delta(x-\lambda)\, dx$$

$$= f(\lambda)\int_{\lambda-\varepsilon}^{\lambda+\varepsilon} \delta(x-\lambda)\, dx = f(\lambda) \tag{A8.20}$$

and

$$\int_{-\infty}^{\infty} f(x)\delta'(x-\lambda)\, dx \equiv \int_{-\infty}^{\infty} f(x)\frac{d}{dx}\delta(x-\lambda)\, dx$$

$$= [f(x)\delta(x-\lambda)]_{-\infty}^{\infty} - \int_{-\infty}^{\infty} f'(x)\delta(x-\lambda)\, dx$$

$$= -f'(\lambda), \tag{A8.21}$$

where in the last reduction we use Eqs. (A8.19) and (A8.20).

We derived Eq. (A8.18) as a necessary condition on a complete orthogonal set. It is also a sufficient condition, because a substitution from Eq.

* See Dirac, P. A. M., *The Principles of Quantum Mechanics*, 4th ed., Oxford Univ. Press, London and New York, 1958, p. 58.

(A8.10) into the right-hand member of Eq. (A8.16) and the use of Eqs. (A8.20) gives

$$\sum_m a_m\phi_m(x) = \int f(x')\left\{\sum_m \phi_m^*(x')\phi_m(x)\right\} dx'$$

$$= \int f(x')\delta(x' - x)\, dx' = f(x). \qquad \text{(A8.22)}$$

Thus, Eq. (A8.18) is a necessary and sufficient condition for the completeness of an orthonormal set of functions.

EXAMPLE 1. Consider the set of functions

$$\sin x, \quad \sin 2x, \quad \ldots, \quad \sin nx, \quad \ldots. \qquad \text{(A8.23)}$$

Show that they are orthogonal to each other for the range $0 \leq x \leq \pi$, and expand the function

$$f(x) = \pi - x, \qquad 0 < x < \pi, \qquad \text{(A8.24)}$$

in a series using these sine functions. Since for m and n integers

$$\int_0^\pi \sin nx \sin mx\, dx = \left[\frac{\sin(n - m)x}{2(n - m)} - \frac{\sin(n + m)x}{2(n + m)}\right]_0^\pi$$

$$= 0, \qquad \text{if} \quad n \neq m, \qquad \text{(A8.25)}$$

the functions are orthogonal. For $n = m$ the integral becomes

$$\int_0^\pi \sin^2 nx\, dx = \left[\frac{x}{2} - \frac{\sin 2nx}{4n}\right]_0^\pi = \frac{\pi}{2} = N_n^2; \qquad \text{(A8.26)}$$

therefore the set of functions

$$\phi_m(x) = \sqrt{\frac{2}{\pi}} \sin mx, \qquad m = 1, 2, \ldots, \infty, \qquad \text{(A8.27)}$$

is an orthonormal set.

In the expansion of $f(x)$ in accordance with Eqs. (A8.10) and (A8.11), we have

$$a_k = \int_a^b f(x)\phi_k^*(x)\, dx = \sqrt{\frac{2}{\pi}} \int_0^\pi f(x) \sin kx\, dx,$$

and hence, substituting $f(x)$ from Eq. (A8.24),

$$a_k = \sqrt{\frac{2}{\pi}} \int_0^\pi (\pi - x) \sin kx \, dx$$
$$= \sqrt{\frac{2}{\pi}} \left[\frac{-\pi \cos kx}{k} - \frac{\sin kx - kx \cos kx}{k^2} \right]_0^\pi$$
$$= \frac{\pi}{k} \sqrt{\frac{2}{\pi}}. \tag{A8.28}$$

If the set of functions $\phi_m(x)$ in Eqs. (A8.27) is complete for $0 \leq x \leq \pi$, we have

$$f(x) = \pi - x = \sqrt{\frac{2}{\pi}} \sum_{k=1}^{\infty} a_k \sin kx = 2 \sum_{k=1}^{\infty} \frac{\sin kx}{k}. \tag{A8.29}$$

To check the validity of Eq. (A8.29), independently of whether the set (A8.27) is complete or not, we may use Eq. (A8.15). Now one has

$$\int_0^\pi f^2(x) \, dx = \int_0^\pi (\pi - x)^2 \, dx$$
$$= [\pi^2 x - \pi x^2 + \tfrac{1}{3} x^3]_0^\pi = \tfrac{1}{3} \pi^3 \tag{A8.30}$$

and

$$\sum_{k=1}^{\infty} a_k^2 = 2\pi \sum_1^{\infty} \frac{1}{k^2} = 2\pi \frac{\pi^2}{6} = \tfrac{1}{3}\pi^3, \tag{A8.31}$$

since $\sum_{k=1}^{\infty}(1/k^2) = \pi^2/6$ is a well-known series. Thus, the set of functions in (A8.23) appears to be complete,* and Eq. (A8.29) is valid for the interval $0 < x < \pi$.

EXAMPLE 2. Show that the set of functions

$$1, \quad \cos x, \quad \cos 2x, \quad \ldots, \quad \cos nx, \quad \ldots \tag{A8.32}$$

is an orthogonal set over the interval $0 \leq x \leq \pi$ and expand the function

$$f(x) = x^2 - 2\pi x + 2\pi^2/3 \tag{A8.33}$$

in a series of these cosines.

We have

$$\int_0^\pi \cos kx \cos mx \, dx = \left[\frac{\sin (k+m)x}{2(k+m)} + \frac{\sin (k-m)x}{2(k-m)} \right]_0^\pi$$
$$= 0, \quad \text{if} \quad k \neq m, \tag{A8.34}$$

* Completeness requires that the equality hold in (A8.15) for any function $f(x)$.

and thus the functions are orthogonal. In addition, it is seen that

$$\int_0^\pi \cos^2 kx\,dx = \left[\frac{\sin 2kx}{4k} + \frac{x}{2}\right]_0^\pi = \frac{\pi}{2}, \qquad \text{if } k \neq 0,$$

$$= \pi, \qquad \text{if } k = 0; \quad \text{(A8.35)}$$

thus, the set

$$\phi_0(x) = \frac{1}{\sqrt{\pi}}, \quad \phi_k(x) = \sqrt{\frac{2}{\pi}}\cos kx, \qquad k = 1, 2, 3, \ldots, \quad \text{(A8.36)}$$

is an orthonormal set.

In accordance with Eqs. (A8.11) and (A8.36) one has

$$a_0 = \int_0^\pi f(x)\phi_0(x)\,dx = \frac{1}{\sqrt{\pi}}\int_0^\pi (x^2 - 2\pi x + 2\pi^2/3)\,dx = 0 \quad \text{(A8.37)}$$

and

$$a_k = \sqrt{\frac{2}{\pi}}\int_0^\pi (x^2 - 2\pi x + 2\pi^2/3)\cos kx\,dx$$

$$= \frac{4}{k^2}\sqrt{\frac{\pi}{2}}, \qquad \text{for } k \neq 0. \quad \text{(A8.38)}$$

Thus the required series expansion is

$$x^2 - 2\pi x + 2\pi^2/3 = 4\sum_{k=1}^{\infty}\frac{\cos kx}{k^2}, \qquad \text{for } 0 < x < \pi, \quad \text{(A8.39)}$$

provided the set (A8.36) is complete. Since we observe that

$$\int_0^\pi f^2(x)\,dx = \int_0^\pi (x^2 - 2\pi x + 2\pi^2/3)^2\,dx = 4\pi^5/45,$$

while by a well-known sum of an infinite series

$$\sum_{k=1}^{\infty} a_k^2 = 8\pi\sum_{k=1}^{\infty}(1/k^4) = 8\pi(\pi^4/90) = 4\pi^5/45$$

the equality holds in Eq. (A8.15). This means that the set of functions is probably complete and that in any case Eq. (A8.39) is valid.

Equation (A8.39), on putting $x = \pi - y$, becomes

$$y^2 - \pi^2/3 = 4\sum_{k=1}^{\infty}\frac{(-1)^k}{k^2}\cos ky, \qquad 0 \leq y \leq \pi. \quad \text{(A8.40)}$$

Since neither side of this equation is changed by replacing y by $-y$, it holds for the interval $-\pi \leq y \leq \pi$. The original equation, Eq. (A8.39), is therefore valid for $0 < x < 2\pi$.

EXAMPLE 3. Expand the function

$$f(x) = \begin{cases} 1, & \text{for} \quad x < \alpha, \\ 0, & \text{for} \quad x > \alpha, \end{cases} \tag{A8.41}$$

where $0 \leq x \leq \pi$ and $0 \leq \alpha \leq \pi$, first in a series of sines and then in a series of cosines.

In the first case by Eq. (A8.11) we have

$$\begin{aligned} a_k &= \sqrt{\frac{2}{\pi}} \int_0^\pi f(x) \sin kx\, dx = \sqrt{\frac{2}{\pi}} \int_0^\alpha \sin kx\, dx \\ &= \sqrt{\frac{2}{\pi}} \frac{1 - \cos k\alpha}{k}, \end{aligned} \tag{A8.42}$$

so that, provided the set is complete,

$$f(x) = \frac{2}{\pi} \sum_{k=1}^{\infty} \frac{(1 - \cos k\alpha)}{k} \sin kx = \begin{cases} 1, & \text{for} \quad 0 < x < \alpha, \\ 0, & \text{for} \quad \alpha < x < \pi. \end{cases} \tag{A8.43}$$

In the second case one has

$$a_0 = \sqrt{\frac{1}{\pi}} \int_0^\pi f(x)\, dx = \sqrt{\frac{1}{\pi}} \int_0^\alpha dx = \alpha \sqrt{\frac{1}{\pi}}$$

and

$$\begin{aligned} a_k &= \sqrt{\frac{2}{\pi}} \int_0^\pi f(x) \cos kx\, dx = \sqrt{\frac{2}{\pi}} \int_0^\alpha \cos kx\, dx \\ &= \sqrt{\frac{2}{\pi}} \frac{\sin k\alpha}{k}, \qquad k \neq 0, \end{aligned} \tag{A8.44}$$

so that, provided the set is complete,

$$f(x) = \frac{\alpha}{\pi} + \frac{2}{\pi} \sum_{1}^{\infty} \frac{\sin k\alpha}{k} \cos kx = \begin{cases} 1, & \text{for} \quad 0 < x < \alpha, \\ 0, & \text{for} \quad \alpha < x < \pi. \end{cases} \tag{A8.45}$$

To verify that the sets are complete we have to show, for the first case, that

$$\sum_{k=1}^{\infty} a_k^2 = \frac{2}{\pi} \sum_{k=1}^{\infty} \left(\frac{1 - \cos k\alpha}{k}\right)^2 = \int_0^\pi f^2\, dx = \int_0^\alpha dx = \alpha; \tag{A8.46}$$

and, for the second case, that

$$\sum_{k=0}^{\infty} a_k^2 = \frac{\alpha^2}{\pi} + \frac{2}{\pi}\sum_{k=1}^{\infty}\left(\frac{\sin k\alpha}{k}\right)^2 = \alpha. \tag{A8.47}$$

Now by squaring and applying Eqs. (A8.31) and (A8.39) one finds that

$$\sum_{k=1}^{\infty}\left(\frac{1-\cos k\alpha}{k}\right)^2 = \sum_{k=1}^{\infty}\frac{1}{k^2}\left(\frac{3}{2} - 2\cos k\alpha + \frac{1}{2}\cos 2k\alpha\right)$$

$$= \frac{\pi^2}{4} - 2\frac{\alpha^2 - 2\pi\alpha + 2\pi^2/3}{4} + \frac{1}{2}\frac{(2\alpha)^2 - 2\pi(2\alpha) + 2\pi^2/3}{4}$$

This simplifies to

$$\sum_{k=1}^{\infty}\left(\frac{1-\cos k\alpha}{k}\right)^2 = \frac{\pi\alpha}{2}, \tag{A8.48}$$

which agrees with Eq. (A8.46). On the other hand, by again using Eqs. (A8.31) and (A8.39) one determines that

$$\sum_{k=1}^{\infty}\left(\frac{\sin k\alpha}{k}\right)^2 = \sum_{k=1}^{\infty}\frac{1-\cos 2k\alpha}{2k^2}$$

$$= \frac{\pi^2}{12} - \frac{1}{2}\frac{(2\alpha)^2 - 2\pi(2\alpha) + 2\pi^2/3}{4}$$

$$= \frac{\pi\alpha}{2} - \frac{\alpha^2}{2}, \tag{A8.49}$$

which agrees with Eq. (A8.47). Thus, the expansions (A8.43) and (A8.45) are established.

EXAMPLE 4. Prove in general that the sets (A8.27) and (A8.36) are complete for the range $0 < x < \pi$.

We have to show, in the first case, in accordance with Eq. (A8.18) that (see Eqs. (A8.27))

$$\frac{2}{\pi}\sum_{k=1}^{\infty}\sin kx' \sin kx = \delta(x'-x) \tag{A8.50}$$

and, in the second case, that (see Eqs. (A8.36))

$$\frac{1}{\pi} + \frac{2}{\pi}\sum_{k=1}^{\infty}\cos kx' \cos kx = \delta(x'-x). \tag{A8.51}$$

Equations equivalent to these are

$$\frac{2}{\pi}\int_0^{\alpha}\sum_{k=1}^{\infty}\sin kx' \sin kx \, dx' = \int_0^{\alpha}\delta(x'-x)\, dx'$$

$$= \begin{cases} 1, & \text{for } 0 < x < \alpha, \\ 0, & \text{for } \alpha < x < \pi, \end{cases} \quad \text{(A8.52)}$$

and

$$\frac{\alpha}{\pi} + \frac{2}{\pi}\int_0^{\alpha}\sum_{k=1}^{\infty}\cos kx' \cos kx \, dx' = \begin{cases} 1, & \text{for } 0 < x < \alpha, \\ 0, & \text{for } \alpha < x < \pi, \end{cases} \quad \text{(A8.53)}$$

where these equations are required to hold for all α in the interval $0 < \alpha < \pi$. Performing the integrations term by term, we see that the resulting equations are just the Eqs. (A8.43) and (A8.45). Thus the sets are complete.

Over the interval from 0 to 2π, neither the set of sines, nor the sets of cosines, is a complete set. This can be seen, for example, from the fact that

$$\int_0^{2\pi} \sin kx \cos mx \, dx = 0 \quad \text{(A8.54)}$$

for k and m arbitrary integers. This shows that $\cos mx$, for a given m, is orthogonal to $\sin kx$ for all integer values of k and hence that the sines do not form a complete set. By a like argument $\sin kx$, for a given k, is orthogonal to all the cosines and the latter can therefore not form a complete set.

The normalization constants are different for this new interval, as is evident from the equations

$$\int_0^{2\pi} \sin^2 kx \, dx = \int_0^{2\pi} \cos^2 kx \, dx = \pi, \qquad k \neq 0, \quad \text{(A8.55a)}$$

and

$$\int_0^{2\pi} dx = 2\pi. \quad \text{(A8.55b)}$$

Thus, the set of functions

$$\sqrt{\frac{1}{\pi}}\sin kx, \quad \sqrt{\frac{1}{2\pi}}, \quad \sqrt{\frac{1}{\pi}}\cos kx, \qquad k = 1, 2, \ldots, \infty \quad \text{(A8.56)}$$

is an orthonormal set. It can be shown to be a complete set for the range $0 < x < 2\pi$.

Since the set (A8.56) is a complete set, any function $f(x)$ defined over the domain $0 \leq x \leq 2\pi$ can be given the following Fourier series expansion:

$$f(x) = \sum_{k=1}^{\infty} a_k \sin kx + \sum_{k=0}^{\infty} b_k \cos kx, \tag{A8.57a}$$

where

$$a_k = \frac{1}{\pi}\int_0^{2\pi} f(x)\sin kx\, dx, \tag{A8.57b}$$

$$b_0 = \frac{1}{2\pi}\int_0^{2\pi} f(x)\, dx, \tag{A8.57c}$$

$$b_k = \frac{1}{\pi}\int_0^{2\pi} f(x)\cos kx\, dx. \tag{A8.57d}$$

The right-hand side of Eq. (A8.57a) is a periodic function of x with a period of 2π; therefore, we can also interpret that equation as the expansion of a periodic function $f(x)$ having a period of 2π for all values of x.

A function having a period L can be expanded into a similar series. Let $x = Ly/2\pi$; then

$$f(x) = f\left(\frac{Ly}{2\pi}\right) = F(y) \tag{A8.58}$$

and

$$f(x + L) = f\left(\frac{Ly}{2\pi} + L\right) = f\left(\frac{L(y + 2\pi)}{2\pi}\right) = F(y + 2\pi). \tag{A8.59}$$

If $f(x) = f(x + L)$, $F(y) = F(y + 2\pi)$ and $F(y)$ is periodic with a period 2π; therefore, it has an expansion given by Eq. (A8.57a), namely,

$$F(y) = \sum_{k=1}^{\infty} a_k \sin ky + \sum_{k=0}^{\infty} b_k \cos ky.$$

From Eq. (A8.58) we then have

$$f(x) = F(y) = \sum_{k=1}^{\infty} a_k \sin \frac{2\pi kx}{L} + \sum_{k=0}^{\infty} b_k \cos \frac{2\pi kx}{L} \tag{A8.60}$$

as an expansion for a periodic function $f(x)$ of period L. The functions

$$\sin \frac{2\pi kx}{L}, \qquad k = 1, 2, \ldots, \infty, \tag{A8.61a}$$

and

$$\cos \frac{2\pi kx}{L}, \qquad k = 0, 1, 2, \ldots, \infty, \tag{A8.61b}$$

form a complete orthogonal set of functions for the expansion of all periodic functions of period L.

To normalize the functions in (A8.61) we note that

$$\int_0^L \sin^2 \frac{2\pi kx}{L}\, dx = \int_0^L \cos^2 \frac{2\pi kx}{L}\, dx = \frac{L}{2}$$

and

$$\int_0^L dx = L,$$

so that the normalized set is

$$\sqrt{\frac{2}{L}} \sin \frac{2\pi kx}{L}, \qquad k = 1, 2, \ldots, \infty, \tag{A8.62a}$$

$$\sqrt{\frac{2}{L}} \cos \frac{2\pi kx}{L}, \qquad k = 1, 2, \ldots, \infty, \tag{A8.62b}$$

and

$$\sqrt{\frac{1}{L}}\,. \tag{A8.62c}$$

Thus any function $f(x)$ having a period of L can be expanded in the form

$$f(x) = \sum_{k=1}^{\infty} A_k \left(\sqrt{\frac{2}{L}} \sin \frac{2\pi kx}{L} \right) + B_0 \sqrt{\frac{1}{L}} + \sum_{k=1}^{\infty} B_k \left(\sqrt{\frac{2}{L}} \cos \frac{2\pi kx}{L} \right), \tag{A8.63}$$

where by Eq. (A8.11)

$$A_k = \int_0^L f(x) \left[\sqrt{\frac{2}{L}} \sin \frac{2\pi kx}{L} \right] dx, \tag{A8.64a}$$

$$B_k = \int_0^L f(x) \left[\sqrt{\frac{2}{L}} \cos \frac{2\pi kx}{L} \right] dx, \qquad k \neq 0, \tag{A8.64b}$$

$$B_0 = \int_0^L f(x) \left[\sqrt{\frac{1}{L}} \right] dx. \tag{A8.64c}$$

Comparing Eqs. (A8.63) and (A8.64) with Eq. (A8.60), we see that

$$a_k = \sqrt{\frac{2}{L}}\, A_k = \frac{2}{L} \int_0^L f(x) \sin \frac{2\pi kx}{L}\, dx, \tag{A8.65a}$$

$$b_k = \sqrt{\frac{2}{L}}\, B_k = \frac{2}{L} \int_0^L f(x) \cos \frac{2\pi kx}{L}\, dx, \tag{A8.65b}$$

$$b_0 = \sqrt{\frac{1}{L}}\, B_0 = \frac{1}{L} \int_0^L f(x)\, dx. \tag{A8.65c}$$

Since

$$\sin kx = \frac{1}{2i}(e^{ikx} - e^{-ikx}), \qquad \cos kx = \tfrac{1}{2}(e^{ikx} + e^{-ikx}), \qquad 1 = e^0,$$

the series Eq. (A8.57a) can be expressed in terms of the functions e^{ikx}, with $k = 0, \pm 1, \ldots, \pm\infty$. We are thus led to consider the set of functions

$$\phi_k(x) = e^{ikx}, \qquad k = 0, \pm 1, \pm 2, \ldots, \pm\infty, \tag{A8.66}$$

which, by virtue of the completeness of the set in (A8.56), must be a complete set for $0 < x < 2\pi$. Now for these exponential functions we find

$$\int_0^{2\pi} \phi_k^*(x)\phi_m(x)\, dx = \int_0^{2\pi} e^{i(m-k)x}\, dx = \left[\frac{e^{i(m-k)x}}{i(m-k)}\right]_0^{2\pi} = 0, \tag{A8.67}$$

if $m \neq k$. For $m = k$, we obtain

$$\int_0^{2\pi} \phi_k^*(x)\phi_k(x)\, dx = \int_0^{2\pi} dx = 2\pi. \tag{A8.68}$$

Therefore the set

$$\phi_k(x) = \frac{1}{\sqrt{2\pi}} e^{ikx}, \qquad k = 0, \pm 1, \pm 2, \ldots, \pm\infty, \tag{A8.69}$$

is a complete orthonormal set for the interval $0 < x < 2\pi$.

EXAMPLE 5. Expand the function $f(x)$ of Example 3, but with $0 \leq x \leq 2\pi$ and $0 \leq \alpha \leq 2\pi$, into a series of the exponentials given in Eqs. (A8.69).

We have, for $0 \leq x \leq 2\pi$ and $0 \leq \alpha \leq 2\pi$,

$$f(x) = \begin{cases} 1, & \text{for } x < \alpha, \\ 0, & \text{for } x > \alpha; \end{cases}$$

therefore, by Eqs. (A8.11) and (A8.69),

$$a_k = \int_0^{2\pi} \phi_k^*(x) f(x)\, dx = \frac{1}{\sqrt{2\pi}} \int_0^{\alpha} e^{-ikx}\, dx$$

$$= \frac{i}{\sqrt{2\pi}} \frac{e^{-ik\alpha} - 1}{k}. \tag{A8.70}$$

Thus $f(x)$ can be expanded in the form

$$f(x) = \sum_{k=-\infty}^{\infty} a_k\phi_k(x) = \frac{i}{2\pi}\sum_{k=-\infty}^{\infty}\left(\frac{e^{-ik\alpha}-1}{k}\right)e^{ikx}$$

$$= \begin{cases} 1, & \text{for } x < \alpha, \\ 0, & \text{for } x > \alpha. \end{cases} \qquad \text{(A8.71)}$$

The types of series occurring in previous examples, in which the orthogonal functions are the sines, cosines, or exponentials with imaginary exponents, are called *Fourier series.* We now have to generalize the concept of orthogonal functions further.

Instead of the sets such as (A8.1) we shall now consider sets of the type $\phi_\xi(x)$ where ξ is a parameter varying continuously between α and β, $\alpha < \beta$. As an example, we shall consider the set

$$\phi_\xi(x) = \left(\frac{1}{2\pi}\right)^{1/2} e^{i\xi x} \qquad \text{(A8.72)}$$

with x and ξ varying from $-\infty$ to ∞. Such functions are usually not quadratically integrable over the infinite domain of these functions. Thus, in our example we have

$$\int_{-\infty}^{\infty} \phi_\xi^*(x)\phi_\xi(x)\,dx = \frac{1}{2\pi}\int_{-\infty}^{\infty} dx = \infty. \qquad \text{(A8.73)}$$

Equations such as Eq. (A8.4) therefore cannot hold. However, a somewhat different equation, namely,

$$I \equiv \int_{-\infty}^{\infty} \phi_\xi^*(x)\phi_{\xi'}(x)\,dx = \delta(\xi - \xi'), \qquad \text{(A8.74)}$$

with a suitable interpretation, does hold. We shall call a set of functions such as given in (A8.72), with a continuously varying index, orthonormal if they satisfy Eq. (A8.74). Let us examine the set (A8.72) from this point of view. The left-hand member of Eq. (A8.74) then becomes

$$I = \int_{-\infty}^{\infty} \phi_\xi^*(x)\phi_{\xi'}(x)\,dx = \frac{1}{2\pi}\int_{-\infty}^{\infty} e^{i(\xi'-\xi)x}\,dx, \qquad \text{(A8.75)}$$

which we proceed to evaluate. We have

$$I = \frac{1}{2\pi}\int_{-\infty}^{\infty} \{\cos(\xi' - \xi)x + i\sin(\xi' - \xi)x\}\,dx$$

$$= \frac{1}{2\pi}\int_{-\infty}^{\infty} \cos(\xi' - \xi)x\,dx, \qquad \text{(A8.76)}$$

since $\sin(\xi' - \xi)x$ is an odd function of x and must give zero on integration from $-\infty$ to ∞. Letting $y = \xi' - \xi$ and noting that $\cos yx$ is an even function, we have

$$I = \frac{1}{\pi}\int_0^\infty \cos yx\,dx = \frac{1}{\pi}\int_0^\infty \frac{\partial}{\partial y}\left(\frac{\sin yx}{x}\right)dx = \frac{d}{dy}J(y), \quad \text{(A8.77)}$$

where*

$$J(y) = \frac{1}{\pi}\int_0^\infty \frac{\sin yx}{x}\,dx = -J(-y). \quad \text{(A8.78)}$$

The last is a well-known integral,† and one finds that

$$J(y) = \begin{cases} \tfrac{1}{2}, & \text{for } y > 0, \\ -\tfrac{1}{2}, & \text{for } y < 0. \end{cases} \quad \text{(A8.79)}$$

Thus, $I = 0$, unless $y = 0$. On the other hand, we have

$$\int_{-\varepsilon}^{\varepsilon} I\,dy = \int_{-\varepsilon}^{\varepsilon} \frac{dJ}{dy}\,dy = J(\varepsilon) - J(-\varepsilon) = 1; \quad \text{(A8.80)}$$

therefore

$$I = \delta(y) = \delta(\xi' - \xi) = \delta(\xi - \xi')$$

and Eq. (A8.74) is verified.

Another way to verify that the functions (A8.72) form a complete set is to use a condition obtained from Eq. (A8.74) by integrating each side with respect to ξ'. The condition is

$$\int_\lambda^\mu d\xi' \int_{-\infty}^\infty \phi_\xi^*(x)\phi_{\xi'}(x)\,dx = \int_\lambda^\mu \delta(\xi - \xi')\,d\xi'$$

$$= \begin{cases} 1, & \text{for } \lambda < \xi < \mu, \\ 0, & \text{for } \xi < \lambda \text{ or } \mu < \xi, \end{cases} \quad \text{(A8.81)}$$

for arbitrary λ and μ, provided only that $\mu > \lambda$. This condition can readily be shown to be equivalent to Eq. (A8.74).

The left-hand side of Eq. (A8.81) for the above set of functions becomes

$$\frac{1}{2\pi}\int_\lambda^\mu d\xi' \int_{-\infty}^\infty e^{i(\xi'-\xi)x}\,dx = \frac{1}{\pi}\int_\lambda^\mu d\xi \int_0^\infty \cos(\xi' - \xi)x\,dx,$$

* Since it is well known that the formalism we are developing is not mathematically rigorous, involving such things as the Dirac delta function, we shall freely assume the legitimacy of interchanging the order of various operations involving infinite limits, etc.

† See, for example, B. O. Peirce, *A Short Table of Integrals*, 3rd rev. ed., Ginn and Co., Boston, 1929, Eq. 484.

because the imaginary part of $e^{i(\xi'-\xi)x}$ is an odd function. On interchanging the order of the integrations this becomes

$$\frac{1}{\pi}\int_0^\infty dx\int_\lambda^\mu \cos(\xi'-\xi)x\,d\xi' = \frac{1}{\pi}\int_0^\infty\left[\frac{\sin(\mu-\varepsilon)x}{x} - \frac{\sin(\lambda-\xi)x}{x}\right]dx$$
$$= J(\mu-\xi) + J(\xi-\lambda).$$

Satisfaction of Eq. (A8.81) thus requires that

$$J(\mu-\xi) + J(\xi-\lambda) = \begin{cases} 1, & \text{for } \lambda < \xi < \mu, \\ 0, & \text{for } \xi < \lambda \text{ or } \mu < \xi, \end{cases} \tag{A8.82}$$

and this is insured by Eq. (A8.79).

One may ask why we introduced $J(y)$ in Eq. (A8.77) instead of evaluating

$$I = \frac{1}{\pi}\int_0^\infty \cos yx\,dx = \frac{1}{\pi}\lim_{\alpha\to\infty}\int_0^\alpha \cos yx\,dx \tag{A8.83}$$

directly. Attempting this latter procedure, we find that

$$I = \frac{1}{\pi}\lim_{\alpha\to\infty}\left[\frac{\sin yx}{y}\right]_0^\alpha = \frac{1}{\pi y}\lim_{\alpha\to\infty}\sin y\alpha.$$

This limit, however, does not exist. The integral in Eq. (A8.83) thus diverges, not by approaching ∞, but by oscillating as the upper limit goes to infinity. In such a case one can use a procedure which, whenever it works, gives correct results for physical applications. Although it can be justified, we shall give it here without justification.*

If

$$I = \lim_{\alpha\to\infty}\int_0^\alpha f(x)\,dx$$

and this limit does not exist due to the oscillation of the integral as α approaches ∞, then we may take

$$I = \lim_{\beta\to\infty}\frac{1}{\beta}\int_0^\beta d\alpha\int_0^\alpha f(x)\,dx \tag{A8.84}$$

provided this new limit exists.

* It is a generalization of the well-known method of summing divergent series, the so-called (C, 1) method. See E. T. Whittaker and G. N. Watson, *A Course in Modern Analysis*, 4th ed., Cambridge Univ. Press, London and New York, 1940, Sections 8.43 and 9.4.

Applying this to our case, we see that for $y \neq 0$

$$I = \lim_{\beta\to\infty} \frac{1}{\beta} \int_0^\beta d\alpha \left[\frac{1}{\pi} \frac{\sin \alpha y}{y}\right] = \frac{1}{\pi y} \lim_{\beta\to\infty} \frac{1}{\beta} \int_0^\beta \sin \alpha y \, d\alpha$$

$$= \frac{1}{\pi y^2} \lim_{\beta\to\infty} \frac{1 - \cos \beta y}{\beta} = 0.$$

This is in agreement with Eq. (A8.74), since $y = \xi - \xi' \neq 0$. For $y = 0$ we have

$$I = \frac{1}{\pi} \int_0^\infty dx = \infty,$$

which also agrees with Eq. (A8.74). This discussion, however, does not show the nature of the singularity of I at $y = 0$. This we can obtain as follows*:

$$\int_{-\varepsilon}^{\varepsilon} I \, dy = \int_{-\infty}^{\infty} I \, dy = \frac{1}{\pi} \lim_{\beta\to\infty} \int_{-\infty}^{\infty} \frac{(1 - \cos \beta y)}{\beta y^2} \, dy$$

$$= \frac{2}{\pi} \lim_{\beta\to\infty} \int_{-\infty}^{\infty} \frac{\sin^2(\beta y/2) \, dy}{\beta y^2}$$

$$= \frac{2}{\pi} \lim_{\beta\to\infty} \int_0^\infty \frac{\sin^2(\beta y/2) \, d(\beta y/2)}{(\beta y/2)^2}$$

$$= \frac{2}{\pi} \int_0^\infty \frac{\sin^2 z \, dz}{z^2} = 1,$$

which shows that $I = \delta(y)$. Thus, combining Eqs. (A8.74) and (A8.75), we have

$$\int_{-\infty}^{\infty} e^{i(\xi' - \xi)x} \, dx = 2\pi\delta(\xi' - \xi), \tag{A8.85}$$

or, more generally,

$$\int_{-\infty}^{\infty} e^{ipx} \, dx = 2\pi\delta(p). \tag{A8.86}$$

In expanding an arbitrary function $f(x)$ in terms of the set $\{\phi_\xi(x)\}$, Eq. (A8.10) is now replaced by its natural generalization

$$f(x) = \int_\alpha^\beta a(\xi)\phi_\xi(x) \, d\xi. \tag{A8.87}$$

* See B. O. Peirce, *A Short Table of Integrals*, 3rd rev. ed., Ginn and Co., Boston, 1929, Eq. 486.

To obtain the coefficients $a(\xi)$ we multiply both sides by $\phi_{\xi'}^*(x)$ and integrate with respect to x from a to b; then with help of Eqs. (A8.74) and (A8.20) we have

$$\int_a^b \phi_{\xi'}^*(x) f(x)\, dx = \int_a^b \left[\phi_{\xi'}^*(x) \int_\alpha^\beta a(\xi)\phi_\xi(x)\, d\xi\right] dx$$

$$= \int_\alpha^\beta \left[a(\xi) \int_a^b \phi_{\xi'}^*(x)\phi_\xi(x)\, dx\right] d\xi$$

$$= \int_\alpha^\beta a(\xi)\delta(\xi' - \xi)\, d\xi = a(\xi').$$

Thus, the coefficients are found from

$$a(\xi) = \int_a^b \phi_\xi^*(x) f(x)\, dx, \tag{A8.88}$$

which replaces Eq. (A8.11).

The completeness conditions are now

$$\int_a^b f^*(x) f(x)\, dx = \int_\alpha^\beta a^*(\xi) a(\xi)\, d\xi, \tag{A8.89}$$

for every $f(x)$, and

$$\int_\alpha^\beta \phi_\xi^*(x')\phi_\xi(x)\, d\xi = \delta(x' - x). \tag{A8.90}$$

Thus for a set of functions with a continuously varying parameter, Eq. (A8.89) replaces Eq. (A8.15) and Eq. (A8.90) replaces Eq. (A8.18).

For the set of functions in (A8.72) the condition in Eq. (A8.90) becomes

$$\frac{1}{2\pi} \int_{-\infty}^{\infty} e^{i(x - x')\xi}\, d\xi = \delta(x' - x), \tag{A8.91}$$

which is merely a special case of Eq. (A8.86). The set is therefore complete. Thus, an arbitrary function $f(x)$ should be expandible in terms of the functions $e^{ix\xi}$ of this set, and one writes

$$f(x) = \left(\frac{1}{2\pi}\right)^{1/2} \int_{-\infty}^{\infty} a(\xi) e^{ix\xi}\, d\xi, \tag{A8.92}$$

where

$$a(\xi) = \left(\frac{1}{2\pi}\right)^{1/2} \int_{-\infty}^{\infty} f(x) e^{-ix\xi}\, dx. \tag{A8.93}$$

The right-hand side of Eq. (A8.92) is called the *Fourier integral* expansion of $f(x)$, and the pair of functions, such as $f(x)$ and $a(\xi)$, connected by Eqs. (A8.92) and (A8.93), are said to be *Fourier* transforms of each other. To verify Eqs. (A8.92) and (A8.93) we substitute from Eq. (A8.93) into Eq. (A8.92) to obtain

$$\begin{aligned} f(x) &= \frac{1}{2\pi}\int_{-\infty}^{\infty} e^{ix\xi}\, d\xi \int_{-\infty}^{\infty} f(x')e^{-ix'\xi}\, dx' \\ &= \frac{1}{2\pi}\int_{-\infty}^{\infty} f(x')\, dx' \int_{-\infty}^{\infty} e^{i(x-x')\xi}\, d\xi \\ &= \int_{-\infty}^{\infty} f(x')\delta(x'-x)\, dx', \end{aligned}$$

which is obviously correct if we remember the property of the δ-function given in Eq. (A8.20).

Generalizing the procedure to three independent variables, x, y, and z, we find a Fourier integral expansion to be expressible in the form

$$f(\mathbf{r}) = \left(\frac{1}{2\pi}\right)^{3/2} \int a(\boldsymbol{k})\exp(i\boldsymbol{k}\cdot\mathbf{r})\, d^3\boldsymbol{k} \tag{A8.94}$$

where $\mathbf{r} = (x, y, z)$, $\boldsymbol{k} = (\xi, \eta, \zeta)$, and $d^3\boldsymbol{k} = d\xi\, d\eta\, d\zeta$. The Fourier transform of $f(r)$ is then

$$a(\boldsymbol{k}) = \left(\frac{1}{2\pi}\right)^{3/2} \int f(\mathbf{r})e^{-i\boldsymbol{k}\cdot\mathbf{r}}\, dx\, dy\, dz. \tag{A8.95}$$

We can generalize still further and show that

$$f(\mathbf{r}) = \int a(\boldsymbol{k})\phi_{\boldsymbol{k}}(\mathbf{r})\, d^3\boldsymbol{k}, \tag{A8.96}$$

where

$$a(\boldsymbol{k}) = \int f(\mathbf{r})\phi_{\boldsymbol{k}}^*(\mathbf{r})\, dV, \tag{A8.97}$$

provided the set of functions $\phi_{\boldsymbol{k}}(\mathbf{r})$ satisfies the equations

$$\begin{aligned} \int \phi_{\boldsymbol{k}}^*(\mathbf{r})\phi_{\boldsymbol{k}'}(\mathbf{r})\, dV &= \delta(\boldsymbol{k}-\boldsymbol{k}') \\ &\equiv \delta(\xi-\xi')\delta(\eta-\eta')\delta(\zeta-\zeta') \end{aligned} \tag{A8.98}$$

and

$$\begin{aligned} \int \phi_{\boldsymbol{k}}^*(\mathbf{r})\phi_{\boldsymbol{k}}(\mathbf{r}')\, d^3\boldsymbol{k} &= \delta(\mathbf{r}-\mathbf{r}') \\ &\equiv \delta(x-x')\delta(y-y')\delta(z-z'). \end{aligned} \tag{A8.99}$$

The special case of these functions used in Eqs. (A8.94) and (A8.95)

$$\phi_{\mathscr{k}}(\mathbf{r}) = \left(\frac{1}{2\pi}\right)^{3/2} \exp(i\mathscr{k}\cdot\mathbf{r}) \tag{A8.100}$$

satisfies the equations

$$\int \exp[i(\mathscr{k}' - \mathscr{k})\cdot\mathbf{r}]\, dV = (2\pi)^3\delta(\mathscr{k}' - \mathscr{k}) \tag{A8.101}$$

and

$$\int \exp[i(\mathbf{r}' - \mathbf{r})\cdot\mathscr{k}]\, d^3\mathscr{k} = (2\pi)^3\delta(\mathbf{r}' - \mathbf{r}). \tag{A8.102}$$

These are special cases of Eqs. (A8.98) and (A8.99), respectively. Both are also special cases of the equation

$$\iiint_{-\infty}^{+\infty} e^{i\mathbf{P}\cdot\mathbf{Q}}\, dP_x\, dP_y\, dP_z = (2\pi)^3\delta(\mathbf{Q}) \equiv (2\pi)^3\delta(Q_x)\delta(Q_y)\delta(Q_z). \tag{A8.103}$$

Appendix A9

Lorentz Transformation of Charge and Current Densities

In Section 16, Eq. (16.18), we introduced a set of four quantities

$$s_\alpha \sim \left(\frac{\rho v_x}{c}, \frac{\rho v_y}{c}, \frac{\rho v_z}{c}, i\rho\right) \sim \left(\frac{\rho \mathbf{v}}{c}, i\rho\right),$$

which our theory showed had to be a four-vector. We have to prove, however, that the functions ρ and $\rho\mathbf{v}$ satisfying this condition actually exist. We shall do this by showing that ρ and $\rho\mathbf{v}$ corresponding to a system of point particles satisfy our requirements.

The charge density $\rho(\mathbf{r}, t)$ for a system of particles, as an extension of Eq. (7.25), is

$$\rho(\mathbf{r}, t) = \sum_s e_s \delta(\mathbf{r} - \mathbf{r}_s), \tag{A9.1}$$

where e_s is the charge and $\mathbf{r}_s = \mathbf{r}_s(t)$ is the position of the sth particle. In dealing with the current density, etc., it will be found convenient to change our notation somewhat to avoid confusion between the velocities of the particles and the velocity of one coordinate system relative to another. The current density we shall now designate by

$$\mathbf{j} = \frac{1}{c}\rho(\mathbf{r}, t)\mathbf{u}(\mathbf{r}, t), \tag{A9.2}$$

where $\mathbf{u}(\mathbf{r}, t)$ is required to be equal to $\mathbf{u}_s$, the velocity of the sth particle, at $\mathbf{r} = \mathbf{r}_s(t)$. At other points $\mathbf{u}$ is undefined, but this ambiguity does not matter since ρ, and hence $\rho\mathbf{u}/c$, is zero at these points. Thus, after substituting ρ from Eq. (A9.1) into (A9.2), we can use $\mathbf{u}_s$ in place of $\mathbf{u}$; it is thus seen that

$$\mathbf{j} = \frac{1}{c}\sum_s e_s\mathbf{u}_s\delta(\mathbf{r} - \mathbf{r}_s). \tag{A9.3}$$

We shall use the general Lorentz transformation given in Eqs. (11.46), namely,

$$\mathbf{r}' = \mathbf{r} + \frac{\gamma - 1}{v^2}\mathbf{v}\cdot\mathbf{r}\,\mathbf{v} - \gamma t\mathbf{v}, \tag{A9.4a}$$

$$t' = \gamma\left(t - \frac{\mathbf{v}\cdot\mathbf{r}}{c^2}\right), \tag{A9.4b}$$

where

$$\gamma = (1 - v^2/c^2)^{-1/2} \tag{A9.4c}$$

and $\mathbf{v}$ is the velocity of the primed system relative to the unprimed. We wish to show that under these transformations the components of

$$s_\alpha \sim (\mathbf{j}, i\rho) \tag{A9.5}$$

transform as components of a four-vector; i.e.,

$$\begin{aligned} \mathbf{j}' &= \mathbf{j} + \frac{\gamma - 1}{v^2}\mathbf{v}\cdot\mathbf{j}\,\mathbf{v} - \frac{\gamma\mathbf{v}\rho}{c}, \\ \rho' &= \gamma\left(\rho - \frac{\mathbf{v}\cdot\mathbf{j}}{c}\right). \end{aligned} \tag{A9.6}$$

Here ρ' and $\mathbf{j}'$ are defined in the primed system just as ρ and $\mathbf{j}$ are defined in the unprimed; i.e.,

$$\rho' = \sum_s e_s\delta(\mathbf{r}' - \mathbf{r}_s') \qquad \text{and} \qquad \mathbf{j}' = \frac{1}{c}\sum_s e_s\mathbf{u}_s'\delta(\mathbf{r}' - \mathbf{r}_s'). \tag{A9.7}$$

The discussion of Section 11 shows that Eqs. (A9.6) must hold. We must prove, however, that the definitions of Eqs. (A9.1), (A9.3), and (A9.7) are in agreement with Eqs. (A9.6). To do this we shall show that Eqs. (A9.6), with the help of Eqs. (A9.1) and (A9.3), reduce to Eqs. (A9.7).

Equations (A9.6) can be written more fully thus:

$$\mathbf{j}'(\mathbf{r}', t') = \mathbf{j}(\mathbf{r}, t) + \frac{\gamma - 1}{v^2}\,\mathbf{v}\cdot\mathbf{j}(\mathbf{r}, t)\mathbf{v} - \frac{\gamma\mathbf{v}}{c}\,\rho(\mathbf{r}, t),$$

$$\rho'(\mathbf{r}', t') = \gamma\left(\rho(\mathbf{r}, t) - \frac{\mathbf{v}\cdot\mathbf{j}(\mathbf{r}, t)}{c}\right);$$

or, upon substitution for $\rho(\mathbf{r}, t)$ and $\mathbf{j}(\mathbf{r}, t)$ from Eqs. (A9.1) and (A9.3),

$$\mathbf{j}'(\mathbf{r}', t') = \sum_s \frac{e_s}{c}\left[\mathbf{u}_s(t) + \frac{\gamma - 1}{v^2}\,\mathbf{v}\cdot\mathbf{u}_s(t)\,\mathbf{v} - \gamma\mathbf{v}\right]\delta[\mathbf{r} - \mathbf{r}_s(t)],$$

$$\rho'(\mathbf{r}', t') = \sum_s e_s\left[1 - \frac{\mathbf{v}\cdot\mathbf{u}_s(t)}{c^2}\right]\gamma\delta[\mathbf{r} - \mathbf{r}_s(t)], \tag{A9.8}$$

where $(\mathbf{r}, t)$ and $(\mathbf{r}', t')$ are connected by the transformations given by Eqs. (A9.4).

We need now to express the right-hand members of Eqs. (A9.8) as functions of $\mathbf{r}'$ and t'. We can do this by solving Eqs. (A9.4) for $\mathbf{r}'$ and t', which is most readily done by interchanging the primed and unprimed coordinates and replacing $\mathbf{v}$ by $-\mathbf{v}$. We thus have

$$\mathbf{r} = \mathbf{r}' + \frac{\gamma - 1}{v^2}\,\mathbf{v}\cdot\mathbf{r}'\,\mathbf{v} + \gamma t'\mathbf{v},$$

$$t = \gamma\left(t' + \frac{\mathbf{v}\cdot\mathbf{r}'}{c^2}\right). \tag{A9.9}$$

An event along the world-line of the sth particle having coordinates $[\mathbf{r}_s(t), t]$ in the unprimed coordinate system will have coordinates $[\mathbf{r}_s'(t'), t']$ in the unprimed system. Equations (A9.9) thus apply to these coordinates and we may write

$$\mathbf{r}_s(t) = \mathbf{r}_s'(t') + \frac{\gamma - 1}{v^2}\,\mathbf{v}\cdot\mathbf{r}_s'(t')\mathbf{v} + \gamma t'\mathbf{v},$$

$$t = \gamma\left(t' + \frac{\mathbf{v}\cdot\mathbf{r}_s'(t')}{c^2}\right). \tag{A9.10}$$

Let $\mathbf{r} - \mathbf{r}_s(t)$ expressed, however, in the primed coordinates be designated by $\mathbf{R}(\mathbf{r}', t)$; i.e.,

$$\mathbf{R}(\mathbf{r}', t') = \mathbf{r} - \mathbf{r}_s(t). \tag{A9.11}$$

Let $F(\mathbf{r}')$ be an arbitrary function of $\mathbf{r}'$. Then (see Appendix A6) we have

$$\iiint F(\mathbf{r}')\,\delta[\mathbf{r}-\mathbf{r}_s(t)]\,dx'\,dy'\,dz'$$

$$=\iiint F(\mathbf{r}')\,\delta[\mathbf{R}(\mathbf{r}',t')]\,dx'\,dy'\,dz'$$

$$=\iiint F(\mathbf{r}')\,\frac{\partial(x',y',z')}{\partial(R_x,R_y,R_z)}\,\delta(\mathbf{R})\,dR_x\,dR_y\,dR_z$$

$$=\left[F(\mathbf{r}')\,\frac{\partial(x',y',z')}{\partial(R_x,R_y,R_z)}\right]_{\mathbf{R}=0}$$

$$=\left[F(\mathbf{r}')\,\frac{\partial(x',y',z')}{\partial(R_x,R_y,R_z)}\right]_{\mathbf{r}=\mathbf{r}_s(t)}$$

The condition $\mathbf{r}=\mathbf{r}_s(t)$ means that $(\mathbf{r},t)=[\mathbf{r}_s(t),t]$ and hence that the event $(\mathbf{r},t)$ lies on the world-line of the sth particle. Since this coincidence is an invariant concept, we must also have $(\mathbf{r}',t')=[\mathbf{r}_s'(t'),t']$ and thus $\mathbf{r}'=\mathbf{r}_s'(t')$. We conclude that $\mathbf{r}=\mathbf{r}_s(t)$ implies $\mathbf{r}'=\mathbf{r}_s'(t')$ and therefore that

$$\iiint F(\mathbf{r}')\,\delta[\mathbf{r}-\mathbf{r}_s(t)]\,dx'\,dy'\,dz'$$

$$=\left[F(\mathbf{r}')\,\frac{\partial(x',y',z')}{\partial(R_x,R_y,R_z)}\right]_{\mathbf{r}'=\mathbf{r}_s'(t')}$$

$$=\iiint F(\mathbf{r}')\,\frac{\partial(x',y',z')}{\partial(R_x,R_y,R_z)}\,\delta[\mathbf{r}'-\mathbf{r}_s'(t')]\,dx'\,dy'\,dz',$$

and consequently $\delta[\mathbf{r}-\mathbf{r}_s(t)]$, regarded as a function of $\mathbf{r}'$ and t', can be expressed as

$$\delta[\mathbf{r}-\mathbf{r}_s(t)]=\frac{\partial(x',y',z')}{\partial(R_x,R_y,R_z)}\,\delta[\mathbf{r}'-\mathbf{r}_s'(t')]$$

$$=\frac{\delta[\mathbf{r}'-\mathbf{r}_s'(t')]}{J},\tag{A9.12}$$

where

$$J=\frac{\partial(R_x,R_y,R_z)}{\partial(x',y',z')}.\tag{A9.13}$$

Since the δ-functions in Eq. (A9.12) are zero, except for $\mathbf{r}'=\mathbf{r}_s'(t')$, we may take for J without loss of generality its value when

$$\mathbf{r}=\mathbf{r}_s(t)\qquad\text{or}\qquad\mathbf{r}'=\mathbf{r}_s'(t').\tag{A9.14}$$

We shall now proceed to evaluate J.

By Eqs. (A9.11) and (A9.9) we have

$$R_x = x' + \frac{\gamma - 1}{v^2}(x'v_x + y'v_y + z'v_z)v_x + \gamma t'v_x - x_s(t) \quad \text{(A9.15)}$$

with

$$t = \gamma\left(t' + \frac{v_x x' + v_y y' + v_z z'}{c^2}\right); \quad \text{(A9.16)}$$

therefore

$$\frac{\partial R_x}{\partial x'} = 1 + \frac{\gamma - 1}{v^2}v_x^2 - \frac{dx_s(t)}{dt}\frac{\partial t}{\partial x'}$$

$$= 1 + \frac{\gamma - 1}{v^2}v_x^2 - \dot{x}_s(t)\frac{\gamma v_x}{c^2}, \quad \text{(A9.17)}$$

where, as elsewhere, a dot over a letter indicates a time derivative.

Similarly, we have

$$\begin{aligned}
\frac{\partial R_x}{\partial y'} &= \frac{\gamma - 1}{v^2}v_x v_y - \dot{x}_s\frac{\gamma v_y}{c^2}, & \frac{\partial R_x}{\partial z'} &= \frac{\gamma - 1}{v^2}v_x v_z - \dot{x}_s\frac{\gamma v_z}{c^2}, \\
\frac{\partial R_y}{\partial x'} &= \frac{\gamma - 1}{v^2}v_y v_x - \dot{y}_s\frac{\gamma v_x}{c^2}, & \frac{\partial R_y}{\partial y'} &= 1 + \frac{\gamma - 1}{v^2}v_y^2 - \dot{y}_s\frac{\gamma v_y}{c^2}, \\
\frac{\partial R_y}{\partial z'} &= \frac{\gamma - 1}{v^2}v_y v_z - \dot{y}_s\frac{\gamma v_z}{c^2}, & \frac{\partial R_z}{\partial x'} &= \frac{\gamma - 1}{v^2}v_z v_x - \dot{z}_s\frac{\gamma v_x}{c^2}, \\
\frac{\partial R_z}{\partial y'} &= \frac{\gamma - 1}{v^2}v_z v_y - \dot{z}_s\frac{\gamma v_y}{c^2}, & \frac{\partial R_z}{\partial z'} &= 1 + \frac{\gamma - 1}{v^2}v_z^2 - \dot{z}_s\frac{\gamma v_z}{c^2}.
\end{aligned} \quad \text{(A9.18)}$$

Evaluating the determinant in Eq. (A9.13), we obtain

$$J = \gamma\left(1 - \frac{v_x\,\dot{x}_s(t)}{c^2} - \frac{v_y\,\dot{y}_s(t)}{c^2} - \frac{v_z\,\dot{z}_s(t)}{c^2}\right)$$

$$= \gamma\left(1 - \frac{\mathbf{v}\cdot\mathbf{u}_s(t)}{c^2}\right). \quad \text{(A9.19)}$$

We can now substitute from Eqs. (A9.12) and (A9.19) into Eqs. (A9.8), obtaining

$$\mathbf{j}'(\mathbf{r}', t') = \sum_s \frac{e_s}{c}\left[\mathbf{u}_s(t) + \frac{\gamma - 1}{v^2}\mathbf{v}\cdot\mathbf{u}_s(t)\,\mathbf{v} - \gamma\mathbf{v}\right]\frac{\delta[\mathbf{r}' - \mathbf{r}_s'(t')]}{\gamma[1 - \mathbf{v}\cdot\mathbf{u}_s(t)/c^2]} \quad \text{(A9.20)}$$

and

$$\rho'(\mathbf{r}', t') = \sum_s e_s \delta[\mathbf{r}' - \mathbf{r}'_s(t)]. \tag{A9.21}$$

To simplify the first of these equations we observe that Eqs. (A9.4) give

$$d\mathbf{r}'_s(t') = d\mathbf{r}_s(t) + \frac{\gamma - 1}{v^2} \mathbf{v} \cdot d\mathbf{r}_s(t)\, \mathbf{v} - \gamma \mathbf{v}\, dt$$

and

$$dt' = \gamma\left[dt - \frac{\mathbf{v} \cdot d\mathbf{r}_s(t)}{c^2}\right] = \gamma\, dt\left(1 - \frac{\mathbf{v} \cdot \mathbf{u}_s(t)}{c^2}\right);$$

whence by division

$$\frac{d\mathbf{r}'_s(t')}{dt'} \equiv \mathbf{u}'_s(t') = \frac{\mathbf{u}_s(t) + [(\gamma - 1)/v^2]\mathbf{v} \cdot \mathbf{u}_s(t)\, \mathbf{v} - \gamma \mathbf{v}}{\gamma[1 - \mathbf{v} \cdot \mathbf{u}_s(t)/c^2]}. \tag{A9.22}$$

With the help of Eq. (A9.22) the Eqs. (A9.20) and (A9.21) reduce at once to Eqs. (A9.7), and thus s_α, defined by Eqs. (A9.5), (A9.1), and (A9.3), is shown to transform as a four-vector.

Appendix A10

Relationship between Gaussian and MKS (Giorgi) Systems

Table A10.1 permits one to convert any of the formulas in the book, which are expressed in Gaussian units (cgs), into their equivalent in rationalized MKS units. The symbols for length, time, mass, force, and other quantities that are not specifically electromagnetic are not changed. The electromagnetic symbols listed under the heading "Gaussian" are replaced by the corresponding symbols listed under the heading "MKS".

In Table A10.2 the Gaussian unit for each physical quantity listed is equivalent to the number of MKS units listed under the last column.

TABLE A10.1
CONVERSION OF SYMBOLS AND FORMULAS

Physical Quantity	Gaussian	MKS
Electric field	$\mathscr{E}$	$\sqrt{4\pi\varepsilon_0}\mathscr{E}$
Magnetic field	$\mathscr{H}$	$\sqrt{4\pi\mu_0}\,\mathscr{H}$
Scalar potential	φ	$\sqrt{4\pi\varepsilon_0}\,\varphi$
Vector potential	$\mathbf{A}$	$\sqrt{\frac{4\pi}{\mu_0}}\mathbf{A}$
Velocity of light	c	$\frac{1}{\sqrt{\varepsilon_0\mu_0}}$
Charge	q	$\frac{1}{\sqrt{4\pi\varepsilon_0}}q$
Charge density	ρ	$\frac{1}{\sqrt{4\pi\varepsilon_0}}\rho$
Current density	$\rho\mathbf{v}/c = \mathbf{J}$	$\frac{1}{\sqrt{4\pi\varepsilon_0}}(\mathbf{J}/c)$
Polarization	$\mathbf{P}$	$\frac{1}{\sqrt{4\pi\varepsilon_0}}\mathbf{P}$
Magnetization	$\mathbf{M}$	$\sqrt{\frac{\mu_0}{4\pi}}\mathbf{M}$
Electric displacement	$\mathscr{D}$	$\sqrt{\frac{4\pi}{\varepsilon_0}}\,\mathscr{D}$
Magnetic induction	$\mathscr{B}$	$\sqrt{\frac{4\pi}{\mu_0}}\,\mathscr{B}$

TABLE A10.2
CONVERSION OF UNITS[a]

Physical quantity	Symbol	Gaussian units	Equivalent MKS units
Length	ℓ or s	cm	10^{-2} m
Mass	m	gm	10^{-3} kg
Time	t	sec	sec
Force	F	dyne	10^{-5} newtons
Energy, work	E	erg	10^{-7} joules
Power	P	erg/sec	10^{-7} watts
Charge	q	statcoul or esu	$\frac{1}{3} \times 10^{-9}$ coulombs
Charge density	ρ	statcoul/cm^3	$\frac{1}{3} \times 10^{-3}$ coulombs/m^3
Current	I	abampere	10 amperes
Current density	$\rho v/c$	abamp/cm^2	10^5 amp/m^2
Potential	φ	statvolt	300 volts
Electric field	$\mathscr{E}$	statvolt/cm	3×10^4 volts/m
Electric displacement	$\mathscr{D}$	statvolt/cm	$\frac{1}{12\pi} \times 10^{-5}$ coul/m^2
Polarization	P	statcoul/cm^2	$\frac{1}{3} \times 10^{-5}$ coul/m^2
Magnetic field	$\mathscr{H}$	oersted	$\frac{1}{4\pi} \times 10^3$ amp-turn/m
Magnetic induction	$\mathscr{B}$	gauss	10^{-4} webers/m^2
Magnetization	M	gauss	$4\pi \times 10^{-4}$ webers/m^2
Conductivity	σ	l/sec	$\frac{1}{9} \times 10^{-9}$ mhos/m

[a] *References:* (1) J. D. Jackson, *Classical Electrodynamics*, Wiley, New York, 1962, pp. 611–621. (2) D. W. Berreman, *Am. J. Phys.*, 27, No. 1, 44–46 (1959).

Index

C

D

E

www.ingramcontent.com/pod-product-compliance
Lightning Source LLC
Chambersburg PA
CBHW071352170526
45165CB00001B/10
9781456481933